12th Natural Gas Conversion Symposium 2019

San Antonio, Texas, USA
2 – 6 June 2019

ISBN: 978-1-5108-8888-3

Printed from e-media with permission by:

Curran Associates, Inc.
57 Morehouse Lane
Red Hook, NY 12571

Printed by Curran Associates, Inc. (2019)

For permission requests, please contact AIChE
at the address below.

AIChE
120 Wall Street, FL 23
New York, NY 10005-4020

Phone: (800) 242-4363
Fax: (203) 775-5177

www.aiche.org

Additional copies of this publication are available from:

Curran Associates, Inc.
57 Morehouse Lane
Red Hook, NY 12571 USA
Phone: 845-758-0400
Fax: 845-758-2633
Email: curran@proceedings.com
Web: www.proceedings.com

TABLE OF CONTENTS

THE CHEMISTRY AND ENGINEERING OF C1 MOLECULES: METHANE CONVERSION AND BEYOND 1
Enrique Iglesia

ENVIRONMENTAL AND COMMUNITY IMPACTS OF SHALE DEVELOPMENT IN TEXAS: A REPORT FROM THE ACADEMY OF MEDICINE, ENGINEERING AND SCIENCE OF TEXAS 2
David Allen, Christine Ehlig, Brian Stump, Melinda Taylor, Joseph Fitzsimons, Tracy Hester, Ramon Alvarez, Matthew Harrison, Danny Reible, Denny Bullard, Michael Young, John Barton, Cesar Quiroga, Gene Theodori, Omar Garcia, Urs Kreuter

MODULAR CHEMICAL PROCESS INTENSIFICATION: ENABLER FOR GAS CONVERSION 4
James Bielenberg, Ignasi Palou-Rivera

ENHANCED LIGHT ALKANE DEHYDROGENATION VIA COUPLING WITH CHEMICAL LOOPING 5
Casper Brady, Xiaoxia Chang, Eric Steinman, Shoucheng Du, Bingjun Xu

ADVANCED NANOCOMPOSITE MEMBRANES FOR NATURAL GAS PURIFICATION 6
Jaesung Park, Hee Wook Yoon, Benny D. Freeman

DISTRIBUTED NATURAL GAS CONVERSION WITH SCALABLE, PROCESS-INTENSIFIED GAS CLEANUP AND REACTOR TECHNOLOGIES 7
Paul S. Dimick

CATALYTIC DEHYDROAROMATIZATION OF ETHANE IN MICROWAVE REACTOR 9
Jianli Hu

CHALLENGES AND SOLUTIONS TO SCALABLE TESTING OF CATALYSTS FOR CONVERTING SYNGAS TO CHEMICALS AND FUELS 10
Anton J. Nagy, Thomas Turek

INNOVATIVE ZONEFLOW® CATALYST OFFERS UNMATCHED SOLUTIONS IN STEAM REFORMING FOR SYNGAS BASED NG CONVERSION 13
Sanjiv Ratan

ENABLING DISTRIBUTED GAS-TO-LIQUID (GTL) OPPORTUNITIES AT THE MICRO-SCALE 15
Sameer Parvathikar, Pradeepkumar Sharma, Atish Kataria, John R. Carpenter

ENERGY OUTLOOK: ROLE OF NATURAL GAS 17
Vijay Swarup

INNOVATION IN THE CHEMICAL INDUSTRY: IMPLICATIONS AND IMPACTS OF THE POST-SHALE WORLD 18
Sandeep S. Dhingra

NATURAL GAS CONVERSION IN MICROPOROUS CATALYSTS; ZEOTYPES AND METAL ORGANIC FRAMEWORKS 19
Unni Olsbye

HOT OXYGEN BASED SYNTHESIS GAS GENERATION 21
Brad Damstedt, Lawrence Bool III, Ines Stuckert

EVALUATION OF THE ABSORPTION PERFORMANCE ON THE OF ACID GAS REMOVAL FROM NATURAL GAS AND SYNTHETIC GAS PROCESSING 22
Kreangkrai Maneeintr, Thanaphat Luemunkong, Kyuro Sasaki

KINETIC PARAMETERS OPTIMIZATION AND ANALYSIS IN PACKED BED STEAM METHANE REFORMER 25
Saiedeh Arabi, Sandeep Patil, Siddarth Chintamani, Brian Dennis

ENHANCED METHANE REFORMING OVER HIERARCHICAL STRUCTURED SUPPORTED NI BASED CATALYST 26
Anup Tathod, Naseem S. Hayek, David Simakov, Oz M. Gazit

ELUCIDATION OF THE REACTION MECHANISM ON LOW TEMPERATURE DRY REFORMING OF METHANE WITH SUPPRESSING CARBON DEPOSITION IN ELECTRIC FIELD 28
Kazuharu Ito, Tomohiro Yabe, Kensei Yamada, Kota Murakami, Kenta Toko, Takuma Higo, Shuhei Ogo, Yasushi Sekine

LOW TEMPERATURE STEAM REFORMING OF METHANE OVER PD-ZN ALLOY CATALYSTS IN AN ELECTRIC FIELD 29
Maki Torimoto, Danny Harjowinoto, Shuhei Ogo, Shinya Furukawa, Takuma Higo, Tomohiro Yabe, Yasushi Sekine

CO-BASED CATALYST WITH HIGH STABLE ACTIVITY FOR SYNGAS PRODUCTION BY METHANE CONVERSION 30
Sholpan Itkulova, Yerzhan Boleubayev, Kirill Valishevsky

CO RICH SYNGAS PRODUCTION BY NTP ASSISTED NI/γ-AL_2O_3 CATALYST PACKED REACTOR: INFLUENCE OF BASIC PROMOTERS 33
Debjyoti Ray, Piu Chawdhury, Surbahmanyam Ch

THE EFFECT OF BARIUM IN NI-LA BASED CATALYSTS ON DRY REFORMING OF METHANE 34
Soraia Brandão

EFFECT OF MANGANESE AND ZEOLITE COMPOSITION ON ZEOLITE-SUPPORTED NICKEL CATALYSTS FOR DRY REFORMING OF METHANE 37
Aaron Najfach, Catherine Almquist

PLASMA-CAO COUPLING FOR CH_4-CO_2 TRANSFORMATION INTO SYNGAS AND/OR HYDROCARBONS, OXYGENATES. CATALYST STRUCTURAL AND TEXTURAL MODIFICATIONS UNDER PLASMA DISCHARGE 38
Nassim Bouchoul, Elodie Fourre, Jean-Michel Tatibouët, Alysson Duarte, Nathalie Tanchoux, Catherine Batiot-Dupeyrat

DRY REFORMING OF METHANE IN MOLTEN METAL ALLOYS 42
Clarke Palmer, David Chester Upham, Simon Smart, Michael Gordon, Horia Metiu, Eric W. McFarland

A STABILITY ANALYSIS OF ELECTROLESS DEPOSITION-DERIVED NI-PT CATALYSTS FOR THE DRY REFORMING OF METHANE 43
Benjamin T Egelske, Jayson Keels, John R. Monnier, John R. Regalbuto

UTILIZING CO_2 FOR PRODUCTION OF CO RICH SYNTHESIS GAS AT LOW STEAM TO CARBON RATIO 46
Peter M. Mortensen, Martin Østberg, Kim Aasberg-Petersen

METHANE PYROLYSIS ON COMPLEX LIQUIDS: REACTIVE SEPARATION OF HYDROGEN AND SOLID CARBON 49
Eric McFarland, Horia Metiu, Michael Gordon, Michael F. Doherty, Clarke Palmer, Jiren Zeng, Dohyung Kang, Nazanin Rahimi

SELECTIVE PHOTOCATALYTIC CONVERSION OF METHANE INTO CARBON MONOXIDE OVER ZINC-HETEROPOLYACID-TITANIA NANOCOMPOSITES 50
Xiang Yu, Vitaly Ordomsky, Andrei Khodakov

CATALYTIC METHANE STEAM REFORMING AT LOW TEMPERATURE BY SURFACE PROTONICS 54
Yasushi Sekine

ZONEFLOW STRUCTURED CATALYTIC REACTORS FOR STEAM METHANE REFORMING: MULTI-SCALE MODELLING AND PILOT PLANT TESTING 55
Florent Minette, Sanjay Katheria, Jianguo Xu, Sanjiv Ratan, Juray De Wilde

ADAPTING TO CHANGING MARKET CONDITIONS, THE EVOLUTION OF REFORMING TECHNOLOGIES TO MEET TODAY'S HYDROGEN SUPPLY REQUIREMENTS 56
Jeffrey Kloosterman

THE INITIAL STAGE OF METAL DUSTING CORROSION OF INCONEL 601 – EFFECTS OF EXPOSURE CONDITIONS AND NEAR-SURFACE STRUCTURE AND COMPOSITION 57
Hilde Venvik, Xiaoyang Guo, Estelle Marie M. Vanhaecke, Per Erik Vullum, Jianyu Ma, Daham Gunawardana Panditha-Vidana, John C. Walmsley, De Chen

OTM SOLID STATE COMBINED REFORMER 60
Ines Stuckert

COST EFFICIENT PROCESS DESIGNS THROUGH AUTOTHERMAL REFORMING 61
Esben Sorensen

DYNAMICS AND MECHANISM OF CARBON FILAMENT FORMATION DURING METHANE REFORMING ON SUPPORTED NICKEL CATALYSTS 62
Samuel L. Leung, Junmei Wei, William L. Holstein, Miguel Avalos-Borja, Enrique Iglesia

A FIRST PRINCIPLES BASED MICROKINETICS MODEL OF DRY REFORMING OF METHANE OVER NICKEL-BASED BIMETALLIC CATALYTIC SYSTEMS 65
Mohammed Minhaj Ghouri, Nimir O. Elbashir, Shaik Afzal

KINETIC ANALYSIS OF METHANE DRY REFORMING ON RH IN SPATIALLY-RESOLVED OPERANDO-RAMAN ANNULAR REACTOR 67
Gianluca Moroni, Matteo Maestri, Alessandro Donazzi

FUNDAMENTALS OF THE REMOVAL OF TRACE SULFUR COMPOUNDS FROM NATURAL GAS DERIVED STREAMS VIA REACTIVE SORPTION WITH COPPER OXIDE 70
Dante Simonetti, Sara Azzam, Faisal H. Alshafei, Richa Ghosh

SEPARATION OF METHANE AND CARBON DIOXIDE USING ZEOLITE@ZIF-95 COMPOSITE 73
Fangyuan Tian
INTENSIFYING SYNGAS PRODUCTION: EXPERIMENTAL INVESTIGATIONS AND PROCESS ANALYSIS FOR PRODUCTION OF INHERENTLY SEPARATED SYNGAS STREAMS 74
Hari C. Mantripragada, Amey More, Götz Veser
PEROVSKITE NANOCOMPOSITES FOR METHANE PARTIAL OXIDATION AND CO_2 SPLITTING IN A CYCLIC REDOX SCHEME 78
Fanxing Li
CHEMICAL LOOPING REFORMING FOR THE PRODUCTION OF HYDROGEN AND CHEMICALS FROM NATURAL GAS 79
Andrew Tong, Tien-Lin Hsieh, Dikai Xu, Yitao Zhang, Dawei Wang, Sourabh Nadgouda, Mandar Kathe, Fanhe Kong, Liang-Shih Fan
MONODISPERSED NI AND NI-CO NANOPARTICLES CAPSULATED WITH SILICA FOR DRY REFORMING OF METHANE 80
Kyung Soo Park, Jong Wook Bae
DRY REFORMING USING MODIFIED METAL CARBIDES 84
Carlos A. Carrero, RajThakur, Justin Smith
EFFECT OF A-SITE NATURE ON $MNIO_3$ (M= LA, SR AND CE) PEROVSKITE-TYPE PRECURSORS APPLIED ON OXY-DRY REFORMING OF CH_4 85
Soraia Brandão, Breno Cerqueira da Silva Sr., Roberto Batista da Silva Jr., , Pedro H. C. Bastos
COMBUSTION SUPPRESSION IN LOW-TEMPERATURE ELECTRO-CATALYTIC TRI-REFORMING OF METHANE OVER NI-MG SUPPORTED CATALYSTS 89
Tomohiro Yabe, Kensei Yamada, Shuhei Ogo, Yasushi Sekine
COMBINED METHANE DRY REFORMING AND METHANE PARTIAL OXIDIZATION FOR SYNGAS PRODUCTION OVER HIGH DISPERSION NI BASED MESOPOROUS AL2O3 CATALYST 90
Qingxiang Ma, Yunxing Han, Jianli Zhang, Tiansheng Zhao
SHALE GAS & BIOMASS CO-GASIFICATION FOR SYNGAS PRODUCTION 91
Amoolya Lalsare, Ali Sivri, Cosmin Dumitrescu, Jianli Hu
GAS CONVERSION: TECHNOLOGY AT THE TIPPING POINT? 93
David Dankworth
OX-ZEO BIFUNCTIONAL CATALYST CONCEPT ENABLES SYNGAS CONVERSION TO LIGHT OLEFINS AND BEYOND 94
Feng Jiao, Xiulian Pan, Junhao Yang, Gen Li, Na Li, Yuxiang Chen, Xinhe Bao
COOL GTL- A NEW PROCESS FROM GTL 97
Terry Marker, Jim Wangerow, Pedro Ortiz-Toral, Martin Linck, Ron Stanis
ADVANCES IN GAS-TO-LIQUIDS TECHNOLOGY AT SHELL 98
Leendert Bezemer
CARBON MOLECULAR SIEVE (CMS) MEMBRANES AS POTENTIAL GAME-CHANGERS IN OLEFINS UNIT OPERATIONS 100
William J. Koros
ADVANCED POLYMER MEMBRANES FOR GAS SEPARATIONS RELEVANT TO SHALE GAS PROCESSING 101
J.C. Davilla, Alec Bridge, Joan F. Brennecke, Benny D. Freeman
CATALYTIC APPLICATIONS BY USING A HIGHLY CRYSTALLINE FERRIERITE-BASED BIFUNCTIONAL CATALYST FOR SYNGAS (CONTAINING CO_2) CONVERSIONS 102
Jong Wook Bae
NOVEL E-IRON CARBIDE FISCHER-TROPSCH CATALYSTS WITH STABLE AND LOW-CO_2 SELECTIVITY 104
Peng Wang, Wei Chen, Fu-Kuo Chiang, A. Iulian Dugulan, Yuanjun Song, Robert Pestman, Kui Zhang, Jinsong Yao, Bo Feng, Ping Miao, Wayne Xu, Zhuowu Men, Emiel Hensen
UNDERSTANDING AND MODELING THE DEACTIVATION OF AN INDUSTRIAL NI/AL_2O_3 CATALYST FOR CO_2 METHANATION IN A MILLI-STRUCTURED FIXED-BED REACTOR-EXCHANGER 105
Isabelle Champon, Sébastien Thomas, Anne Cecile Roger, Alain Bengaouer, Albin Chaise
A NEW, SUSTAINABLE VISIBLE-LIGHT-DRIVEN METHANOL TO ETHYLENE GLYCOL (MTEG) PROCESS 109
Shunji Xie, Zebin Shen, Jiao Deng, Pu Guo, Qinghong Zhang, Haikun Zhang, Chao Ma, Zheng Jiang, Jun Cheng, Dehui Deng, Ye Wang

THE COMBINED ROLE OF CO AND WATER ON THE STABILITY OF COBALT FISCHER-TROPSCH CATALYSTS 112
Moritz Wolf, Nico Fischer, Michael Claeys
WATER AS KEY TO ACTIVITY AND SELECTIVITY IN CO FISCHER-TROPSCH SYNTHESIS 115
Erling Rytter, Øyvind Borg, Nikolaos Tsakoumis, Bjørn Christian Enger, Anders Holmen
DEACTIVATION OF THE COBALT FISCHER-TROPSCH CATALYST - A KINETIC STUDY 120
Ljubisa Gavrilovic, Jan Brandin, Anders Holmen, Hilde Venvik, Rune Myrstad, Kumar Ranjan Rout, Erling Rytter, Magne Hillestad, Edd Blekkan
CONVERSION OF NATURAL GAS TO ENERGY THROUGH PEMFC TO APPLY AT COMMERCIAL APPLICATION 124
Abhi Soni, Hiralkumar Morker, Naim Memon, Vaishali Umrigar, Sopariwala Vishnu Manoj
DEVELOPMENT OF A HIGHLY EFFICIENT AMMONIA SYNTHESIS USING BASE METALS IN AN ELECTRIC FIELD 128
Kota Murakami, Yuta Tanaka, Ryuya Sakai, Tomohiro Yabe, Shuhei Ogo, Yasushi Sekine
ANALYSIS AND OPTIMIZATION OF GAS TO LIQUID (GTL) PROCESS USING UNCONVERTED SYNGAS RECYCLE 129
Saiedeh Arabi, Sandeep Patil, Brian Dennis
CFD STUDY OF LOW TEMPERATURE FISCHER-TROPSCH REACTOR IN JET FUEL APPLICATION 130
Saiedeh Arabi, Sandeep Patil, Brian Dennis
CATALYST AND PROCESS DEVELOPMENT FOR REACTIONS WITH GAS RECYCLE: CHALLENGES AND OPPORTUNITIES 131
Peter Kolb
EFFECT OF POTASSIUM ON HIGHLY DISPERSED IRON NANOPARTICLES SUPPORT ON CARBON FOR CO HYDROGENATION 133
Joakim Tafjord
KINETICS FOR COBALT FISCHER-TROPSCH SYNTHESIS; A MULTI-DIMENSIONAL TASK 137
Magne Hillestad, Erling Rytter, Mohammad Ostadi
DEVELOPMENT OF HIGHLY SELECTIVE, WATER TOLERANT COBALT CATALYSTS FOR FISCHER-TROPSCH SYNTHESIS 140
Denzil Moodley, Thys Botha, Jana Potgieter, Michael Claeys, Eric van Steen
CO-NI SUPPORTED ON REDUCIBLE OXIDES AS HIGHLY ACTIVE AND SELECTIVE FISCHER-TROPSCH CATALYSTS 141
Carlos Hernández Mejía, Jessi E.S. van der Hoeven, Petra E. de Jongh, Krijn de Jong
IMPROVED CO/TI CATALYSTS BY DIRECT REDUCTION OF COBALT NITRATE AND THE INCORPORATION OF PROMOTERS 145
Mohammad Mehrbod, Michela Martinelli, Annabelle Martino, Burtron H. Davis, Donald Cronauer, A. Jeremy Kropf, Christopher Marshall, Gary Jacobs
OPTIMIZING OXYGENATES PRODUCTION IN THE FISCHER TROPSCH REACTION OVER ALKALI-PROMOTED CO/MN CATALYSTS 148
Norbert Kruse, Yizhi Xiang, Libor Kovarik
MOF-MEDIATED SYNTHESIS OF HIGHLY ACTIVE FE- AND CO-BASED FTS CATALYSTS 152
Freek Kapteijn, Giusy Petty, Christian van der Sande, Patricia S Garcia Guerrero, Lide Oar-Arteta, Tim A. Wezendonk, Xiaohui Sun, Jorge Gascon
METHANOL SYNTHESIS FROM CO/CO_2/H_2 AND CO/H_2 OVER CU/ZNO/AL_2O_3 AND CU/CEO_2 CATALYSTS 154
Leon G.A. van de Water, Sam Wilkinson, Brendon Miller, Mark. J Simmons, E. Hugh Stitt, Michael J. Watson
CO_2 CONVERSION TO MEOH-DME FUELS:CATALYTIC AND TECHNOLOGICAL ASPECTS 158
Francesco Frusteri, Catia Cannilla, Fabio Costa, Aldo Mezzapica, Giuseppe Bonura
CARBON DIOXIDE HYDROGENATION INTO METHANOL WITH CUO-ZNO-ZRO_2 CATALYST PREPARED BY A CONTINUOUS FLOW MICROFLUIDIC-ASSISTED SYNTHESIS 160
L'hospital Valentin, Parkhomenko Ksenia, Christophe Serra, Anne Cecile Roger
OXYGENATE FORMATION OVER K/MO_2C CATALYSTS - ROLE OF PREPARATION AND PROMOTION 164
Wijnand Marquart, David J. Morgan, Graham J. Hutchings, Michael Claeys, Nico Fischer
HIGH THROUGHPUT TESTING OF CATALYSTS WITH FAST DEACTIVATION FOR METHANOL-TO-HYDROCARBONS (MTH) 168
Marius Kirchmann, Alfred Haas, Christoph Hauber
SELECTIVE CONVERSION OF SYNGAS TO CHEMICALS AND FUELS ENABLED BY BIFUNCTIONAL OX-ZEO CATALYSTS 170
Feng Jiao, Junhao Yang, Gen Li, Na Li, Yuxiang Chen, Xiulian Pan, Xinhe Bao

INTERMEDIATE PRODUCT REGULATION IN TANDEM SYNGAS CONVERSION TO LIQUID FUELS 171
Gonzalo Prieto
BIFUNCTIONAL CATALYSIS FOR DIRECT TRANSFORMATIONS OF SYNGAS AND CARBON DIOXIDE INTO LOWER OLEFINS AND AROMATICS 173
Xiaoliang Liu, Mengheng Wang, Jincan Kang, Kang Cheng, Qinghong Zhang, Ye Wang
TRANSIENT KINETICS OF THE FISCHER-TROPSCH REACTION ON COBALT CATALYSTS 177
Wei Chen, Robert Pestman, Emiel Hensen
SOME ENDURING MECHANISTIC PUZZLES IN THE FISCHER-TROPSCH SYNTHESIS REVISITED 182
David Hibbitts, Jianwei Liu, Hale Ay, Brett Loveless, Matthew Neurock, Enrique Iglesia
CO AND H_2 ADSORPTION ON CO(11-20) - A COMBINED EXPERIMENTAL AND THEORETICAL INVESTIGATION 183
Hilde Venvik, Marie Døvre Strømsheim, Mari Helene Farstad, Ingeborg-Helene Svenum, Mehdi Mahmoodinia, Kees-Jan Weststrate, Anne Borg
MECHANISTIC UNDERSTANDING OF METHANOL CARBONYLATION: INTERFACING HOMOGENEOUS AND HETEROGENEOUS CATALYSIS VIA CARBON SUPPORTED IR-LA 186
Alyssa Hensley, Jianghao Zhang, Ilka Vinçon, Xavier Isidro Pereira Hernández, Diana Tranca, Gotthard Seifert, Jean-Sabin McEwen, Yong Wang
COBALT NANOCRYSTALS AS MODEL CATALYSTS FOR THE FISCHER-TROPSCH SYNTHESIS 192
Tom W. van Deelen, Krijn de Jong
ZEOLITE CATALYSTS AS ENABLING MATERIALS FOR CARBON MANAGEMENT 195
Ulrich Müller, S. Altwasser, Craig Cain-Borgman, A. Gordillo, A. Parvulescu, L. Karwacki
DIRECTLY PRODUCING LIGHT AROMATICS FROM SYNGAS OVER MODIFIED CEZROX/ZSM-5 BI-FUNCTIONAL CATALYSTS 197
Hualong Xu
SOLDERING PROMOTERS FOR IRON AND COBALT FISCHER-TROPSCH CATALYSTS 200
Bang Gu, Vitaly Ordomsky, Mounib Bahri, Ovidiu Ersen, P.a. Chernavskii, Andrei Khodakov
PROMOTED COBALT METAL CATALYSTS SUITABLE FOR THE PRODUCTION OF CHEMICALS AND FUELS FROM NATURAL GAS 202
Jingxiu Xie, Pasi Paalanen, Tom W. van Deelen, Bert M. Weckhuysen, Manuel J. Louwerse, Krijn de Jong
TWO-DIMENSIONAL GRAPHENE-ASSISTED FORMATION OF THE CYLINDRICAL IRON CARBIDE NANOCAPSULES FOR FISCHER-TROPSCH SYNTHESIS 204
Feng Jiang
INCREASED CO_2 HYDROGENATION TO LIQUID PRODUCTS USING PROMOTED IRON CATALYSTS 205
Wilson D. Shafer, Gary Jacobs, Uschi M. Graham, Hussein H. Hamdeh, Burtron H. Davis
AIR LIQUIDE RETHINKS THE PROCESS FOR TRANSFORMING CO2 TO METHANOL 206
Timm Schuhmann, Stéphane Haag, Holger Schlichting, Nga Thi Quynh Do, Martin Gorny, Veronika Gronemann, Tobias Oelmann, Bryce A. Williams
BEHAVIOUR OF CUZNAL AND CUZNZR FERRIERITE-BASED CATALYSTS TO PRODUCE DME BY CO_2 HYDROGENATION 208
Fabio Salomone, Giuseppe Bonura, Francesco Frusteri, Samir Bensaid, Raffaele Pirone
DIRECT CO_2 HYDROGENATION INTO LIQUID FUELS AND CHEMICALS WITH HIGH SELECTIVITY 212
Peng Gao Sr., Hui Wang, Yuhan Sun
CATALYTIC HYDROGENATION OF CO_2 TO ISOPARAFFINS OVER FE-BASED MULTIFUNCTIONAL CATALYSTS 213
Jian Wei, Ruwei Yao, Jian Sun, Qingjie Ge
ZEOLITE A CATALYSTS REDUCED BY $NABH_4$ FOR HYDROFORMYLATION OF 1-HEXENE 216
Yuan Fang, Guohui Yang, Yoshiharu Yoneyama, Noritatsu Tsubaki
DIRECTLY CONVERTING CARBON DIOXIDE TO LINEAR I ± -OLEFINS ON BIO-PROMOTED CATALYSTS 217
Lisheng Guo, Jian Sun, Qingjie Ge, Noritatsu Tsubaki
INVESTIGATION OF THE ACTIVE PHASES IN ALKALI METAL PROMOTED MOLYBDENUM SULFIDE CATALYSTS FOR THE SYNTHESIS OF METHANETHIOL FROM SYNGAS AND HYDROGEN SULFIDE 218
Miao Yu, Nikolay Kosinov, Emiel Hensen, Wei Chen

IMPACT OF CATALYST DEACTIVATION ON A CARBON DIOXIDE METHANATION "POWER-TO-GAS" PROCESS: DEACTIVATION KINETIC STUDY AND REACTOR MODELLING 222
Eduard Alexandru Morosanu, Albin Chaise, Alain Bengaouer, Samir Bensaid

CO HYDROGENATION OVER NI CATALYSTS SUPPORTED ON MAL_2O_4@AL (M=MN, MG, ZN) 224
Thien An Le, Ji Eun Kim, Jong Kyu Kang, Eun Duck Park

MECHANISTIC INSIGHTS IN THE METHANOL-TO-HYDROCARBON PROCESS USING ADVANCED AND COMBINED OPERANDO SPECTROSCOPY 225
Sophie H. van Vreeswijk, Joris G. Goetze, Bert M. Weckhuysen

KINETIC PARAMETERS OPTIMIZATION OF FISCHER-TROPSCH PROCESS FOR OIL PRODUCTION AND EXPERIMENTAL VALIDATION 229
Saiedeh Arabi, Sandeep Patil, Ashkan Akbariyeh, Siddarth Chintamani, Brian Dennis

EVALUATING CU/BEA CATALYST PERFORMANCE WITH TECHNO-ECONOMIC ANALYSIS TO DEVELOP A MARKET-RESPONSIVE BIOREFINERY CONCEPT AROUND THE CONVERSION OF METHANOL TO HIGH-OCTANE HYDROCARBONS 230
Daniel Ruddy, Jesse E. Hensley, Connor Nash, Eric Tan, Earl Christensen, Carrie A. Farberow, Joshua A. Schaidle

CATALYTIC OPTIONS FOR METHANE VALORIZATION CATALYST PERFORMANCE EVALUATION USING PARALLEL FIXED BED REACTOR SYSTEMS AND DATA DRIVEN CATALYST DEVELOPMENT 232
Roel Moonen, Amy Triesscheijn, Erik-Jan Ras, Carlos Ortega

ONE-PASS TUNABLE FORMATION OF CHEMICALS OR SUPER-CLEAN FUELS VIA SYNGAS CONVERSION: PRECISE DESIGN OF MULTIPLE FUNCTIONAL CATALYSTS 234
Noritatsu Tsubaki

BIFUNCTIONAL CATALYSTS FOR SELECTIVE CONVERSION OF SYNGAS BEYOND FISCHER-TROPSCH SYNTHESIS 237
J. Lennart Weber, Daniel Martinez, Carmen Martos, Arturo Vizcaino, Petra E. de Jongh, Krijn de Jong

DIRECT CONVERSION OF SYNGAS TO LIGHT OLEFINS: CONSTRUCTION OF REACTION NETWORK AND DESIGN OF OXIDE-ZEOLITE BIFUNCTIONAL CATALYSTS 241
Yangdong Wang, Junjie Su, Haibo Zhou, Wenqian Jiao, Su Liu, Dong Wang, Weimin Yang, Zaiku Xie, Mingyuan He

CONVERSION OF SYNTHESIS GAS TO LIGHT OLEFINS: IMPACT OF HYDROGENATION ACTIVITY ON PROCESS SELECTIVITY OVER CR-ZN AND CU-ZN WITH SAPO-34 242
Alexey Kirilin, Joseph DeWilde, Adam Chojecki, Vera Santos, Kinga Scieranka, Andrzej Malek

THE EFFECTS OF INTRAPARTICLE DIFFUSION PHENOMENA ON DIMETHYL ETHER DIRECT SYNTHESIS 245
Simone Guffanti, Carlo Giorgio Visconti, Gianpiero Groppi

DESIGNING A HIERARCHICAL NANOSHEET ZSM-35 ZEOLITE TO REALIZE MORE EFFICIENT ETHANOL SYNTHESIS FROM DIMETHYL ETHER AND SYNGAS 247
Xiaobo Feng, Guohui Yang, Yoshiharu Yoneyama, Noritatsu Tsubaki

ADOPTION OF HIGHLY CONDUCTIVE OPEN CELL FOAMS AS AN EFFECTIVE SOLUTION TO ENHANCE THE HEAT TRANSFER PERFORMANCES OF PACKED-BED FISCHER-TROPSCH REACTORS 248
Laura Fratalocchi, Carlo Giorgio Visconti, Luca Lietti, Gianpiero Groppi, Enrico Tronconi

FISCHER-TROPSCH SYNTHESIS ON FREEZE-CAST HYBRID-BACKBONE MESO-MACROPOROUS MICROMONOLITH CATALYSTS: RECONCILING HIGH PORE TRANSPORT RATES WITH IMPROVED REACTION HEAT MANAGEMENT 252
Kai Jeske, Jonglack Kim, Valentina Nese, Jochen Joos, Nicolas Duyckaerts, Norbert Pfänder, Gonzalo Prieto

CHALLENGES AND OPPORTUNITIES IN DIRECT CONVERSION OF METHANE TO HIGHER VALUE PRODUCTS 255
Benjamin A. McCool, Randall Meyer

MICROKINETIC MODELING OF NON-OXIDATIVE METHANE COUPLING TO ETHYLENE OVER IRON/SILICA CATALYST 256
Hilal Ezgi Toraman, Konstantinos Alexopoulos, Dionisios G. Vlachos

ELECTROCHEMICAL OXIDATION OF METHANE TO OLEFINS IN HF/HSO_3F 257
Yuxin Fang, Christopher G. Arges, John C. Flake, Craig Plaisance

SYNERGY OF THIN CATALYTIC LAYERS AND DBD PLASMA FOR COUPLING OF METHANE 258
Nuria Garcia-Moncada, Leon Lefferts

STRATEGIES FOR VALORIZATION OF METHANE: A COMPUTATIONAL PERSPECTIVE 260
Ambarish Kulkarni

DIRECT METHANE CONVERSION TO OLEFINS AND AROMATICS UNDER NON-OXIDATIVE CONDITION 261
Xinhe Bao, Jianqi Hao, Guangzong Fang, Xiaoguang Guo, Pierre Schwach, Xiulian Pan
CATALYSIS OF LIQUID-METAL INDIUM FOR DIRECT DEHYDROGENATIVE CONVERSION OF METHANE INTO HIGHER HYDROCARBONS 263
Yuta Nishikawa, Ayumi Nakaya, Hitoshi Ogihara, Yuki Ohtsuka, Akira Nakayama, Jun-ya Hasegawa, Shoji Iguchi, Ichiro Yamanaka
METHANE DEHYDROAROMATIZATION IS MEDIATED BY SURFACE CARBON 267
Nikolay Kosinov, Emiel Hensen
ON THE DYNAMIC NATURE OF MO SITES FOR METHANE DEHYDROAROMATIZATION 269
Freek Kapteijn, Ina Vollmer, Bart van der Linden, Samy Ould-Chikh, Irina Yarulina, Alma I. Olivos Suarez, Jorge Gascon
DIRECT METHANE CONVERSION TO VALUE-ADDED PRODUCTS OVER FE-FER CATALYSTS 273
Guangyu Zhao, Michael Stockenhuber, Eric M. Kennedy, Adesina Adesoji A., Meng-Jung Li, Matthew Drewery
CONTROL OF KINETIC AND DIFFUSIVE LENGTH-SCALES DURING ABSORPTIVE HYDROGEN REMOVAL FOR ENHANCED AROMATICS YIELD IN METHANE DEHYDROAROMATIZATION ON MO/H-ZSM-5 CATALYSTS 277
Neil Razdan, Anurag Kumar, Aditya Bhan
A STABLE MO@HZSM-5 CATALYST FOR DIRECT CONVERSION OF METHANE TO AROMATICS 283
Yi Liu, Tianyun Wang, Franklin (Feng) Tao, Yi Zhang
ACTIVITY AND SELECTIVITY OF OXIDATIVE COUPLING OF METHANE ON DOPED LA2O3 CATALYSTS: A DENSITY FUNCTIONAL THEORY STUDY 286
Zhiqiang Wang, Xueqing Gong
BIFUNCTIONAL CATALYSIS INVOLVING SURFACE COUPLING FOR LOW TEMPERATURE OXIDATIVE COUPLING OF METHANE 287
Zhinian Li, Lei He, Shenliang Wang, Shihui Zou, Liping Xiao, Jie Fan
MICROWAVE-ASSISTED NATURAL GAS CONVERSION IN PRESENCE OF LOW RANK COAL 293
Victor Abdelsayed, Mark W. Smith, Dushyant Shekhawat, Sonia Hammache
RESEARCH ON THE STABILIZER TO NON-CATALYTIC COMBUSTION REACTION IN OXIDATIVE COUPLING OF METHANE 294
Li Wei, Qingrui Zhao, Mingsen Zhang
CATALYTIC DEHYDROAROMATIZATION OF METHANE: STRATEGIES FOR ENHANCING THE YIELD TO AROMATICS AND IMPROVING COKING RESISTANCE 295
Sheima J. Khatib, Apoorva Sridhar, Mustafizur Rahman
LOW TEMPERATURE SELECTIVE OXIDATION OF METHANE USING GOLD-PALLADIUM BASED CATALYSTS 296
Nishtha Agarwal, Graham J. Hutchings, Qian He, Christopher Kiely, Sultan Althahban, Simon Freakley, Nicholas Dummer
SELECTIVE OXIDATION OF METHANE TO METHANOL WITH MOLECULAR OXYGEN USING AQUEOUS AU-PD COLLOIDS AT LOW TEMPERATURE 297
N. Agarwal, S. Freakley, Sultan M. Althahban, C.J. Kiely, G.J. Hutchings
SELECTIVE OXIDATION, REFORMING AND COMBUSTION OF METHANE OVER NIO/NI ON CERIA-ZIRCONIA 298
Yimeng Lyu, Jennifer Jocz, Rui Xu, Eli Stavitski, Carsten Sievers
DRIFTS STUDIES ON ZEOTYPES: TOWARDS FUNDAMENTAL UNDERSTANDING OF THE DIRECT CONVERSION OF METHANE TO METHANOL 299
Simone Creci, Per-Anders Carlsson, Magnus Skoglundh
SELECTIVE OXIDATION OF METHANE TO METHANOL BY FE AND CU-EXCHANGED ZEOLITES: A SPECTROSCOPIC INVESTIGATION 301
Dieter Plessers
IDENTIFICATION OF THE REACTION PATHWAY AND ACTIVE SITE FOR THE CATALYTIC CONVERSION OF METHANE TO METHANOL OVER COPPER-EXCHANGED ZEOLITES 304
Kimberly Dinh, Mark Sullivan, Pedro Serna, Randall Meyer, Mircea Dinca, Yuriy Roman
METHANE TO METHANOL CONVERSION OVER CU-ZEOLITES 305
Elisa Borfecchia, Dimitrios K. Pappas, Michael M. Dyballa, Kirill A. Lomachenko, Andrea Martini, Gloria Berlier, Bjornar Arstad, Carlo Lamberti, Silvia Bordiga, Unni Olsbye, Pablo Beato, Stian Svelle
BINUCLEAR TRANSITION METAL ION CENTERS IN ZEOLITES FOR SELECTIVE OXIDATION OF METHANE 307
Jiri Dedecek, Edyta Tabor, Mariia Lemishka, Kinga Mlekodaj, Stepan Sklenak, Zdenek Sobalik

IMPACT OF AI DISTRIBUTION IN THE CU-EXCHANGED CHA-TYPE ZEOLITE ON THE CATALYTIC PERFORMANCE IN CH_4 CONVERSION 309
Toshiyuki Yokoi, Ryota Osuga, Yusuke Kunitake, Junko N. Kondo

SELECTIVE OXIDATION OF METHANE TO METHANOL ON CU/SBA-15 VIA A STEPWISE MANNER 311
Ha V. Le

MECHANISTIC INSIGHTS ON THE CONVERSION OF METHANE INTO METHANOL OVER CU-ZSM-5 ZEOLITE 315
Jian-Feng Wu, Xu-Dong Gao, Long-Min Wu, Si-Min Yu, Shi Bai, Wei Wang

ACTIVATION OF OXYGEN - KEY TO THE OXIDATIVE COUPLING OF METHANE (OCM) REACTION 317
Lukas Thum, Prof. Dr. Reinhard Schomäcker, Annette Trunschke, Robert Schlögl

EXPERIMENTAL AND THEORETICAL MECHANISTIC INSIGHTS INTO THE OXIDATIVE COUPLING OF METHANE WITH SOFT OXIDANTS 318
Tobin J. Marks, Shanfu Liu

CATALYST DEVELOPMENT FOR AUTOTHERMAL ADIABATIC OPERATION OF OXIDATIVE COUPLING OF METHANE 319
Wugeng Liang, Luanyi Li, Hector Perez, Sarsani Sagar, Tian Gu, David West, Balakotaiah Vemuri

CRYSTALLIZATION OF METAL OXIDES WITH WELL-DEFINED MORPHOLOGIES FOR CONVERTING METHANE TO HIGH-VALUE CHEMICALS 320
Mariano Susman, Hien N. Pham, Abhaya Datye, S. Chinta, Jeffrey D. Rimer

LOW-TEMPERATURE ONE-STEP CONVERSION OF METHANE TO METHANOL AND ACETIC ACID WITH SUPPORTED RHODIUM SINGLE CATIONS USING MOLECULAR OXYGEN 321
Mengwei Li, Junjun Shan, Maria Flytzani-Stephanopoulos

DIRECT GAS-PHASE CONVERSION OF METHANE INTO METHANOL WITH WATER 323
Jong Kyu Kang, Sae Ha Lee, Yu-ri Jeong, Eun Duck Park

SELECTIVE OXIDATION OF METHANE TO OXYGENATES IN AQUEOUS HYDROGEN PEROXIDE USING BIMETALLIC AU-PD SUPPORTED BY DIFFERENT NANOSTRUCTURED MATERIALS 324
Saeed Alshihri

HIGH EFFICIENCY GAS-TO-LIQUIDS THROUGH CONDENSATION OF SMALL HYDROCARBONS USING CORONA DISCHARGE PROCESSES 325
Yu Miao, Alexandre Yokochi, Goran Jovanovic, Nick AuYeung, Annette von Jouanne, Ryan Collin, Ian Reddick, Adam Shareghi, Andrew Traverso

AN AUTOMATICALLY GENERATED PRESSURE-DEPENDENT PAH MECHANISM FOR RICH-METHANE COMBUSTION 326
Te-Chun Chu, Zachary J. Buras, Patrick Oßwald, Mengjie Liu, Mark J. Goldman, William H. Green

CATALYTIC CONVERSION OF METHANE INTO OXYGENATED COMPOUNDS OVER $H_4PMO_{10}WVO_{40}$ CATALYST 327
Djaouida Allam

MILD OXIDATION OF METHANE TO C_1-C_2 OXYGENATES OVER MONONUCLEAR COMPLEXES REACTION-INDUCED ON ACTIVATED CARBON 328
Yunjie Ding

SINGLE RHODIUM ATOMS ANCHORED IN MICROPORES FOR EFFICIENT TRANSFORMATION OF METHANE TO METHANOL AND ACETIC ACID UNDER MILD CONDITIONS 329
Yu Tang, Franklin (Feng) Tao, Yuting Li

A STUDY OF SELECTIVE METHANE CHLORINATION REACTION USING SOLID SUPERACID METAL OXIDE CATALYSTS 330
Jip Kim, Youngmin Kim, Kyoung-Su Ha, Ho-Jeong Chae

HISTORICAL REVIEW AND RECOMMENDATIONS ON THE OXIDATIVE COUPLING OF METHANE 331
Anne Gaffney

NEW PERSPECTIVES TO "OLD" CATALYSTS: IMPACT OF REACTOR OPERATING MODE ON THE PERFORMANCE RANKING FOR OXIDATIVE COUPLING OF METHANE 332
Laura Pirro, Pedro S.F. Mendes, Ana Obradovic, Bart D. Vandegehuchte, Guy B. Marin, Joris W. Thybaut

PROCESS INTENSIFICATION FOR OXIDATIVE COUPLING OF METHANE 334
Laurien A. Vandewalle, Manuel Nunez Manzano, Kevin M. Van Geem, Guy B. Marin

PACKED FOAMS: A NOVEL REACTOR CONFIGURATION FOR INTENSIFICATION OF THE METHANE STEAM REFORMING PROCESS 336
Riccardo Balzarotti, Matteo Ambrosetti, Tommaso Selleri, Alessanda Beretta, Gianpiero Groppi, Enrico Tronconi

DENSITY FUNCTIONAL THEORY ANALYSIS OF PROPANE DEHYDROGENATION ON PALLADIUM ALLOYS 339
Ranga Rohit Seemakurthi, Yinan Xu, Brandon C. Bukowski, Fabio H. Ribeiro, Jeffrey Greeley

SUBSURFACE EFFECTS ON SUPPORTED 2 NM PT_3MN NANOPARTICLES FOR CATALYTIC PROPANE DEHYDROGENATION 341
Zhenwei Wu, Brandon C. Bukowski, Zhe Li, Cory Milligan, Lin Zhou, Tao Ma, Yue Wu, Yang Ren, Fabio H. Ribeiro, W. Nicholas Delgass, Jeffrey Greeley, Guanghui Zhang, Jeffrey T. Miller

METAL PHOSPHIDES FOR SELECTIVE ALKANE DEHYDROGENATION 344
Jessica A. Muhlenkamp, Jeonghyun Ko, William F. Schneider, Jason C. Hicks

REGENERABLE SINGLE ATOM CATALYSTS FOR PROPANE DEHYDROGENATION 346
Griffin A. Canning, Ryan Alcala, Haifeng Xiong, Megan Senn, Nicole LiBretto, Jeffrey T. Miller, Fabio H. Ribero, Abhaya Datye

IMPROVED STABILITY OF PTSN FOR PROPANE DEHYDROGENATION BY BORON-MODIFIED ZRO2 SUPPORT 347
Zhonghai Ji, Xiulian Pan, Xinhe Bao

BORON-BASED MATERIALS AS HIGHLY SELECTIVE CATALYSTS FOR THE OXIDATIVE DEHYDROGENATION OF ALKANES 349
Ive Hermans

MIXED OXIDE REDOX CATALYSTS FOR CHEMICAL LOOPING OXIDATIVE DEHYDROGENATION OF LIGHT ALKANES 350
Fanxing Li

STABILITY OF SUPPORTED METAL OXIDE CATALYSTS FOR THE NON-OXIDATIVE DEHYDROGENATION OF PROPANE: AN IN SITU/OPERANDO RAMAN STUDY 351
Carlos A. Carrero, Jorge Moncada, Lindsey Parsons

SUPPORTED LIQUID METALS - A NEW CLASS OF ROBUST DEHYDROGENATION CATALYSTS 353
Nicola Taccardi, Mathias Grabau, Jonas Debuschewitz, Monica Distaso, Marco Brandl, Rainer Hock, Florian Maier, Christian Papp, Jannis Erhard, Christian Neiss, Wolfgang Peukert, Andreas Gorling, Moritz Wolf, Marco Haumann, Hans-Peter Steinruck, Peter Wasserscheid

NEW INSIGHT INTO ACTIVATION OF LIGHT PARAFFIN: A CASE STUDY OF ELECTROCHEMICAL DEPROTONATION OF ETHANE AT REDUCED TEMPERATURES 355
Dong Ding, Wei Wu, Lucun Wang

DOW FLUID CATALYTIC DEHYDROGENATION FOR ON-PURPOSE PROPYLENE PRODUCTION 356
Lin Luo, Matt Pretz

DECIPHERING THE REACTION NETWORK OF THE SELECTIVE OXIDATION OF BUTANE TO MALEIC ANHYDRIDE OVER $(V_{1-X}W_X)OPO_4$ CATALYSTS 357
Jingxiu Xie, Felix Pohl, Christian Schulz, Knut Wittich, Ralph Kraehnert, Benjamin Frank, Robert Glaum, Frank Rosowski

LIGHT ALKENE OLIGOMERIZATION FOR SHALE GAS UPGRADING: A MICROKINETIC MODEL 359
Sergio Vernuccio, Elsa Koninckx, Elizabeth E. Bickel, Han-Ting Tseng, Rajamani Gounder, Fabio Ribeiro, Linda J. Broadbelt

MICROKINETIC MODELING OF OXIDATIVE DEHYDROGENATION OF ETHANE TO ETHYLENE AT HIGH TEMPERATURES 362
Hilal Ezgi Toraman, Dionisios G. Vlachos

HIGH-PERFORMANCE NI-FOAM-STRUCTURED NB_2O_5-NIO NANOCOMPOSITE CATALYSTS FOR OXIDATIVE DEHYDROGENATION OF ETHANE TO ETHYLENE 363
Zhiqiang Zhang, Guofeng Zhao, Jian Zhu, Ye Liu, Yong Lu

OXIDATIVE DEHYDROGENATION OF PROPANE OVER PT-SN/SI-BETA CATALYSTS: KEY ROLE OF PT-SN INTERACTION 364
Lanlan Sun

A FRESH LOOK AT NI-BASED CATALYSTS FOR THE OXIDATIVE DEHYDROGENATION OF ETHANE 365
Kara Stowers, Justin Park, Kelsey Canizales, Morris D. Argyle

STATUS AND PROSPECTS OF THE HALOGEN-MEDIATED CATALYTIC PROCESSES FOR NATURAL GAS UPGRADING 366
Guido Zichittella, Vladimir Paunovic, Javier Perez-Ramirez

CONVERSION OF METHANE, ETHANE AND PROPANE TO FORM ALKYL ESTERS: OXY-ESTERIFICATION AS A STRATEGY FOR SELECTIVE PARTIAL OXIDATION 368
T. Brent Gunnoe, John Groves, William A. Goddard III, Nichole Schwartz, Nicholas Boaz, Steven Kalman, Jonathan Goldberg, George Fortman, Ross Fu, Robert Nielsen

CO-PROMOTING EFFECTS OF MG AND K IN ETHYLENE OXYCHLORINATION 369
Hongfei Ma, Endre Fenes, Kumar R. Rout, Terje Fuglerud, De Chen

SELECTIVE OXIDATION OF PROPANE TO PROPYLENE IN THE PRESENCE OF HCL OVER CEO2 AND NIO-MODIFIED CEO2 NANOCRYSTALS 372
Jincan Kang, Quanhua Xie, Huamin Zhang, Jun Cheng, Qinghong Zhang, Ye Wang

HALOGEN-DEPENDENT SURFACE CONFINEMENT GOVERNS SELECTIVE ALKANE FUNCTIONALIZATION TO OLEFINS 376
Guido Zichittella, Matthias Scharfe, Begoña Puértolas, Vladimir Paunovic, Patrick Hemberger, Andras Bodi, László Szentmiklósi, Núria López, Javier Pérez-Ramírez

ALKANE DEHYDROGENATION WITH LEWIS-BRONSTED ACID PAIRS IN GA/H-ZSM-5 379
Lingli Ni, Moritz Schreiber, Ricardo Bermejo-Deval, Johannes A. Lercher

ETHANE OXIDATIVE DEHYDROGENATION ON M1 PHASE MIXED METAL OXIDES 381
Prashant Deshlahra, Annamalai Leelavathi, Yilang Liu

INTRODUCTION OF SB INTO TRIGONAL MO3VOX OXIDE AND ITS CRYSTAL TRANSFORMATION TO ORTHORHOMBIC MO3VOX BY HEAT TREATMENT 383
Satoshi Ishikawa, Shoma Inukai, Toyokazu Tanabe, Wataru Ueda

ELUCIDATING THE ROLE OF CO2 IN THE SOFT OXIDATIVE DEHYDROGENATION OF PROPANE OVER CERIA-BASED CATALYSTS 386
James H. Carter, Ewa Nowicka, Christian Reece, Sultan Althahban, Khaled Mohammed, Simon Kondrat, David J. Morgan, Qian He, David Willock, Stanislaw Golunski, Christopher J Kiely, Graham J. Hutchings, Nicholas Dummer

CO_2 ASSISTED OXIDATIVE DEHYDROGENATION OF ETHANE TO ETHYLENE OVER FEO_X-BASED CATALYSTS 391
Charis Loizidis, Eleni Heracleous, Angeliki Lemonidou

EFFECTS OF BRØNSTED ACID SITE DENSITY AND PROXIMITY IN MFI AND CHA ZEOLITES ON LIGHT ALKANE ACTIVATION AND LIGHT ALKENE OLIGOMERIZATION CATALYSIS 396
Philip M. Kester, Young Gul Hur, Elizabeth E. Bickel, Fabio H. Ribeiro, Jeffrey T. Miller, Rajamani Gounder

HIGH AND STABLE ACTIVITY OF DEALUMINATED ZN/H-ZSM-5 FOR ETHANE DEHYDROAROMATIZATION 398
Hikaru Saito, Satoshi Inagaki, Qian Han, Ryota Terunuma, Tomohiro Yabe, Shuhei Ogo, Yoshihiro Kubota, Yasushi Sekine

PARAFFIN ACTIVATION AND CONVERSION TECHNOLOGY (PACT), A NEW FLEXIBLE AND ROBUST TECHNOLOGY FOR THE CONVERSION OF C5+/NATURAL GASOLINE TO HIGHER VALUED PRODUCTS 399
Jens Michael Poulsen, Anthony Baldridge

ETHENE DIMERIZATION MEDIATED BY THE COORDINATION-INSERTION MECHANISM AT NICKEL CATIONS SUPPORTED ONTO MOLECULAR SIEVES 400
Ravi Joshi, Guanghui Zhang, Jeffrey T. Miller, Rajamani Gounder

OBSERVATION OF PT MIGRATION ON PT/ZSM-5 CATALYST FOR ETHANE AROMATIZATION 402
Hui Wang, Junjun Shan, Lisa Nguyen, Joshua Miles, Hua Liu, Fu-Kuo Chiang, Jihong Cheng

ORIGIN OF SELECTIVITY IN THE DIRECT PROPYLENE EPOXIDATION REACTION USING MOLECULAR OXYGEN 404
Janvit Teržan, Petar Djinovic, Janez Zavasnik, Iztok Arcon, Gregor Zerjav, Matjaz Spreitzer, Albin Pintar

THE ROLE OF PHOSPHORUS ON VANADIA CATALYST FOR THE SELECTIVE OXIDATION OF N-BUTANE TO MALEIC ANHYDRIDE 408
Jingxiu Xie, Christian Schulz, K. Knemeyer, R. Naumann d'Alnoncourt, Benjamin Frank, Ralph Kraehnert, Frank Rosowski

IDENTIFYING THE ACTIVE SITE FOR THE V/P MIXED OXIDES 410
Gang Fu Sr.

COMBINED SOLID- AND GAS PHASE KINETIC MODELLING: ETHYLENE OXYCHLORINATIONOVER K-PROMOTED CU/AL2O3 CATALYST 412
De Chen, Terje Fuglerud, Endre Fenes, Hongfei Ma, Magnus Rønning, Samuel Regli, Yanying Qi, Kumar Rout

UNDERSTANDING THE REDOX REACTIONS OF NI AND CO DURING CATALYTIC COMBUSTION OF METHANE 417
Shirley E. Liland, Kumar R. Rout, Endre Fenes, De Chen

ATOMICALLY RESOLVED IN SITU INVESTIGATION OF CU_2O REDUCTION WITH METHANOL 419
Hao Chi, Meng Li, Matthew Curnan, Christopher M. Andolina, Judith Yang, Götz Veser, Stephen House

NICKEL-MANGANESE BIMETALLIC CATALYSTS FOR SELECTIVE HYDROGENATION OF CO_X IN THE PRESENCE OF LIGHT HYDROCARBONS: PROMOTIONAL EFFECT OF MANGANESE 422
Vahid Shadravan, Matthew Drewery, Meng-Jung Li, Eric M. Kennedy, Michael Stockenhuber
THE FEASIBILITY OF THE THERMOCATALYTIC CONVERSION OF CO2 INTO SYNTHETIC NATURAL GAS (SNG) 428
Sogol Mottaghi-Tabar, Robert Currie, Yichen Zhuang, David Simakov
HIGH-EFFICIENT CHEMICAL/ELECTROCHEMICAL CH_4 CONVERSION 430
Yongdan Li
EFFECT OF NI-FE SEGREGATION ON THE DEACTIVATION BEHAVIOR OF CO-PRECIPITATED $NIFEALO_X$ CATALYSTS IN THE CO_2 METHANATION REACTION 433
Thomas Burger, Stefan Ewald, Klaus Köhler, Olaf Hinrichsen
STRONG NI-MNO SYNERGY IN THE SABATIER REACTION 437
Emiel Hensen, Wilbert Vrijburg
PREPARATION AND TESTING OF RU/AL_2O_3 CATALYSTS FOR A COMPACT-SCALE FIXED BED CO_2 METHANATION REACTOR 439
Alessandro Porta, Leonardo Falbo, Carlo Giorgio Visconti, Luca Lietti, Claudia Bassano, Paolo Deiana
ADVANCED CHEMICAL LOOPING CONCEPT FOR ENERGY AND FUELS APPLICATIONS 443
Guy B. Marin, Vladimir V. Galvita
CONCEPTUAL DESIGN OF A DUAL FLUIDIZED BED SYSTEM FOR AN INTENSIFIED STEAM METHANE REFORMING CONCEPT COUPLED WITH CA-NI LOOPING 445
Theodoros Papalas, Andy Antzara, Angeliki Lemonidou
METHANE OXY-REFORMING COUPLED WITH A WATER GAS SHIFT MEMBRANE REACTOR FOR PURE HYDROGEN PRODUCTION 449
Francesco Basile, Gabriele Centi, Andrea Fasolini, Salvatore Abate
SYNERGY EFFECT OF TWO SETS OF SING-ATOM SITES (NI_1 AND RU_1) ON CEO_2 FOR REFORMING CH_4 452
Franklin (Feng) Tao, Yu Tang, Yu Chen, Ben deGlee, Meilin Liu
NANO-TAILORING NICKEL CATALYST SURFACES FOR CO_2 REFORMING OF METHANE 453
Anuj Prakash, Shaik Afzal, Anjaneyulu Chatla, Patrick Littlewood, Nelly Catherin, Peter C. Stair, Nimir Elbashir
DIRECT CONVERSION OF BIOGAS INTO SYNTHETIC NATURAL GAS (SNG) OVER THE 0.1-0.5WT% RU/AL2O3 CATALYST 455
Yichen Zhuang, Sogol Mottaghi-Tabar, David Simakov
LOW-TEMPERATURE CO_2 METHANATION CATALYST FOR POWER-TO-GAS 457
Audrey Waldvogel, Sébastien Thomas, Anne Cecile Roger
ADVANCED REACTOR TECHNOLOGY FOR CO_2 UTILIZATION IN METHANE REFORMING 462
Mohamed Sufiyan Challiwala, Hanif Choudhury, Nimir O. Elbashir, Shaik Afzal
PHOTOCATALYTIC UPGRADE OF CO_2 TO SYNTHETIC NATURAL GAS 465
Steven Bardey, Audrey Bonduelle-Skrzypczak, Antoine Fécant, Valérie Caps, Valérie Keller
GREEN METHANATION OF CO USING NEW CO-PT@FLG NANOCOMPOSITES 466
Loïc Michel, Cerise Robert, Sécou Sall, Valérie Caps
MULTIFUNCTIONAL CATALYTIC MATERIALS WITH CO_2 CAPTURE AND OXYGEN TRANSFER FUNCTIONALITIES FOR HIGH-PURITY H_2 PRODUCTION VIA SORPTION ENHANCED CHEMICAL LOOPING STEAM METHANE REFORMING 468
Andy Antzara, Eleni Heracleous, Angeliki Lemonidou
LOW TEMPERATURE METHANE STEAM REFORMING IN A HYDROGEN-SELECTIVE MEMBRANE REACTOR 473
Francesco Basile, Gabriele Centi, Andrea Fasolini, Salvatore Abate
OXYGEN REMOVAL FROM NG BY METHANE OXIDATION AT LOW TEMPERATURE 476
Carla Queiroz, Valéria Vicentini, Holli Garrett, Chandra Ratnasamy, Greg Cullen
UNDERSTANDING THE ROLE OF TITANIUM IN THE STABILITY OF SUPPORTED PALLADIUM CATALYSTS FOR THE OXIDATION OF VENTILATION AIR METHANE 478
Matthew Drewery, Hadi Hosseiniamoli, Meng-Jung Li, Eric M. Kennedy, Adesina Adesoji A., Michael Stockenhuber
APPLICATION OF CHEMICAL LOOPING IN FISCHER-TROPSCH GAS-TO-LIQUIDS PROCESSES: PERFORMANCE AND ECONOMIC ANALYSIS 483
Vishnu Priya Ramakrishnan, Omar Basha
INNOVATIVE PROCESSES DESIGN FOR VALUE-ADDED PRODUCTS FROM NATURAL GAS LIQUIDS IN SHALE GAS 484
Yiru Li, Zewei Chen, Wasiu P. Oladipupo, Edwin Andres Rodriguez Gil, Jeffrey J. Siirola, Fabio Ribeiro, Rakesh Agrawal

EFFECTS OF MORPHOLOGY AND DOPANTS ON THE CO_2 CAPACITY OF NANOFIBROUS CALCIUM-OXIDE BASED MATERIALS FOR SORPTION-ENHANCED STEAM METHANE REFORMING 485
Dante Simonetti, Luke Minardi, Faisal H. Alshafei, Zubin Mishra
PEI IMPREGNATED MESOPOROUS CARBON SPHERES AS IDEAL SORBENT FOR POST-COMBUSTION CO_2 CAPTURE FROM NATURAL GAS POWER PLANT 486
Qingjun Chen Sr., De Chen, Kumar Rout, Siyu Wang
FUTURE ASPECTS OF SYNGAS CONVERSION 490
Frits Dautzenberg
SELECTIVE OXIDATION AND OXIDATIVE DEHYDROGENATION OF ETHANE AND PROPANE 491
Srikant Gopal, Mohammed Al-Hazmi, Pankaj S. Gautam
TECHNO-ECONOMIC ANALYSIS OF DRM+COSORB INTEGRATED PROCESS FOR HIGH H_2/CO SYNGAS PRODUCTION 494
Shaik Afzal, Debalina Sengupta, Mahmoud El-Halwagi, Nimir Elbashir
SYSTEMS ANALYSIS OF LIGHT ALKANE RESOURCES FOR FUELS AND PETROCHEMICALS MANUFACTURE 497
Sean E. DeRosa, Yosuke Kimura, Gary McGaughey, Elena C. McDonald-Buller, David Allen, Mark A. Stadtherr
ENHANCING ASSET DESIGN, PERFORMANCE AND OPERATION MANAGEMENT THROUGH RESERVOIR-PRODUCTION-PROCESSING FACILITY INTEGRATED MODELING 498
Carlos Yengle, Hemant Kumar
MODEL-BASED OPTIMIZATION OF AN ALKENE OLIGOMERIZATION PROCESS TO PRODUCE HIGH OCTANE FUELS 500
Hari S. Ganesh, Sergio Vernuccio, Michael Baldea, Thomas F. Edgar, Linda J. Broadbelt, Mark A. Stadtherr, David Allen
IDENTIFICATION OF OPTIMAL GAS CONVERSION STRATEGIES USING A PROCESS SYSTEMS ENGINEERING APPROACH 501
Patrick Linke, Dhabia Al-Mohannadi, Shaza Shehab, Razan Ahmad
A GENERAL EVALUATION FRAMEWORK FOR DIRECT NONOXIDATIVE CONVERSION STRATEGIES OF METHANE TO AROMATICS 502
Kefeng Huang, Christos T. Maravelias
OPTIMAL ALLOCATION OF MODULAR AND TRANSPORTABLE METHANE PROCESSING UNITS FOR USE IN OIL AND GAS FIELDS 503
R. Cory Allen, Styliani Avraamidou, Douglas L. Allaire, Mahmoud M. El-Halwagi, Efstratios N. Pistikopoulos
CENTER FOR INNOVATION ON NEW ENERGIES -CINE- A NEW RESEARCH EFFORT FOR METHANE CONVERSION USING ELECTROCHEMICAL TECHNOLOGIES 505
Andre Ferlauto, Estevam Spinacé, Reginaldo Muccillo, Daniel de Florio, Almir Oliveira Neto, Elisabete Santiago, Thiago Lopes, Jorge Vaz, Eliana Muccillo, Fabio Fonseca
PROMOTION OF AMMONIA SYNTHESIS BY SURFACE PROTONICS AT LOW TEMPERATURE 506
Yuta Tanaka, Kota Murakami, Ryuya Sakai, Tomohiro Yabe, Shuhei Ogo, Yasushi Sekine
TPSR STUDIES: THE EFFECTS OF PRETREATMENT TECHNIQUES ON SILICA SUPPORT FOR COBALT FISCHER-TROPSCH SYNTHESIS CATALYSTS 507
Chike G. Okoye-Chine
STRUCTURE-PERFORMANCE RELATIONSHIPS ON CO-BASED FISCHER-TROPSCH SYNTHESIS: THE IMPACT OF H2-CO-H2 ACTIVATION TREATMENT 508
Nikolaos Tsakoumis, Eleni Patanou, Rune Myrstad, Erling Rytter, Edd Blekkan
NEW INSIGHTS INTO THE CATALYTIC PERFORMANCE OF NA-PROMOTED FE/AL_2O_3 CATALYST IN FISCHER-TROPSCH SYNTHESIS 512
Aliu A. Adeleke, Xinying Liu, Michela Martinelli, Diane Hildebrandt, Burtron H. Davis
LOW-TEMPERATURE CATALYTIC OXIDATIVE COUPLING OF METHANE OVER CE-W-O SYSTEM CATALYSTS IN THE ELECTRIC FIELD 513
Yuna Takeno, Shuhei Ogo, Ayaka Sato, Tomohiro Yabe, Yasushi Sekine
TUNGSTEN POLYOXIDE CATALYSTS FOR OXIDATIVE CONDENSATION OF METHANE 514
Tolkyn S. Baizhumanova, Svetlana A. Tungatarova, Zauresh T. Zheksenbaeva, Kaysar Kassymkan, Gulnar N. Kaumenova, Rabiga O. Sarsenova
PERFORMANCE OF CE-ADDED NA_2WO_4 CATALYST FOR DIRECT CONVERSION OF METHANE TO ETHYLENE 517
Heeyeon Kim, Jang-won O, Guk-hyun Kwon, Dae-keun Lee, Ji-haeng, Yu
NA_2WO_4/MN/SIC CATALYST FOR OXIDATIVE COUPLING OF METHANE 518
Ji Eun Kim, Suji Yoon, La Hee Park, Jeong-Myeong Ha, Eun Duck Park

TUNING CATALYST ELECTRONIC STRUCTURE & O_2 ACTIVITY FOR OCM REACTIONS 519
Shanti Kiran Nayak, Angelica Benavidez, Lok-kun Tsui, Fernando Garzon

THE HIERARCHICAL POROUS W-MN/SIO_2 CATALYST FOR OXIDATIVE COUPLING OF METHANE 521
Ning Zhang, Jian Yang, Huahua Zhao, Liang Yan, Huanling Song, Lingjun Chou

THE CATALYTIC PERFORMANCE OF THE MN-NA-W/SIO_2 CATALYST IN OXIDATIVE COUPLING OF METHANE: SURFACE STRUCTURE AND FURTHER ADDITIVES 524
Naseem S. Hayek, Oz M. Gazit

EFFECT OF THE MODIFICATION OF PD/CEO_2 CATALYST ON STEAM REFORMING OF METHANE AT LOW TEMPERATURE IN AN ELECTRIC FIELD 527
Ayako Takahashi, Reona Inagaki, Takuma Higo, Shuhei Ogo, Yasushi Sekine

PROMISING COMBINED PYROCHLORE-PEROVSKITE CATALYSTS TO SYNGAS PRODUCTION 528
Adriana P. Ramon, Elisabete M. Assaf, José M. Assaf

SYNGAS GENERATION BY CFD SIMULATION OF KINETIC MODEL AND HEAT TRANSFER IN STEAM METHANE REFORMER 529
Saiedeh Arabi, Sandeep Patil, Brian Dennis

C-H ACTIVATION CHEMISTRY AT METAL PHOSPHIDE SURFACES 530
Jeonghyun Ko, Jessica A. Muhlenkamp, William F. Schneider, Jason C. Hicks

EFFECT OF BA ADDITION ON THE CATALYTIC ACTIVITY OF GA-AL2O3 FOR ETHANE DEHYDROGENATION 531
Yukiko Hosono, Hikaru Saito, Hirofumi Seki, Shun Maeda, Shuhei Ogo, Kunihide Hashimoto, Yasushi Sekine

THE ROLE OF ACTIVATED CARBON SUPPORTED CEO_X AND VO_X PHASES DURING OXIDATIVE DEHYDROGENATION OF PROPANE TO PROPYLENE USING CO2 AS A SOFT OXIDANT 532
Petar Djinovic, Janez Zavasnik, Ivo Jerman, Gregor Zerjav

DRY REFORMING AND OXIDATIVE CONVERSION OF METHANE ON COMPOSITE MATERIALS PREPARED BY SELF-PROPAGATING COMBUSTION SYNTHESIS AND IMPREGNATION METHOD 536
Svetlana A. Tungatarova, Galina Xanthopoulou, Tolkyn S. Baizhumanova, Zauresh T. Zheksenbaeva, Manapkhan Zhumabek, Gulnar N. Kaumenova

TRI-REFORMING OF METHANE - OPERATIONAL STUDIES OVER A NI/$CEZRO_2$/$MGAL_2O_4$ CATALYST 540
Ananda V.P. Lino, Elisabete M. Assaf, Jose M. Assaf

INDIUM OXIDE AS PROMOTER IN CU/ZRO_2 CATALYSTS FOR CO_2 HYDROGENATION TO METHANOL 542
Marco A. Rossi, Jose M. Assaf, Elisabete M. Assaf

SORPTION-ENHANCED WATER-GAS SHIFT TECHNOLOGY PLATFORM FOR COST EFFECTIVE CO_2 REDUCTION IN THE IRON & STEEL INDUSTRY 544
Leon G.A. van de Water, Liliana Lukashuk, H.A.J. van Dijk, Paul D. Cobden, Magnus Lundqvist, Giampaolo Manzolini, Letitia Petrescu, Siebe van der Veer, Luca Mancuso, Jeremy Johns, David Bellqvist

SPECTROSCOPIC CHARACTERIZATION OF MONOMERIC AND DIMERIC COPPER SPECIES IN CHABAZITE ZEOLITES FOR PARTIAL METHANE OXIDATION TO METHANOL 548
Laura N. Wilcox, Ishant Khurana, Hui Li, Christopher Paolucci, William F. Schneider, Rajamani Gounder

A STUDY ON THE CHARACTERISTICS OF METHANE CHLORINATION BY CRYSTALLINE / AMORPHOUS CARBON 550
Jip Kim, Youngmin Kim, Kyoung-Su Ha, Ho-Jeong Chae

LIQUID-PHASE OXIDATION OF METHANE INTO OXYGENATES WITH O_2 551
Jong Kyu Kang, Eun Duck Park

ETHENE DIMERIZATION OVER ZR-BASED METAL-ORGANIC FRAMEWORK (UIO-67) FUNCTIONALIZED WITH NI AND BIPYRIDINE 552
Mustafa Komurcu, Andrea Lazzarini, Gurpreet Kaur, Silvia Bordiga, Karl Petter Lillerud, Unni Olsbye

EXPERIMENTAL AND COMPUTATIONAL FLUID DYNAMICS STUDY OF TWO-STAGE NATURAL GAS COMBUSTION PYROLYSIS REACTOR IN HIGH-VALUED CHEMICAL PRODUCTION: THE EFFECT OF FEEDSTOCK-TO-FUEL RATIO AND FEEDSTOCK COMPOSITIONS 555
Lei Chen, Sreekanth Pannala, Balamurali Nair, Istvan Lengyel, Byeongjin Baek, David West

Author Index

557826 The Chemistry and Engineering of C_1 Molecules: Methane Conversion and Beyond

Monday, June 3, 2019: 8:30 AM
Texas Ballroom EF (Grand Hyatt San Antonio)

Enrique Iglesia, Chemical and Biomolecular Engineering, University of California at Berkeley, Berkeley, CA

The recurring interest in advantaged feedstocks as precursors to chemicals and fuels brings us today, once again, to scientific and engineering matters pertaining to the conversion of molecules without C-C bonds, such as methane, methanol, and dimethyl ether, and other small unreactive components in natural gas. Such C_1 molecules, especially methane, pose formidable thermodynamic and kinetic challenges that bring significant complexity and costs into the practice of their chemical transformations. These challenges, along with those inherent in preserving C-C bonds, are also evident in the activation of C-H bonds in small alkanes, a chemical event required for their conversion and functionalization. This lecture attempts to gather some unifying concepts and, in doing so, seeks to chart feasible paths between feedstocks and specific target products in order to guide our search for the most attractive conversion strategies. These underlying concepts include: (i) the ubiquitous thermodynamic constraints and the essential limitations of catalysis in circumventing them; (ii) the pre-eminence of process simplicity and inexpensive oxidants over the allure and glamour of more direct conversions; (iii) the enduring fundamental principles of kinetic and thermodynamic protection and how to best implement them in engineering practice (iv) the coupling of separations with catalytic reactions, of multiple catalytic functions in tandem sequences, and of endothermic and exothermic transformations either at the molecular scale or at least within thermal conduction distances; (v) the prevalence of kinetic bottlenecks in forming the first C-C bond from C_1 molecules and protecting them in larger alkanes; (vi) the counterintuitive benefits of very reactive hydrogen abstractors and of high temperatures in disfavoring the activation of weaker C-H bonds; and (vii) the emergence of C_2 conversion platforms, as the limits of C_1 chemistries are circumvented (or accepted) and as condensate streams from shale gas or ethanol feedstocks from biogenic origins emerge as advantaged feedstocks. The conclusions garnered through chemistry and engineering concepts along this journey are sobering, as the magnitude of the challenges seems to warrant; they are essential, however, in guiding our search for opportunities.

Environmental and Community Impacts of Shale Development in Texas: A Report from the Academy of Medicine, Engineering and Science of Texas

Monday, June 3, 2019: 10:15 AM - 12:30 PM
Texas Ballroom D (Grand Hyatt San Antonio)

Description:
The biggest change in the global oil and gas industry during the past decade has been the proliferation of horizontal drilling and multi-stage hydraulic fracturing. Improvements in many aspects of the technologies and materials used in the horizontal drilling and hydraulic fracturing processes have opened up vast shale deposits that previously were not viable economically for oil and gas production. A significant portion of this major energy development and technological breakthrough since the mid-2000s has taken place in Texas. Today, Texas produces more crude oil than any other state, and is responsible for more than one-third of the nation's total oil production. At the same time, there is opposition to this expansion in many places, including some U.S. states and some nations. However, hydraulic fracturing for shale development will continue to be an important and likely growing part of the Texas and United States energy production portfolio. The Academy of Medicine, Engineering and Science of Texas (TAMEST) convened a task force to prepare a report on the Texas shale development experience. This session will i) introduce TAMEST and the process used to prepare the report; ii) present the major findings and recommendations; and iii) report advances since the report's release.

Sponsor:
Featured and Special Sessions
Chair:
David Allen **Email:** allen@che.utexas.edu

10:15 AM
Introduce TAMEST - David T. Allen

10:25 AM
Overview - Christine Ehlig Economides

10:35 AM
Geology and Earthquake Activity - Brian Stump, Kris J. Nygaard, Craig Pearson

10:45 AM
Land Resources - Melinda Taylor, Joseph Fitzsimons, Tracy Hester

10:55 AM
Air Quality - David Allen, Ramón Alvarez, Matthew Harrison

11:05 AM
Water Quantity and Quality - Danny Reible, Denny Bullard, Michael Young

11:15 AM
Transportation - John Barton, Cesar Quiroga

11:25 AM
Economic and Social Impacts - Gene Theodori, Omar Garcia, Urs Kreuter

11:35 AM
Updates Since the Report's Release

11:50 AM
Panel Discussion

See more of this Group/Topical: General Submissions
<< Previous Session | Next Session >>

553045 Modular Chemical Process Intensification: Enabler for Gas Conversion

Monday, June 3, 2019: 5:12 PM
Republic ABC (Grand Hyatt San Antonio)

James Bielenberg and Ignasi Palou-Rivera, RAPID, New York, NY

Process intensification and modular technologies (MCPI) are critical enabling technology areas in the area of gas conversion. While this has always been the case, the recent changes in the supply landscape for light gases (particularly in the US) and advances in technology have made potential for modular and intensified processes greater than ever. However, there have been very limited commercial deployments of MCPI. The AIChE-led Rapid Advancement in Process Intensification Deployment (RAPID) Manufacturing Institute is focused on fostering the development, demonstration, and commercial deployment of modular process intensification technology leading to dramatic reductions in energy consumption, capital investment, and operating cost. In this talk, the intrinsic hurdles in gas conversion technologies will be discussed, with a particular focus on how modular and intensified processes can be used to address these challenges. A brief overview of current RAPID projects related to gas conversion will also be included.

551532 Enhanced Light Alkane Dehydrogenation via Coupling with Chemical Looping

Monday, June 3, 2019: 5:15 PM
Republic ABC (Grand Hyatt San Antonio)

Casper Brady[1], Xiaoxia Chang[2], Eric Steinman[3], Shoucheng Du[4] and Bingjun Xu[1], (1)Center for Catalytic Science and Technology, Department of Chemical and Biomolecular Engineering, University of Delaware, Newark, DE, (2)Key Laboratory for Green Chemical Technology of Ministry of Education, School of Chemical Engineering and Technology, Tianjin University, Tianjin 300072, China, (3)University of Delaware, Newark, DE, (4)Chemours Engineering Technology, The Chemours Company, Wilmington, DE

Catalytic dehydrogenation reactions for light alkane upgrading are limited by equilibrium at most industrially relevant temperatures. These thermodynamic limitations can be mitigated via removal of hydrogen using reactive separation techniques. Typical reactive separation techniques include the use of hydrogen permeable membranes1, selective hydrogen combustion2, and selective hydrogen adsorption3. We propose a novel reactive separation technique for the removal of hydrogen during alkane dehydrogenation based on chemical looping. In our proposed process a dehydrogenation reactor is coupled with a chemical looping reactor for selective hydrogen oxidation and a high temperature water adsorbent to remove the produced water. In our initial work we demonstrate the viability of this concept for use with methane dehydroaromatization by a combination of single stage dehydroaromatization experiments and metal oxide redox experiments in quartz tube microreactors4. In our current work we apply a newly constructed novel microreactor, capable of recirculating heated gas through 3 separate reactors as well as intermittently regenerating catalyst beds and re-oxidizing metal oxide beds, to investigate

coupling methane dehydroaromatization to chemical looping in more realistic process conditions. In the future we intend to apply this unique design to other dehydrogenation reactions including propane dehydrogenation, ethane dehydrogenation, and methane dehydrogenation to ethylene.

References

[1] R. W. Borry; E. C. Lu; Y.H. Kim; E. Iglesia, *Stud. Surf. Sci. Catal.*, **119** (1998), 403.

[2] J.G. Tsikoyiannis; D.L. Stern; R.K. Grasselli, *J. Catal.*, **184** (1999) 77.

[3] A. Kumar; K. Song; L. Liu; Y. Han; A. Bhan, *Angew. Chem.,* (2018). [4] C. Brady; B. Murphy; B. Xu, *ACS Catal.*, **7** (2017) 3924.

551527 Advanced Nanocomposite Membranes for Natural Gas Purification

Monday, June 3, 2019: 5:18 PM
Republic ABC (Grand Hyatt San Antonio)

Jaesung Park, McKetta Department of Chemical Engineering, The University of Texas at Austin, Austin, TX, **Hee Wook Yoon**, McKetta Department of Chemical Engineering, University of Texas at Austin, Austin, TX and Benny D. Freeman, Mcketta Department of Chemical Engineering, Center for Energy and Environmental Resources, and Texas Materials Institute, The University of Texas at Austin, Austin, TX

Using membrane technology for removal of CO_2 from natural gas is a proven technology with a track record spanning several decades. Compared with amine absorption, membrane units often operate with lower energy input, and they have a smaller footprint than conventional amine absorption units. However, conventional commercial membrane materials do not offer high-enough CO_2 flux and selectivity to compete effectively with amine absorption.

In the present work, we report highly permeable polymer nanocomposites applied to natural gas purification. Using a polymer of intrinsic microporosity (PIM), which contains ethanoanthracene (EA) and Tröger base (TB) units as control material (PIM-EA(Me_2)-TB), we design and synthesize a version of this PIM without methyl side chains, PIM-EA(H_2)-TB. Nanocomposites are made by combining PIM-EA(H_2)-TB with 10 wt. % of porous aromatic framework-1 (PAF-1) nanoparticles (which is a self-assembled microporous polymer with repeating units comprising aromatic carbon rings arranged in a tetrahedron). ^{13}C solid state nuclear magnetic resonance (NMR) spectra and the cross-sectional scanning electron microscope (SEM) demonstrated that incorporation of PAF-1 into PIM-EA(H_2)-TB yields a phase separated system. Small angle X-ray scattering (SAXS) and wide-angle X-ray scattering (WAXS) were used to characterize the presence of PAF-1 impact on the spacing between PIM-EA(H_2)-TB chains, suggesting that PAF-1 drastically enhances *d*-spacing in PIM-EA(H_2)-TB. Pure gas permeability for CO_2 and CH_4 was measured as a function of upstream pressure at 35 . Incorporation of PAF-1 substantially increases pure CO_2 permeability of PIM-EA(H_2)-TB, but decreases pure CO_2/CH_4 selectivity, presumably attributed to free volume increase imparted by PAF-1 in PIM-EA(H_2)-TB matrix. Permeation of mixtures of CO_2 and CH_4 (50:50 mol. %) are presented alongside the measurements of pure components to characterize the performance of these materials under simulated natural gas conditions.

551528 Distributed Natural Gas Conversion with Scalable, Process-Intensified Gas Cleanup and Reactor Technologies

Monday, June 3, 2019: 5:21 PM
Republic ABC (Grand Hyatt San Antonio)
Paul S. Dimick, IntraMicron, Inc., Auburn, AL

Shale gas is a distributed resource that is typically gathered and processed at centralized facilities. The high capital cost of the gathering, collection, and treatment infrastructure required to produce and treat shale gas results in the largest reserves (with a value high enough to justify the capital expenditure) being produced while smaller reserves are commonly avoided until the production risk is low enough to justify their production. This is particularly the case if there is any significant cleanup burden associated with these resources (ie CO_2, H_2S, etc.). One avenue to effectively utilize the smaller reserves is to treat the gas in a distributed manner and or raise its value by converting it directly to value-added products with more favorable transportation economics. For example, shale gas can be converted directly into easily transportable liquid products or pipeline spec natural gas can be produced from sour gas or acid gas. This approach is not commonly taken because traditional processing methods and approaches do not scale down efficiently to allow this to be done in an economical manner, and traditional processes do not have sufficient automation and robustness to withstand operation in the remote and sometimes harsh environments. Effective Implementation of process intensification (PI) and modular chemical process intensification (MCPI) technologies will open the door for distributed gas cleanup and conversion.

IntraMicron has developed several PI and MCPI technologies which can be applied to enable the distributed production and conversion of shale gas. Since effective gas cleanup and conditioning is required for these distributed conversion operations, IntraMicron has developed a modular, scalable catalytic desulfurization process, SourCat™, that was successfully demonstrated at an Eagle Ford Shale Well head site in the Fall of 2018 and is applying its Microfibrous Entrapped Sorbent (MFES) technology to modular PSA-based CO_2 separation systems for distributed acid gas treatment. Both of these technologies, which have the opportunity to supplant traditional amine-based separations followed by post treatment, are the subject of research grants from the RAPID Institute for Process Intensification. Initial TEAs have demonstrated significant cost savings for these processes compared to traditional approaches even at scales as low as 1 MMSCFD. Once contaminants are removed from the natural gas, converting it to value added components is commonly done with a series of catalytic processes that are either highly exothermic or endothermic. IntraMicron's Microfibrous Entrapped Catalyst (MFEC) technology provides the process simplicity of a fixed bed reactor with up to 250 times the heat transfer and 10 times the mass transfer of traditional fixed bed catalyst systems. This enhanced heat transfer typically enables 20% of the catalyst to be used at similar or greater activity with improved selectivity. MFEC can be prepared with any catalyst in the appropriate size range. For highly exothermic processes the high radial thermal conductivity of MFEC enables a 16x increase in the reactor tube area (i.e 1" to 4" for FTS) to minimize part count and simplify reactor construction. Since convective heat transfer does not need to be the primary driver for reactor design, the reactors can be operated in different regimes (higher conversion, low recycle) that significantly reduce balance of plant to minimize CAPEX and OPEX. This technology has been applied to various GTL approaches and chemistries. The approach is also well suited to accommodate many of the new direct conversion chemistries that are being developed. Several scenarios for the implementation of MFEC for distributed gas conversion will be presented.

551529 Catalytic Dehydroaromatization of Ethane in Microwave Reactor

Monday, June 3, 2019: 5:24 PM
Republic ABC (Grand Hyatt San Antonio)

Jianli Hu, Department of Chemical and Biomedical Engineering, West Virginia University, Morgantown, WV

Ethane is separated from shale gas to minimize condensation in natural gas pipeline. As an effective way to utilize ethane, direct non-oxidative ethane dehydroaromatization has been studied for decades. Conventionally, high reaction temperature (550-600°C) is required to activate the hydrocarbon molecules. In this study, a microwave catalytic reactor system is used to activate ethane at temperature 200°C lower than conventional catalytic approach. The objective of this study is to elucidate reaction mechanism in microwave-assisted ethane dehydroaromatization. The study consists of synergistically integrating microwave reaction chemistry with novel heterogeneous catalysis that selectively activates natural gas through microwave irradiation. The challenges in C-H bon activation is overcome by a combination of changing in reaction pathways, novel microwave reactor and catalysts. Specifically, the focus of the study is to ascertain reaction pathways and the catalytic function of catalyst that are sensitive to electromagnetic energy. One of the mechanisms by which the catalyst and reacting species can interact with the microwave field and provide energy to the reaction is by relaxation processes, such as dipolar or Debye processes, which involve the coupling of the radiation with dipoles in the solid catalyst. These dipoles can be defect sites (i.e. atomic vacancies) in the catalysts or dangling bonds on the surface of catalysts. From the standpoint of catalysis, dipoles on the surface can be reactant or products that would be susceptible to selective bond activation effects, which in turn can affect reaction rates.

Using ethane as the feedstock, aromatization reaction was studied in a variable frequency microwave reactor in the presence of metal promoted zeolite catalyst. The significance of microwave frequency, reaction temperature and residence time were investigated respectively. Aromatic formation can be achieved at temperature lower than 400°C. Transmission electron microscope with energy dispersive X-ray spectroscopy, thermogravimetric analysis and other technologies were applied to characterize the metal doped catalyst. Energy economy is also calculated in this study to evaluate the feasibility of microwave operation for commercial application.

551645 Challenges and Solutions to Scalable Testing of Catalysts for Converting Syngas to Chemicals and Fuels

Monday, June 3, 2019: 5:30 PM
Republic ABC (Grand Hyatt San Antonio)

Anton J. Nagy, ILS-Integrated Lab Solutions GmbH, Berlin, Germany and Thomas Turek, Clausthal University of Technology, Institute of Chemical and Electrochemical Process Engineering, Clausthal-Zellerfeld, Germany

Challenges and Solutions to Scalable Testing of Catalysts for Converting Syngas to Chemicals and Fuels

Dr. Anton J. Nagy[1]* (underline presenting author) and Prof. Dr. T. Turek[2]

[1] ILS-Integrated Lab Solutions GmbH, Berlin (Germany)

[2]TU Clausthal, Institut für Chemische und Elektrochemische Verfahrenstechnik,

Clausthal-Zellerfeld (Germany)

*anton.nagy@integratedlabsolutions.com

Introduction

The recent increase in availability of short chain alkanes derived from the US Fracking gas initiative combined with worldwide recognition of the need to reduce CO_2 emissions to mitigate climate change has resulted in ever-increased interest in converting synthesis gas to chemicals and/or fuels. This presentation will highlight a number of new systems developed by ILS for performing such tests and also provide real-life examples of pitfalls to avoid in GTL catalyst testing.

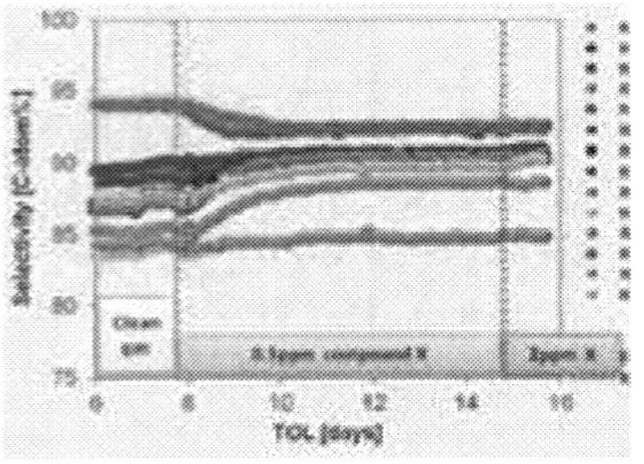

GTL Types

Fischer Tropsch

Fischer Tropsch catalyst for low-temperaure cobalt-catalyzed FT are done industrially in either parallel fixed-bed reactors or bubble-column reactors. At the lab scale there serious challenges in testing these catalysts[1]. We will address a number of these issues in the presentation pointing out a number of practicable solutions including:

Figure 1. Example Data for Long-Term Parallel FT Testing

1. Isothermicity issues: Despite what is often published we have found that even small-diamter, SiC thinned catalysts tested with 2mm inner diameters show deviation

from true temperature.

2. Axial Oxidation Gradients: Product water formed during reaction results in oxidation of the lower catalyst bed and potential falsification of results

3. Slurry CSTR. Vs. Fixed-Bed: The rapid response of small fixed-bed lab reactors makes them powerful tools for studying process variations that only appear after long time-on-streams in larger s lurry reactors.

4. Pore-diffusion limitations: Fixed-bed reactor testing of high-alpha FT catalysts almost all exhibit pore-diffusion limitations. This can be overcome by operating reactors at similar conversion levels as long as product spectrums are similar.

CO2/CO Methanation

CO and CO_2 methanation are already industrially-relevant processes used for example for converting syngas generated from coal to clean fuels for city power and fuel requirements.

Challenges in high-throughput testing of extremely effective nickel-based methanation catalysts will be shown. For example; the need to operate at extremely high flowrates to obtain data far enough from thermodynamic equilibrium.

Methanation catalysts for use as energy storage via hydrogenation of CO and CO_2 using green-powered electrolysis units is a very exciting topic at the moment. These catalysts are exposed to large swings in redox atmosphere as the availability of H_2 varies as a function of the electricity available.

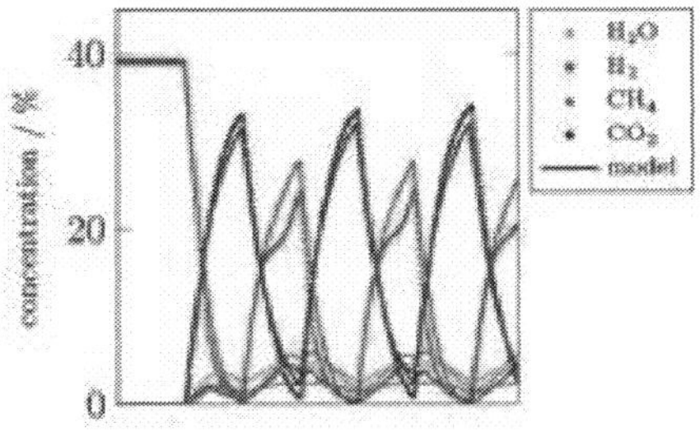

The ILS-Berty internal recycle reactor is a next-generation Berty derived from the exclusive licensing agreement that ILS has with the Friedrich-Alexander-University in Erlangen, Germany. The use of a novel, electromagnet makes possible stirring rates of 10.000rpm and completely eliminates the need for an external motor with rotating parts. The high speeds and lack of moving parts ensures a broader pressure operating regime and virtually no downtime. This is particularly advantageous for high-temperature, low-to-medium pressure gas-phase reactions.

Figure 2. Instationary CO_2 Methanation in ILS Berty Reactor

Recent data from the Turek group isusing a Berty-type CSTR from ILS to study the instationary methanation kinetics of CO_2. The pulse RTD results clearly show the ideal backmixing of the ILS Berty reactor. Initial transient kinetic studies performed for CO_2 methanation in the CSTR show the transient kinetic behavior to be very different than that predicted by more traditional steady-state models.

Methanol Synthesis

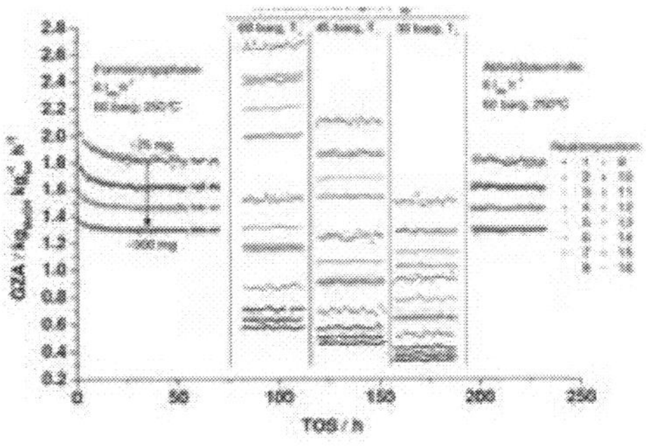

Methanol synthesis is a reaction of major industrial relevance. We will present the modern high-throughput fixed-bed methanol synthesis testing units provided by ILS which allow independent temperature control in order to do rapid parallel kinetic testing. The ability to have temperature control on the order of +/- 0,5°C was critical for these units which have been in operation at Clariant for over 10 years now. Challenges like carbonyl poisoning and effective measurement of low-concentrations of byproducts will be addressed.

Figure 3. Results of 16-Paralle Methanol Synthesis Catalyst Testing[2]

References

1. Sie S.T., Miniaturization of hydroprocessing catalyst testing systems: Theory and practice, *AICHE Journal,* , Vol 42, Issue 12, Dec 1996

2. Dissertation. T. Henkel, Modellierung von Reaktion und Stofftransport in geformten Katalysatoren am Beispiel der Methanolsynthese, TU Munich 2011

545626 Innovative Zoneflow® Catalyst Offers Unmatched Solutions in Steam Reforming for Syngas Based NG Conversion

Monday, June 3, 2019: 5:36 PM
Republic ABC (Grand Hyatt San Antonio)
Sanjiv Ratan, ZoneFlow Reactor Technologies LLC, Windsor, CT

Natural gas has been a prime, rather increasingly preferred contender for meeting current and future energy demand based on its abundance, lower carbon intensity and modes of distribution. With the advent of shale gas, it has not only driven a quick switchover from high carbon fuel sources like coal anf fuel oil to NG but also offered ample opportunities and potential for monetizing it through conversion to Syngas as a building block. Moreover curtailment of flare gas also offers lot of opportunities and incentives for gas conversion. The syngas based wide spectrum ranges from major petro-chemicals to fertilizers and futuristic clean power with concerted carbon capture.

The predominant technology for Syngas generation over all these years has been steam reforming of natural gas commonly called steam-methane reforming (SMR). This mature technology, however, has seen only incremental improvements in its catalyst design in aiming for better performance and operating life whihc otherwise plays a crucial role in its thermal design, overall performance, reliability and cost-effectiveness.

The conventional "pellet" steam reforming catalysts suffer from few inherent limitations and/or operational deficiencies. These mainly relate to a) uneven random packing voidage leading to undesired flow and temperature mal-distribution, b) catalyst attrition and breakage from thermal cycling leading to increasing pressure drop and related capacity limitations and c) pore-diffusion limited reaction path leading to curtailed intrinsic activity. Past attempts to develop structured steam reforming catalysts to overcome some of these limitations have suffered from the innate problems around differential expansion gaps leading to feed by-passing and disturbed boundary layer endotherm.

ZoneFlow™ Reactor Technologies LLC (ZFRT) has developed proprietary, radically innovative and differentiating structured catalyst systems called ZoneFlow™ (ZF) Reactors. These not only overcome the above-mentioned deficiencies but also offer an unmatched combination of higher heat transfer, lower pressure drop, higher geometric surface area (GSA), more uniform flow and heat distribution and long-term structural integrity compared to the current state-of-the-art pellet catalyst.

ZF Reactor's unique structure also imparts a high degree of proximal flexibility thus eliminating the differential thermal expansion gaps, while also having high (metal substrate) physical strength against breakage over its lifetime. Its nested modules provide much more geometric surface than pellets, and take away the diffusion-limited access to the layered catalyst thus resulting in much higher catalyst effectiveness.

ZF Reactors also offer the advantage that they can be loaded horizontally, which allows for in-situ 'convective pre-reforming' in the reformer convection section. Furthermore its annular casing design directly enables regenerative or "bayonet" reformer designs, while overcoming the major concerns and risks associated with the current pellet-based designs. The ZF Reactor product range covers radiant steam reforming, convective pre-reforming as well as recuperative reforming. Their performance parameters have been validated through advanced CFD modeling and bench scale test rigs, to be further confirmed in upcoming pilot plant in Belgium.

ZF Reactor designs can be customized for required optimization between heat transfer and pressure drop, and to fit in any commercial SMR tube size. These features of the ZF Reactor can translate into attractive benefits to the end-user ranging from relieving "stressed" reformers, debottlenecking and capacity upgrades as well as new plant SMRs. The use of ZF reactors enables significant capital and operating cost reduction

This paper will provide status update on ZF technology developments and also the upcoming pilot plant progress with its extensive test program under near commercial operating conditions as well as various case studies covering capacity revamps and new plant for enhancing the economics and reliability of steam reforming based syngas generation for Natural Gas conversion and monetization.

546153 Enabling Distributed Gas-to-Liquid (GTL) Opportunities at the Micro-Scale

Monday, June 3, 2019: 5:39 PM
Republic ABC (Grand Hyatt San Antonio)
Sameer Parvathikar, Pradeepkumar Sharma, Atish Kataria and John R. Carpenter, RTI International, Research Triangle Park, NC

Advancements in drilling technologies have provided beneficial energy sources but also contributed to significant flaring of natural gas. Similarly, more environmental pushes to deal with waste streams has increased the production of biogas. These sources of gas many times are smaller than most conventional conversion technologies at micro-scales of less than 1 MMSCFD. This increased availability of distributed small gas volumes has driven pushes for the development of small-scale technologies for its conversion to various products as a way to both monetize these streams and assuage environmental concerns from their emission. One of the biggest challenges in the development of small-scale technologies is overcoming the economic advantages resulting from the large scales in traditional processes. RTI has developed and demonstrated an internal combustion-engine based Microreformer as a syngas generator from natural gas. In this study, we evaluate the feasibility and economics of using this technology for monetizing stranded gas resources by producing liquid chemicals using two case studies demonstrated at pilot scale – methanol production and an advanced syngas-to-liquids (STL) technology.

In the first case study, RTI's Microreformer technology uses commercially available internal combustion engine blocks, thereby countering the economies of scale with economies of mass manufacturing. RTI's engine-based, impurity-tolerant Microreformer has been demonstrated at near-deployment scale, to process 50,000 SCFD natural gas feed. The use of air as an oxidant further reduces the cost of the system. We have demonstrated the operation of this air-blown non-catalytic partial oxidation (POx) system for over 250 hours cumulatively. Methane conversions > 90% were achieved, with very low oxygen slip (<0.5%) resulting in syngas with H_2:CO of 1.6. Techno-economic analysis of the Microreformer showed that the cost of syngas at the small-scale was comparable to costs from conventional technologies at scales much larger in size.

Methanol production was integrated with the engine to demonstrate the viability of producing liquid fuels or chemicals. The use of air as an oxidant in the engine results in a diluted syngas stream. A 1 barrel/day (BPD) methanol skid was designed based on a commercial methanol synthesis catalyst and built to use a 10% slip stream of the engine exhaust. The methanol skid included gas conditioning and compression prior to a 2-stage synthesis reactor with interstage cooling. The process also included a hydrogen membrane to recycle tail gas back to the process. The data collected was used to validate techno-economic analyses of methanol production from the Microreformer at a 300,000 SCFD scale. The production cost was calculated to be ~$1/gal. This cost shows that while additional savings might be needed to make distributed methanol cost competitive with large-scale facilities, the Microreformer can be attractive for markets where gas resources are stranded.

A second example of a fuel production technology is an advanced syngas-to-liquids (STL) system RTI recently demonstrated at a pilot scale. Working with a vendor, a small-footprint 1BPD Fischer-Tropsch (FT) skid was constructed. The system is based on compact, heat exchanger reactors that consist of alternating channels of catalysts and heat transfer fluid, and hence provide excellent temperature control of the exothermic reaction. Downstream of the reactors are a set of product collection vessels, hot and cold traps, to condense organic hydrocarbons. The compact nature of the reactors and the excellent heat transfer play a crucial role in modularizing and lowering the capex of the small-scale FT system. While this FT skid was not directly integrated with the Microreformer, its operation was tested with a natural gas POx system at a partner's test site. In particular, the system was tested with diluted syngas simulating air-blown POx at Standard FT synthesis conditions. High CO conversions were observed, and periodic organic product sampling revealed that 95% of the hydrocarbons had a carbon number below C-25 with a mean between 10-12. Distillation testing according to ASTM D-1160 showed that the majority of the FT product consisted of diesel and lighter hydrocarbons. The minimal wax production from the system further reduces plant complexity and reduces capex, and hence makes it attractive for stranded gas applications. Based on the operation of the 1 BPD pilot skid, RTI has developed detailed designs for a 12.5 BPD commercial-scale modular skid. Techno-economic analysis of the integration of the advanced STL process with the engine-based Microreformer has also shown promising results.

RTI's Microreformer and other modularized technologies provide a unique, cost-effective technologies for the utilization of distributed micro-scale gas resources. In the case of the Microreformer, the use of a mass-manufactured engine greatly reduces the capex of the system. Low replacement costs and times mean that these systems can operate with a low opex and very little down-time if needed. RTI is currently evaluating this technology in several distributed gas markets in addition to flared/associated gas, particularly hog farms in Eastern North Carolina, where increasing environmental concerns with swine lagoons have led to widespread nuisance lawsuits. RTI is also assessing other biogas sources including waste water treatment plants, crop residues and landfills. These markets can have far-reaching impact, offering owners/operators/farmers an incentive to implement solutions that offer an additional revenue stream, while providing adequate waste management, and most importantly reinvigorating non-exportable manufacturing jobs in economically depressed rural America.

558680 Energy Outlook: Role of Natural Gas

Tuesday, June 4, 2019: 8:30 AM
Texas Ballroom EF (Grand Hyatt San Antonio)
Vijay Swarup, ExxonMobil Research and Engineering Company, Clinton, NJ

With an increasing global population and developing middle class, more and more people around the world will be looking for affordable, reliable energy solutions to support and improve their standard of living. Many factors, including technology, environmental, economic and political to name a few will help shape and impact the decision and choices that are made. Understanding the role natural gas will play in solving the dual energy challenge of providing scalable energy solutions while reducing emissions to address the risk of climate change is a goal of ExxonMobil scientists and engineers.

555981 Innovation in the Chemical Industry: Implications and Impacts of the Post-shale World

Wednesday, June 5, 2019: 8:45 AM
Texas Ballroom EF (Grand Hyatt San Antonio)
Sandeep S. Dhingra, Dow, Houston, TX

Shale gas created euphoria within the USGC petrochemical industry, providing preferred feedstocks at advantaged prices at a time when most forecasts predicted the exhaustion of North American supplies and the death of the industry. Euphoria has given way to the hard work of completing the announced projects as market dynamics erode the economic advantage of light gas resources. The chemical industry faces new opportunities and challenges as we enter the post-shale years. Monomers are now made on-purpose, not as byproducts. Concerns over sustainability create pressure to reduce CO_2 emissions and to reduce environmental impact of the plastics that are the industry's largest products. The time is ripe for new developments in process, products and applications that will drive our industry forward in the post-shale years.

556708 Natural Gas Conversion in Microporous Catalysts; Zeotypes and Metal Organic Frameworks

Thursday, June 6, 2019: 8:45 AM
Texas Ballroom EF (Grand Hyatt San Antonio)

Unni Olsbye, Department of Chemistry, University of Oslo, OSLO, Norway

Since the first Natural Gas Conversion Symposium in Auckland, New Zealand, in 1987, the natural gas conversion area has evolved from a prospective field of intensive research efforts, to a mature field with major industrial implementation. In recent years, ethane-rich shale gas offered opportunities to successfully refurbish industrial naphta crackers into ethane crackers, which concomitantly intensified the search for alternative routes to propene and butene production.

This lecture will focus on two such routes that are currently under study in our group; the methanol to olefins (MTO) reaction, and the ethene oligomerization reaction. Both reactions are catalyzed by Brønsted acid sites in zeotype materials (including zeolites), and ethene oligomerization is further promoted by Ni sites ion-exchanged onto the lattice.

During the past decade, significant progress was made in the fundamental understanding of zeotype-catalyzed processes, owing to a combination of mechanistic studies, synthesis of materials with systematically modified properties (pore and cavity diameter, crystal size and morphology, acid-site strength, density and distribution) and improved operando and ex situ characterization techniques [1].

Considering the MTO reaction, the mechanism of alkene conversion to (poly-)aromatic compounds, and furthermore; the site selectivity of zeotype catalysts, is currently a hot topic. The first part of the lecture will be dedicated to recent advances in this research field [2].

The second part of the lecture will focus on the ethene oligomerization reaction. Industrial processes are currently carried out over homogeneous Ni-complex catalysts, and many studies addressed Ni-containing zeotypes as potential heterogeneous analogues [3]. An alternative approach, metal organic frameworks (MOFs), offers the possibility of anchoring Ni to an organic entity, similar to those found in commercial, homogeneous catalysts [4]. As one of a few examples where MOF catalysts are used to convert gas phase molecules, current insight in Ni-MOFs versus Ni-zeolites as ethene oligomerization catalysts will be presented.

A long-term goal of our research activity is to develop guiding principles for tailoring microporous catalysts towards selective applications. Experimental determination of intrinsic kinetic parameters is a prerequisite of this approach. However, it is a daunting task to obtain experimental data for individual reactions in zeotypes, due to the plethora of reactions that may take place simultaneously inside the porous network. For this reason, we are currently exploring the use of the Temporal Analysis of Products (TAP) technique to measure intrinsic kinetics of zeotype-catalyzed reactions [5]. This unique apparatus operates under vacuum conditions, and favors monomolecular interaction with surface sites. The status and prospects of such studies will be addressed in the final part of the lecture.

References

[1] Olsbye et al. Angew. Chemie Int. Ed., 2012, 51, 5810; Chem. Soc. Rev. 2015, 44, 7155; Ilias and Bhan, ACS Catalysis, 2013, 3, 18; Yarulina et al. Nature Catal. 2018, 1, 2398; Gounder and Iglesia, Acc. Chem. Res. 2012, 45, 229; Sarazen et al. J. Catal. 2016, 344, 553.

[2] Martinez-Espin et al.; J. Catal., 2017, 349, 136; Cat.Sci.Technol. 2017, 7, 2700; ACS Catalysis, 2017, 7, 5173; Yarulina et al, Nature Chemistry, 2018, 10, 809; Morten et al. ChemPhysChem. 2018, 19, 484.

[3] Hulea ACS Catalysis, 2018, 8, 3263; Brogaard and Olsbye ACS Catalysis 2016, 6, 1205.

[4] Canivet, JACS 2013, 135, 4195; Metzger JACS, 2017, 139, 757; Liu JACS, 2018, 140, 11174; Kømurcu et al (*in preparation*).

[5] Brogaard et al J. Catal., 2014, 314, 159; Redekop et al. (*in preparation*).

545196 Hot Oxygen Based Synthesis Gas Generation

Monday, June 3, 2019: 5:12 PM
Texas Ballroom A (Grand Hyatt San Antonio)

Brad Damstedt[1], Lawrence Bool III[1] and **Ines Stuckert**[2], (1)Industrial Applications R&D, Praxair, Tonawanda, NY, (2)R&D, Praxair, Inc, Tonawanda, NY

Similar to well-known partial oxidation (POx) systems, Praxair has developed a synthesis gas generation method utilizing hot oxygen technology, improving on conventional oxygen systems. This paper reviews hot oxygen technology and its application to natural gas derived synthesis gas generation.

A hot oxygen burner (HOB) produces a reactive jet by combusting a small amount of fuel with an excess of oxygen and accelerating the combustion products through a nozzle. The resulting jet is very reactive due to high oxygen concentration, high temperature and exceptional mixing capability. Advantages achieved using an HOB system include simplified reactor design and the ability to engineer a syngas generation system for relatively small production applications while maintaining high performance levels.

Praxair has experimentally investigated HOB syngas generation at lab and pilot scales, at pressures ranging up to 400 psig, with feedstock variation including steam and CO2 addition. A broad spectrum of bench and pilot scale results are presented including syngas production efficiency, oxygen usage, product H2:CO ratio and quantitative measurements of minor species including soot, methane, acetylene, ethylene, ammonia and hydrogen cyanide. Experimental results are compared to detailed modeling of the hot oxygen process.

545542 Evaluation of the Absorption Performance on the of Acid Gas Removal from Natural Gas and Synthetic Gas Processing

Monday, June 3, 2019: 5:15 PM
Texas Ballroom A (Grand Hyatt San Antonio)

Kreangkrai Maneeintr, Department of Mining and Petroleum Engineering, Faculty of Engineering, Chulalongkorn University, Bangkok, Thailand, Thanaphat Luemunkong, Department of Chemical Engineering, Faculty of Engineering, Chulalongkorn University, Bangkok, Thailand and Kyuro Sasaki, Department of Earth Resources Engineering,, Faculty of Engineering, Kyushu University, Fukuoka, Japan

Abstract

1. Introduction

Due to the high energy consumption and decreasing oil supply, the alternative fuels such as natural gas and biomass can be considered as the main sources of energy in the future. Currently, one of the technologies to convert such biomass to more valuable products, syngas, is gasification process. However, these sources including natural gas contain acid gases like carbon dioxide (CO2) and hydrogen sulfide (H2S). These gases are required to remove because of the decrease in product heating values, operational problems, corrosion and environmental concern. The requirement for the minimum level of contaminants in the gas cleaning of CO2 and H2S concentrations from many criteria are less than 5% and 10 ppb, respectively. The basic tools to do preliminary study is to apply process simulation. Nowadays the effective technologies used for acid gas removal is absorption both physical and chemical absorptions. The chemicals commercially used for acid gas removal are monoethanolamine (MEA) and methyldiethanolamine (MDEA) for chemical absorption and dimethylether of polyethylene glycol (DMPEG) or Selexol for physical absorption process. The advantages of these chemicals are that they can capture CO2 and H2S at the same time with various conditions; then they can be used in single acid gas removal unit. Therefore, the data for solubility, physical properties and so on are available from literature. Much of the experimental work on acid gas removal has been studied. The simulation work of acid gas removal from natural gas and gasification processes is very interesting.
Therefore, the objective of this work is to simulate and compare the absorption process for acid-gas treating by using various solutions at different operating conditions in order to meet the requirement of product gas.

2. Simulation

The simulation on acid gas cleaning process by means of absorption column with structured packing is investigated. The study used the ASPEN Plus software package to the process modelling based on aqueous MEA solution for gas processing in natural gas. In this work, the effects of solution concentration, liquid flow rate, temperature and pressure for chemicals used are investigated for the optimal design in the future.

2.1 Simulation of Physical Solvent

The physical solvent used is Selexol. Because there is no chemical reaction, this solvent combined less strongly with CO_2 and H_2S. The equilibrium concentration of the absorbing agent in the liquid phase is strongly dependent on the partial pressure of that component in the gas phase. The advantage of using such solvents is that these gases can be separated from them in the regenerator by reducing the pressure, resulting in much lower energy consumption.

2.2 Simulation of Chemical Solvent

For chemical absorption, the chemical reactions illustrating the process have been used in simulation. However, the reactions can be simplified by assuming the following conditions:

1. The equilibrium curve is linear over the range of concentrations encountered.
2. The partial pressure of the inert gas is essentially constant over the length of the column.
3. The solute contents of gaseous and liquid phases are sufficiently low that the partial pressure and liquid concentration values may be assumed proportional to the corresponding values when expressed in terms of moles of solute per mole of inert gas (or of solvent).

3. Results and Discussion

The absorption column is practical for acid gas removal. It is used for simulation to study the effect of parameters such as types of absorbing agents, concentration of solutions and liquid flow rate on CO_2 and H_2S removal from feed gas. The simulation results are based on the data of experimental work from the previous study at 0.9-2.5 MPa absolute and 25-40 °C and concentration of inlet acid gases.

3.1 Effect of Chemicals Used

The effects of the chemicals (MEA, MDEA and Selexol) on acid gas cleaning are studied for CO_2 and H_2S at the various conditions. The simulation results show that MDEA and MEA are excellent to remove CO_2 and MDEA. Also, Selexol is very outstanding in capturing H_2S. Moreover, the product gas consists of 3.64 vol% CO_2 and 27.24 ppb H_2S for MEA, 0.74% CO_2 and 0.02 ppb H_2S for MDEA and 7.45% CO_2 and 1.12×10^{-6} ppb H_2S for Selexol. From the requirement, the product gas should have less than 5% for CO_2 and 10 ppb for H_2S. Therefore, MDEA is the suitable chemical for this application because it can capture both gases at the required level.

3.2 Effect of Liquid Flow Rate in the Packed Column

It is evident that the liquid flow rate has a direct relationship with the efficiency of acid gas removal in that the efficiency increases as the liquid flow rate increases. This can be explained by the fact that applying a higher liquid flow rate to the system results in a larger driving force for acid gas removal, and consequently a higher rate of removing acid gas by the chemical per unit of time. This leads to a higher overall mass transfer coefficient that is directly proportional to the mass transfer values.

From the viewpoint of liquid phase, at a higher liquid flow rate, the amount of free-amine molecules in the system becomes larger or the system has more capacity to absorb acid gas from the gas phase, thus enhancing the mass transfer. However, an increase in solution flow rate leads to a higher circulation rate and regeneration costs; thus, may not improve the overall system efficiency. Maximizing liquid flow rates may not lead to optimum operating conditions. Therefore, an optimum flow rate has to be determined.

3.3 Effect of Solution Concentration

The effect of solution concentration is an interesting aspect since it affects the design of the process. The results of solution concentration of MEA at 3, 4 and 5 M are studied for both CO2 and H2S removal. At flow rate of 150 kmol/h, the amount of CO2 left in the product gas for 3M MEA is 5.03% and that for H2S is 39.6 ppb. For MDEA solution, the corresponding amount of CO2 remained in the product gas for 3M is found to be 4.00% and that for H2S is found to be 0.05 ppb. Also, the results of MDEA can meet the requirement for the criteria for the amount of acid gas in the product gas. From the results, it is clear that the efficiency of the process can increase as solution concentration increases because the amount of free molecules has increased per unit volume, resulting in more acid gases removed. However, the higher concentration can cause more equipment corrosion and the cost of chemical used as well as liquid viscosity. However, for the case of Selexol, there is no data for effect of concentration because normally Selexol is used as relatively pure chemical.

3.4 Effect of Temperature

The temperatures used in this study are 25 and 40 °C and the effects of system temperature are simulated for all chemicals. At flow rate of 150 kmol/h, MDEA provides better performance among them. In addition, theoretically, at higher temperature, gas has lower solubility because of higher kinetic energy to escape to gas phase. However, these two temperatures show little difference for solubility. Therefore, the practical operating temperature can be at 40 °C especially for the tropical countries.

3.5 Effect of Pressure

The effects of absolute pressure at 0.1, 0.9 and 2.5 MPa in the column are reported for both CO2 and H2S removal with MEA. It is evident that solubility of gases in solution increases at higher pressure corresponding to Henry's law. Therefore, the operation at 2.5 MPa offers the best solubility for MEA especially at the flow rate of 150 kmol/h. However, the increasing pressure means that extra equipment will be needed and leads to higher capital cost for acid gas removal.

4. Conclusion

Among MEA, MDEA and Selexol, MDEA is the most effective chemical to capture both CO2 and H2S at the studied conditions. The effects of liquid flow rate, solution concentration, temperature and pressure on removal performance are also studied and investigated. From the results, it is evident that the higher liquid flow rate can remove more gas because it has a higher driving force because of the higher amount of free amine molecules per unit of time. Consequently, a higher rate of removing acid gases by the chemicals can be achieved.

For higher solution concentration, the amount of free-amine molecules in the system becomes larger or the system has more capacity to absorb acid gases. However, it is noted that increasing solution concentration or liquid flow rate can lead to a higher circulation rate, regeneration costs, more equipment corrosion, and cost of chemical used. In addition, high pressure and low temperature can provide higher solubility for acid gases but these conditions need some equipment to fulfill their requirement and the cost will be higher.

545709 Kinetic Parameters Optimization and Analysis in Packed Bed Steam Methane Reformer

Monday, June 3, 2019: 5:18 PM
Texas Ballroom A (Grand Hyatt San Antonio)

Saiedeh Arabi, Sandeep Patil, Siddarth Chintamani and Brian Dennis, Mechanical & Aerospace Engineering, The University of Texas at Arlington, Arlington, TX

Steam Methane Reformer (SMR) has been in use for several years for generation of H_2 and Syngas needed for different applications. Production of oil is one of such application, where this gas is further fed to Fischer-Tropsch process.

Lab scale experiment has been setup to study kinetics of steam methane reforming over catalyst to obtain kinetic data of reactor modelling. The overall reactions in SMR process is to form CO, H_2, CO_2 and H_2O include number of equations. However, according to thermodynamic analysis of previous studies, only the three important equations are considered in this study. The reaction equilibrium constants of these three equations are evaluated from literature,

Reaction rate expressions for SMR is determined based on the detailed studies of possible mechanisms and appropriate rate controlling steps, which have been presented in this research. It is widely accepted that both methane and steam are adsorbed on the catalyst with dissociation and the surface reactions producing CO and CO_2 are the rate controlling steps. Considering information in the literature and thermodynamics analysis, appropriate forms of reaction rate expressions are considered for the kinetic model. There are four adsorption constants and three rate constants in rate expressions where the temperature dependency of each was estimated. The kinetic parameters were determined by minimizing the squared sum of the error between experimental data and predicted results from catalyst bed model. The values of the parameters for optimal fit were determined using nonlinear regression coupled with nonlinear ordinary differential equation solver.

545035 Enhanced Methane Reforming over Hierarchical Structured Supported Ni Based Catalyst

Monday, June 3, 2019: 5:30 PM
Texas Ballroom D (Grand Hyatt San Antonio)

Anup Tathod[1], **Naseem S. Hayek**[1], David Simakov[2] and Oz M. Gazit[1], (1)Chemical Engineering, Israel Institute of Technology - Technion, Haifa, Israel, (2)Chemical Engineering, University of Waterloo, Waterloo, ON, Canada

The reforming of methane, using CO_2 as an oxidizer, to make syngas (H_2 and CO) is an intriguing route to produce value added chemicals. For this reaction to become viable the catalyst needs to be based on nonprecious transition metals that show high activity and good resistance to deactivation induced by metal sintering and coke formation. Supported nickel catalysts were identified as a cost effective and active catalyst for MDR. However, the high concentrations of carbon species in MDR and the absence of water makes carbon deposition on the nickel catalyst the thermodynamically favored path. This stimulates the use of even higher reaction temperatures, which was shown thermodynamically to limit carbon formation. Unlike most rare earth metals, for which the T_{Tam} (Tammann temperature) is above the desired reaction temperature (600 °C - 800 °C), Ni has a lower T_{Tam} (590 °C), making it highly prone to sintering. Supporting nano-sized Ni catalysts on a strongly interacting support (SIS) will produce a stable catalyst but, may also block important active sites. In contrast, the same nano-sized Ni particles supported on a weekly interacting support (WIS) will be less stable but, will also be more active. This highlights the challenge of balancing the degree of interaction between the metal catalyst and the underlying support material, which is pivotal for obtaining optimal catalytic performance.

To mitigate this challenge, we used a "surface phase oxide" in the form of MgAl hydrotalcite nanosheets (<2 nm thick) as an interlayer SIS to mediate the interaction between nickel nanoparticles and an underlying WIS (i.e. ZrO_2 or SiO_2). We show that this hierarchical configuration, in which the Ni is supported only on the hydrotalcite SIS nanosheets, the obtained reaction performance, for an unoptimized reaction conditions, enhanced the conversion by up to 2-fold, as compared to Ni supported on ZrO_2, and by up to 15-fold as compared to Ni on bulk Mg-Al hydrotalcite. Moreover, selectivity of the hierarchical catalyst was similar to that of the Ni on ZrO_2, despite the fact that the Ni phase had virtually no physical interaction with the underlying ZrO_2. We found that the interaction of the Ni nanoparticles with the hydrotalcite surface phase oxide maintained a stable 2 nm Ni catalyst for a total of 250 h on stream. H_2-TPR analysis showed that the presence of the SIS surface phase oxide increased the Ni reduction temperature above that of the underlying ZrO_2 but kept it below that of Ni supported on bulk hydrotalcite. To gain better understanding into the mechanism of enhanced catalytic performance we used a combination of PXRD, SAXS, Cryo-TEM, H_2-TPR, STEM-EDS, and XPS in conjunction with the results for the catalytic performance and kinetic analysis. The combined results show that the amplified catalytic performance is strongly correlated with the modification of the metal-support interaction (MSI) by the MgAl hydrotalcite nanosheets interlayer. The changes in MSI interactions is postulated to arise either due to the ability of the Ni sites to electronically communicate with the underlying ZrO_2 or due to the effect of the new interface between the hydrotalcite nanosheets and the underlying ZrO_2. Molecular level studies into these effects are being conducted using atomic friction force microscopy, in situ-spectroscopy analysis and temperature programmed adsorption measurements.

545383 Elucidation of the Reaction Mechanism on Low Temperature Dry Reforming of Methane with Suppressing Carbon Deposition in Electric Field

Monday, June 3, 2019: 5:33 PM
Texas Ballroom D (Grand Hyatt San Antonio)

Kazuharu Ito, Tomohiro Yabe, Kensei Yamada, Kota Murakami, Kenta Toko, Takuma Higo, Shuhei Ogo and Yasushi Sekine, Applied Chem., Waseda Univ., Tokyo, Japan

Dry reforming of methane (DRM) has recently attracted a significant interest, producing syngas with a low H_2:CO ratio, which is adequate for C1 chemistry such as Fischer-Tropsch reaction, hydroformylation and carbonylation reactions.
$CH_4 + CO_2 \rightarrow 2CO + 2H_2$ $\Delta H^0{}_{298} = 247.2$ kJ mol^{-1} (1)
However, this reaction requires higher temperatures above 1073 K because of a large endothermic reaction, resulting in carbon deposition and requiring for sophisticated reactors.
Recently, we reported that DRM over Ni/La-ZrO_2 was significantly assisted at low temperature regions (423 K) by applying an electric field. The operation of DRM at low temperatures greatly suppresses carbon deposition on the catalyst and the catalytic activity is stably maintained. Furthermore, the kinetic analysis revealed that CH_4 conversion, which is the rate-determining step in conventional catalyst system, is enhanced by the electric field and the reaction rate was regulated by CO_2 dissociation.
To elucidate the reaction mechanism of DRM in the electric field, transient isotopic experiments using oxygen isotope ($^{18}O_2$) was conducted. Results showed that redox mechanism using lattice oxygen is important for CO_2 activation in the electric field. It is assumed that lattice oxygen (Olat) will be consumed by hydrocarbon species (CH_x) which is stem from CH_4 and supplied CO_2 will re-oxidize the defect.
$CH_4 \rightarrow CH_x + (4-x)/2\ H_2$ (2)
$CH_x + O_{lat} \rightarrow CO + x/2\ H_2 + V^{\cdot\cdot}{}_o$ (3)
$CO_2 + V^{\cdot\cdot}{}_o \rightarrow CO + O_{lat}$ (4)
Herein, the effects of surface defect on the catalytic activity of La-ZrO_2 were considered using DFT calculations. Results clarified that the formation of carbonate or bicarbonate, which are common intermediates formed during CO_2 conversion, occurs favorably at oxygen vacancy site. These results suggest that CO_2 conversion was assisted by promoting the formation of oxygen vacancy in the electric field.

545388 Low Temperature Steam Reforming of Methane over Pd-Zn Alloy Catalysts in an Electric Field

Monday, June 3, 2019: 5:36 PM
Texas Ballroom D (Grand Hyatt San Antonio)

Maki Torimoto[1], Danny Harjowinoto[1], Shuhei Ogo[1], Shinya Furukawa[2], Takuma Higo[1], Tomohiro Yabe[1] and Yasushi Sekine[1], (1)Applied Chem., Waseda Univ., Tokyo, Japan, (2)Hokkaido Univ., Sapporo, Japan

In recent years, world-wide demand for H_2 has been increasing because of its potential for alternative energy source. While steam reforming of methane (SR) is the most common process for H_2 production, it requires high temperature and pressure to dissociate the strong C-H bond of methane. Therefore, a new process that can be operated under milder condition is desired. Steam reforming of methane in an electric field (Electreforming: ER) proceeds at low temperature at which conventional catalytic reaction (SR) hardly proceeds. Our previous works revealed that reaction mechanism for ER was different from that for SR, and surface protonics induced by the electric field plays key role to promote reaction at low temperature. However, the role of supported metals on ER activity has not been investigated. In this study, to elucidate effects of supported metals on ER activity, we prepared Pd-Zn alloy catalysts supported on $Ce_{0.5}Zr_{0.5}O_2$ (CZO) and performed activity tests with or without an electric field. The prepared catalysts were also characterized by X-ray diffraction (XRD), X-ray absorption fine structure (XAFS), and Infrared spectroscopy (IR). From activity tests over Pd/CZO and Zn/CZO, it is considered that Pd is the active site for both ER and SR because Zn/CZO showed almost no activity with or without an electric field. Therefore, we evaluated activity of Pd-Zn alloy catalysts by calculating TOF per Pd atom. As the Zn amount increased, TOF for ER increased while TOF for SR decreased. XAFS and IR measurements revealed that adjacent Pd sites decreased with increased the Zn content and the electric density of Pd over Pd-Zn/CZO increased by the electron transfer from Zn to Pd. From these results, it is considered that increasing the electron density of Pd atom by the alloying with Zn contributes to the high ER activity of Pd-Zn/CZO catalysts.

545569 Co-Based Catalyst with High Stable Activity for Syngas Production By Methane Conversion

Monday, June 3, 2019: 5:39 PM
Texas Ballroom D (Grand Hyatt San Antonio)

Sholpan Itkulova, Yerzhan Boleubayev and Kirill Valishevsky, D.V.Sokolsky Institute of Fuel, Catalysis&Electroc, Almaty, Kazakhstan

Abstract: Natural gas is a prospective source for replacement of the traditional oil to produce synthetic oil fractions via syngas. This work deals with the production of synthesis gas by dry and bireforming of methane over the Co-based catalysts supported on a matrix composed of alumina modified with zirconia and rare-earth metal oxides. The multicomponent 5%Co-M/Al_2O_3-ZrO_2-R catalysts showed the high stable activity during its long-term continuous testing for 400 hours. The catalysts were characterized by a number of physico-chemical methods.

Introduction

Lately, there has been considerable interest in the development of more efficient processes to generate syngas and hydrogen. Syngas, a mixture of H_2 and CO in any ratios has been considered as a potential and promising sustainable energy from an economic and environmental point of view. Syngas is an intermediate in the multi-tonnage production of fuels and chemicals, including ammonia, methanol, dimethyl ether, oxo-synthesis and Fischer–Tropsch (F-T) fuels and can be produced by three methane reforming processes, including steam (SRM, Eq.1) or dry reforming (DRM, Eq.2) and partial oxidation (POM, Eq.3), or also by any combination of these reactions – bireforming (BRM), tri-reforming, autho-thermal reforming of methane (ATR).

$CH_4 + H_2O \leftrightarrow CO + 3H_2$ $\Delta H^o = 206$ kJ/mol (Eq. 1)

$CH_4 + CO_2 \leftrightarrow 2CO + 2H_2$ $\Delta H^o = 247$ kJ/mol (Eq. 2)

$CH_4 + 1/2\ O_2 \leftrightarrow CO + 2H_2$ $\Delta H^o = -38$ kJ/mol (Eq. 3)

The ratio of H_2/CO is of great importance for the subsequent application of syngas. The required H_2/CO ratio of syngas depending on the target processes can be achieved by the combined steam-dry reforming of methane so-called bireforming of methane (BRM). This is a feasible process to adjust the H_2/CO ratio of syngas with various feed ratios of H_2O and CO_2.

The carbonization of catalysts for these processes is a main challenge, which is not yet solved and requires the development of effective catalysts resistant to coke formation.

This work deals with the design and synthesis of the high active and stable catalyst based on Co and supported on a matrix composed of alumina modified with zirconia and rare-earth metal oxides for dry and bireforming of methane to produce synthesis gas with different ratio depending on a steam amount added.

Experimental

The multicomponent 5%Co-M/Al_2O_3-ZrO_2-R catalysts, where M is a noble metal and R is Ce or/and La were synthesized by impregnation and tested in dry and steam reforming of methane. The processes were carried out in a quarts flow reactor supplied with programmed heating, controlled feed velocity, a programmable syringe pump, and GCs. The dry (DRM) and bireforming (BRM) reforming of methane have been studied under atmospheric pressure, varying gas hourly space velocity (GHSV) within 1000-3000 h^{-1} with using a feed with a ratio of CH_4/CO_2=1:1 and varying temperature within 300-800°C. The 0.5-1.0 vol. parts of steam has been added to a feed to provide bireforming of methane. To elucidate the stability the 5%Co-M/Al_2O_3-ZrO_2-CeO_2-La_2O_3 catalyst was continuously tested for a long-term period: 400 hours in DRM and 100 hours bireforming of methane. The catalysts were characterised by BET, SEM, TEM, and TPR methods.

Results/Discussion

The catalysts perform a high activity in both dry and steam reforming of methane. Methane is almost completely converted at 750-800°C depending on the catalyst composition and process type. Adding steam in amount 10-50 vol.% to methane always leads to decrease in temperature of methane complete conversion and enriching syngas with hydrogen over the synthesized catalysts.

The 5%Co-M/Al_2O_3-ZrO_2-CeO_2-La_2O_3 catalyst showed a high stable activity for all period of testing in both processes: 400 hours in DR and 100 hours in steam reforming of methane. During 400 hours of DRM process over the catalyst under conditions t=700°C, P=1 atm, GHSV=1000 h^{-1}, CH_4:CO_2=1:1, methane conversion was varied within 96.0-96.6%, while the extent of carbon dioxide conversion was insignificantly less and varied within 90.4-92.3%. The ratio of H_2/CO in the syngas formed was 0.9. In steam reforming of methane (CH_4:CO_2:H_2O=1:1:0.5), H_2/CO ratio was increased to 1.3.

A high dispersed state of metals and the effect of the second metal on reducibility of Co was observed by TEM and TPR.

Conclusion

The 5%Co-M/Al_2O_3-ZrO_2-R catalysts developed performs the high activity and stability in syngas production from methane. The synthesis gas composition can be controlled by adjusting the composition of the feed.

Acknowledgments

The work was financed by the Ministry of Education and Science of the Republic of Kazakhstan (Program number # PCF_BR05236739).

545409 CO Rich Syngas Production By Ntp Assisted Ni/γ-Al_2O_3 Catalyst Packed Reactor: Influence of Basic Promoters

Monday, June 3, 2019: 5:42 PM
Texas Ballroom D (Grand Hyatt San Antonio)

Debjyoti Ray, Piu Chawdhury and Surbahmanyam Ch, Chemistry, Indian Institute of Technology Hyderabad, Sangareddy, India

CO rich syngas production by NTP assisted Ni/γ-Al_2O_3 catalyst packed reactor: Influence of basic promoters

Debjyoti Ray, Piu Chawdhury, Ch. Subrahmanyam*

Department of Chemistry, Indian Institute of Technology Hyderabad, Telangana 502 285, India, Email: csubbu@iith.ac.in

Dry reforming of methane (DRM) was carried out in a packed bed dielectric barrier discharge (DBD) reactor. DBD plasma alone suffers from poor energy efficiency. Hence a combination of heterogeneous catalyst with plasma is necessary. During the present study, the discharge gap of 6 mm was filled with Ni/MO_Al_2O_3 (M= Ce, Mg) catalyst.[1] Ni/γ-Al_2O_3 is known to have superior activity for DRM reaction and syngas production. But only Ni/Al_2O_3 catalyst shows poor activity and fast deactivation due to coke formation. Hence CeO2 or MgO has been introduced as a promoter to suppress the coke formation. Catalysts were prepared by wet impregnation method and were characterized by PXRD, BET, TPD (NH_3 and CO_2), TGA to bring out structure-function correlations and to notify the active species generated. Typical observations indicated that MgO integrated catalyst improved the conversion of the reactants as well as the selectivity of the products. The best syngas ratio was observed with Ni/MgO_Al_2O_3 catalyst.

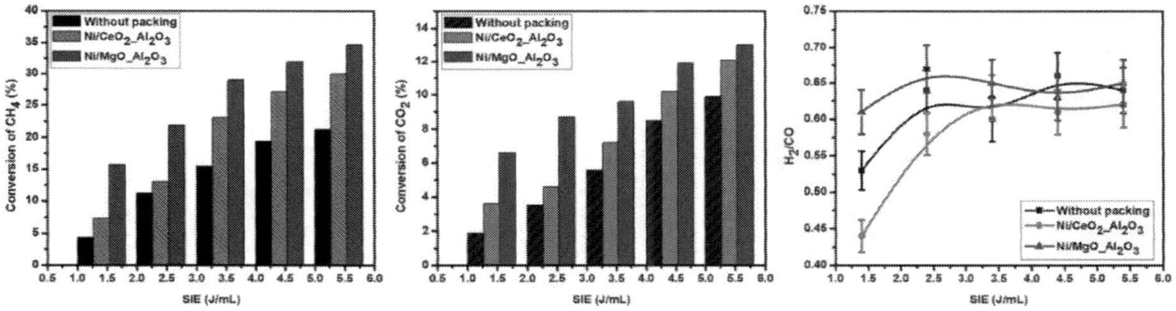

[1] X. Tu et al., *Appl. Catal., B*, 2018, *224*, 469.; Sk. Mahammadunnisa et al., *Energy Fuels*, 2013, *27*, 4441.; D. Ray et al., *Catal. Today,* 2018, *309*, 212.

541327 The Effect of Barium in Ni-La Based Catalysts on Dry Reforming of Methane

Tuesday, June 4, 2019: 10:15 AM
Texas Ballroom D (Grand Hyatt San Antonio)

Soraia Brandão, Inorganic Chemistry, Universidade Federal da Bahia, Salvador, Brazil

1. Introduction

Many studies have been conducted about Dry Reforming of Methane (DRM) mechanism and is a consensus that this reaction follows a bifunctional mechanism where CH_4 is activated preferentially on metallic sites and CO_2 is activated by the support [1, 2]. Since the catalysts applied on this reaction are prone to deactivate, the support exhibits a fundamental role.

Several studies on the literature have proposed that the nature of the support effects on the mechanism of carbon species oxidation. For example Ni-based catalyst supported on CeO_2, La_2O_3, CaO, MgO and BaO are alternatives to inhibit the deactivation by carbon deposition. However, for nickel based perovskites, few investigations have been done with barium as a promoter [3].

Therefore, the aim of this study was evaluating Ni-La based catalysts and the impact of barium on the DRM reaction. It was investigated the effect of the synthesis method (nickel perovskite reduction or wetness impregnation) on Ni/La_2O_3 catalysts properties as well as evaluate the influence of barium species on carbon deposition.

2. Experimental

In this work perovskite precursors $La_{1-x}Ba_xNiO_3$ (x = 0.0, 0.05, 0.1 and 0.2) were prepared using citrate method. The samples were characterized by X-ray fluorescence (XRF), nitrogen adsorption, temperature-programmed reduction (TPR), temperature-programmed oxidation (TPO), thermogravimetric analysis (TGA) and scanning electron microscopy (SEM). *In situ* X-ray diffraction experiments were conducted at the XPD beam line using a radiation with $\lambda = 1.2397Å$ at the Brazilian Synchrotron Light Source at Campinas (LNLS). DRM reaction was carried out in a stainless-steel tubular reactor at 973K during 20 h under 100 mL min^{-1} 25% CH_4/ 25% CO_2/ $Ar_{(balance)}$. The gas effluent was analyzed using a GC Shimadzu 2014 equipped with TCD detector and column Carboxen 1010. Before the test the samples were pre-treated as follows: heated (10°C min^{-1}) under H_2 flow (30 mL min^{-1}) to 1073 K during 1 h. Also aging tests were performed. The used samples were characterized by TGA and SEM to determine the amount and morphology of the coke formed.

3. Results

The results of *in situ* XRD experiments are shown in Figure 1 (a) and (b) and the results of DRM aging tests are presented in Figure 2.

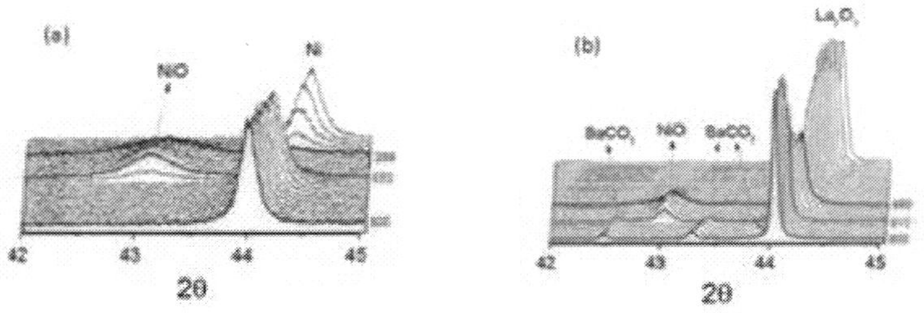

Figure 1: *In situ*]XRD under CH_4 and CO_2: (a) $LaNiO_3$ and (b) $La_{0.8}Ba_{0.2}NiO_3$

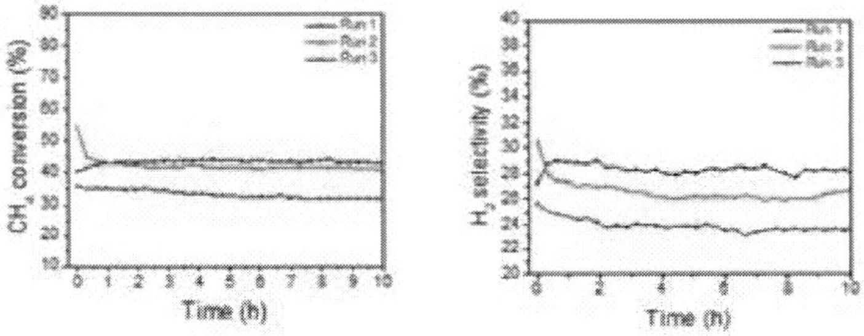

Figure 2: $La_{0.8}Ba_{0.2}NiO_3$ aging tests at 973 K: CH_4 conversion and H_2 selectivity curves.

Table 1 – Carbon deposited after DRM catalytic test determined by TPO.

Catalyst	$mg_{carbon}/g_{catalyst}$
$LaNiO_3$	1.64
$La_{0.95}Ba_{0.05}NiO_3$	0.08
$La_{0.9}Ba_{0.1}NiO_3$	0.16
$La_{0.8}Ba_{0.2}NiO_3$	0.03

4. Discussion

In situ XRD showed the phase evolution when the catalysts were exposed under CH_4/CO_2 mixture from room temperature to 800 ºC. For all catalysts, metallic nickel was oxidized to nickel oxide when exposed to reaction atmosphere. This event suggests that CO_2 present on the feed was responsible for the nickel sites oxidation. Comparing the catalysts obtained via $LaNiO_3$ and $La_{0.8}Ba_{0.2}NiO_3$ (fig. 1a and 1b) it was verified that the nickel sites present on $LaNiO_3$ catalyst are more susceptible to oxidation. At 289 ºC, the intensities of the diffraction lines associated to La_2O_3 phase decreased in intensity and the peak (111) attributed to metallic nickel disappeared, characterizing the total nickel oxidation to NiO. At the same temperature, the peaks characteristics of NiO (111) and (200), started to appear, reaching a maximum at 463 ºC. Moreover, at 463 ºC, the peak (111) attributed to metallic nickel becomes detectable, suggesting the catalyst was being regenerated. For the catalyst obtained from $La_{0.8}Ba_{0.2}NiO_3$, the oxidation event takes place at higher temperatures, starting from 432 ºC and ending at 610 ºC. Nonetheless, the nickel sites were not totally oxidized, suggesting a partial oxidation process, which was characterized only by the decreasing of the metallic nickel peak (111). This behavior considered that $BaCO_3$, during reduction process, may cover the nickel sites, at least partially, blocking such sites and making them less susceptible to oxidation.

Barium carbonate and carbon species can interact to form a carbonate-carbon complex that decomposes rapidly to carbon monoxide. If this process is taken into account, it is possible to explain the higher resistance of barium-content catalysts to carbon deposition observed in table 1. It is worth mentioning the CO_2 present on the feed may promote the $BaCO_3$ regeneration, since this phase is present in all temperature range in which the catalyst was exposed to CH_4/CO_2.

5. Conclusions

In situ XRD under reaction atmosphere suggested all catalysts were oxidized to NiO when exposed to reaction atmosphere and regenerated increasing the temperature. Nevertheless, barium catalysts showed more resistant to oxidation. Aging tests and TPO analysis suggested barium addition promotes higher resistance to catalysts deactivation since $BaCO_3$ can act as a carbon oxidant in reaction conditions. Also, those catalysts may provide other reaction surface regeneration pathways.

References

[1] S. Dama, S.R. Ghodke, R. Bobade, H.R. Gurav, S. Chilukuri, Appl. Catal. B Environ. 224 (2018) 146–158.

[3] D. Pakhare, J. Spivey, Chem. Soc. Rev. 43 (2014).

[2] L.M.T.S. Rodrigues, R.B. Silva, M.G.C. Rocha, P. Bargiela, F.B. Noronha, S.T. Brandao, Catal. Today. 197 (2012) 137–143.

544160 Effect of Manganese and Zeolite Composition on Zeolite-Supported Nickel Catalysts for Dry Reforming of Methane

Wednesday, June 5, 2019: 2:51 PM
Republic ABC (Grand Hyatt San Antonio)

Aaron Najfach and **Catherine Almquist**, Chemical, Paper, Biomedical Engineering, Miami University, Oxford, OH

Dry reforming of methane (DRM) is a reaction of interest for utilizing two greenhouse gases, carbon dioxide and methane, for syngas production. Catalysts are used to facilitate DRM as high temperatures are required. Noble metals are chosen for their effectiveness; however, these metals are expensive. Nickel as a substitute for DRM catalysts has interest for its low relative cost while still exhibiting effectiveness. Carbon deposition from side reactions is a concern as it causes catalysts fouling. To address this, addition of a second metal has shown success at facilitating the consumption of built-up carbon. Manganese has shown promise as an inexpensive second metal for this purpose. This study examines the effect of manganese loading on nickel catalysts and the effect different zeolites have on DRM. Samples were characterized with BET analysis, X-Ray Diffraction (XRD), Scanning Electron Microscopy (SEM), and Thermogravimetric Analysis (TGA). Catalysts were tested for activity and carbon deposition. It was found that the addition of manganese has a beneficial effect on reducing catalyst fouling. It was also found, over 24 hours, zeolite type had some effect on catalyst activity and may influence carbon formation. The role of manganese and sodium on zeolite destruction is also discussed.

545671 Plasma-CaO Coupling for CH_4-CO_2 transformation into Syngas and/or Hydrocarbons, Oxygenates. Catalyst Structural and Textural Modifications Under Plasma Discharge

Tuesday, June 4, 2019: 11:03 AM
Texas Ballroom D (Grand Hyatt San Antonio)

Nassim Bouchoul[1], Elodie Fourre[2], Jean-Michel Tatibouët[2], Alysson Duarte[3], Nathalie Tanchoux[3] and **Catherine Batiot-Dupeyrat**[2], (1)University of poitiers, Poitiers, France, (2)University of Poitiers, Poitiers, France, (3)University of Montpellier, Montpellier, France

Introduction

Non-thermal plasma is considered as an attractive alternative method for methane and carbon dioxide transformation to overcome carbon deposition limitation in classical catalytic processes. Many reactions are initiated at low temperature under the plasma discharge leading to valuable chemicals such as hydrocarbons or oxygenates [1]. However the selectivity towards targeted products is often difficult to achieve due to the occurrence of a large number of reactions in gaseous phase. The coupling of catalyst and plasma was proposed to overcome this drawback [2]. In many studies, the selected catalysts were similar to those used in conventional thermal catalytic processes, such as Ni/Al_2O_3, but it was shown that the presence of conductive Ni active sites decreased the electric field strength and consequently the electron density reducing thus reactant conversion [3]. The combination of non-thermal plasma with different metal oxides was investigated, while alumina possesses no catalytic activity in the thermal dry reforming of methane reaction, a significant improvement in CH_4 and CO_2 conversion over Al_2O_3 was observed when coupled with plasma [4]. The combination of plasma and catalyst requires the development of specific materials able to transform reactants present under active species form, such as excited species, radicals and ions. In this respect, the influence of calcium oxide located in a packed bed DBD reactor was studied. This material was chosen due to its relatively low dielectric constant (9-30), since as shown by Wang et al. [5] a too high dielectric constant of the packing material limits the discharge to the contact point of the catalyst beads.

- **Experimental part**

The reaction was performed at room temperature and atmospheric pressure in a coaxial dielectric barrier discharges (DBD) reactor. The non-thermal plasma reactor was composed of an alumina tube (ID: 4mm; ED: 6mm), a stainless steel electrode inside the reactor (1.0mm) and a copper electrode wrapped around the quartz tube (100 mm long). Helium, methane and carbon dioxide were flown through the plasma reactor at a total flow rate of 40mL.min^{-1} using different ratio CH_4/CO_2 (from 0.5 to 2) with a constant concentration in He: 75%, corresponding to a contact time of 1.6s.

A sinusoidal supply of power was applied across the electrodes (TG1010A Aim-TTi, Thurlby Thandar Instruments Brand). The discharge power, calculated from the Lissajous figures, was fixed at 8 watts (frequency at 800Hz and voltage at 13.5kV).

The oxide used in this study we prepared according to the alginate way after optimization of the synthesis procedure allowing to obtain CaO after calcination at 600°C with a high purity [6].

2 Results and discussion

2.1 Influence of the grain size and role of CaO material

CaO oxide was sieved to diameters d1: 250-355µm, d2: 355-650µm, d3:650-800 and d4: 800-1000µm. The only parameters that are affected by the grain size are the accessible surface, the average size of space and the number of contact point between the particles. Results showed that conversion of CO_2 and CH_4 increased with decreasing particle size. Kasinathan et al. [7] obtained a similar effect for CH_4 conversion into C2, C3 and C4 hydrocarbons over MgO/Al_2O_3. They conclude that the smaller the catalyst particle, the more diffuse the plasma streamers on the catalyst surface.

The selectivity to products depends on the particle size. Ethane formation is favored when the biggest particles are used, corresponding to the largest space between grains, suggesting that the recombination of CH_3 radicals to form C_2H_6 is favored in gaseous phase and not at the surface of the solid. As expected, CO is the main product formed, while the amount of oxygenated products remains low whatever the grain size. The carbon balance is always higher than 90%. It appears that the presence of CaO does not lead to significant catalytic effect in the experimental conditions used. It is believed that the reactivity is governed by the characteristics of the discharge at the surface of the packing material possessing a low dielectric constant (≈11 [8]). CaO is characterized by the presence of meso and macropores with porous volume of 0.01 cm^3/g and an average pore diameter of 10 nm with a specific surface area of 15 m^2/g. The formation of micro-discharges in pores is possible when their diameter is larger than the Debye length. This value depends on electron density and temperature in the plasma streamer but is typically in the order of 100 nm to 1µm. According to calculations of A. Bogaerts group, plasma can penetrate in pores within 50nm but at very short time [9, 10]. Consequently, using CaO, the penetration of micro-discharges within pores can be excluded.

2.2. Influence of the ratio CH_4/CO_2

The carbon balance and the selectivity to CO is maximum with an excess of CO_2. Besides CO and hydrocarbons, oxygenated products (formaldehyde, methanol, acetaldehyde and ethanol) were quantified, but with a low selectivity. An opposite trend is visible concerning formaldehyde and methanol formation: CH_3OH is slightly favored when an excess of CO_2 is used, while the formation of HCOH is favored in an excess of methane. These results are in agreement with the reported research of De Bie et al. using a one-dimensional fluid model [11]. The dominant reaction pathways was through the following three-body reaction between CH_3 and OH radicals: $CH_3 + OH + M \rightarrow CH_3OH + M$. It is also supported by the work of Wang et al. based on density functional theory (DFT) calculations [12] and Istadi using a hybrid artificial neural network – genetic algorithm technique [13]. The production of formaldehyde comes from the reaction between CO_2 and CH_2 radicals and acetaldehyde is obtained by O atom (from CO_2 dissociation) reaction with ethyl radical [11].

2.3 Influence of temperature

The results obtained for grain size: 355 µm < d < 650 µm, P= 8W, from 100 to 300°C and a CH_4/CO_2 ratio of 2 show that methane conversion is relatively stable during time on stream and is closed to 20% whatever the reaction temperature, while for CO_2 the conversion is strongly increased at the beginning of the experiment performed at 300°C, reaching 40%, then decreasing regularly with time and stabilizing at around 15% after 30 minutes. The carbon balance follows an opposite trend suggesting a reaction of CO_2 with the basic CaO material. The selectivity to the main products: CO and C_2H_6 is not modified by the increase in the reaction temperature. Little modifications are observable on oxygenates, an increase in the reaction temperature lead to a decrease of the selectivity to methanol and formaldehyde while the selectivity to acetaldehyde is not strongly modified.

2.3 Characterization of the oxide after reaction at different temperatures

No significant differences (from TGA analysis) are observed before reaction and after reaction at room temperature and 100°C. The weight loss between 500-700°C is attributed to the presence of stable carbonate species corresponding to 7 monolayers.

When the reaction is performed at 200°C, TGA profile exhibits an important weight loss between 300-500°C (6.1%) corresponding to an endothermic peak. It is attributed to the decomposition of $Ca(OH)_2$. The formation of calcium hydroxide can result from H_2O dissociation to H^+, OH^- in gas phase. After reaction at 300°C, the presence of carbonate oxide is preponderant. This result shows, for the first time, that CO_2 or excited CO_2 under plasma discharge can diffuse from the gas phase deep into porous CaO.

Conclusion

This study shows that plasma-catalysis can be considered as an alternative method for the production of oxygenated compounds and CO_2 valorization, despite the low selectivity to methanol. CaO particle size, CH_4/CO_2 ratio or temperature had little effect on methanol yield. However, the presence of the catalyst in the plasma discharge led to the formation of carbonate oxide, proof of the CO_2 diffusion into CaO pores.

References

[1] B. Eliasson, W. Egli, U. Kogelschatz, Pure Appl. Chem. 66(6) (1994) 1275-1286

[2] E.C. Neyts, K. Ostrikov, M.K. Sunkara, A. Bogaerts, Chem. Rev. 115 (2015) 13408-13446

[3] X. Tu, H.J. Gallon, M. V. Twigg, P. A. Gorry, J.C. Whitehead, J.Phys. D: Appl. Phys. 44 (2011) 274007-274017

[4] D. Yap, J.M. Tatibouët, C. Batiot-Dupeyrat, Catal. Today, 299 (2018) 263-271

[5] W. Wang, H.H. Kim, K. Van Laer, A. Bogaerts, Chem. Eng. J 334 (2018) 2467-2479

[6] Y. Monakhova, P. Agulhon, F. Quignard, N. Tanchoux, D. Tichit, Catal. Today, 189 (2012) 28-34

[7] P. Kasinathan, S. Park, W.C. Choi, Y.K. Hwang, J.S. Chang, Y.K. Park, Plasma Chem. Plasma process. 34 (2014) 1317-1330

[8] J. Robertson, Eur. Phys. J. Appl. Phys. 28 (2004) 265-291

[9] Y.R. Zhang, K.Van Laer, E. C. Neyts, A. Bogaerts, Appl. Catal. B: Env 185 (2016) 56-67

[10] Q.Z. Zhang, A. Bogaerts, Plasma Sources Sci. technol. 27 (2018) 35009-35019

[11] C. De Bie, J. Van Dijk, A. Bogaerts, J. Phys. Chem. C 119 (2015) 22331-22350

[12] J.G. Wang, C. Liu, B. Eliassion, Energy Fuels, 18 (2004) 148-153

[13] I. Istadi, N.A.S. Amin, Chem. Eng. Sci. 62 (2007) 6568-6581

545795 Dry Reforming of Methane in Molten Metal Alloys

Tuesday, June 4, 2019: 11:27 AM
Texas Ballroom D (Grand Hyatt San Antonio)

Clarke Palmer, University of California, Santa Barbara, CA, David Chester Upham, Chemical Engineering, Stanford University, Stanford, CA, Simon Smart, University of Queensland, Australia, Brisbane, Australia, Michael Gordon, Department of Chemical Engineering, University of California, Santa Barbara, CA, Horia Metiu, Department of Chemistry and Biochemistry, University of California - Santa Barbara, Santa Barbara, CA and Eric W. McFarland, Chemical Engineering, University of California, Santa Barbara, Santa Barbara, CA

Dry reforming of methane (DRM) has had limited practical application due to the relatively low H_2:CO product ratio and challenges in catalyst coking. Syngas with a H_2:CO ratio of 2:1 is required for oxo-alcohol and hydrocarbon production; however, the DRM processes explored thus far produce H_2:CO ratios of at most 1:1. The low ratio is a result of the trade-off between the thermodynamics of the reforming reaction and coking kinetics. By combining dry reforming of methane with methane pyrolysis to produce H_2-rich syngas and solid carbon in a single catalytic molten metal-based reactor, high H_2:CO ratios can be obtained and catalyst deactivation from coke is eliminated. In molten metal bubble column reactors, the low-density solid carbon is continually removed from the high-density melt and methane is activated within the rising bubble with a continuously-renewed metal surface. DRM is demonstrated using a molten alloy of 65 mole % nickel – 35 mole % indium with 92% CO_2 conversion and 55% CH_4 conversion to produce a H_2:CO ratio of 1.2:1. The system produces a nanocrystalline graphitic carbon product as measured by Raman spectroscopy. The variability of the syngas H_2:CO ratio with changes in the feed ratios of methane to carbon dioxide, the reaction temperature, and the residence time is demonstrated.

545435 A Stability Analysis of Electroless Deposition-Derived Ni-Pt Catalysts for the Dry Reforming of Methane

Tuesday, June 4, 2019: 11:51 AM
Texas Ballroom D (Grand Hyatt San Antonio)

Benjamin T Egelske, Jayson Keels, John R. Monnier and John R. Regalbuto,
Chemical Engineering, University of South Carolina, Columbia, SC

Dry reforming of methane (DRM) by CO to form CO and H is an opportunity to blend sustainable initiatives with inplace hydrocarbon infrastructure. To overcome sluggish reactivity methane reforming must take place at temperatures exceeding 700 °C which leads to rapid deactivation of current Ni-based catalyst technology.

Past studies have shown improved stability and performance when Group VIII metals are paired with Ni, but these results often suffer from two deficiencies. First, samples have been prepared using traditional dry impregnation which randomizes metal-metal contact making it difficult to determine correlations between catalyst composition and performance [1]. Secondly, evaluation data is reported over a wide range of temperatures often below 700 °C resulting in a lack of industrial relevance [1].

In this study Electroless Deposition (ED), where one metal is placed directly and only on a second metal, is used to prepare a series of Ni-Pt catalysts. Evaluations were performed at both low temperature (<= 600 °C) and high temperature (700 °C) conditions in a continuous flow reactor. Low temperature data agrees with the promotional effects previously reported in literature but at high temperature conditions performance is reversed, i.e. all bimetallic samples exhibit lower conversion and higher deactivation rates compared to monometallic Ni.

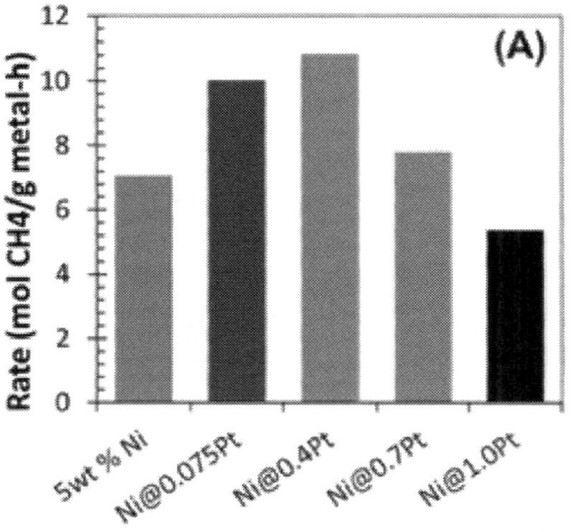

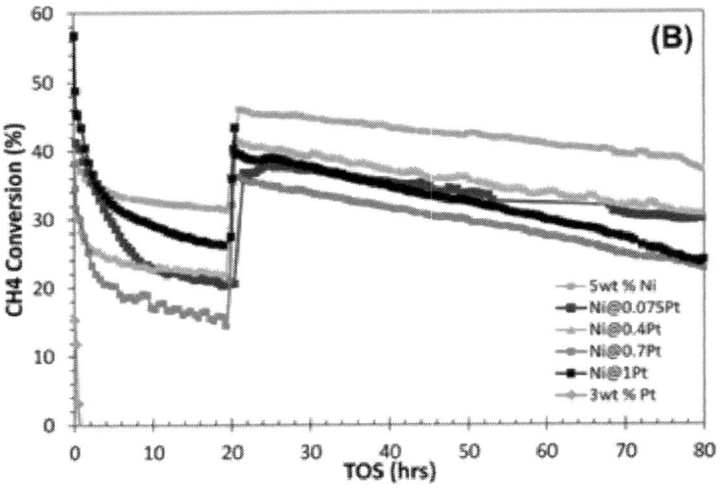

Figure 1: Activities of Ni-Pt catalysts at 550 °C (A) and 700°C (B). All experiments performed at identical conditions so instantaneous CH conversion is proportional to reaction rate.

X-ray Diffraction data suggests high temperature deactivation results from Pt-assisted methane decomposition due to separation of the Ni-Pt alloy above 630 °C. Pre and post reaction STEM imaging showed a slight growth in monometallic Ni particles but large Pt clusters and the presence of whisker coke for bimetallic samples. TPO analysis on spent samples shows a direct relationship between Pt loading and surface carbon. These results suggest that coke formation is the primary cause for deactivation with only small effects from sintering of Ni.

An analysis of low temperature kinetics indicates the presence of a compensation effect or the offset of apparent activation energies by pre-exponential factors up to an isokinetic temperature at which point the order of catalytic performance is reversed [2]. For the Ni-Pt series experimental data suggests a link between the calculated isokinetic temperature of 624 °C and the historically reported phase separation point at 630 °C [3]. The findings in this study highlight the benefits of evaluating catalysts of known compositions and the importance of including practical reaction conditions in scientific studies.

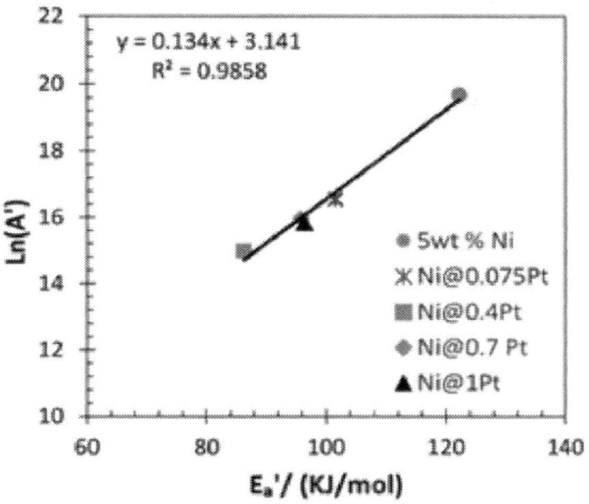

Figure 2: Ni-Pt compensation plot under stable operation (T <= 600 °C).

[1] A. N. Şener, M. E. Günay, A. Leba, and R. Yıldırım, "Statistical review of dry reforming of methane literature using decision tree and artificial neural network analysis," *Catalysis Today,* vol. 299, pp. 289-302, 2018.

[2] G. C. Bond, M. A. Keane, H. Kral, and J. A. Lercher, "Compensation Phenomena in Heterogeneous Catalysis: General Principles and a Possible Explanation," *Catalysis Reviews,* vol. 42, no. 3, pp. 323-383, 2000.

[3] H. Okamoto, "Ni-Pt (Nickel-Platinum)," *Journal of Phase Equilibria and Diffusion,* vol. 31, no. 3, pp. 322-322, 2010.

545647 Utilizing CO_2 for Production of CO Rich Synthesis Gas at Low Steam to Carbon Ratio

Tuesday, June 4, 2019: 1:54 PM
Texas Ballroom D (Grand Hyatt San Antonio)

Peter M. Mortensen, Martin Østberg and Kim Aasberg-Petersen, Haldor Topsoe, Kgs. Lyngby, Denmark

Introduction

Synthesis gas production is one of the largest industries in the world with its development stemming back from 1930s. Important bulk chemicals as hydrogen, ammonia, and methanol are produced on the basis of this. Haldor Topsoe A/S has many years of experience in reforming, with the first start-up of a steam reformer designed by Haldor Topsoe A/S dating back to 1956.

Reforming of methane with mixtures of steam and carbon dioxide (in the following referred to as CO_2-reforming) is an environmentally interesting process as it in theory offers a way of using CO_2, which in many industries is considered as a waste product and in parallel is a polluting greenhouse gas. Overall, CO_2-reforming is one of few chemical processes which can be designed with an overall negative CO_2-emission, as the process can be designed to utilize more CO_2 than what is produced from the process. The process could therefore play an important role in combination with CO_2 capture technologies, which are receiving much attention currently.

When performing CO_2-reforming with a large import of CO_2, CO will make up a significant fraction of the produced synthesis gas and H_2/CO ratios in the range from 0.5-3 can be produced more or less directly from the process. The large amounts of CO can be used within the chemical industry for production of higher alcohol, synthetic fuels, or acetic acid, among others, but especially the polymer industry has an increasing demand for CO which is an essential component in production of polycarbonates and polyurethane, among others.

One of the primary tasks in the development of CO_2-reforming is to find operating conditions in combination with a suitable catalyst to avoid carbon formation. Carbon formation in a steam reforming plant is dictated by thermodynamics, and in a typical design of a reformer, it is a requirement that there is no affinity for carbon formation anywhere in the catalyst bed. This means that the process gas will have to be balanced with water in order to circumvent the carbon formation area. In the case of the classical fired reformer, the process gas enters the reformer at 400–500°C while leaving the reformer at maybe 950°C (not experiencing temperatures above 1000°C). Thus, when designing a reformer, there must not be an affinity for carbon formation of the equilibrated gas anywhere between 400–1000°C. This criterion can be used to evaluate the carbon limit of the reformer.

Nickel, cobalt, and noble metal catalysts have been tested as catalyst for CO_2-

reforming, where nickel has been the most investigated system, as this is a relatively cheap catalyst in comparison to noble metals. Noble metal catalysts typically have a lower affinity for carbon formation and a higher activity for steam methane reforming, however, the markedly higher raw material prices of these material makes them unattractive for extensive use in industrial scale. Consequently, nickel based catalyst are established as the preferred catalyst for industrial use, especially for the conventional steam methane reforming (SMR) reaction, as a Pareto optimality between catalyst price and performance. A large incentive towards continued use of nickel as reforming catalyst therefore also exists.

New high temperature reactor for CO production

In the current work, a new reactor configuration is presented where hot CO_2 is added directly downstream a reformer and equilibrated in an adiabatic reformer, a so-called Adiabatic POst Convertor (APOC). This concept utilizes the high temperature of the principle reformers product gas and detailed understanding of the underlying thermodynamic and kinetic mechanisms to circumvent the carbon formation area. This allow for a method to tailor the product toward practically any H_2/CO ratio without the risk of carbon formation, while still using a cost-efficient nickel based catalyst.

By placing the APOC directly downstream a reformer allows for operating the reformer at its absolute minimum steam to carbon ratio, because no CO_2 will be added to this. This reformer can be practically any known reforming technology including the well-known tubular reformer or the autothermal reformer. Additionally, the introduction of the APOC can be flexibly introduced for both revamps of existing reforming plants as well as new build plants.

Bench scale experiments were done to evidence the operation and the catalyst performance of the APOC. These experiments showed that operation at what would correspond to a steam to carbon ratio of 1 is possible, without the risk of carbon formation. These results were included in process studies, which showed that easy integration of the APOC in a reforming plant can be achieved, where the highlights of the technology is:

- The size of the primary reformer can be reduced by up to ca. 25% by including the APOC. Or the capacity of an existing plant can be boosted accordingly.
- The carbon for the CO product to a larger extent stems from the CO_2 feed and the plant consequently has a large netto use of CO_2.
- No potential for carbon formation is found inside the reactor.
- Practically any H_2/CO ratio can be supplied in the product synthesis gas.

Conclusion

In conclusion, the new high temperature CO production reactor (APOC) is found as a promising technology for producing synthesis gas with a high content of CO at a low practical steam to carbon ratio. The technology can be used to retro-fit an existing production towards more CO production or for boost in capacity, but can also be included in new projects directly and is an excellent match in cases where excess

CO_2 is available.

544595 Methane Pyrolysis on Complex Liquids: Reactive Separation of Hydrogen and Solid Carbon

Tuesday, June 4, 2019: 2:18 PM
Texas Ballroom D (Grand Hyatt San Antonio)

Eric McFarland[1], Horia Metiu[2], Michael Gordon[1], Michael F. Doherty[3], Clarke Palmer[4], Jiren Zeng[4], Dohyung Kang[1] and Nazanin Rahimi[1], (1)Department of Chemical Engineering, University of California, Santa Barbara, CA, (2)Department of Chemistry and Biochemistry, University of California - Santa Barbara, Santa Barbara, CA, (3)Chemical Engineering, University of California, Santa Barbara, CA, (4)University of California, Santa Barbara, CA

Steam methane reforming is the most economical means of producing H_2 and CO_2 at large scale; however, if, in the future, large segments of society are willing to pay a significant price for CO_2 emissions reduction while natural gas remains relatively inexpensive, H_2 produced from natural gas pyrolysis would offer the most cost competitive alternative. Specific compositions of molten metal alloys and/or complex molten salt mixtures at high temperature (~1000 C) are shown to have high activity for methane decomposition and high selectivity for molecular hydrogen and solid carbon as the only products. When the physical properties of the carbon and the liquids are selected optimally, solid carbon produced from pyrolysis in the high temperature melts can be conveniently separated in bubble column reactors. In molten alloys of Ni and Bi or mixtures of $MnCl_2$ and KCl hydrogen and graphitic carbon products are produced at high conversion and high selectivity and are separable from the melt. Solid catalysts (including Ni/Al_2O_3) suspended in specific melts are also shown to be continuously reactivated as the high temperature liquid serves as a solvent to remove carbonaceous surface deposits prior to the irreversible formation of coke. Single pass methane conversion of over 98% to molecular H_2 at over 98% selectivity is demonstrated in several molten metal alloys and molten salt mixtures. Fundamentals of C-H bond activation on melt surfaces will be discussed for metals and salts and distinctions made as to how the subsequent reaction pathways can produce different pyrolysis pathways and different types of solid carbons. The practical problem of producing a commercially acceptable carbon product will be discussed along with the heat transfer and materials of construction challenges.

545615 Selective Photocatalytic Conversion of Methane into Carbon Monoxide over Zinc-Heteropolyacid-Titania Nanocomposites

Tuesday, June 4, 2019: 2:42 PM
Texas Ballroom D (Grand Hyatt San Antonio)

Xiang Yu, Vitaly Ordomsky and Andrei Khodakov, Univ. Lille, CNRS, Centrale Lille, ENSCL, Univ. Artois, UMR 8181 - UCCS - Unité de Catalyse et Chimie du Solide, F-59000 Lille, France, Villeneuve d'Ascq, France

Selective photocatalytic conversion of methane into carbon monoxide over zinc-heteropolyacid-titania nanocomposites

Xiang Yu, Vitaly Ordomsky, Andrei Khodakov

Univ. Lille, CNRS, Centrale Lille, ENSCL, Univ. Artois, UMR 8181 - UCCS - Unité de Catalyse et Chimie du Solide, F-59000 Lille, France

In recent years, methane has become increasingly abundant due to the development of shale gas fields and other cost efficient feedstocks. Methane is also considered as one of the greenhouse gases with a global warming potential 50 times higher than carbon dioxide [1]. Methane activation is therefore a formidable challenge for catalysis [2,3]. High reaction temperatures (>700°C), low selectivity to the target products and often high CO_2 production are major drawbacks of the conventional thermocatalytic approach.

Photocatalysis, which uses sunlight, has been shown very promising for water decomposition and environmental remediation. Photocatalysis has been also considered as one of pathways to break the thermodynamic barrier [4]. Only a limited number of examples of methane photocatalytic conversion have been available in the literature. Carbon monoxide is a very important compound and a building block in chemical industry. It is utilized as a feedstock in the production of chemicals ranging from acetic acid to polycarbonates and polyurethane intermediates. Syngas, which is a mixture of carbon monoxide and hydrogen, is a valuable feedstock for manufacturing methanol, hydrocarbon fuels, numerous oxo-alcohols and aldehydes. CO is also an important reducing agent and it is used for manufacturing pure metals and in particular iron, cobalt and nickel.

Chemical utilization of vast fossil and renewable feedstocks of methane remains one the most important challenges of modern chemistry. Herein, we report direct selective photocatalytic conversion of methane into carbon monoxide under ambient conditions with the selectivity to CO higher than 80% and only marginal CO_2 production: $2CH_4 + 3O_2 \rightarrow 2CO + 4H_2O$

A series of catalysts were developed on the basis of metals, $H_3PW_{12}O_{40}$ heteropolyacids (HPW) and TiO_2 (P25). The Zn-HPW/TiO_2 system exhibited exceptional photocatalytic activity in selective carbon monoxide production from methane (Figure 1). In the composite catalysts tungstophosphoric acid constitutes a thin layer of 1-2 nm over titanium oxide, while zinc species which are involved in methane photocatalytic oxidation, are highly dispersed on the tungstophosphoric layer. Zinc species seem to play an important role in methane activation [5,6]. The role of the TiO_2 semiconductor is primarily assigned to the absorption of light with generation of electron and hole pairs. The electron and hole pairs can then migrate through HPW, which also has properties of semiconductor, to the zinc species, which are active in methane selective photooxidation. Another important role of HPW is relevant to enhancing Zn dispersion and mobility in the catalysts. Importantly, methane activation and reaction are carried out at ambient temperature. High carbon monoxide yields (up to 3-4% in a single batch experiment) and extended catalyst stability achieved in this process make it practical for industrial applications. To the best of our knowledge, the present work presents the first example of utilizing photocatalysis for methane selective oxidation into carbon monoxide.

The conducted in-situ investigation of the reaction mechanism is indicative of the zinc reduction by methane with important modifications of the catalyst structure. Exposure to oxygen leads to subsequent regeneration of the composite catalyst according to Mars-Van Krevelen mechanism [6,7]. And we found that the process involves formation of the surface methoxy-carbonate as reaction intermediate (Figure 2).

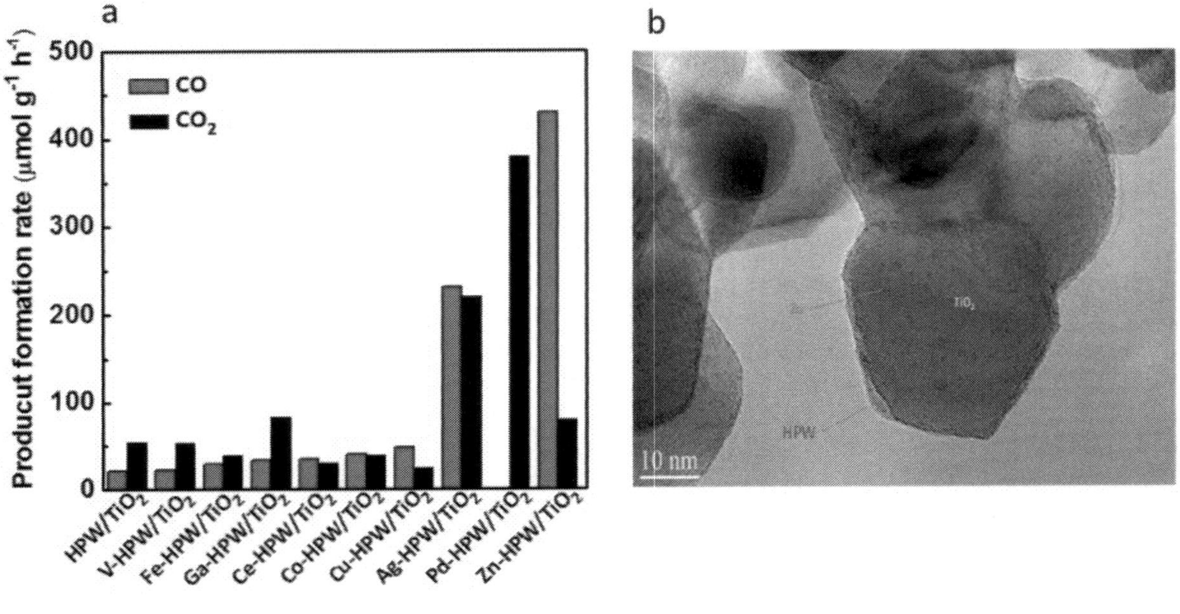

Fig. 1 Performances of various samples for photocatalytic conversion of CH_4 (a) and TEM image of the catalyst Zn-HPW/TiO_2

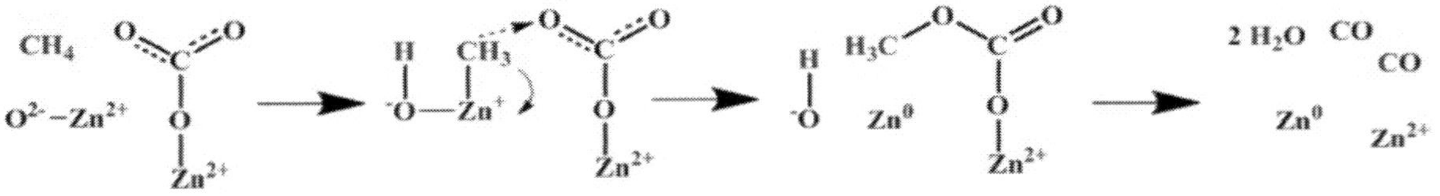

Fig. 2 Mechanism of photocatalytic conversion of CH_4.

[1] Nisbet, E.G., Dlugokencky, E.J., Bousquet, P., Methane on the rise--again. Science 343, 493-495 (2014).

[2] Arakawa, H., Aresta, M., Armor, J.N., Barteau, M.A., Beckman, E.J., Bell, A.T., et al. Catalysis research of relevance to carbon management: progress, challenges, and opportunities. Chem. Rev. 101, 953-996 (2001).

[3] Cargnello, M., Jaén, J.D., Garrido, J.H., Bakhmutsky, K., Montini, T., Gámez, J.C., et al. Exceptional activity for methane combustion over modular Pd@CeO_2 subunits on functionalized Al_2O_3. Science 337, 713-717 (2012).

[4] Yuliati, L, Yoshida, H. Photocatalytic conversion of methane. Chem. Soc. Rev. 37, 1592-1602 (2008).

[5] Bae, K.-L., Kim, J., Lim, C.K., Nam, K.M., Song, H. Colloidal zinc oxide-copper (I) oxide nanocatalysts for selective aqueous photocatalytic carbon dioxide conversion into methane. Nature Commun. 8, 1156 (2017).

[6] Ali, A.M., Emanuelsson, E.A., Patterson, D.A. Photocatalysis with nanostructured zinc oxide thin films: The relationship between morphology and photocatalytic activity under oxygen limited and oxygen rich conditions and evidence for a Mars Van Krevelen mechanism. Appl. Catal. B 97, 168-181 (2010).

[7] Ali, A.M., Emanuelsson, E.A., Patterson, D.A. Conventional versus lattice photocatalysed reactions: Implications of the lattice oxygen participation in the liquid phase photocatalytic oxidation with nanostructured ZnO thin films on reaction products and mechanism at both 254 nm and 340 nm. Appl. Catal. B 106, 323-336 (2011).

545384 Catalytic Methane Steam Reforming at Low Temperature By Surface Protonics

Tuesday, June 4, 2019: 3:06 PM
Texas Ballroom D (Grand Hyatt San Antonio)
Yasushi Sekine, Applied Chem., Waseda Univ., Tokyo, Japan

Generally, methane steam reforming is conducted at high temperatures (923–1123 K) with Ni catalysts because of thermodynamic equilibrium and because of methane molecule stability. However, such high reaction temperatures raise some issues such as the necessity for expensive materials that can tolerate at high temperatures, catalyst deactivation, and complex processes with multiple heat-exchangers. Lowering steam reforming temperatures not only has the potential to improve the hydrogen production efficiency, but also has high potential for waste heat recovery.

We have found that a steam reforming process in an electric field, Electreforming (ER), shows high activity even at a low temperature of 423 K. For ER, Pt or Pd catalyst supported on CeO_2-based oxide are an effective catalyst. Activities increased drastically by the application of an electric field with low electric power consumption of 130 MJ/kg-H_2. However, the reason why the reaction proceeded with application of the electric field was uncertain. So, the aims of this study are to elucidate the electric field effects on catalytic methane steam reforming and the reaction mechanism for Electreforming. To achieve these aims, kinetic analyses, in particular with isotopes, and operando-IR measurements were conducted with Pd catalyst supported on CeO_2. Results show that the application of the electric field promoted surface protonics. Moreover, it enabled catalytic steam reforming even at the low temperature of 423 K.

Application of the electric field increased the activity drastically. To elucidate the effects of the electric field on methane steam reforming, kinetic investigations and operando-DRIFTS measurements were conducted.
Results of our kinetic analyses demonstrated that the water pressure dependency of the reaction rate increased during the application of the electric field. However, the apparent activation energy decreased with the electric field, especially at lower reaction temperatures, indicating that the reaction mechanisms with and without the electric field differ considerably.

Results of operando-DRIFTS revealed that surface proton conduction via adsorbed water on the catalyst surface occurred with the electric field, known as the Grotthuss mechanism. Furthermore, our operando analyses from the viewpoint of an inverse kinetic isotope effect revealed that methane was activated by proton collision derived from the Grotthuss mechanism. Furthermore, ER proceeds mainly at the interface between Pd and CeO_2. Therefore, the surface protonics by the application of electric field serve an important role in the enhancement of catalytic methane steam reforming at a low reaction temperatures of 423 K.

545757 Zoneflow Structured Catalytic Reactors for Steam Methane Reforming: Multi-Scale Modelling and Pilot Plant Testing

Tuesday, June 4, 2019: 4:00 PM
Texas Ballroom D (Grand Hyatt San Antonio)

Florent Minette[1], Sanjay Katheria[2], Jianguo Xu[3], **Sanjiv Ratan**[3] and **Juray De Wilde**[4], (1)Materials and Process Engineering (IMAP), Universite catholique de Louvain (UCL), Louvain-la-Neuve, Belgium, (2)Materials and Process Engineering (IMAP), Université catholique de Louvain, Louvain-la-Neuve, Belgium, (3)ZoneFlow Reactor Technologies LLC, Windsor, CT, (4)Materials and Process Engineering (IMAP), Université Catholique de Louvain (UCL), Louvain-la-Neuve, Belgium

The paper addresses the multi-scale modelling and pilot plant testing of ZoneFlow structured catalytic reactors for steam methane reforming (zoneflowtech.com). Compared to conventional packed beds, ZoneFlow reactors offer reduced pressure drop, improved heat transfer and increased catalyst effectiveness. Aspects discussed include the catalyst, kinetic modelling, reactor modelling and required scale-bridging strategies, and pilot plant testing and multi-scale model validation.

Wash coated or inter-diffused catalyst coatings (Alloy Surfaces, Co. Inc.) can be applied. For the kinetic modelling, experiments with a micro-packed bed reactor were carried out. Plug flow, isothermal operation, small pressure drop and absence of interfacial and intra-catalyst transport limitations are ensured by careful reactor design and selection of the operating conditions. Model discrimination and parameter estimation are based on regression and physico-chemical and statistical testing. The concept of rate determining step was applied to derive the Hougen-Watson Langmuir-Hinschelwood type rate equations. The latter allow accounting for the details of the reaction mechanism while facilitating integration at the catalyst and reactor scale. A total of 30 possible sets of rate equations were tested. To describe intra-catalyst transport, a continuum description of the catalyst coating was adopted and the effective transport properties determined.

A CFD reactor model was developed to account for the complex fluid dynamics. To account for the effects of turbulence, the RANS approach with effective transport properties and the standard k-epsilon turbulence model with wall functions were adopted. Cold flow pressure drop tests in a wide range of flow rates allowed independently determining the turbulence model parameters. Thermal conduction in the reactor tube and in the reactor internals coated with catalyst and radiative heat transfer were accounted for, the latter applying the Rosseland-WSGGM approach. The CFD code was coupled with the calculation of the intra-catalyst diffusion-reaction calculation, using the independently determined reaction kinetics.

Finally, cold, hot and SMR pilot plant test are presented that allow validation of different aspects of the multi-scale reactor model and its application for the design and optimization at the commercial scale.

546497 Adapting to Changing Market Conditions, the Evolution of Reforming Technologies to Meet Today's Hydrogen Supply Requirements

Tuesday, June 4, 2019: 4:24 PM
Texas Ballroom D (Grand Hyatt San Antonio)
Jeffrey Kloosterman, Process Gases Technology - Syngas, Air Products, Allentown, PA

For more than 30 years, Air Products has been supplying hydrogen to customers. Over that time market drivers have shifted, and Air Products adapted designs and operating facilities to meet those challenges. Whether that was with an innovative cycle design, recuperative reforming, or capturing CO_2 for beneficial use, Air Products has led the way in bringing cutting edge technology to provide customers with reliable and cost-effective hydrogen solutions. Remaining competitive in the market place is the life blood of any of the major industrial gas companies. Air Products' evolution with the market to provide innovative solutions in the past, present, and future will be presented. Examples of innovations in our existing fleet and current state of our offerings will be discussed including our Port Arthur facility where CO_2 is captured from a steam methane reformer. Bringing technical innovations to our fleet of more than 60 SMRs will be covered with a focus on how our approach delivers solutions to our customer base.

Steam methane reforming has evolved over time from refinery operated, basic reformer furnaces with relatively poor efficiencies to today's over the fence highly heat integrated offerings from industrial gas companies. Early on, reductions in natural gas usage per H_2 has driven the industry to conducting analysis of heat exchanger network design. Requirements on export steam purities have further tested the ability of suppliers to provide steam turbine quality steam for customers at an incremental cost. Steam to carbon ratios and elimination of waste heat from the plant have generated further changes to design conditions and flowsheets. The advancement of purification technologies and hydrogen recovery have generated constraints on the reforming process design and operability. Environmental regulations create opportunity for improvement, as well as a challenge to the economic design of the plant. Fundamentals are at the heart of providing the insight necessary for generating technoeconomic solutions. The application of heat integration, use of CFD models, imaging, internet of things, and detailed optimization and process monitoring software have driven innovation and built our fleet.

546828 The Initial Stage of Metal Dusting Corrosion of Inconel 601 – Effects of Exposure Conditions and Near-Surface Structure and Composition

Tuesday, June 4, 2019: 4:48 PM
Texas Ballroom D (Grand Hyatt San Antonio)

Hilde Venvik[1], Xiaoyang Guo[1], Estelle Marie M. Vanhaecke[1], Per Erik Vullum[2], Jianyu Ma[1], Daham Gunawardana Panditha-Vidana[3], John C. Walmsley[4] and De Chen[1], (1)Chemical Engineering, Norwegian University of Science and Technology, Trondheim, Norway, (2)SINTEF Industry, Trondheim, Norway, (3)Yara ASA, Porsgrunn, Norway, (4)Materials Science, University of Cambridge, Cambridge, United Kingdom

Fe, Ni and Co are known as catalysts for carbon formation. They are also main constituent elements of industrial alloys due to e.g. temperature properties. Alloys based on these metals are therefore susceptible to metal dusting corrosion; a catastrophic degradation phenomenon that proceeds by a gradual breakdown of the material into a powdery mixture of graphite, carbide and metal particles. It constitutes a major issue in the synthesis gas production, where (heat exchange) surfaces are extensively exposed to carbon-saturated gaseous environments with low partial pressures of oxygen and/or steam in a critical temperature range of 400–900 °C [1].

Cr_2O_3 is known as impermeable to carbon and remains stable in carbonaceous atmospheres down to very low oxygen partial pressure [2]. One measure to prevent metal dusting is the incorporation of chromium (Cr) in the alloy to facilitate formation of a protective oxide layer at the surface. The inhibition of carbon formation that a dense Cr_2O_3 layer that prohibits access to the bulk matrix can be formed and maintained during industrial operation.

The overall objective of our investigations is to provide a better general understanding and prediction of carbon formation phenomena leading to metal dusting, as well as enabling alloy selection and alloy pre-treatment procedures that minimize metal dusting corrosion. We have previously demonstrated in a qualitative way that the tendency to form solid carbon on the Inconel 601 alloy sample surface is strongly linked to the pre-oxidation conditions, through the composition and structure of the Cr-rich oxide layer formed [3]. The carbon formation was catalyzed by Fe and/or Ni (alloy) particles, seemingly originating from the reducible phases presence within this layer. Moreover, we used exposures to a model syngas at varying temperature to show that features resembling the so-called pitting observed in alloys applied in industrial process equipment could be formed within relatively short exposures [4]. In this study, by varying the exposure time we assess the point at which the reducible phases are initiating the first carbon formation as well as how the accelerated pitting may occur.

Inconel 601 alloy samples were first subjected to a polishing and oxidation procedure intended to yield a consistent starting point. Two types of gas mixtures were used for the carburizing experiments at 750 °C; one representing infinite carbon activity and consisting of 10% CO in Ar at 1bar (I-a_C), the other simulating syngas at industrially relevant conditions, i.e. a finite low carbon activity mixture of H_2 (25%), CO (20%), CO_2 (15%), H_2O (10%) and Ar (30%) at 20 bar total pressure (FL-a_C).

The resulting samples were characterized by a range of techniques (in an attempt to relate the carbon formation propensity and potential progression of the metal dusting to the initial structure and composition, as well as the development of the metal/oxide matrix during the carburizing exposures; i.e. field emission gun scanning electron microscopy (FE-SEM), Auger and Raman spectroscopy, x-ray diffraction (XRD) and, finally, transmission electron microscopy (TEM), energy-dispersive x-ray spectroscopy (EDS) and electron energy loss spectrometry (EELS) analysis of near-surface cross-section lamellas prepared by focused ion beam (FIB) milling.

The nature of the surface oxide layer, i.e. microstructure and composition as resulting from the initial polishing to uncover the bulk and subsequent oxidation in steam at elevated temperature, represents the starting point of this investigation. Combined, the information from XRD, Raman spectroscopy and Auger depth profiling shows that very thin oxide scale has formed after oxidization in 10% steam in Ar at540 °C that contains Cr_2O_3 as well as (Ni, Fe, Cr)$_3O_4$ spinel.

Carbon filaments are found on the surface for nearly all I-a_C and FL-a_C exposure times (1 to 20h), with density, length diameter increasing with time. Metal particles are found at the tip of the filament as well as along the center axis as the length increases, typical of a Ni/Fe catalyzed growth mode. The induction period appeared, however, longer under FL-a_C at 20 bar than under I-a_C at 1 bar, but eventually the carbon filaments were usually larger in diameter in the former case.

In Raman spectroscopy comparison of the so-called *G* (1577-1608 cm^{-1}), *D* (1326-1328 cm^{-1}) and *D'* (1600-1610 cm^{-1}) bands can be used to investigate the structural properties of the carbon on the surface [5-8]. The results imply that, while some disordered and amorphous carbon may synthesize first, the overall crystallinity of the carbon formed is improved by prolonged exposure to CO only (I-a_C). When it comes to the FL-a_C condition increase in the I_D/I_G value implies that there is no improved crystallinity of the growing carbonaceous deposits. Raman and XRD jointly indicate that the pre-oxidized alloy samples continue to develop near-surface oxides phases during exposure to the – essentially reducing - carburizing gas mixture at 750 °C. But while Cr_2O_3 becomes predominant for I-a_C, the major additional phase observed after FL-a_C exposure is (Ni, Fe, Cr)$_3O_4$ spinel. Such spinels are thermodynamically less stable and can also be spalled more easily as a result of stress [9].

The TEM/EDS/EELS cross-section analysis of the near-surface region reveals that the changes in structure and composition have penetrated significantly deeper into the alloy matrix for the simulated synthesis gas exposure than for the CO only case (

I-a_C), with clear evidence of pitting. The synthesis gas thus seem to facilitate continuous formation of (Ni, Fe, $Cr)_3O_4$ species that can be subsequently reduced to metallic (Ni, Fe) catalyst particles, while under CO a thin Cr_2O_3 layer can be maintained even if carbon forms on the surface. The results also reveal how the underlying alloy grain structure affects the propensity to form carbon on the surface. A fine-grain structure is believed able to provide easy-diffusion paths such as grain boundaries, sub-boundaries and dislocations resulting in faster formation of the protective Cr-rich oxide layer [10, 11].

[1] M. Holland and H. De Bruyn, *International journal of pressure vessels and piping*. 66 (1996) 125-133.

[2] Z. Zeng et al., *Corrosion* 60 (2004) 632-642.

[3] P. D. S. Gunawardana, H. Venvik, and J. Walmsley, "Investigation of the initial stage of metal dusting corrosion in the conversion of natural gas to synthesis gas," in *CORROSION 2013*, 2013: NACE International.

[4] P. D. S. Gunawardana, T. T. M. Nguyen, J. C. Walmsley, and H. J. Venvik, "Initiation of metal dusting corrosion in conversion of natural gas to syngas studied under industrially relevant conditions," *Industrial & Engineering Chemistry Research*, vol. 53, no. 5, pp. 1794-1803, 2013.

[5] M. M. Lucchese *et al.*, "Quantifying ion-induced defects and Raman relaxation length in graphene," *Carbon*, vol. 48, no. 5, pp. 1592-1597, 2010.

[6] L. G. Cançado *et al.*, "Quantifying defects in graphene via Raman spectroscopy at different excitation energies," *Nano letters*, vol. 11, no. 8, pp. 3190-3196, 2011.

[7] I. Childres, L. A. Jauregui, W. Park, H. Cao, and Y. P. Chen, "Raman spectroscopy of graphene and related materials," New developments in photon and materials research, vol. 1, 2013.

[8] R. Saito, M. Hofmann, G. Dresselhaus, A. Jorio, and M. Dresselhaus, "Raman spectroscopy of graphene and carbon nanotubes," Advances in Physics, vol. 60, no. 3, pp. 413-550, 2011.

[9] M. W. Edwards and N. S. McIntyre, "Gas Phase Initial Oxidation of Incoloy 800 Surfaces," *Oxidation of metals*, vol. 79, no. 1-2, pp. 179-200, 2013.

[10] H. Grabke, E. Muller-Lorenz, S. Strauss, E. Pippel, and J. Woltersdorf, "Effects of grain size, cold working, and surface finish on the metal-dusting resistance of steels," *Oxidation of Metals*, vol. 50, no. 3, pp. 241-254, 1998.

[11] N. Parimin, E. Hamzah, and A. Amrin, "Effect of grain size on the isothermal oxidation of superalloys," *Advances in Environmental Biology*, vol. 7, no. 12, pp. 3720-3726, 2013.

544483 OTM Solid State Combined Reformer

Tuesday, June 4, 2019: 5:12 PM
Texas Ballroom D (Grand Hyatt San Antonio)
Ines Stuckert, R&D, Praxair, Inc, Tonawanda, NY

Reactively-driven oxygen transport membranes (OTM) that utilize oxygen ion and electron-conducting ceramic materials are a breakthrough technology for high-temperature processes, such as chemicals and liquid fuels production and oxyfuel combustion. OTM reformers can efficiently convert natural gas to mixtures of hydrogen and carbon monoxide (known as syngas) to make higher value products. Our OTM technology is a step-change process intensification of the front end of a gas to liquids cycle, combining the oxygen supply and reforming in one compact step. The OTM system is modular and scalable down to very small systems without losing its capital cost and operating efficiency advantages, and can produce liquid fuels with a 30% lower greenhouse gas footprint compared to conventional technologies. The performance and reliability of this technology has been demonstrated through long term lab scale and pilot system testing. We have increased our membrane manufacturing capacity to supply the requirements of the first commercial systems.

545801 Cost Efficient Process Designs through Autothermal Reforming

Tuesday, June 4, 2019: 5:36 PM
Texas Ballroom D (Grand Hyatt San Antonio)

Esben Sorensen, Technical Sales, Chemicals, Haldor Topsoe, Lyngby, Denmark

The Syncor™ Technology developed and industrialized by Haldor Topsoe early in this century has proven to be an amazing way to produce syngas for various petrochemical applications. The Syncor™ technology features autothermal reforming which enables very cost effective steam reforming at a steam-carbon ratio as low as 0.6. The technology was originally developed for GTL processes and it has proven to be extremely successful for that application. There are currently four Syncor™ reactors in operation each with a capacity corresponding to production of 17,000 bpd Fischer-Tropsch liquids. Subsequently - early in this decade, Haldor Topsoe launched a methanol process based on the Syncor™ Reactor and showed significant advantages of this process lay-out relative to the established processes which rely on traditional steam methane reforming technology. This lay-out is used both for methanol processes and gasoline processes which are based on the MTG route. One example of such a process is the Turkmengas 15,500 bpd TIGAS facility scheduled to come on-line early 2019. Recently, the Syncor™ technology was applied for a 35 MM SCFD CO production facility with co-production of Hydrogen which is also anticipated to be started up in 2019. The Syncor™ technology has an excellent cost scaling factor and hence has the potential to become the preferred technology for gas producers wherever there is a need for large scale production of both CO and Hydrogen. The presentation discusses some of the features of the Syncor™ technology and explains why the technology is so versatile for various petrochemical processes.

550876 Dynamics and Mechanism of Carbon Filament Formation during Methane Reforming on Supported Nickel Catalysts

Wednesday, June 5, 2019: 10:30 AM
Texas Ballroom D (Grand Hyatt San Antonio)

Samuel L. Leung[1], Junmei Wei[1], William L. Holstein[1], Miguel Avalos-Borja[2] and Enrique Iglesia[1], (1)Department of Chemical and Biomolecular Engineering, University of California, Berkeley, Berkeley, CA, (2)Department of Advanced Materials, Instituto Potosino de Investigación Cientifica y Tecnológica, San Luis Potosí, Mexico

Dynamics and Mechanism of Carbon Filament Formation during Methane Reforming on Supported Nickel Catalysts

Samuel L. Leung[1], Junmei Wei[1], William L. Holstein[1], Miguel Avalos-Borja[2], Enrique Iglesia[1]

[1] Department of Chemical and Biomolecular Engineering, University of California – Berkeley, Berkeley, California 94720, United States

[2] Department of Advanced Materials, Instituto Potosino de Investigación Cientifica y Technológia, San Luis Potosí, Mexico

Keywords: Methane reforming, carbon deposition, carbon filaments, nickel catalyst

1. Introduction

Carbon filaments tend to form on Ni-based catalysts during the steady-state catalysis of CH_4 reforming reactions, thus necessitating H_2O/CH_4 or CO_2/CH_4 ratios in excess of those required by stoichiometry in order to suppress coking. Bulk diffusion of carbon from exposed surfaces to those in contact with the growing filament has been proposed to limit the rate of filament growth. This diffusive flux depends, in turn, on the gradient in the thermodynamic activity of carbon between metal surface exposed to gaseous reactants ($a_{C,s}$) and that in contact with the filament ($a_{C,f}$) [1]. The rate of filament formation therefore reflects a value of $a_{C,s}$ that is determined by the relative rates of the elementary steps that form and remove C-atoms during steady-state CH_4 reforming catalysis and thus also on the structure and composition of the active surfaces. Sperle et al. previously used similar arguments determine a critical H_2O/CH_4 ratio below which carbon growth would not occur [2]. This study describes a mechanistic basis for interpreting CH_4 reforming rates, which accounts also for the activity of surface carbon during steady-state CH_4 reactions, over a broad range of conditions relevant to the practice of reforming catalysis.

2. Experimental/methodology

Experiments were performed on 7% wt. and 15% wt. Ni/MgO samples prepared by incipient wetness impregnation [3]. The dynamics of carbon deposition were measured during CH_4 reforming reactions using a tapered element oscillating microbalance under controlled reactive and hydrodynamic environments at 843-973 K. The value of $a_{C,s}$ was varied through systematic changes in the inlet CH_4, CO_2, CO, H_2, and H_2O concentrations and the chemical conversion for each inlet gas composition. The effluent was continuously monitored by on-line mass spectrometry to concurrently determine CH_4 reforming rates. High-resolution transmission electron microscopy (TEM) was used to determine Ni particle sizes and the morphology of the carbon deposits formed after exposure to reactive environments.

3. Results

The mechanistic interpretation of CH_4 reforming rates on Ni-based catalysts, previously derived from isotopic and theoretical inquiries [3], shows that $a_{C,s}$ values are proportional to $P_{CO}P_{CH4}/P_{CO2}$ (χ) (or equivalently to $P_{H2}P_{CH4}/P_{H2O}$ (j), because of the prevalent water-gas shift equilibration) during CH_4-H_2O and CH_4-CO_2 reforming at conditions far removed from its thermodynamic equilibrium. For each Ni/MgO sample and temperature, the carbon morphology and carbon deposition rate were solely determined by the magnitude of χ (or j). Three regimes, leading to distinct carbon morphologies, were evident from TEM analysis with increasing χ or j values for 7% wt. Ni/MgO: minimal formation of carbon deposits (regime I, $\chi < 1.1$ kPa), formation of filamentous carbon without detectable changes in reforming turnover rates (regime II, $1.1 < \chi < 4.2$ kPa), and formation of encapsulating carbon shells with a concomitant decrease in reforming rates (regime III, $\chi > 4.2$ kPa). These regimes also aligned with trends noted in steady-state carbon deposition rates (Figure 1). Carbon deposition rates were consistently higher on 11 nm than on 5.4 nm average diameter Ni particles and decreased with increasing temperature for a given value of χ.

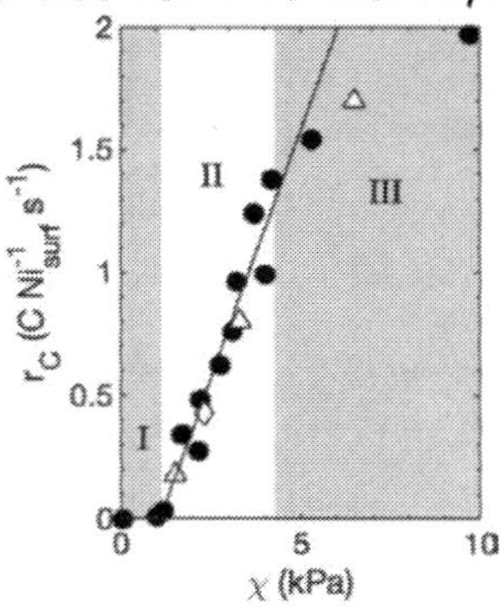

Figure 1. C deposition rates on 7% wt. Ni/MgO at 873 K. CH_4/CO_2 (●), CH_4/H_2O (△), and CH_4/CO_2 with H_2 (◇) reactions.

4. Discussion

Carbon deposition rates are negligible in regime I because filament growth cannot occur until the carbon activity in the Ni particle exceeds that required for nucleation of a carbon filament phase. The activity of carbon filaments can therefore be estimated from the threshold for filament growth. The linear dependence of carbon deposition rates on $a_{C,s}$ in regime II is consistent with the proposed diffusion model, while the lower than expected rates in regime III from extrapolation of rates in regime II is caused by the encapsulation and subsequent deactivation of the Ni clusters. CH_4 turnover rates and carbon deposition rates were not influenced by the formation of carbon filaments in regime II because such growth of filaments continuously depletes the carbon pool, maintaining carbon activity below supersaturation levels, and does not hinder access by gas-phase reactants to active Ni surfaces. Carbon deposition rates increased with increasing Ni crystallite size, reflecting lower thermodynamic activities for the larger diameter filaments that tend to form on larger clusters [4], leading to steeper carbon activity gradients. Carbon deposition rates decreased with increasing temperature for a given value of χ, reflecting the higher activation energy for the removal of surface carbon than for its deposition [3, 5] and therefore the more rapid removal of surface carbon and thus lower values of $a_{C,s}$ at higher temperatures.

5. Conclusion

Carbon diffusion and filament growth is driven by a gradient in thermodynamic activity across the metal catalyst particle. The carbon activity at the gas-metal interface is described by the pressure ratio χ or j at conditions before equilibration is achieved, and carbon deposition rates are a single valued function of these parameters on a given supported Ni catalyst and temperature. These results provide a framework for determining CH_4 reforming conditions at which $a_{C,s}$ can be kept below threshold values that lead to carbon deposition.

6. References

[1] Holstein, W. L., *J Catal.* 152 (1995) 42-51.

[2] Sperle, T., Chen, D., Lodeng, R., Holmen, A., *Appl Catal a-Gen.* 282 (2005) 195-204.

[3] Wei, J., Iglesia, E., *J Catal.* 224 (2004) 370-383.

[4] Rostrup-Nielsen, J. R., *Steam Reforming Catalysts*. Teknisk Forlag: 1975.

[5] Snoeck, J. W., Froment, G. F., Fowles, M., *Ind Eng Chem Res.* 41 (2002) 4252-4265.

Financial support from BP as part of the Methane Conversion Cooperative Research Program at the University of California at Berkeley and from MEXUS grant CN-02-76 are gratefully acknowledged.

545533 A First Principles Based Microkinetics Model of Dry Reforming of Methane over Nickel-Based Bimetallic Catalytic Systems

Wednesday, June 5, 2019: 10:54 AM
Texas Ballroom D (Grand Hyatt San Antonio)

Mohammed Minhaj Ghouri, Chemical Engineering, Texas A&M University at Qatar, Doha, Qatar, Nimir O. Elbashir, TEES Gas & Fuels Research Center, Texas A&M at Qatar, Doha, Qatar and **Shaik Afzal**, Chemical Engineering Program, Texas A&M University at Qatar, Doha, Qatar

Dry reforming of methane (DRM) reaction utilizes CO_2, a major greenhouse gas to convert natural gas (mainly composed of methane) to synthesis gas, an important feedstock which could further be converted into valuable chemicals and cleaner fuel. This reaction presents a clear incentive in terms of its positive impact on the global environment and thus it has gained increasing attention from the scientific community lately. The superior catalytic activity of Nickel based catalysts and also their relatively lower costs make them the most promising catalyst for this reaction. However, these catalysts are also subjected to rapid deactivation owing to their high susceptibility to coke formation and filamentous carbon deposition. This severe catalyst deactivation is one of the major drawbacks that is obstructing the widespread commercialization of DRM. Several alternative catalysts have been explored for this reaction, including noble metals such as Rh and Pt. Even though these metals are found to be more reactive, as well as more resistant to carbon deposition, their high cost generally hinders their use [1]. Many approaches have been proposed in the literature to tackle the issue of rapid coking of the catalysts. One of the techniques among these is to modify the electronic structure of the host catalytic system by its doping with foreign transition metal atoms [2].

In the current work, we present our exhaustive work on the solid state density functional theory (DFT) model results to study theoretically the wide network of elementary reactions comprising the DRM reaction on facets of pure nickel catalyst such as Ni(111) surface. Calculations were performed using rev-PBE as exchange-correlation functional within the generalized gradient approximation (GGA) as implemented in the software VASP. Adsorption energies were calculated for all the DRM reaction intermediate species and then subsequently the activation barriers were calculated for all the elementary reactions in the DRM cycle. The catalytic activity of these pure nickel surfaces in terms of DRM reaction rate are then compared to the rates obtained on various transition metal doped bimetallic systems such as Cu/Ni, Fe/Ni and Co/Ni systems. Electronic structure analysis of various catalytic surfaces is performed in terms of the d-band theory of catalysis.

We then proceed to construct a comprehensive microkinetic model of DRM reaction mechanism consisting of more than 30 elementary reactions on each of the above-mentioned bimetallic catalysts. Our results help to improve our mechanistic understanding of DRM reaction on bimetallic catalysts. They indicate that the catalyst stability is greatly improved by a transition metal doped bimetallic nickel based surfaces. Furthermore, our DFT results for the predicted performance of the Ni-Cu system have been confirmed experimentally under typical DRM reaction conditions.

References:

1. Pakhare, D. and J. Spivey, *A review of dry (CO2) reforming of methane over noble metal catalysts.* Chemical Society Reviews, 2014. **43**(22): p. 7813-7837
2. Fan, C.; Zhu, Y.; Xu, Y.; Zhou, Y.; Zhou, X. and Chen, D; *Origin of synergistic effect over Ni-based bimetallic surfaces: A density functional theory study.* The Journal of Chemical Physics, 2012. 137: 014703

545255 Kinetic Analysis of Methane Dry Reforming on Rh in Spatially-Resolved Operando-Raman Annular Reactor

Wednesday, June 5, 2019: 11:18 AM
Texas Ballroom D (Grand Hyatt San Antonio)

Gianluca Moroni, Matteo Maestri and Alessandro Donazzi, Laboratory of Catalysis and Catalytic Processes - Dipartimento di Energia, Politecnico di Milano, Milano, Italy

Methane Dry Reforming (MDR) represents an effective route for the valorization of bio-gas from biomass anerobic digestion to syngas. Metal-based catalysts are efficient for this reaction but are severely affected by carbon coking which eventually leads to catalyst deactivation. Hence, understanding the mechanism of carbon formation and its kinetic consequences in MDR is of primary importance to improve the catalytic materials and optimize the operating conditions.

In this study, we performed a combined spectroscopic-kinetic analysis by using an Operando Raman Annular reactor[1]. The experimental tests, carried out under nearly-isothermal conditions, were aimed at the analysis of the effect of dilution in MDR. We performed three different tests at constant space velocity, catalyst load (4% w/w of Rh on α-alumina) and CO_2/CH_4 ratio (2). The outlet gas composition was analyzed between 300 to 750°C, with steps of 50°C. The experimental results showed that higher reactant concentration negatively affects the activity of the system (Figure 1a).

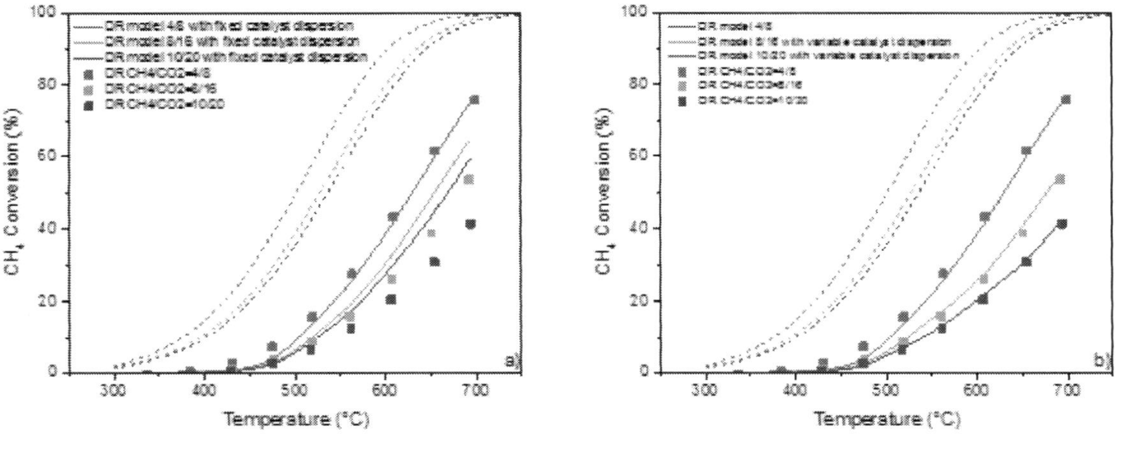

Figure 1 – *CH_4 conversion during DR experiements with different dilution in the T range 300-750°C. a) Comparison between experimental data (dots) and microkinetic model predictions with fixed catalyst dispersion (solid lines). b) Comparison between experimental data (dots) and microkinetic model predictions with variable catalyst dispersion (solid lines).*

The measured data were then compared with the predictions of a microkinetic model in order to relate the operando information to the mechanistic interpretation of the working catalyst[2,3]. In particular, the Reaction Path Analysis (RPA) showed that the mechanism consists of two different routes, one for the CH_4 and the other for CO_2. CH_4 dehydrogenates till the formation of C ad-atoms at the catalyst surface. C atoms are then oxidized to CO by OH. CO_2 decomposes to O, which is involved in the formation of OH, and CO. By performing a sensitivity analysis, the RDS of the whole process is identified as the dehydrogenation of CH_3 to CH_2.

The microkinetic model is able to predict with good agreement the experimental results for the test with a 4% of CH_4 in the inlet stream but it only qualitatively reproduced the decrease in CH_4 conversion due to the increase of reactants concentration. The analysis of the surface coverage predicted by the microkinetic model shows that an increase of reactant concentration results in a decrease of the free-active sites, thus affecting the rate of the RDS and the activity of the whole system. The mismatch between the model predictions and the experimental data appeared to be related to a phenomenon that involves the catalyst surface. To clarify this point, we performed long term MDR tests by keeping constant the T at 600°C and the CH_4 inlet concentration at 8%. The CH_4 conversion was monitored for more than 2 hours, showing a decreasing trend (Figure 2a). At the same time, we performed spatially resolved Raman analyses in order to monitor the catalyst surface (Figure 2b).

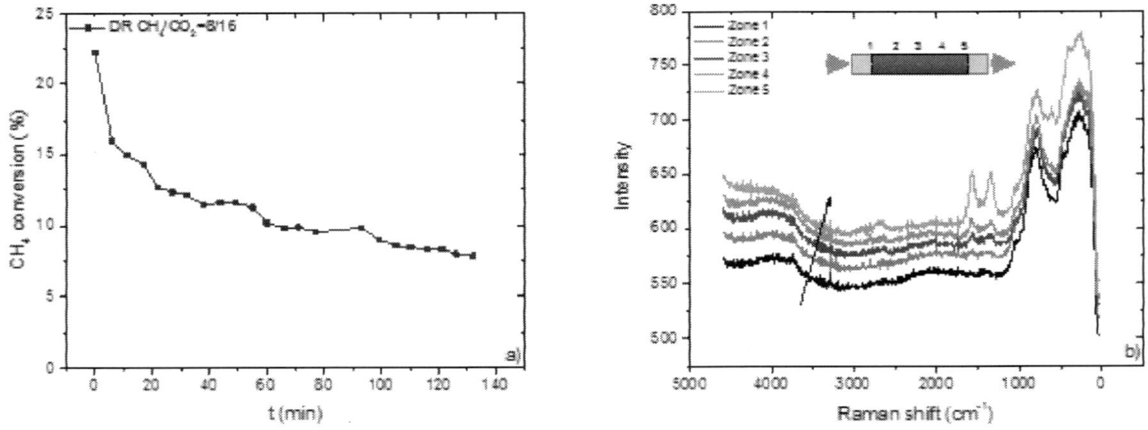

Figure 2 – *a) CH_4 conversion during DR run with CH_4/CO_2 = 8/16 performed at 600°C; b) Spatially Resolved Raman analysis performed during DR CH_4/CO_2=8/16: spectra recorded starting from the beginning (zone 1) to the end (zone 5) of the catalyst bed.*

Two peaks, related to graphitic (G peak at 1580 cm^{-1}) and disordered (D peak at 1350 cm^{-1}) carbonaceous species, appeared in the Raman spectra. Thanks to the spatially resolved analysis, it was also evident that carbon is not uniformly present along the axis of the catalyst, but more deposits accumulated at the end of the 2 cm catalyst bed compared to the inlet, which maintained almost clean. This result suggests that formation of the stable carbonaceous species are stratified along the reactor length and they tend to accumulate towards the outlet of the reactor. CH_4 is responsible for the formation of C intermediates while CO_2 provides OH for converting it. However, CO_2 is only weakly adsorbed at the surface and as CO_2 decreases with the time on stream CH_4 dehydrogenation prevails over the OH formation, which is not longer equilibrated. As a consequence, the C-species are not oxidized by OH and tend to organize at the surface. This leads to the formation and accumulation of stable C-deposits at the surface, covering the active sites and thus reducing the conversion of CH_4. The nature of these stable C-deposits is confirmed by the fact that agreement between the model and experiment is recovered by imposing that the availability of catalytst free sites (Rh) decreases as a result of a coverage of the active sites by C-deposits (Figure 1b).

To further assess this mechanistic picture we considered an additional test by boosting the formation of OH species via the co-feed of a small amount of O_2. In particular, the time evolution of the CH_4 conversion at 600°C with $CO_2/CH_4=1$ showed a decreasing trend due to catalyst deactivation. The same test was repeated adding to the inlet stream 0.1% of O_2. The decay of the conversion is substantially reduced. The model clarifies that this small amount of oxygen is converted to H_2O at the very entrance of the reactor and the water produced provides an additional route for OH formation, which is independent from the concentration of CO_2. Consequently, OH is not only related to the CO_2 activation ot CO and O and thus it sustains the conversion of C to CO instead to C-deposits. This is further confirmed by Raman spectra, which provide no evidence of formation of stable C-deposits at the surface.

In conclusion, operando spatially-resolved Raman-spectroscopy was applied to the study of catalyst deactivation in MDR reaction. In particular, this methodology helps revealing that the deactivation of the catalyst is mainly due to the accumulation of C-deposits at the catalyst surface caused by the failure of CO_2 activation route in MDR mechanism. By means of spatially-resolved Raman analyses, it was possible to observe that these C-deposits are not uniformly present along the axis of the catalyst bed, but their formation and accumulation are faveored towards the outlet. Moreover, these pieces of information has allowed us to extend the microkinetic model to the description of the relevant routes which lead to the formation of stable C-despositsat the catalyst surface.

References

1. Maghsoumi, A.; Ravanelli, A.; Consonni, F.; Nanni, F.; Lucotti, A.; Tommasini, M.; Donazzi, A.; Maestri, M. *React. Chem. Eng.* 2017, *2*, 908-918.

2. Maestri, M.; Vlachos, D.G.; Beretta, A.; Groppi, G.; Tronconi, E. *J. Catal.* 2008, *259*, 211-222.

3. Maestri, M.; Livio, D.; Beretta, A.; Groppi, G. *Ind. Eng. Chem. Res.* 2014, *53 (27)*, 10914–10928.

545433 Fundamentals of the Removal of Trace Sulfur Compounds from Natural Gas Derived Streams Via Reactive Sorption with Copper Oxide

Wednesday, June 5, 2019: 11:42 AM
Texas Ballroom D (Grand Hyatt San Antonio)

Dante Simonetti[1], Sara Azzam[2], Faisal H. Alshafei[1] and Richa Ghosh[1], (1)Chemical and Biomolecular Engineering, University of California, Los Angeles, Los Angeles, CA, (2)Chemical and Biomolecular Engineering, UCLA, Los Angeles, CA

Hydrocarbon streams derived from natural gas and oil processing are often contaminated by sulfur containing compounds, such as hydrogen sulfide and alkyl-mercaptans, which affect the quality and specifications of the product streams, pose a corrosion hazard in transfer pipelines and equipment, and conflict with health and environmental regulations. Several removal techniques have been studied and implemented in industrial purification processes ranging from gas-alkaline liquid absorption and physisorption on molecular sieve materials. Chemical reaction of sulfur molecules with copper oxide is an attractive purification strategy (especially for dilute concentrations) because the reactions are thermodynamically favorable at low temperature and the sulfur is sequestered as solid Cu-sulfur products. This study aims to elucidate the fundamentals and mechanistic details of these gas-solid reactions. Specifically, the results from this study led to semi-empirical correlations between reaction conditions (contaminant concentration, temperature, and presence of other contaminants) and the reaction rates and saturation capacities of the solid phase. Moreover, the effect of structural differences resulting from different preparation methods and starting precursors on the performance of copper oxide adsorbents in reactive sorption experiments was investigated to develop a structure-function relationship that can help in future design and tuning of adsorbent properties. A combination of thermodynamic calculations, fixed bed sorption experiments, and advanced characterization techniques are used to achieve this goal.

Apparent H_2S uptake kinetics and contaminant capacities were determined from contaminant breakthrough profiles collected from fixed beds of sorbents. The breakthrough curves were analyzed using irreversible adsorption models based on linear driving force approximations. This assumption leads to an analytical solution to the contaminant mass balance that consists of two parameters: the saturation concentration of contaminant in solid phase and an apparent sorption rate parameter. The rate parameter can be de-convoluted into contributions from bulk diffusion, pore diffusion, and reaction-diffusion phenomena within the solid. The effect of agglomerate size was probed using a commercially produced Cu-Zn-Al oxide material. As the agglomerate size increased from 600 to 1000 microns, the sorption rate parameter decreased from 1.04×10^{-4} s^{-1} to 6.05×10^{-5} s^{-1}. The value of the sorption parameter for agglomerates larger than 550 microns were only weakly dependent on temperature and pressure and were in close agreement with theoretically calculated values of the contribution from diffusion within the agglomerate macropores. For agglomerate sizes less than 500 microns, the rate parameter is less sensitive to decreasing agglomerate size, and the rate parameter eventually becomes independent of particle size at 196 microns. The results from these agglomerate size studies indicate that bulk and pore diffusion are the dominant resistances to H_2S uptake for agglomerates that are larger than ~500 microns, however, as agglomerate size decreases, bulk and pore diffusion resistances become negligible compared to reaction-diffusion phenomena within the copper oxide and copper sulfide phases.

The effects of varying reaction conditions (temperature, pressure, inlet concentration of H_2S) on the rate parameter were also studied. Varying the total pressure (1-2 atm) while maintaining constant temperature and inlet concentration had negligible effect on the rate parameter while increasing the inlet concentration (930-2200 ppm) of H_2S led to an increase in the parameter from 3.74×10^{-5} s^{-1} to 1.1×10^{-4} s^{-1}. Semi-empirical relations between concentration and the reaction rate parameter were developed. The rate parameter has an apparent 0.5 order with respect to inlet H_2S concentration. The saturation capacity remained constant at an average of 12.8 wt% (g of H_2S/100 g of sorbent) over a concentration range of 930-2200 ppm. Moreover, increasing temperature (21-150°C) increased the saturation capacity from 11 wt% to 42 wt%. The rate parameter exhibits an Arrhenius-type dependency on temperature with a pseudo-activation energy of 18 kJ/mol.

To test the applicability of linear driving force models for the reactive sorption of other sulfur containing compounds, reactive sorption experiments were conducted using methyl and ethyl mercaptan at varying temperatures (21-200°C) on the commercially produced Cu-Zn-Al material. Unlike the CuO-H_2S reactions that yielded the same products throughout the experiments' duration and with varying conditions, CuO-mercaptan experiments resulted in various products. Examining the concentration profiles of different species identified the dominant reaction/s under given conditions. At room temperature, methyl and ethyl mercaptans were physically adsorbed by the aluminum base of the materials and breakthrough happened fast with no other sulfur containing compounds detected in effluent streams which suggests that, although thermodynamically favored, the rate of formation of dimethyl and diethyl sulfide at room temperature is too low for any reaction to be detected. At 120°C the methyl mercaptan reactive sorption experiment initially produces water and a high molecular weight sulfur containing compound which indicates the reaction: $2CH_3SH + 2\ CuO$ à $Cu_2O + H_2O + CH_3S\text{-}SCH_3$. Over time, the water concentration in the effluent decreases with a corresponding increase in CH_3SCH_3 and methyl mercaptan which suggests that the reaction $2CH_3SH + CuO$ à $CH_3SCH_3 + CuS$ is taking place at a low rate. The spent sample of this experiment showed a color change from black to orange which indicates the formation of Cu_2O via $2CH_3SH + 2\ CuO$ à $Cu_2O + H_2O + CH_3S\text{-}SCH_3$. Increasing temperature to 200°C yielded CH_3SCH_3 and CH_3SH throughout the reaction duration which suggests that at 200°C the $2CH_3SH + 2\ CuO$ à $Cu_2O + H_2O + CH_3S\text{-}SCH_3$ reaction pathway is both kinetically and thermodynamically favored.

Fresh and spent nanostructures were analyzed via XRD and SEM to identify composition, crystallite agglomerate size, and crystallite size (using the Sherrer

Equation). XRD patterns of fresh CuO nanostructures showed the characteristic peaks of tenorite, the monoclinic structure of CuO, for both nanofibers and nanoparticles regardless of precursor. Furthermore, nanofibers and nanoparticles consisted of CuO crystallites of similar size (16-18 nm). XRD analysis of the commercial materials also indicated monoclinic CuO with ~20 nm crystallites. SEM images revealed that the nanofibers exist as approximately rectangular solid shaped agglomerates of crystallites that are 250-300 nm in size and arranged in a one-dimensional stacking pattern. Nanoparticles and commercial Cu-Zn-Al oxide also consisted of 250-300 nm sized crystallite agglomerates, however, these agglomerates are packed into sphere like structures that are 5-10 microns in size.

To further probe the reaction between H_2S and CuO at the structural/morphological level, H_2S removal using CuO nanofibers and nanoparticles was compared. Nanofibers (230 nm diameter) and nanoparticles exhibited similar maximum capacities (2.4 wt%), however, sorption rate parameters for nanofibers (3.0×10^{-3} s^{-1}) were 1.7 times larger than those for nanoparticles (1.7×10^{-3} s^{-1}). This faster sorption rate is likely the result of the higher intra-agglomerate void fraction in the nanofibers compared to the nanoparticles. Moreover, the effects of changing nanofiber diameter on sorption rate and saturation capacity was studied. It was found that an increase of diameter (230-600 nm) resulted in a drop in the saturation capacity (2.4-0.3 wt%) and a decrease in the sorption rate parameter from 3.0×10^{-3} s^{-1} to 1.4×10^{-4} s^{-1}. This decrease can be explained by the decrease in the intra-agglomerate void space which decreases accessibility to the reactive domains. The temperature effects were probed in CuO nanofibers and nanoparticles in similar fashion to the commercial Cu-Zn-Al. Increasing the temperature from 21°C to 200°C increased the saturation capacity for both 600 nm nanofibers (from 0.3 wt% to 38.8 wt%) and CuO nanoparticles (from 2.4 wt% to 40 wt%) very close to the stoichiometric limit of 42 wt% for complete conversion of CuO to CuS. Moreover, the concentration effects on rate parameter and saturation capacity were probed in CuO nanoparticles. In the range of 250-1000 ppm, the increase in inlet concentration increased the rate parameter from 6.95×10^{-5} to 1.6×10^{-3} while for higher concentrations (1000-2000 ppm) the dependency becomes weaker and the rate parameter eventually becomes invariant to concentration.

The results from these studies indicate that equations derived from linear driving force approximations can be used as predictive models for the removal of sulfur compounds via reactive sorption with CuO. Mass transfer is the predominant resistance to contaminant removal for large particles of sorbent material, and thus, reaction rate parameters for linear force driving models can be calculated from theoretical relationships. Reaction-diffusion phenomena within the solid are more complex and require semi-empirical relationships to determine rate parameters for linear driving force models. Studies of CuO from different synthesis methods reveals that the fundamental kinetics and mechanism of these gas-solid reactions are highly sensitive to the nanostructure of CuO.

545608 Separation of Methane and Carbon Dioxide Using Zeolite@ZIF-95 Composite

Wednesday, June 5, 2019: 12:06 PM
Texas Ballroom D (Grand Hyatt San Antonio)

Fangyuan Tian, Chemistry and Biochemistry, California State University Long Beach, Long Beach, CA

Methane, as a greenhouse gas, is detrimental to the environment. With the increasing number of municipal solid waste (MSW) landfills, human-related methane emission level has kept rising in the last five decades. Meanwhile, methane is the primary component in natural gas, which can be utilized as an alternative fuel source. One of the major challenges in MSW controlling process is to purify methane from landfill gas (LFG), a mixture of CH_4 (~55%), CO_2 (~45%), and other gas molecules.In this study, a composite material contains zeolite and zeolitic imidazolate framework (zeolite@ZIF-95) was synthesized via a one-pot solvothermal approach. Commercial zeolite 4A as nuclei promoted the heterogeneous crystal growth during the synthesis process. The resulted composite material was characterized by power X-ray diffraction (XRD) and infrared spectroscopy (IR) to confirm its crystal structure and chemical composition. A layered microstructure of the synthesized zeolite@95 composite was observed using scanning electron microscope (SEM). Moreover, zeolite@ZIF-95 exhibited the highest BET surface area compared to pristine zeolite and ZIF-95 crystal, confirmed by the nitrogen gas sorption isotherm analysis at 77K. Gas sorption of CO_2 and CH_4were carried out on zeolite 4A, ZIF-95, and zeolite@ZIF-95 composite at 273 K and 298 K, respectively. The gas adsorption selectivity toward CO_2 and CH_4 at different temperatures were performed based on single-component gas adsorption isotherms with the aid of the Ideal Adsorbed Solution Theory (IAST). The selectivity of CO_2/CH_4 on zeolite@ZIF-95 has been improved to 20.1 (volume ratio per gram of adsorbate), comparable to other porous metal-organic frameworks (MOFs). But the cost of making zeolite@ZIF-95 composite is less expensive than synthesizing MOFs. Overall, zeolite@ZIF-95 exhibits great potential as a low-cost selective adsorbate in landfill gas separation. Furthermore, the versatile and large pores of ZIF-95 in the composite can be used as catalyst carriers, which provides numerous applications for heterogeneous catalytic conversion of hydrocarbons.

545450 Intensifying Syngas Production: Experimental Investigations and Process Analysis for Production of Inherently Separated Syngas Streams

Wednesday, June 5, 2019: 1:30 PM
Texas Ballroom D (Grand Hyatt San Antonio)

Hari C. Mantripragada[1], Amey More[1] and Götz Veser[2], (1)Department of Chemical and Petroleum Engineering, University of Pittsburgh, Pittsburgh, PA, (2)Chemical Engineering, University of Pittsburgh, Pittsburgh, PA

The continuing growth in world population along with a rapidly rising standard of living in the developing world is fueling a continuously growing demand for energy and bulk chemicals. At the same time, ever increasing concerns regarding climate change urge a transformation in how we produce this energy and chemicals without dramatically increasing the energy and environmental footprint of both industries. In this context, the recent expansion of economically exploitable natural gas reserves—the "shale gas boom"—is both a huge opportunity and a major concern unless handled responsibly.

One emerging "clean energy" technology that enables fossil fuel utilization with minimal carbon footprint is chemical looping combustion (CLC). In CLC, an oxide—typically an oxidized transition metal—is used as an oxygen source for fuel combustion in a "reducer" reactor, generating a stream of CO_2 and H_2O, which can be easily separated via condensation to generate a sequestration-ready CO_2 stream. After combustion, the reduced metal is then regenerated with air in an "oxidizer" reactor and recycled back to the reducer, completing the materials "loop." Thus, CLC bypasses most of the problems of conventional fuel combustion, including NO_x formation and CO_2 separation from dilute gaseous streams, and is currently emerging as a favored solution for large-scale clean combustion applications.

However, "chemical looping" is in fact a rather flexible platform technology for oxidation reactions well beyond combustion, including CO_2 utilization and natural gas upgrading to value-added chemicals [1-4]. Our own group has demonstrated the use of chemical looping for partial oxidation and reforming of methane to synthesis gas (a mixture of CO & H_2), the most complex and costly step in the (indirect) upgrading of methane to valuable chemicals. We have demonstrated that pure H_2 streams can be produced by replacing air as oxidant with H_2O, resulting in chemical looping steam reforming of methane (CLSR) [5], or with CO_2, enabling CO_2 activation via reduction to CO [6, 7]. Furthermore, by carefully adjusting the fuel conversion in the "reducer" reactor, syngas is possible via chemical looping partial oxidation of methane (CLPOM)[8,9] or chemical looping dry reforming (CLDR) [10], i.e. the chemical looping variants of the conventional methane partial oxidation and dry reforming processes.

The attractiveness of these chemical looping syngas production processes could be further increased by making the syngas ratio (the H_2:CO ratio in the product stream) adjustable, thus adding flexibility to the downstream use of the produced syngas—such as high purity H_2 for fuel cells or ammonia production, or pure CO streams for carbonylation reactions. We are demonstrating that such a flexible syngas production is indeed achievable via a modification of these looping processes that results in the production of fully separated H_2 and CO effluent streams. In this redesigned two-step process, CH_4 is first converted to a pure hydrogen stream and solid carbon via thermocatalytic cracking over a metal catalyst in a "cracker" reactor. The deposited carbon is then selectively oxidized in an oxidizer reactor using CO_2 as oxidant, resulting in the production of high-purity CO streams and regenerating the metal which is recycled back to the cracker reactor. The metal is thus periodically cycled between the two reactors carrying carbon between the two half steps, i.e. the "oxygen carrier" that characterizes chemical looping processes effectively acts like a "carbon" carrier between the two reaction half steps.

This process configuration thus combines thermocatalytic methane cracking, which has been studied extensively over the past several decades [11-13], with CO_2 activation, resulting in the formation of inherently separated syngas streams, while at the same time addressing some of the key shortcomings of conventional methane cracking, such as low CH_4 conversion, rapid catalyst deactivation, and large temperature excursions during carbon burn-off in the oxidation half cycle. The required cracking step can be carried out efficiently using ceria-supported Ni carriers. Utilizing a reducible support with strong metal support interactions (i.e. ceria) allows efficient production of carbon (and hydrogen) at elevated temperatures without rapid catalyst deactivation and thus high fuel conversion at high throughput, making the process well suited for scalable, decentralized application. Similarly, the deposited carbon can be effectively removed by reaction with either O_2 or CO_2 to regenerate the active Ni species. While the use of oxygen as oxidant is favorable from an energetic point of view, it limits the selectivity of the oxidation reaction and yields a mixed CO/CO_2 product stream. In contrast, using CO_2 as a "soft" oxidant not only allows for CO_2 activation via reduction to CO (enhancing the total CO yield from the process), but also improves the overall process selectivity by selectively oxidizing carbon while keeping Ni in its metallic state, which is active and selective for catalytic cracking. However, it also renders the process overall endothermal and hence requires the use of an additional heat source, either via combustion of additional methane or combustion of part of the product stream.

Based on our experimental results, we are finally evaluating various process configurations for this intensified, flexible syngas process using a systems-level analysis. The overall system is designed to enable a desired syngas production rate while maximizing yield and minimizing overall energy consumption and CO_2 emissions. The model involves a detailed chemical looping reactor model, incorporated into the systems-level process model [14]. Reactor-level mass and energy balance calculations are performed to calculate the effect of operating conditions such as flow rates, solids conversion, and temperature on the final product composition, selectivity and yields and solids inventory. The systems-level model includes thermodynamic calculations of individual process components such as blowers, compressors, and heat recovery equipment. Using sensitivity analysis on the process parameters, we will identify operating conditions to optimize process performance and compare results to conventional alternatives such as SMR and POX processes.

[1] S. Bhavsar, M. Najera, R. Solunke, G. Veser, Chemical looping: To combustion and beyond, Catalysis Today 228 (2014) 96–105.

[2] B. Moghtaderi, Review of the Recent Chemical Looping Process Developments for Novel Energy and Fuel Applications, Energy & Fuels, 26 (2012) 15-40.

[3] L.-S. Fan, L. Zeng, W. Wang, S. Luo, Chemical looping processes for CO2 capture and carbonaceous fuel conversion - prospect and opportunity, Energy & Environmental Science, 5 (2012) 7254-7280.

[4] L.S. Fan, L. Zeng, S. Luo, Chemical looping technology platform, AIChE Journal, 61 (2015) 2-22.

[5] R.D. Solunke, G. Veser, Hydrogen production via chemical looping steam reforming in a periodically operated fixed-bed reactor, Industrial & Engineering Chemistry Research, 49 (2010) 11037-11044.

[6] M. Najera, R. Solunke, T. Gardner, G. Veser, Carbon capture and utilization via chemical looping dry reforming, Chemical Engineering Research and Design, 89 (2011) 1533-1543.

[7] S. Bhavsar, M. Najera, G. Veser, Chemical Looping Dry Reforming as Novel, Intensified Process for CO2 Activation, Chemical Engineering & Technology, 35 (2012) 1281-1290.

[8] A. More, S. Bhavsar, G. Veser, Iron–Nickel Alloys for Carbon Dioxide Activation by Chemical Looping Dry Reforming of Methane, Energy Technology, 10 (2016) 1147-1157.

[9] A. More, G. Veser, Physical mixtures as simple and efficient alternative to alloy carriers in chemical looping processes, AIChE Journal, (2016).

[10] S. Bhavsar, G. Veser, Chemical looping beyond combustion: production of synthesis gas via chemical looping partial oxidation of methane, RSC Advances, 4 (2014) 47254-47267.

[11] A.M. Amin, E. Croiset, W. Epling, Review of methane catalytic cracking for hydrogen production, International Journal of Hydrogen Energy, 36 (2011) 2904-2935.

[12] Y. Li, D. Li, G. Wang, Methane decomposition to COx-free hydrogen and nano-carbon material on group 8–10 base metal catalysts: A review, Catalysis Today, 162 (2011) 1-48.

[13] U.P.M. Ashik, W.M.A. Wan Daud, H.F. Abbas, Production of greenhouse gas free hydrogen by thermocatalytic decomposition of methane – A review, Renewable and

Sustainable Energy Reviews, 44 (2015) 221-256.

[14] H.C. Mantripragada, E.S. Rubin, Performance model for evaluating chemical looping combustion (CLC) processes for CO2 capture at gas-fired power plants. Energy Fuels 30 (2016) 2257-2267.

545559 Perovskite Nanocomposites for Methane Partial Oxidation and CO_2 Splitting in a Cyclic Redox Scheme

Wednesday, June 5, 2019: 1:54 PM
Texas Ballroom D (Grand Hyatt San Antonio)

Fanxing Li, Department of Chemical and Biomolecular Engineering, North Carolina State University, Raleigh, NC

The current study explores a methane partial oxidation (POx) and CO_2 splitting scheme assisted by industrial waste heat or solar-thermal energy. Specifically, we report mixed oxide nanocomposites as highly effective redox catalysts to facilitate the abovementioned redox scheme. Close to 100% CO_2 to CO conversion was consistently achieved over 30 redox cycles. In this process, methane is introduced as an oxygen "sink" to promote the reduction of the redox catalyst followed with re-oxidation via CO_2 splitting. Separate streams of Fischer-Tropsch/methanol ready syngas and near pure CO were produced from methane and CO_2.

A number of mixed metal oxides with varying structures (Perovskite, Ruddlesden-Popper, and Fluorite) and compositions were studied. It was determined that single phase perovskite and Ruddlesden-Popper (R –P) materials exhibited excellent initial performance but deactivated over multiple redox cycles. Nanocomposites of inert ceramic support and perovskite/R-P materials, however, exhibit excellent recyclability, activity, and selectivity. The effect of support was subsequently determined by detailed characterization of nanocomposites both during and after the redox reactions. Up to 96% syngas selectivity in the methane partial oxidation step was achieved, in addition to close to complete conversion of CO_2 to CO. The corresponding CO productivity and production rate were about 7 times higher than those in state-of-the-art solar-thermal CO_2 splitting processes, which are carried out at significantly higher temperature. In addition to optimizing the bulk structure and compositions of these mixed oxides, surface promotions were attempted to enhance the activity of the redox catalysts for methane activation. The resulting redox catalysts exhibited satisfactory activity/selectivity for methane POx and CO_2-splitting at temperatures lower than 600 °C. Integration of solar-thermal or waste heat and the ability to produce methanol ready syngas and concentrated CO can potentially simply the production of chemicals derived from methanol carbonylation (e.g. acetic acid and ethylene glycol), resulting in reduced energy consumptions and CO_2 emissions.

547089 Chemical Looping Reforming for the Production of Hydrogen and Chemicals from Natural Gas

Wednesday, June 5, 2019: 2:18 PM
Texas Ballroom D (Grand Hyatt San Antonio)

Andrew Tong, Tien-Lin Hsieh, Dikai Xu, Yitao Zhang, Dawei Wang, Sourabh Nadgouda, Mandar Kathe, Fanhe Kong and Liang-Shih Fan, William G. Lowrie Department of Chemical and Biomolecular Engineering, The Ohio State University, Columbus, OH

Though the use of the oxidation-reduction chemical looping reaction pathway for producing chemical products and industrial gases dates back over 100 years, no chemical looping technology is in practice today at the commercial production scale due, in part, to limitations in the performance and recyclability of the redox material used as the oxygen carrier, the reactor design, and the interconnecting reactor components used. The chemical looping reaction pathway offers many advantages for increased material, heat, and functional step integration. The present paper summarizes the chemical looping processes The Ohio State University (OSU) is developing for syngas production of chemical product applications and hydrogen production for industrial gas supply. The iron titanium complex metal oxide (ITCMO) developed is capable of sustaining over 3,000 redox cycles without loss in strength and reactivity. The syngas chemical looping process, developed for high purity hydrogen generation from fossil fuels using a counter current moving bed reducer design, was tested at the 250 kW_{th} capacity for approximately 1,000 hours. The use of chemical looping to produce syngas is advantageous as it eliminates the need for an air separation unit. The type of metal oxides used as the active oxygen carrier also dictate the purity and quality of the syngas produced. A concurrent moving bed reducer reactor is used to for precise control of the oxidation state of the oxygen carrier throughout the reactor to maximize natural gas conversion and syngas purity. Multiple cases of the OSU chemical looping processes for syngas production are considered using ASPEN Plus process modelling to determine optimum process configurations that efficiently converts natural gas to syngas for producing methanol or other chemical products with minimal to negative CO_2 emissions. Experimental studies in the 15 kWth sub-pilot scale and 1.5 kWth bench-scale chemical looping reactors confirm the flexibility of the co-current moving bed design and ITCMO oxygen carrier for producing syngas with a wide range of CO/H_2 ratios using natural gas, CO_2 and steam as reactants. Results from the experimental studies and economic assessments of the chemical looping processes for syngas production will be presented.

545469 Monodispersed Ni and Ni-Co Nanoparticles Capsulated with Silica for Dry Reforming of Methane

Wednesday, June 5, 2019: 2:45 PM
Republic ABC (Grand Hyatt San Antonio)
Kyung Soo Park and **Jong Wook Bae**, School of Chemical Engineering, Sungkyunkwan University (SKKU), Suwon, Kyonggi-do, Korea, Republic of (South)

Abstract: To figure out the intrinsic effect of Co to the enhanced activity and stability in Ni catalysts for the dry reforming of methane (DRM), the nanoreactor of Ni-Co@SiO_2 catalysts were synthesized, which are composed of Ni-Co nanoparticles encapsulated in the SiO_2 shell. The steric hindrance to the sintering was induced from the silica shell where the nanoparticles could retain their pristine size at the high temperature dry reforming of methane without serious deactivation as supported Ni catalysts showed. The optimum molar ratio of Ni/Co was determined in terms of DRM activity and stability. Also mechanisms of synergetic effect of Ni and Co are proposed by the various characterizations such as TPR, in-situ IR, EXAFS and others.

Keywords: Nanoreactor, Ni-Co nanoparticles, SiO_2 coating, Core-shell catalysts, Dry reforming of methane (DRM), Oxophilic cobalt

1. **Introduction**

Dry reforming of methane (DRM) is nowadays spotlighted in environmentally and industrially since it could convert two main greenhouse gases, methane and carbon dioxide which are responsible for global warming and could produce syngas ratio close to 1, which is favorable for the further conversion to the synfuels. In addition the shale revolution, which increased the production of natural gases tremendously, made the process more feasible which converting the C1 gases to liquid fuels and chemicals such as methanol and synthetic gasoline **[1]**.

Two main reactions in DRM process as follow: Methane decomposition, Boudouard reaction.

(R1) Methane decomposition : CH_4 -> C(s) + $2H_2$
(R2) Boudouard Reaction : 2CO = C(s)+CO_2

During the reaction, it is pointed that the CH_4 is dissociated to be hydrogen and solid carbon on the active metal, which can be gasified by the CO_2 molecule to 2 molecules of CO. Since this solid carbon can deposit on the surface of active metals, it can harm the stable production of the syngas by catalyst deactivation and reactor blocking. The DRM process is usually done at high temperature of 700 – 900 ^oC to accelerate the gasification of the coke so that the long life time of the catalysts is granted.

Generally Ni catalysts are favored which is far cheaper than the precious metal based ones with comparable activity on the reforming processes. Major drawback of the Ni catalysts is the process stability. Since Ni nanoparticles are easily sintered to make bigger agglomerates, where the coke can easily deposit onto, the Ni catalysts are easily deactivate in the high temperature DRM condition **[2]**. Efforts are being made by many researchers continuously to increase the stability of Ni catalysts. One of the promoters can be used for the Ni catalysts is Co, which is known to promote the activity and stability of the catalysts with proper amount of introduction to the catalysts. Co itself is not active for the reforming process as Ni is, however the oxophilic nature of Co which will easily utilize the surface oxygen to facilitate the gasification of the coke during the reaction leading to the promotion of activity and stability of the catalysts **[3]**. Some researchers noted that the synergetic effects of Ni-Co are induced from the formation of Ni-Co alloy where the Ni can be electron-rich from the electron donation of Co upon alloy formation **[4]**. However the formation of Ni-Co alloy was hardly observed directly instead of only ex-situ characterizations which cannot simulate the characteristics of catalysts situated in the high temperature DRM reaction.

Via the rational design of nanoreactor, which preserving the state of metal nanoparticles composed of Ni and Co during the DRM reaction, it is possible to observe the synergetic effect of Ni and Co which can be hardly confirmed in the conventional supported catalysts due to its heterogeneity nature. From the encapsulation of Ni-Co alloyed nanoparticles in the silica shell, it is possible to lessen the phenomena from sintering problems, metal-support interactions and coke deposition whereas the pristine synergetic effect of Ni-Co can be directly observed.

In this research, nano-reactor of Ni-Co@SiO_2 multi core-shell catalysts were prepared and evaluated in terms of DRM reaction to confirm the synergetic effect of Ni-Co nano-alloys and to figure out the origin of the effects by means of various characterizations.

2. **Experimental**

Ni-Co@SiO_2 multi core-shell catalysts are prepared by two main steps of nanoparticle synthesis and encapsulation of the nanoparticles with SiO_2.

Ni and Ni-Co nanoparticles are prepared from the thermal decomposition of precursors by hot injection method. Firstly, regarding to the targeted Ni/Co ratio, the determined amounts of nickel acetylacetonate and cobalt acetylacetoate of 2 mmol in total were dissolved in the solution of 3 mL oleylamine and determined amounts of oleic acid and trioctylphosphine. The solution was degassed with bubbling of UHP N_2 at 110 ^oC for more than 0.5 h. After the 25 mL of oleylamine solvent was heated to the injection temperature ranged among 240 - 250 ^oC as targeted Ni/Co ratio with bubbling of N_2 in the oil bath, the degassed precursor solution is rapidly injected to the solvent to nucleate the Ni and Co at the same time. After the temperature of the mixture reached to the injection temperature, the temperature was maintained for appropriate time to grow the nanoparticles sized about 6 nm in diameter. The mixture was cooled in the room temperature until the temperature reached to 100 ^oC where the 40 mL of ethanol was added to the mixture to separate the nanoparticles from the solvent. The nanoparticles are collected and washed via the centrifugation with ethanol as washing agent. The washed nanoparticles are dispersed in the cyclohexane for further coating of SiO_2.

The encapsulation of nanoparticles was done via the reverse micelle method. Nanoparticles-dispersed cyclohexane and the solution composed of IGEPAL CO-630 and 35 mL of cyclohexane were sonicated for more than 20 min. Then, the solutions are mixed together and stirred vigorously overnight to reach the equilibrium of ligand exchange from OAc and TOP to CO-630 on the surface of nanoparticles. 2mL of tetraethoxy silane (TEOS) was injected slowly as a silica source right after the NH_4OH was added drop by drop to hydrolyze the TEOS. After the reaction mixture was stirred for 72 h, the prepared nanoparticles were collected by means of centrifugation following the addition of 20 mL methanol to separate the particles from cyclohexane. The collected particles are further washed with hexane and methanol subsequently for few times followed by drying at 80 °C convection oven overnight. Dried ones were crushed and calcined at the 450 °C and 800 °C subsequently for 0.5h and 2h respectively at a ramping rate of 1°C/min. The catalytic activity of Ni-Co@SiO_2 catalysts were reduced in-situ at the 800 °C for 1 h and tested in a 3/8 inch fixed-bed reactor at the following reaction conditions; T = 800 °C, P = 0.1 MPa, gas hourly space velocity (GHSV) = 36,000 $L/kg_{cat}\cdot h$ and the feed molar ratio of $CH_4/CO_2/N_2$=1/1/2. The effluent products were analyzed using online gas chromatography equipped with TCD.

The catalysts are characterized with various tools such as XRD, TPR, XPS, TEM, ICP-AES, In-situ IR, XAFS and so on to clarify the mechanism of the synergetic effect of Ni and Co for the DRM reaction.

3. Results and discussion

The Ni-Co nanoparticles with different molar composition of Ni and Co were successfully synthesized to have the monodispersed sizes of 5-6 nm of diameters. Since the injection temperature was set to decompose Ni and Co at the simultaneously but to attain the small sized nanoparticles less than 10 nm, the ratio of Ni/Co was a lot bigger than as targeted. Still it was possible to increase the contents of Co in the particles gradually with increasing the amount of the cobalt precursor put. After the silica coating process, the nanoparticles retained their sizes after the high temperature reduction at the temperature of 800 °C with successfully structured as multi core-shell Ni-Co@SiO_2 nanocatalysts. However the catalytic activity and stability towards DRM is changing with the molar ratio of Ni/Co. Especially the specific rates of conversion with regard to the amount of metals showed the volcano shaped trends where the optimum existed. The research is still under investigation. Up to now the ratio of Ni/Co is supposed to be related to the conversion of CO_2, accelerating the decomposition of CH_4 on the Ni since it promotes the gasification of carbons to CO. With the combination of the results of In-situ IR and other characterization tools, it would be possible to disclose the main roles of Co used with Ni and the phenomena involved in the DRM process.

References

1. Ning Wang, Kui Shen, Lihong Huang, Xiaopeng Yu, Weizhong Qian, Wei Chu. *ACS Catal.* 3 (**2013**) 1638.
2. Liu Chang-jun, Jingyun Ye, Jiaojun Jiang, Yunxiang Pan. *ChemCatChem* 3 (**2011**) 529.
3. Weifeng Tu, Mireille Ghoussoub, Chandra Veer Singh, Yahuei Cathy Chin. *J. Am. Chem. Soc.* 139 (**2017**) 6928.
4. Xingyuan Gao, Zhenwei Tan, Kus Hidajat, Sibudjing Kawi. *Catal. Today* 281 (**2018**) 250.

549464 Dry Reforming Using Modified Metal Carbides

Wednesday, June 5, 2019: 2:48 PM
Republic ABC (Grand Hyatt San Antonio)
Carlos A. Carrero, Department of Chemical Engineering, Auburn University, Auburn, AL

Dry Reforming using Modified Metal Carbides

Raj Thakur, Justin Smith, Carlos A. Carrero*

† Department of Chemical Engineering, Auburn University, Auburn, Alabama 36830, United States

* cac0134@auburn.edu

Our research focusses on the conversion of methane (CH_4) and carbon dioxide (CO_2) to syngas, so-called dry reforming of methane (DRM). Despite of the challenges associated to the high endothermicity of this reaction, DRM produces H_2 and CO in a 1:1 ratio which is optimal to produce, for instance, liquids via the Fischer-Tropsch process.

In this study, prior performing the dry reforming reaction, a series of catalysts comprised of bulk and supported vanadium oxide were converted to vanadium carbide. We developed an approach to convert the metal oxide precursors in the carbide form by exploring various conditions and precursors to then correlate the selected parameters with the structural and catalytic properties of the final carbide materials. We characterize the materials, prior, during, and after the carburization via various *ex situ*, *in situ*, and *operando* techniques, such as Raman spectroscopy, XRD, XPS, and SEM. Our preliminary kinetic studies show that by tuning the conditions in the carburization process, the catalytic properties of the resulted vanadium carbide changes. These differences are more evident in terms of the stability of the catalysts, or in other words, in terms of the type and amount of carbon produced during the reaction. We aim to maximize the activity, selectivity in terms of CO/H_2 ratio, and stability towards syngas of vanadium carbide catalysts.

545484 Effect of a-Site Nature on $MNiO_3$ (M= La, Sr and Ce) Perovskite-Type Precursors Applied on Oxy-Dry Reforming of CH_4

Wednesday, June 5, 2019
Texas Ballroom Prefunction Area (Grand Hyatt San Antonio)

Soraia Brandão[1], Breno Cerqueira da Silva Sr.[2], Roberto Batista da Silva Jr.[2] and , Pedro H. C. Bastos[2], (1)Inorganic Chemistry, Universidade Federal da Bahia, Salvador, Brazil, (2)Inorganic Chemistry, Universidade Federal da Bahia -UFBA, SALVADOR, Brazil

1. Introduction.

Nickel-based catalysts show excellent activity towards C-C, C-O and C-H bond cleavage, therefore, because of its lower price and higher abundance compared to noble metals, these materials are widely employed in syngas generation from methane reforming processes [1]. However, nickel catalyst shows high deactivation rates due to coke deposition [2]. In this context, perovskite-type precursors appear as a valid solution to this problem, due to their particular ability to produce nickel dispersed particles over an oxide support after H_2 pre-treatment, increasing catalyst stability [3]. Many studies on the literature have reported that the nature of A and B-site cation on perovskites strongly influences its catalytic properties. Particularly, cerium and strontium substituted lanthanum-nickel perovskites have shown interesting properties that worth deeply investigation. In this sense, this work aims to study the effect of Ce, La and Sr over nickel-based catalyst precursors on Oxy-dry reforming.

2. Experimental.

Perovskite precursors $MNiO_3$ (M= La, Sr and Ce) were prepared using citrate method and calcined at 1073°C for 5h. After that, the materials were characterized using temperature-programmed reduction (TPR), X-ray diffraction (XRD), *in-situ* X-ray diffraction at reducing atmosphere (XRD-H_2), Temperature-programmed surface reaction (TPSR) and Thermo-programmed CH_4-cracking reaction (TPCR-CH_4). Oxy-dry reaction was conducted in a tubular quartz reactor at 1073K during 15h, with 15 mg of catalyst, under a continuous feed of 70 $mL.min^{-1}$ in a $CH_4/CO_2/O_2$ 4:2:1 flow. The samples were pretreated under H_2 atmosphere (30 $mL.min^{-1}$) at 1073 K during 1h before TPSR and TPCR experiments.

3. Results and Discussion.

The XRD and TPR results, for the samples, are shown in Figure 1.

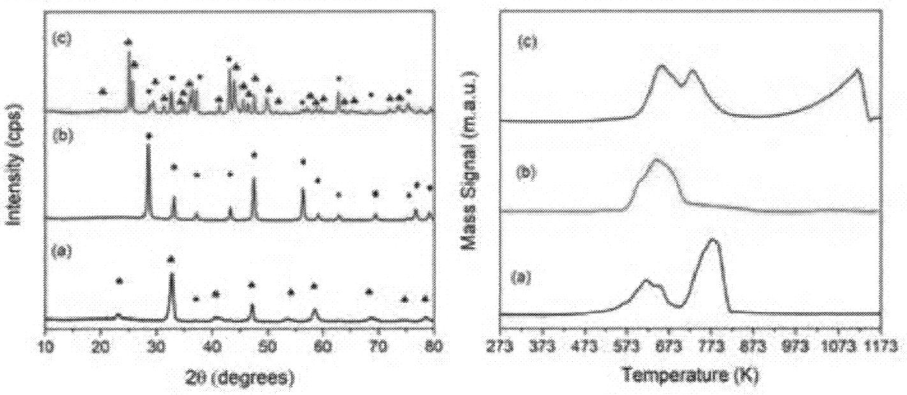

Figure 1: X-ray diffraction (Left) and Thermo Programmed Reduction (Right) profiles of the samples a) $LaNiO_3$ b) $CeNiO_3$ c) $SrNiO_3$. ♥ $Sr_2Ni_2O_5$; ♦ CeO_2; ♣ $SrCO_3$; • NiO; ♠ $LaNiO_3$

The XRD results have shown that the sample $LaNiO_3$ presented the formation of perovskite phase (ICDD-01-079-2451), along with NiO segregate (ICDD-00-002-1216) while, the precursor $CeNiO_3$ were composed basically by CeO_2 (ICDD-03-065-5923) and NiO (ICDD- 01-075-0269) oxides. Meanwhile, the perovskite $SrNiO_3$ exhibited a complex diffraction profile, with lines relative to $SrCO_3$ (ICDD-00-005-0418), NiO (ICDD-01-078-0423) and $Sr_2Ni_2O_5$ (ICDD-00-028-1242). The results indicated a low solubility of Ce and Sr cations in the perovskite MNO_3 structures, since the first one stabilizes as Ce^{4+} in the fluorite structure CeO_2 because of the high calcination temperatures employed in the perovskite synthesis [4]. While the latter is thermodynamically more stable as a +2 cation than a +3 cation and tends to form carbonates in presence of organic matter, like citrates.

In fact, the presence of these phases is in accordance with the TPR profiles of the samples. In the case of the reaction using $LaNiO_3$, three-steps can be clearly observed, relative to the sequential reduction of Ni^{3+} to Ni^{2+} followed by Ni^{2+} to Ni^0. An intermediary peak was attributed to NiO, which was reduced to Ni^0, consuming H_2. Similarly, the $SrNiO_3$ sample showed three H_2 consumption events. Both were attributed to Ni^{2+} reduction to Ni^0 at lower temperature, being the first attributed to Ni^{2+} from NiO phase, and the later to the decomposition of $Sr_2Ni_2O_5$ to Ni^0 and SrO. The last peak, according to *in-situ* XRD studies is relative to $SrCO_3$ reduction, generating SrO and CO_2. On the other hand, $CeNiO_3$ presented a reduction in two steps due to the reduction of NiO and partial reduction of Ce-based oxides.

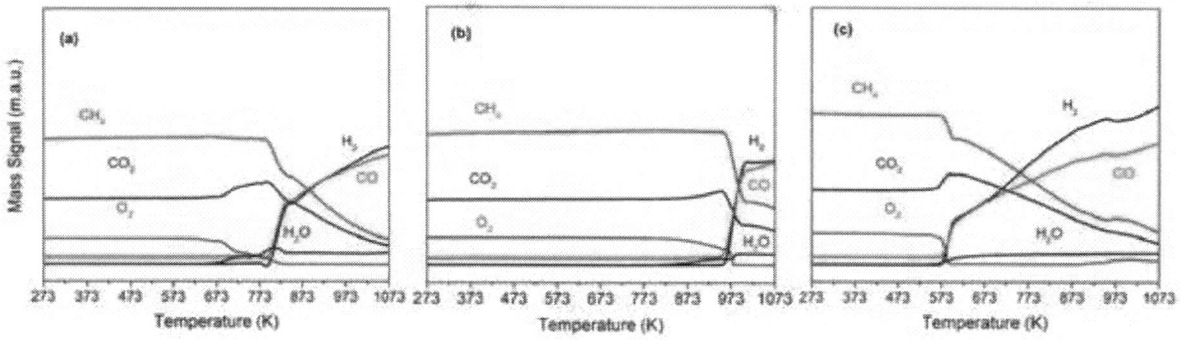

Figure 2. Thermo programmed superficial oxy-dry reactions profiles for a) $LaNiO_3$; b) $SrNiO_3$; c) $CeNiO_3$.

Previously the TPSR analyses (Fig. 2), methane cracking tests were performed. The presence of cerium seemed to decrease the activation energy of C-H bond cleavage, leading to lower reaction temperatures, while the strontium addition has promoted an activity decrease. Indeed, the Oxy-CO_2 reactions (1 atm, 1073 K, 15 h), after H_2-reduction pretreatment, were according to these results and the methane and carbon dioxide conversions had the following decreasing order: $CeNiO_3 > LaNiO_3 > SrNiO_3$, which can be attributed to the possible coverage of Ni^0 sites by $SrCO_3$ [4] detected via XRD and TPR analyses. No deactivation was observed over $CeNiO_3$, possibly due to the inherent oxygen mobility of Ceria-based catalysts [4].

TPSR studies indicated a reaction scheme with an initial combustion-reforming of methane, followed by dry reforming of methane leading to the formation of syngas. The ignition temperature of the combustion following the order: $CeNiO_3 > LaNiO_3 > SrNiO_3$. This indicates the superior ability of cerium oxides to promote oxygen-based reactions. Moreover, $CeNiO_3$ presented also a short combustion step, almost instantly followed by the dry reforming reaction, comparing with $LaNiO_3$. This may be associated with the higher reducibility of NiO in relation with the perovskite oxides. The selectivity to H_2 and CO were not affected by the A-site perovskite substitution. Deactivation of the catalysts was not observed after 15 h on-stream.

4. Conclusions.

XRD studies indicated the presence of perovskite-type structure only on La-containing precursor. Instead, the other two precursors revealed by XRD a mixture of segregate solids. All materials were active to syngas production and the A-site substitution caused significant variation on C-H cleavage activation rate, following the order: $CeNiO_3 > LaNiO_3 > SrNiO_3$. No significant deactivation by carbon deposition was verified after 15h on stream.

5. References.

[1] Xue Z., Shen Y., Zhu, S. Promoting effects of lanthanum oxide on the NiO/CeO_2 catalyst for hydrogen production by autothermal reforming of ethanol. Catal. Comm.

108 (2018) 12-16.

[2] Castro T., Rabelo C., Noronha F. Study of the performance of Pt/Al_2O_3 and Pt/CeO_2/Al_2O_3 catalysts for steam reforming of toluene, methane and mixtures. Catal. T. (http://dx.doi.org/10.1016/j.cattod.2017.05.067).

[3] Santos, M. Noronha, F., Brandao, S. Perovskite as catalyst precursors in the partial oxidation of methane: The effect of cobalt, nickel and pretreatment. Catal. T. 299, (2018) 229-241.

[4] Yang E., Noh Y., Perovskite as catalyst precursors in the partial oxidation of methane: The effect of cobalt, nickel and pretreatment. Catal. T. (http://dx.doi.org/10.1016/j.cattod.2017.03.050)

545632 Combustion Suppression in Low-Temperature Electro-Catalytic Tri-Reforming of Methane over Ni-Mg Supported Catalysts

Wednesday, June 5, 2019
Texas Ballroom Prefunction Area (Grand Hyatt San Antonio)
Tomohiro Yabe, Kensei Yamada, Shuhei Ogo and Yasushi Sekine, Applied Chem., Waseda Univ., Tokyo, Japan

For reduction of automobile fuel consumption, a novel method for fuel reforming uses waste heat and exhaust gas components to regenerate the exhaust loss energy. We have investigated tri-reforming of CH_4 (TRM) using H_2O, CO_2, O_2, and CH_4 as model gas. TRM comprises three reactions: CH_4 steam reforming (SRM), CH_4 dry reforming (DRM), and CH_4 combustion (MC). To be suppressed to recover the enthalpy, SRM and DRM are expected to proceed preferentially to CH_4 combustion (exothermic).

Recently, our group reported that Co-impregnated catalysts of Ni and Mg on La-ZrO_2 (LZO) showed high reforming activity, suppressed CH_4 combustion, and achieved enthalpy recovery at 473 K by applying an electric field.

In this study, to elucidate the causes of high reforming activity and CH_4 combustion suppression over NiO-MgO co-supported catalysts in the TRM reaction in an electric field at 473 K, the reactivity and structure of NiO-MgO/LZO catalysts with or without an electric field was investigated using various experiments such as SRM activity and MC activity, and various characterization such as H_2-TPR, XRD, XPS, XAFS. As a result, we found that Mg addition to 10wt%Ni/LZO catalyst suppresses methane combustion, while SRM proceeds well by virtue of surface protonics in the electric field. NiO-MgO solid solution formed on the support played an important role in combustion suppression. Ni^{2+} on 10wt%Ni/10wt%Mg/LZO catalyst was more cationic than that on 10wt%Ni/LZO catalyst.

546105 Combined Methane Dry Reforming and Methane Partial Oxidization for Syngas Production over High Dispersion Ni Based Mesoporous Al2O3 catalyst

Wednesday, June 5, 2019
Texas Ballroom Prefunction Area (Grand Hyatt San Antonio)

Qingxiang Ma, **Yunxing Han**, **Jianli Zhang** and Tiansheng Zhao, Ningxia University, Yinchuan, China

Ni based mesoporous Al_2O_3 catalysts were prepared by simple sol-gel method, and were employed as catalyst in combined methane dry reforming (DRM) and methane partial oxidization (POM) reaction. The catalysts were characterized by transmission electron micrographs (TEM), N_2 physical adsorption-desorption, H_2 chemisorption and temperature programmed reduction (H_2-TPR), X-ray diffraction (XRD) and Raman spectroscopy. The TEM and N_2-physisorption characterization showed that the catalysts had typical mesoporous structure. The catalytic activity tests results indicated that the catalysts prepared by the one-pot method had better catalytic performance for combined methane dry reforming and methane partial oxidation reaction. The TEM and Raman analysis of used catalysts showed that combined DRM with POM could effectively suppress the formation of carbon on catalysts surface.

545754 Shale Gas & Biomass Co-Gasification for Syngas Production

Wednesday, June 5, 2019
Texas Ballroom Prefunction Area (Grand Hyatt San Antonio)

Amoolya Lalsare[1], Ali Sivri[2], Cosmin Dumitrescu[2] and Jianli Hu[3], (1)Chemical & Biomedical Engineering, West Virginia University, Morgantown, WV, (2)Mechanical and Aerospace Engineering (MAE), West Virginia University, Morgantown, WV, (3)Department of Chemical and Biomedical Engineering, West Virginia University, Morgantown, WV

According to the U.S Energy Information Administration, the production of Appalachian shale gas grew manifold from 7.8 Bcf/d in 2012 to 23.8 Bcf/d in 2017. Thus Appalachian shale gas which is abundant in West Virginia has been the key component of U.S's ever increasing natural gas production. With the cost of U.S natural gas hovering just over $2.5/MMBtu, utility of natural gas as a primary feedstock for fuels and value added chemicals has tremendous potential. At the same time lingo-cellulosic biomass generated from the agriculture, dairy, poultry, and related activities has potential to transform the transportation, energy, and value-added chemicals industry in a major way. The complexity and variation in biomass feedstock and its lower heating value compared to coal and natural gas limits extensive use of biomass as a sustainable feedstock. Ligno-cellulosic biomass being a complex array of aromatic rings and unsaturated carbons, it needs upgradation before it can be used as fuel or feedstock for chemical synthesis.

Thus biomass gasification is a commercially viable and economic pathway to produce syngas from biomass which can be used as feedstock for Fischer-Tropsch or related processes for chemical synthesis. However, high oxygen content and aromatic nature of ligno-cellulosic biomass yields low quality syngas with H_2:CO ratio of about 1.2 to 1.4.

Thus we propose co-gasification of shale gas or essentially methane – biomass without pyrolysis or partial-gasification of methane to CO, CO_2 but via activation of the methane species and dissociation of the C-H bond. This study aims to perform an extensive reaction kinetics and mechanistic study of a possible hydrogen transfer between methane hydrogen and biomass unsaturated carbon. The co-gasification of biomass and methane will be performed at different reaction conditions of temperature, equivalence ratio, heating rate etc. Methane – biomass co-gasification will be studied under catalytic and non-catalytic environment to observe the effect of metal-doped zeolite ZSM-5 like catalyst on activation of methane and co-gasification of methane-biomass.

We have developed two reactor setups to test the gasification reaction kinetics. A fixed bed reactor setup and a bench scale bubbling fluidized bed reactor. Initially Thermo-gravimetric Analyzer (TGA) – Mass Spectrometry setup will be used for parametric reaction kinetics study of biomass-methane co-gasification and the optimized reaction conditions will be tested on the two gasifier reactor setups mentioned above. Hardwood pellet biomass will be used in this study and methane will be used initially for mechanistic studies which will be later replaced by commercial natural gas.

546098 Gas Conversion: Technology at the Tipping Point?

Monday, June 3, 2019: 10:15 AM
Texas Ballroom EF (Grand Hyatt San Antonio)

David Dankworth, Corporate Strategic Research, ExxonMobil Research and Engineering Co., Annandale, NJ; Andlinger Center for Energy & Environment, Princeton University, Princeton, NJ

As natural gas becomes more and more abundant, the potential supply exceeds the demand expected from power and heat applications. Methane could be the ideal raw material for production of cleaner-burning, lower carbon fuels. We have not so far seen a wave of investments in GTL projects, even for higher margin scenarios such as flaring avoidance. The reason is simple: current technologies are too capital intensive relative to other options, and have less than stellar carbon-efficiency. Technology developers and researchers have not been standing still, however. Over the past 5-10 years, several potentially step-out new technologies have passed into the later stages of development and/or demonstration. At the same time, several new scientific leads have emerged in the academic literature which could lead to further advances in cost and yield. These new and improved approaches, deployed in the right combinations, could bring gas conversion into competition with other gas monetization options.

This keynote address will discuss the current state of art in gas conversion technologies, including syngas generation and conversion as well as direct methane conversion pathways. The emerging new technologies and ideas will be examined in terms of what they may bring in terms of cost reduction and yield improvement. Finally, the remaining gaps and areas where further R&D is needed to arrive at the "tipping point" will be outlined for the community. The talk should provide a excellent scene set for the topics and ideas to be presented in the rest of the meeting.

Note: This initial abstract is a placeholder. A more specific abstract text will be provided later.

545983 OX-Zeo Bifunctional Catalyst Concept Enables Syngas Conversion to Light Olefins and Beyond

Monday, June 3, 2019: 10:54 AM
Texas Ballroom EF (Grand Hyatt San Antonio)

Feng Jiao, Xiulian Pan, Junhao Yang, Gen Li, Na Li, Yuxiang Chen and Xinhe Bao,
Dalian Institute of Chemical Physics, Chinese Academy of Sciences, Dalian, China

Title

OX-ZEO bifunctional catalyst concept enables syngas conversion to light olefins and beyond

Authors:

Feng Jiao, XiulianPan,* Junhao Yang, Gen Li, Na Li, Yuxiang Chen, Xinhe Bao*

State Key Laboratory of Catalysis and Dalian National Laboratory for Clean Energy Dalian Institute of Chemical Physics, Chinese Academy of Sciences, Dalian, 116023, China.

*Xiulian Pan: panxl@dicp.ac.cn; *Xinhe Bao: xhbao@dicp.ac.cn

Abstract text

Light olefins, include ethylene, propylene and butylene, are important chemical building blocks which are widely used for producing plastics, cosmetics, and drugs. Syngas, as a platform of coal, natural gas and biomass utilization, can be transformed to light olefins via traditional Fischer-Tropsch synthesis (FTS).[1] However, hydrocarbon distributions of FTS follows the so-called Anderson-Schulz-Flory (ASF) distribution, according to which the selectivity of C_2-C_4 hydrocarbon does not exceed 58%.[1-3] Therefore, to precisely control the C-C coupling as well as suppress the methane formation and over-hydrogenation of olefins are key challenges in this filed.

We recently reported a bifunctional catalyst design concept of OX-ZEO (Oxide-Zeolite) for syngas conversion. For example, a mix of $ZnCrO_x$ metal oxides and mesopore SAPO-34 (MSAPO) leads to a light olefin selectivity up to 80% in hydrocarbons at CO conversion of 17%.[4] This selectivity is much higher than the upper limit of C_2-C_4 according to the ASF model and the best iron and cobalt based FTS catalysts reported so far. We demonstrated that this OX-ZEO concept turns syngas conversion into a tandem reaction, i.e. activating CO and H_2, generating intermediates over the partially reduced metal oxides, and C-C coupling within the pores of zeolites.[4] Thus, the product selectivity can be manipulated with shape-selective zeolites by changing the acidity and topology of zeolites.[4-6] In addition, it is important to match the two components properly in terms of the compositions, structures and intimacy to achieve an optimum catalytic activity and selectivity. This concept is further corroborated by substituting the zeolite component with mordenite, which steers syngas conversion selectively toward ethylene. The selectivity of ethylene alone in hydrocarbons reaches as high as 83%.[6] If the zeolite component is ZSM-5, syngas then can be selectively converted to aromatics and the selectivity reaches 73%.[5] Furthermore, our recent results shown that this concept can be extended for direct synthesis of liquid fuels and oxygenates from syngas. In addition, it was demonstrated recently that this concept is also applicable to CO_2 hydrogenation to form a variety of chemicals.[7-10] Although the mechanisms how the two functionalities cooperate and how the product selectivity is controlled are not well understood yet, the increasing number of publications in CO and CO_2 hydrogenation[4-14] reveals the versatility of this concept and may provide a third alternative for fabrication of basic chemicals from non-petroleum carbon resources such as coal, natural gas and biomass. In this presentation, detailed discussion will be given about the OX-ZEO catalyst design concept and the possible reaction mechanism.

References

1. Torres Galvis, H. M.; de Jong, K. P., Catalysts for Production of Lower Olefins from Synthesis Gas: A Review. *ACS Catal.* **2013,** *3* (9), 2130-2149.
2. Torres Galvis, H. M.; Bitter, J. H.; Khare, C. B.; Ruitenbeek, M.; Dugulan, A. I.; de Jong, K. P., Supported iron nanoparticles as catalysts for sustainable production of lower olefins. *Science* **2012,** *335* (6070), 835-8.
3. Zhong, L.; Yu, F.; An, Y.; Zhao, Y.; Sun, Y.; Li, Z.; Lin, T.; Lin, Y.; Qi, X.; Dai, Y.; Gu, L.; Hu, J.; Jin, S.; Shen, Q.; Wang, H., Cobalt carbide nanoprisms for direct production of lower olefins from syngas. *Nature* **2016,** *538* (7623), 84-+.
4. Jiao, F.; Li, J.; Pan, X.; Xiao, J.; Li, H.; Ma, H.; Wei, M.; Pan, Y.; Zhou, Z.; Li, M.; Miao, S.; Li, J.; Zhu, Y.; Xiao, D.; He, T.; Yang, J.; Qi, F.; Fu, Q.; Bao, X., Selective conversion of syngas to light olefins. *Science* **2016,** *351* (6277), 1065-8.
5. Yang, J.; Pan, X.; Jiao, F.; Li, J.; Bao, X., Direct conversion of syngas to aromatics. *Chem. Commun.* **2017,** *53* (81), 11146-11149.
6. Jiao, F.; Pan, X. L.; Gong, K.; Chen, Y. X.; Li, G.; Bao, X. H., Shape-Selective Zeolites Promote Ethylene Formation from Syngas via a Ketene Intermediate. *Angewandte Chemie-International Edition* **2018,** *57* (17), 4692-4696.
7. Gao, P.; Li, S.; Bu, X.; Dang, S.; Liu, Z.; Wang, H.; Zhong, L.; Qiu, M.; Yang, C.; Cai, J.; Wei, W.; Sun, Y., Direct conversion of CO2 into liquid fuels with high selectivity over a bifunctional catalyst. *Nat. Chem.* **2017,** *9* (10), 1019-1024.
8. Li, Z.; Wang, J.; Qu, Y.; Liu, H.; Tang, C.; Miao, S.; Feng, Z.; An, H.; Li, C., Highly Selective Conversion of Carbon Dioxide to Lower Olefins. *ACS Catal.* **2017,** *7* (12), 8544-8548.
9. Liu, X.; Wang, M.; Zhou, C.; Zhou, W.; Cheng, K.; Kang, J.; Zhang, Q.; Deng, W.; Wang, Y., Selective transformation of carbon dioxide into lower olefins with a bifunctional catalyst composed of ZnGa2O4 and SAPO-34. *Chem. Commun.* **2018,** *54* (2), 140-143.

10. Ni, Y. M.; Chen, Z. Y.; Fu, Y.; Liu, Y.; Zhu, W. L.; Liu, Z. M., Selective conversion of CO2 and H-2 into aromatics. *Nat. Commun.* **2018,** *9*.
11. Zhu, Y. F.; Pan, X. L.; Jiao, F.; Li, J.; Yang, J. H.; Ding, M. Z.; Han, Y.; Liu, Z.; Bao, X. H., Role of manganese oxide in syngas conversion to light olefins. *ACS Catal.* **2017,** *7* (4), 2800-2804.
12. Cheng, K.; Gu, B.; Liu, X.; Kang, J.; Zhang, Q.; Wang, Y., Direct and highly selective conversion of synthesis gas into lower olefins: design of a bifunctional catalyst combining methanol synthesis and carbon-carbon coupling. *Angew. Chem. Int. Ed.* **2016,** *55* (15), 4725-8.
13. Cheng, K.; Zhou, W.; Kang, J.; He, S.; Shi, S.; Zhang, Q.; Pan, Y.; Wen, W.; Wang, Y., Bifunctional Catalysts for One-Step Conversion of Syngas into Aromatics with Excellent Selectivity and Stability. *Chem* **2017,** *3* (2), 334-347.
14. Zhang, P.; Tan, L.; Yang, G.; Tsubaki, N., One-pass selective conversion of syngas to para-xylene. *Chemical Science* **2017,** *8* (12), 7941-7946.

551531 Cool GTL- A New Process from GTL

Monday, June 3, 2019: 11:18 AM
Texas Ballroom EF (Grand Hyatt San Antonio)

Terry Marker, Jim Wangerow, Pedro Ortiz-Toral, Martin Linck and **Ron Stanis**, Gas Technology Institute, Des Plaines, IL

GTI is currently working to develop Cool GTL, a new processes for the conversion of natural gas to high quality liquids.

Cool GTL can be used to convert high CO_2 content natural gas, biogas streams, and gas produced from gas plants to liquids. In Cool GTL we convert high CO_2 feeds to synthesis gas at a 2/1 H_2/CO ratio using a novel CO_2/steam reforming catalyst which has high stability. The 2/1 H_2/CO synthesis gas ratio is always achieved, for a variety of levels of feed CO_2 content, by adjusting the amount of water added according to the amount of CO_2 in the feed. The reformer catalyst has been tested for 500 hours and shows no deactivation. The product from the reformer then goes directly to a low temperature bubbling fluid bed Fischer Tropsch reactor which makes hydrocarbon liquids from the synthesis gas. A unique catalyst formulation enables the fluid bed to make no wax. The fluid bed allows excellent temperature control. Cool GTL is under development but still requires longer integrated testing.

Currently Cool GTL is being developed at GTI through DOE funding in the project "Cool GTL for the Production of Jet Fuel from Biogas". Although this project uses biogas as a feed, the technology is equally applicable to any natural gas stream. In this DOE project, our goal is to make 100 gallons of high quality jet fuel. This project will also demonstrate Cool GTL catalyst and process stability through long term, 24 hour operation. Once the DOE project is complete, we believe the Cool GTL process will be ready for larger scale commercial deployment in multiple applications.

544681 Advances in Gas-to-Liquids Technology at Shell

Monday, June 3, 2019: 11:42 AM
Texas Ballroom EF (Grand Hyatt San Antonio)

Leendert Bezemer, Shell Global Solutions International B.V., Amsterdam, Netherlands

Advances in Gas-to-Liquids Technology at Shell

G.L. Bezemer

[1] Shell Global Solutions International BV, Grasweg 31, 1031 HW Amsterdam, The Netherlands

Introduction

Shell researchers have been studying the Fischer-Tropsch (FT) reaction for more than 40 years. This contributed to commercial Gas-to-Liquids (GTL) plants in Malaysia and Qatar and ongoing developments for new GTL prospects at various locations. Shell Middle Distillate Synthesis (Shell MDS) in Bintulu, Malaysia, started up in 1993 with a current capacity of 14,700 bbl/d synthetic products. Pearl GTL is situated in Ras-Laffan, Qatar, and started up in 2011. It is the largest GTL plant in the world with a production capacity of 140,000 bbl/d synthetic products. In Shell, research to further improve the economical and technical aspects of the GTL process has been a continuous activity. This has resulted in improved technology for each of the main process steps. In this talk FT related developments will be discussed first. In both Shell MDS Bintulu and Pearl GTL, multi-tubular fixed bed technology is used for the Heavy Paraffin Synthesis (HPS). This technology has proven scale-up benefits and several other differentiating features compared to other reactor types such as slurry reactors. Second main topic will be improvements made in the Heavy Paraffin Conversion (HPC) technology. Afterwards the product slate of the Shell GTL process will be discussed in conjunction with the technology. Finally, an update will be given on the implications of these and other improvements on new GTL plant designs.

HPS catalyst improvements

In Shell MDS Bintulu and Pearl GTL, a multitude of HPS reactors are installed where the chemical conversion of gas into liquids takes place. The catalyst in the reactors can be reactivated in situ several times before it has to be unloaded and reclaimed. The design and early operation of Shell MDS Bintulu was based on the first-generation Fischer-Tropsch catalyst. The second-generation FT catalysts were implemented from 2000 onwards. In the design of Pearl GTL, the full potential of this improved catalyst could be harvested, being a doubled production rate per unit catalyst volume, a heavier product distribution with higher liquid selectivity and lower CO_2 make.

Since the start-up of Pearl, catalyst development has continued resulting in third-generation catalysts with even further improved liquid selectivity and reduced CO_2

production. Besides that, catalyst stability was further increased. Introduction of these third-generation catalysts started in 2014 in Pearl GTL and Shell MDS followed in 2015. In the lecture the performance of next generation HPS catalyst will be discussed and some of the underlying capabilities that were behind the development will be highlighted.

HPC catalyst improvements

The conversion of raw FT product into final products is done in the Liquid Processing Unit (LPU). The LPU can be regarded as the refinery part of the GTL plant. The molecular structure of FT wax is however quite different compared to the crude oil diet of a normal refinery and hence dedicated catalysts and process units were developed for this purpose. In the HPC reactor (de)hydrogenation, isomerisation and cracking reactions are taking place. The noble metal based HPC catalyst designed for Pearl GTL provided excellent product quality. In the lecture improvements in the HPC catalyst system will be discussed with examples shown of stability and selectivity improvements that were enabled by fundamental understanding of the process.

GTL products developments

Shell's FT synthesis process results in very clean paraffinic hydrocarbons. The unique nature of the molecules makes them very suitable in specialty applications. The very heavy wax produced by the HPS allows the conversion into lubricant baseoils with superior properties compared to oil derived baseoils available in the market. Shell baseoils are used in premium lubricants and enable improved fuel economy during driving. In the lecture a comparison will be made with different baseoil types and the basis of the enhanced performance will be discussed.

Scalable GTL design for new prospects

With the Shell MDS and the Pearl GTL projects, Shell has proven GTL on 15 and 140 kbpd scale. This experience enables the selection of a scale (hence: "Scalable GTL") that gives the optimum balance of economy of scale versus project complexity and resource requirements. The Scalable GTL design It is based on the latest technology improvements in various areas (gasification, HPS, HPC) and has an optimized cost based on competitive scoping, supply chain transformation & efficient execution. In the lecture the interaction between technology steps and the resulting design will be highlighted.

Conclusions

Almost 100 years after its initial discovery, the Fischer-Tropsch process still receives a lot of attention in academia and industry. In the lecture, four areas where recently significant improvements have been made will be show cased alongside an overview of the Gas-to-Liquids process deployed in Shell MDS Bintulu and Pearl GTL.

549752 Carbon Molecular Sieve (CMS) Membranes as Potential Game-Changers in Olefins Unit Operations

Monday, June 3, 2019: 12:06 PM
Texas Ballroom EF (Grand Hyatt San Antonio)
William J. Koros, School of Chemical & Biomolecular Engineering, Georgia Institute of Technology, Atlanta, GA

Shale gas has opened the door to new resources for olefin production, but current separation technology still relies primarily upon energy-intensive distillation processes for key separations to take advantage of these resources. This presentation will explain why carbon molecular sieve membranes in asymmetric hollow fiber forms can provide significant positive changes for olefins production. The state of the art, including fundamentals related to this exciting technology will be presented. Challenges and opportunities in implementing the technology as the large scale needed to impact olefins production will also be considered.

549956 Advanced Polymer Membranes for Gas Separations Relevant to Shale Gas Processing

Monday, June 3, 2019: 1:30 PM
Texas Ballroom EF (Grand Hyatt San Antonio)

J.C. Davilla[1], **Alec Bridge**[1], Joan F. Brennecke[2] and Benny D. Freeman[3], (1)The University of Texas at Austin, Austin, TX, (2)McKetta Department of Chemical Engineering, University of Texas at Austin, Austin, TX, (3)McKetta Department of Chemical Engineering, The University of Texas at Austin, Austin, TX

The availability of large amounts of shale gas has stimulated the search for processes that could upgrade these light hydrocarbon resources to potentially more valuable chemicals and fuels. Novel processes will demand advances in catalysis and separations. Membranes may well play a significant role in separations associated with shale gas conversion. This presentation will outline a potentially new approach, based upon the CISTAR process, for conversion of C_2 and C_3 alkane hydrocarbons to olefin analogs useful for further processing into, for example, chemicals. Separations of interest for this application include H_2 removal from mixtures of olefins and paraffins as well as olefin/paraffin separation. Based upon process considerations, some of these separations may be more efficiently performed at temperatures substantially above ambient. Recent results on thermally stable membrane materials, including those employing ionic liquids to assist in separation properties, will be discussed, as well as potential new directions for materials research in this space.

544682 Catalytic Applications By Using a Highly Crystalline Ferrierite-Based Bifunctional Catalyst for Syngas (containing CO_2) Conversions

Monday, June 3, 2019: 2:09 PM
Texas Ballroom EF (Grand Hyatt San Antonio)
Jong Wook Bae, School of Chemical Engineering, Sungkyunkwan University (SKKU), Suwon, Kyonggi-do, Korea, Republic of (South)

A gas-phase carbonylation of dimethyl ether (DME) to methyl acetate (MA) on the zeolites has been largely investigated to produce the value-added useful chemical intermediates **[1]**. Although a lot of solid-acid zeolites such as MFI, BEA, and MOR can be applied for DME carbonylation to produce MA, DME conversion on those solid-acid catalysts was relatively lower and deactivated significantly **[2]**, where DME can be mainly produced by CO and CO_2 hydrogenation using bifunctional hybrid Cu-ZnO-Al_2O_3/zeolites **[3,4]**. Based on our previous results **[5,6]**, DME carbonylation was investigated by using a lab-made FER with high crystallinity to prohibit the coke depositions effectively for a stable long-term activity in terms of crystallinity and dispersion of active metallic sites on FER-based heterogeneous catalysts. The seed-derived ferrierite (FER@FER) with a high crystallinity was applied for a gas-phase DME carbonylation to methyl acetate (MA), and it was further applied for the direct syngas conversions through CO and (or) CO_2 hydrogenation to DME on the bifunctional Cu-ZnO-Al_2O_3/FER (CZA/FER). The observed higher catalytic activity and stability were mainly attributed to the larger amounts of active Brønsted acid sites with less defected acidic sites of the FER@FER. The typical gas-phase carbonylation reaction of DME to MA was carried out by using the FER zeolite having different Si/Al molar ratios of 10.4 – 12.5 as well as high crystallinity synthesized by using various zeolite seed materials such as the MOR, ZSM-5 and USY. The enhanced crystallinity of the FER prepared by simply using the FER seed (FER@FER) having newly formed mesopore structures was responsible for an increased amount of the active Brønsted acid sites. The highly crystalline FER@FER revealed the suppressed depositions of aromatic coke precursors due to less presence of defect sites. Compared to other zeolite seed-derived FER zeolites, the less amount of defect sites (extra-framework Lewis acidic Al species) on the FER@FER was successfully controlled. The active Brønsted acid sites for DME carbonylation reaction were originated from the preferential formations of the stable tetrahedral Al sites (especially, T2 sites of the Al-O-Si-O-Al framework of FER) on the 8 and 10-membered ring channels of the FER@FER. The FER@FER synthesized with different FER-seed content without using any organic structure-directing agent (OSDA) showed a highly active and stable activity for carbonylation of dimethyl ether (DME) to methyl acetate (MA) at an optimal amount of previously synthesized FER-seed with 15wt%. The FER-seed played an important role to maximize the number of Bronsted acid sites with less coke deposition by concomitant decrease of defect sites through its recrystallization possibly, where the defected Lewis acidic sites on the FER surfaces can be responsible for the preferential coke depositions as well as for the retarded CO insertion rate by preventing the formation of surface acetyl groups. In addition, the selective synthesis of ethanol from syngas through multi-step reactions via methyl acetate (MA) intermediate is possible using a highly crystalline FER and hybridized bifunctional CZA/FER as well. The extent of crystallinity of the FER played a crucial role for catalytic activities of syngas conversion to DME, DME carbonylation to MA as well as MA hydrogenation to ethanol. The CZA/FER can be utilized to convert MA to ethanol with a similar equilibrium yield of ~42 mol%. The highly crystalline of FER@FER, which was synthesized by using a FER seed, showed a higher DME conversion to MA and stability due to the abundant presence of Bronsted acid sites with less defect sites.

References

1 J. Liu, H. Xue, X. Huang, P.H. Wu, S.J. Huang, S.B. Liu, W. Shen, Chin. J. Catal. 31 (2010) 729.

2 A. Bhan, A.D. Allian, G.J. Sunley, D.J. Law, E. Iglesia, J. Am. Chem. Soc. 129 (2007) 4919.

3 J.W. Jung, Y.J. Lee, S.H. Um, P.J. Yoo, D.H. Lee, K.W. Jun, J.W. Bae, Appl. Catal. B: Environ. 126 (2012) 1.

4 H. Ham, J. Kim, S.J. Cho, J.H. Choi, D.J. Moon, J.W. Bae, ACS Catal. 6(9) (2016) 5629.

5 S.Y. Park, C.H. Shin, J.W. Bae, Catal. Commun. 75 (2016) 28.

6 H. Ham, J. Kim, J.H. Lim, W.C. Sung, D.H. Lee, J.W. Bae, Catal. Today 303 (2018) 93.

545566 Novel e-Iron Carbide Fischer-Tropsch Catalysts with Stable and Low-CO_2 selectivity

Monday, June 3, 2019: 2:33 PM
Texas Ballroom EF (Grand Hyatt San Antonio)

Peng Wang[1,2], Wei Chen[2], Fu-Kuo Chiang[1], A. Iulian Dugulan[3], Yuanjun Song[4], Robert Pestman[2], Kui Zhang[1], Jinsong Yao[1], Bo Feng[1], Ping Miao[1], Wayne Xu[1], Zhuowu Men[5] and Emiel Hensen[2], (1)National Institute of Clean-and-Low-Carbon Energy, Beijing, China, (2)Schuit Institute of Catalysis, Department of Chemical Engineering and Chemistry,, Eindhoven University of Technology, Eindhoven, Netherlands, (3)Fundamental Aspects of Materials and Energy Group, Delft University of Technology, Delft, Netherlands, (4)School of Mathematics and Physics, University of Science and Technology Beijing, Beijing, China, (5)National Institute of Clean-And-Low-Carbon Energy, Beijing, China

Fischer-Tropsch (FT) synthesis represents one of the most advanced technologies to convert fossil and renewable carbon feedstock into liquid fuels. The technology is well established for the conversion of synthesis gas derived from natural gas. Currently, significant efforts are underway to realize coal-to-liquids (CTL) processes in certain parts of the world where coal is a cheap and abundant resource. Compared to natural-gas based FT, CTL processes face an important challenge: since the H/C ratio of coal is too low, the process requires the water-gas shift (WGS) reaction to increase the H_2/CO ratio of the FT feedstock. This leads to significant CO_2 emissions. In essence, this leads to the requirement to couple CTL processes to CO_2 capture. Typically, conventional Fe-based FT catalyst converts 30% of feeding CO into by-products CO_2 in FT units. Taking into account the whole CTL process, decreasing CO_2 release in FT unit will dramatically reduce the pilot costs and enhance productivity.

To overcome this challenge, we present a completely novel and scalable preparation method of stable and phase-pure ε(')-iron carbide Fischer-Tropsch catalysts with a near-zero intrinsic (primary) CO_2 selectivity. In situ characterization techniques are employed to clarify the formation and phase composition of ε(′)-carbide and the catalytic performance of these novel catalysts was evaluated under industrial FT conditions. Importantly, no catalyst deactivation was observed and the ε(′)-carbide phase was intrinsically stable under these conditions, even after a high temperature excursion. This is contrary to current belief that ε(′)-carbide is not stable under industrial FT conditions. These results shed a completely new view on the utility of ε(′)-carbide, which has hitherto been disregarded in favor of Hägg carbide as the stable and catalytically active phase. We unequivocally show that ε(′)-carbide can convert syngas to liquid fuels without CO_2 as a primary product. Thus, we achieved to prepare a catalyst with low CO_2-selectivity.

This work is published as *Science Advances* recently, DOI: 10.1126/sciadv.aau2947.

546535 Understanding and Modeling the Deactivation of an Industrial Ni/Al_2O_3 Catalyst for CO_2 Methanation in a Milli-Structured Fixed-Bed Reactor-Exchanger

Monday, June 3, 2019: 2:57 PM
Texas Ballroom EF (Grand Hyatt San Antonio)

Isabelle Champon[1,2,3], Sébastien Thomas[2], Anne Cecile Roger[2], Alain Bengaouer[1] and Albin Chaise[1], (1)LITEN, CEA, Grenoble, France, (2)ICPEES, Université de Strasbourg, Strasbourg, France, (3)ADEME, Angers, France

Understanding and modeling the deactivation of an industrial Ni/Al_2O_3 catalyst for CO_2 methanation in a milli-structured fixed-bed reactor-exchanger

Isabelle CHAMPON, Sébastien THOMAS, Anne-Cécile ROGER, Alain BENGAOUER, Albin CHAISE

The integration of renewable energy into the electric grid underlines the issue of the storage of an excess and intermittent electricity production. Power-to-gas technology is one of the solutions to overcome this problem. The catalytic reaction of interest is here the Sabatier reaction, to produce Substitute Natural Gas (SNG) from hydrogen, synthetized by water electrolysis and carbon dioxide. This process allows also the CO_2 valorization if the CO_2 comes from the biomass or is catch from industrial smoke. Regardless the reactor, the deactivation of the catalyst happened with time-on-stream and is reflected by the loss of the catalytic activity and selectivity.

The reactor studied here is a milli-structured plate reactor-exchanger filled with a fixed bed of 14-17% nickel-alumina industrial catalyst powder and cooled by a thermal oil which circulates on both sides of the catalytic bed. The purpose of this work is to study the deactivation of the catalyst in order to implement the deactivation model in the chemical kinetics.

First, some reactive tests are performed on the industrial fresh catalyst using the reactor previously described, which allows tests at the scale of several grams of catalyst. The purpose of these tests is to grasp the evolution of the catalyst performances regarding the operating temperature, pressure and time-on-stream for a stoichiometric ratio $H_2/CO_2 = 4$. Over 1000 h on-stream, the loss of CO_2 conversion with time-on-stream shows 2 zones: in the first one, the deactivation rate is much faster than in the second one.

Sintering of the nickel particles is one of the main routes of deactivation[1]. The thermal degradation of the catalyst is studied by submitting the catalyst at different atmospheres between 450 and 600 °C for several hours. The post-experiment catalysts are characterized by H_2-TPD analysis to evaluate the surface area of metallic nickel and TEM analysis to confirm the dispersion and particle size of nickel particles. Sintering laws of nickel surface area decline are obtained. Under hydrogen at 600 °C, over 300 h, an order of sintering of five is calculated by the simple power law[2], meaning an Ostwald ripening process as the main route of sintering. In the same conditions, a third order generalized power law[3] is achieved. However, under hydrogen at 450°C, results show that the catalyst is quite stable over 300 h.

Knowing that water accelerates the sintering of metal[4], and because water is one of the products of methanation reaction, the thermal degradation of the catalyst is studied in atmosphere $H_2/H_2O/N_2$. Results are in accordance with expectation, they are shown in Figure 1. The evolution of the nickel surface area with time on stream under hydrothermal deactivation is shown Figure 2 for a flow rate of 50 Nml/min of H_2, 25 Nml/min of N_2 and 14 Nml/min of H_2O. More than half of the surface of Ni° is lost during 100 hours of sintering in these conditions. The different power laws obtained throughout this hydrothermal study will be used to model the deactivation inside the reactor, (see below).

In parallel, the kinetic laws of the three reactions happening in CO_2 methanation are established: direct way (CO_2 methanation) and indirect way (Reverse Water Gas Shift - RWGS and CO methanation) on the industrial fresh catalyst. The scale of this study is the milligram of catalyst, so the catalytic bed is diluted in SiC in order to obtain a plug flow reactor. After finding conditions in which the internal and external diffusional limitations can be neglected (modifying the reactive flow rate, the catalyst mass, and the particle diameter), the influence of each component on each reaction is studied at low conversion and for different temperatures to build a kinetic law. When the pressure of one component is varied, all the other are kept constant, and the total flow is compensated with argon to always have the same residence time inside the reactor.

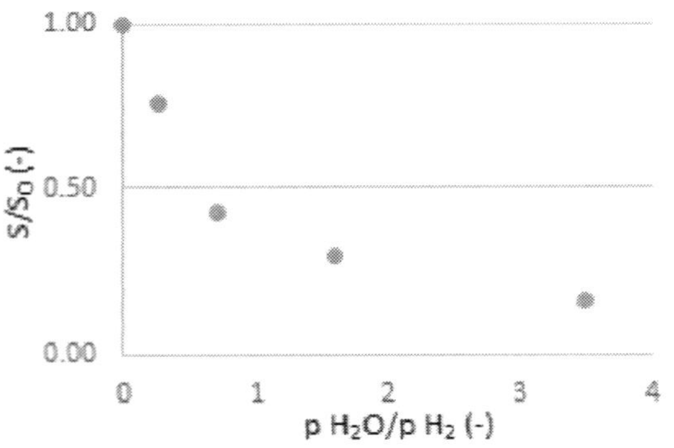

Figure 1 : Influence of water on the evolution of Ni surface area for 20 h on stream at 600°C

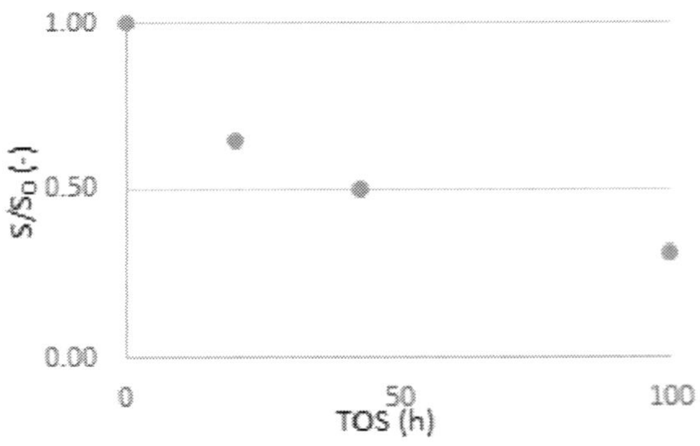

Figure 2 : Evolution of the Ni surface area with time on stream for $p_{H2O}/p_{H2} = 0,3$ at 600°C

The conclusion of the kinetic study is that hydrogen has a stronger influence on the kinetic of CO_2 methanation than CO_2. Concerning the products of the CO_2 methanation, methane has no effect on the kinetics of CO_2 methanation, whereas water inhibits the reaction. For the kinetics of RWGS, same conclusions are made for CO_2 and H_2O as the CO_2 methanation, except for H_2 which has no effect on the

RWGS kinetics. Concerning the CO methanation, both a strong influence of CO and H_2 is seen on the CO methanation kinetic.

Different kinetic laws of the literature are tested by linearization at low conversion[5]–[8]. Finally, kinetic and adsorption parameters are obtained at 350°C, 400°C and 450°C. Assuming Arrhenius laws, the pre-exponential factors, activation energy and heat of adsorption are calculated. Some tests are performed at higher conversion rate to adjust the parameters previously identified. Parity plots of experimental and modeled values are drawn and are shown in Figure 3.

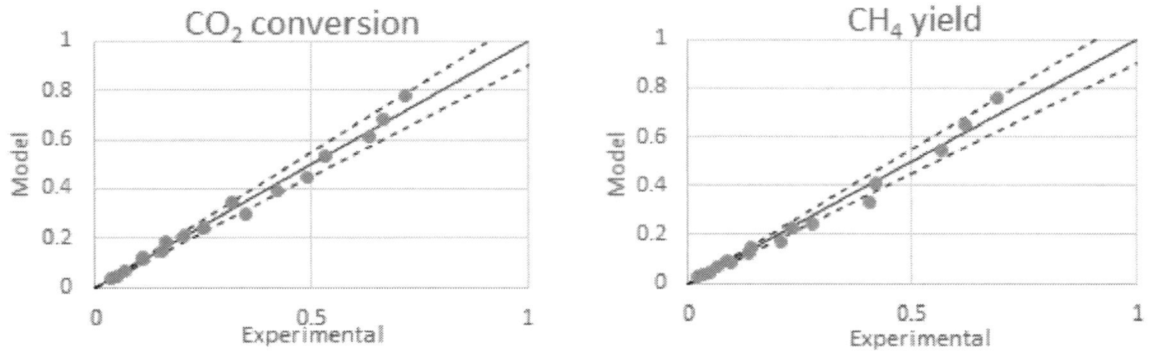

Figure 3 : Parity plots

The same experimental tests are applied on partially deactivated catalyst to see the evolution of the kinetic and adsorption parameters. The different samples of post-mortem catalyst are those obtained after the hydrothermal deactivation, and those obtained after deactivation in reactive conditions. The final purpose is to establish a model of kinetics taking into account the deactivation of the catalyst. The validation of this coupled model of kinetic-deactivation is done on COMSOL Multiphysics, with a solid-gas heterogeneous 1D model to represent all the phenomena inside the reactor and based on mass, energy and momentum balances in the gas phase and mass and energy balances for the solid phase[9].

In conclusion, with the results of experimental tests performed at different scales: from the milligram-scale of catalyst to several grams-scale of catalyst, a coupled model of kinetic/deactivation is established to model the behavior of a CO_2 methanation reactor.

Keywords — Power-to-gas, CO_2 methanation, deactivation, kinetic model, sintering

[1] C. H. Bartholomew, "Mechanisms of catalyst deactivation," *Applied Catalysis A: General*, vol. 212, no. 1, pp. 17–60, 2001.

[2] E. Ruckenstein and B. Pulvermacher, "Kinetics of crystallite sintering during heat treatment of supported metal catalysts" *AIChE Journal*, vol. 19, no. 2, pp. 356–364, 1973.

[3] C. H. Bartholomew, "Sintering kinetics of supported metals: new perspectives

from a unifying GPLE treatment" *Applied Catalysis A: General*, vol. 107, no. 1, pp. 1–57, 1993.

[4] J. Sehested, "Sintering of nickel steam-reforming catalysts: effects of temperature and steam and hydrogen pressures" *Journal of Catalysis*, vol. 223, no. 2, pp. 432–443, 2004.

[5] G. Weatherbee, "Hydrogenation of CO_2 on group VIII metals II. Kinetics and mechanism of CO_2 hydrogenation on nickel" *Journal of Catalysis*, vol. 77, no. 2, pp. 460–472, 1982.

[6] J. Xu and G. F. Froment, "Methane steam reforming, methanation and water-gas shift: I. Intrinsic kinetics" *AIChE Journal*, vol. 35, no. 1, pp. 88–96, 1989.

[7] C. Wheeler, "The water–gas-shift reaction at short contact times," *Journal of Catalysis*, vol. 223, no. 1, pp. 191–199, 2004.

[8] F. Koschany, D. Schlereth, and O. Hinrichsen, "On the kinetics of the methanation of carbon dioxide on coprecipitated $NiAl(O)_x$" *Applied Catalysis B: Environmental*, vol. 181, pp. 504–516, 2016.

[9] R. Try, A. Bengaouer, P. Baurens, and C. Jallut, "Dynamic modeling and simulations of the behavior of a fixed-bed reactor-exchanger used for CO_2 methanation" *AIChE Journal*, vol. 64, no. 2, pp. 468–480, 2018.

545517 A New, Sustainable Visible-Light-Driven Methanol to Ethylene Glycol (MTEG) Process

Monday, June 3, 2019: 3:21 PM
Texas Ballroom EF (Grand Hyatt San Antonio)

Shunji Xie[1], Zebin Shen[1], Jiao Deng[1], Pu Guo[1], Qinghong Zhang[1], Haikun Zhang[1], Chao Ma[2], Zheng Jiang[3], Jun Cheng[1], Dehui Deng[4] and Ye Wang[1], (1)Xiamen University, Xiamen, China, (2)Hunan University, Hunan, China, (3)Shanghai Synchrotron Radiation Facility, Shanghai, China, (4)Dalian Institute of Chemical Physics, Chinese Academy of Sciences, Dalian, China

1. Introduction

Methanol can be derived from a variety of carbon resources via synthesis gas process, such as natural gas or shale gas, coal, biomass and carbon dioxide, and is an abundant and renewable one-carbon (C_1) building block.[1] Several processes such as methanol-to-olefins (MTO) and methanol-to-gasoline (MTG) have been developed for the C–C coupling of methanol into chemicals, but these processes usually show limited selectivity to a specific product.[2] Traditionally, the conversion of methanol involves the activation of its O–H or C–O bond. The system that can selectively activate the unreactive C–H bond of methanol with hydroxyl group intact and form C–C bond is rare.[3] Here, we present the first visible-light-driven dehydrogenative coupling of methanol to ethylene glycol (MTEG) process (Eq. 1).[4]

$$2CH_3OH \rightarrow HOCH_2CH_2OH + H_2 \qquad (1)$$

2. Experimental

Photocatalytic reactions were carried out in a sealed quartz reactor. Reaction conditions: solution, 76 wt% CH_3OH + 24 wt% H_2O, 5.0 cm^3; atmosphere, N_2; light source, 300 W Xe lamp.

3. Results & Discussion

Eq. 1 cannot proceed via conventional catalysis because of the thermodynamic limitation. Photocatalysis is a promising strategy for C–C coupling reactions. We investigated many typical semiconductor photocatalysts for the conversion of CH_3OH. However, over most semiconductors examined, instead of EG, HCHO was formed as a major carbon-based product (Table 1), suggesting that the O–H bond in CH_3OH is easier to be activated. We discovered that a few metal sulfides such as ZnS and CdS could catalyze the formation of EG, and CdS showed better selectivity under visible-light irradiation. Then, we investigated the effect of various co-catalysts. We found that the loading of MoS_2 nanofoam onto CdS nanorods could remarkably enhance the formation rate of EG. The rates of EG and H_2 formations over MoS_2-foam/CdS were about 24 and 16 times higher than those over CdS alone (Table 1). We performed characterizations and found that MoS_2 nanofoam with abundant edge sites accelerates the photocatalytic activity for EG formation by both providing H_2-evolution active sites and enhancing the transfer of photogenerated electrons and hole (Figure 1). The mechanistic studies reveal that CdS-based photocatalyst is quite unique in the preferential activation of the C–H bond in methanol without affecting the O–H group, forming EG via •CH_2OH radical intermediate. High selectivity (90%) and yield (16%) of EG were obtained by designing a process-intensified reactor with EG separation capability.

Table 1. Catalytic performances of typical semiconductor photocatalysts for the conversion of methanol.

Catalyst	Formation rate (mmol g_{cat}^{-1} h^{-1})							e^- / h^+	Selectivity (%)		
	EG	HCHO	HCOOH	CO	CO_2	H_2	CH_4		**EG**	HCHO	HCOOH
					UV-Vis light						
TiO_2	**0**	1.6	0.11	0.16	0.042	2.0	0.053	0.91	**0**	84	5.6
ZnO	**0**	3.0	0.038	0.23	0.028	3.1	0.14	0.90	**0**	91	1.2
g-C_3N_4	**0**	0.79	0.33	0.11	0	1.5	0.039	0.92	**0**	64	27
ZnS	**1.3**	2.2	0.067	0.083	0	3.4	0.087	0.92	**54**	43	1.3
					Visible light						
ZnS	**0**	0	0	0	0	0	0	–	**–**	–	–
Cu_2O	**0**	0.46	0	0	0	0.42	0	0.91	**0**	100	0
Bi_2S_3	**0**	0.13	0.017	0.023	0	0.19	0	0.91	**0**	77	10
CuS	**0**	0.11	0.013	0	0	0.13	0	1.0	**0**	89	11
CdS	**0.46**	0.38	0	0	0	0.75	0	0.90	**71**	29	0
MoS_2-foam/CdS	**11**	2.5	0	0	0	12	0	0.92	**90**	10	0

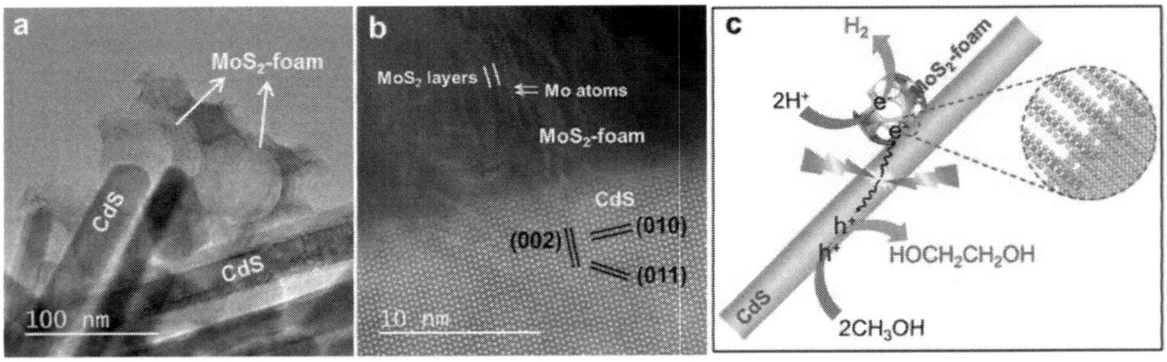

Figure 1. Structural properties of the MoS_2-foam/CdS catalyst. **a,** TEM image. **b,** High-resolution HAADF-STEM image. **c,** Schematic illustration of MoS_2-foam/CdS for photocatalytic synthesis of EG and H_2 from CH_3OH.

4. Conclusions

We have discovered that the CdS-based photocatalytic system is quite unique in the preferential activation of the C–H bond in methanol for subsequent coupling to EG. The loading of MoS_2 significantly improves the formation of EG. The present visible light-driven MTEG process not only offers an atom-efficient method for the synthesis of EG under mild conditions but also opens up a new avenue for preferential C–H bond activation without affecting other functional groups in the same molecule.

References

1. G. A. Olah, *Angew. Chem. Int. Ed.* **52**, 104-107 (2013).

2. U. Olsbye, S. Svelle, M. Bjørgen, P. Beato, Ton V. W. Janssens, F. Joensen, S. Bordiga, K. P. Lillerud, *Angew. Chem. Int. Ed.* **51**, 5810-5831 (2012).

3. S. Zhang, F. Zhang, Y. Tu, *Chem. Soc. Rev.* **40**, 1937-1949 (2011).

4. S. Xie, Z. Shen, J. Deng, P. Guo, Q. Zhang, H. Zhang, C. Ma, Z. Jiang, J. Cheng, D. Deng, Y. Wang, *Nat. Commun.* **9**, 1181 (2018).

545228 The Combined Role of CO and Water on the Stability of Cobalt Fischer-Tropsch Catalysts

Monday, June 3, 2019: 4:00 PM
Texas Ballroom EF (Grand Hyatt San Antonio)

Moritz Wolf, Nico Fischer and **Michael Claeys**, University of Cape Town, Cape Town, South Africa

Supported cobalt crystallites in the nanometer range are being applied industrially in the Fischer-Tropsch synthesis (FTS).[1] Small crystallites have been reported to show a very different resistance with regard to deactivation compared to larger cobalt particles.[2,3] This observation has been attributed to differences in surface energy.[2] The Fischer-Tropsch (FT) product water has been reported to have a severe effect on the deactivation of cobalt-based FT catalysts.[1-8] Three mechanisms have been proposed in literature (see figure 1): oxidation of metallic cobalt to FT inactive cobalt oxide,[2,4,7] hydrothermal sintering of cobalt particles resulting in a loss of specific surface area,[6,8] and formation of FT inactive cobalt-support compounds.[5,7]

An *in situ* magnetometer at the University of Cape Town [3,7-9] was specifically designed for research on such phase transformations in FT catalysts under fully relevant working conditions for the FTS. These phase changes can be identifies via changes of magnetic properties, e.g. from ferromagnetic to anti-ferromagnetic species in the event of metallic cobalt forming oxidic cobalt species. Furthermore, the special magnetic properties of nanoparticles allow for a qualitative and, in some cases even, quantitative analysis of crystallite size changes. Cobalt oxide nanoparticles were prepared separately allowing the application of powerful techniques for the controlled synthesis of well-defined nanoparticles with narrow size distributions. These well-defined nanoparticles were subsequently supported, reduced and tested under various conditions. A series of model catalysts consisting of nanoparticles with well-defined average sizes ranging from 3 to 10 nm on Al_2O_3 or SiO_2 was exposed to different ratios of the partial pressures of the FT product H_2O to H_2 and/or syngas.[3,7]

Co-feeding of H_2O with synthesis gas mimics high FT conversion levels while the actual CO conversion was below 10%. Both model catalyst systems displayed a size-dependent oxidation behaviour with smaller nanoparticles being less stable than larger ones [3,7] as predicted for the oxidation to CoO.[2] However, the formation of cobalt-support compounds, namely cobalt aluminates and silicates, coincided with the formation of CoO.[7] Thus, in addition to these metal oxides, solid carbon spheres were utilised as a support in order to limit the formation of metal-support compounds. The effect of CO and H_2O on the deactivating Co oxidation reaction was tested in H_2O-rich environments simulating high FT conversion levels. Importantly, the reactant CO could be identified as the major oxygen source during water-induced oxidation of cobalt to cobalt oxide, while the oxidation of Co by H_2O-derived oxygen is seemingly kinetically hindered. In another experiment, a newly developed Co/C model catalyst with well-defined cobalt nanoparticles on exfoliated graphite has been developed for testing the stability of well-defined monodisperse Co nanoparticles in H_2O-rich atmospheres. An earlier hypothesised inter-connection between hydrothermal sintering and oxidation [6] of cobalt nanoparticles was captured *in situ*. Utilisation of non-functionalised carbon as support material allowed for the monitoring of the sintering process as the pristine carbon provides limited stabilisation due to weak metal-support interactions.

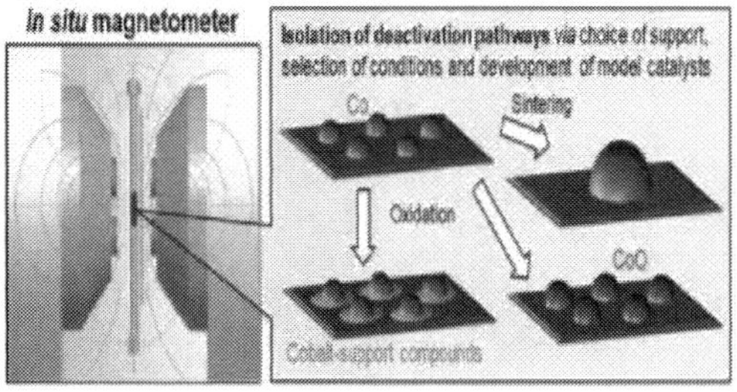

Figure 1: Characterisation of deactivation mechanisms of cobalt-based Fischer-Tropsch model catalysts under high conversion levels utilising an *in situ* magnetometer.

In our studies we could show that deactivation of cobalt based catalysts strongly depends on reaction conditions during the Fischer-Tropsch synthesis. Importantly, there appears to be an interplay between carbon monoxide and water regarding catalyst stability.

[1] M.E. Dry, *Stud. Surf. Sci. Catal.* **152** (2004) 533.

[2] E. van Steen, M. Claeys, M.E. Dry, J. van de Loosdrecht, E.L. Viljoen, J.L. Visagie, *J. Phys. Chem. B.* **109** (2005) 3575.

[3] N. Fischer, B. Clapham, T. Feltes, E. Van Steen, M. Claeys, *Angew. Chem. Int. Ed.* **53** (2014) 1342.

[4] J. van de Loosdrecht, B. Balzhinimaev, J.-A. Dalmon, J.W. Niemantsverdriet, S.V. Tsybulya, A.M. Saib, P.J. van Berge, J.L. Visagie, *Catal. Today* **123** (2007) 293.

[5] N.E. Tsakoumis, J.C. Walmsley, M. R_nning, W. van Beek, E. Rytter, A.J. Holmen, *J. Am. Chem. Soc.* **139** (2017) 3706.

[6] M. Sadeqzadeh, J. Hong, P. Fongarland, D. Curulla-Ferr_, F. Luck, J. Bosquet, D. Schweich, A.Y. Khodakov, *Ind. Eng. Chem. Res.* **51** (2012) 11955.

[7] M. Wolf, H. Kotz_, N. Fischer, M. Claeys, *Faraday Discuss.* **197** (2017) 243.

[8] M. Claeys, M.E. Dry, E. van Steen, P.J. van Berge, S. Booyens, R. Crous, P.

van Helden, J. Labuschagne, D.J. Moodley, A.M. Saib, *ACS Catal.* **5** (2015) 841.

[9] M. Claeys, E. van Steen, J.L. Visagie, J. van de Loosdrecht, US Patent 8,733,118 B2 (2014).

545411 Water As Key to Activity and Selectivity in Co Fischer-Tropsch Synthesis

Monday, June 3, 2019: 4:24 PM
Texas Ballroom EF (Grand Hyatt San Antonio)

Erling Rytter[1], Øyvind Borg[2], Nikolaos Tsakoumis[3], Bjørn Christian Enger[4] and Anders Holmen[1], (1)Chemical Engineering, Norwegian University of Science and Technology, Trondheim, Norway, (2)Equinor, Trondheim, Norway, (3)Department of Chemical Engineering, NTNU, Trondheim, Norway, (4)SINTEF, Trondheim, Norway

Water as key to activity and selectivity in Co Fischer-Tropsch synthesis

Erling Rytter [a,b], Øyvind Borg [a,c], Nikolaos E. Tsakoumis [a], Bjørn Christian Enger, [b] and Anders Holmen [a]

[a] Department of Chemical Engineering, Norwegian University of Science and Technology (NTNU), N-7491 Trondheim

[b] SINTEF Industry, N-7465 Trondheim, Norway

[c] Present address: Equinor Research Center, Trondheim, Norway <>1. Introduction

Supported cobalt catalysts are widely studied and applied for conversion of synthesis gas in low-temperature (< 250 °C) Fischer-Tropsch (FT) synthesis. These catalysts provide high activity, high selectivity to long chain paraffins and low-water-gas shift activity. It is common practice to add a noble metal promoter, typically platinum or rhenium, to optimize the performance of the catalyst. Both cobalt and the promoter are rather expensive, and optimal usage in the reactor is of crucial importance. For all catalyst systems and process conditions, water added to the feed and indigenous water produced during synthesis result in higher selectivity to higher hydrocarbons. The C_{5+} selectivity increases with conversion due to a positive effect of produced water. Apparently, there is moderate effect of *produced* water on the activity. Nevertheless, in a recent paper it was suggested that water plays a key role in activation of CO by providing hydrogen to the oxygen atom.[1] Adsorbed water presumably interacts with CO and lowers the energy barrier for CO activation, thereby enhancing the surface coverage of polymerization intermediates. *Added* water to the syngas may increase or decrease the catalyst activity.[2] More inert supports and large pore size supports have a positive response.[3]

The present work concentrates on Re-promoted Co-catalysts on γ-alumina supports with large variations in pore sizes and pore size distributions,[4] as well as on α-alumina supports, and the effect of water. Differences in pore characteristics result in large variations in measured cobalt particle sizes. Effect of water is studied by adding water vapor during the FT-reaction. An attempt is made to unravel the relationship between pore structure and cobalt size with water response and selectivity. Most data reported are for lab. scale fixed-bed reactors. <>2. Results

Thirteen γ-alumina supports with different pore sizes and pore size distributions, and three supports prepared by calcining γ-alumina at temperatures between 1130 and 1140 C to give α-alumina contents between 84 and 100%, have been impregnated with cobalt and rhenium. The catalysts were tested for Fischer-Tropsch synthesis (FTS) under dry and enhanced water vapor pressure conditions. It is shown that there is a positive correlation between pore size and cobalt crystallite size of the reduced catalyst after incipient wetness impregnation as long as the pore size distribution is sufficiently narrow. There is a concurrent increase in selectivity to C_{5+} with pore diameter due to higher chain propagation probabilities α_{3+}. Higher α values are ascribed to larger cobalt crystallites that promote CO activation resulting in higher surface coverage of C_{Hx} monomers. Adding water vapor to the syngas feed has a strong positive effect on selectivity to higher hydrocarbons and activity. Water imposes a significant enhancement of α_1 that is attributed to higher relative cobalt surface coverage of H_2O(OH-) and C_{Hx} relative to hydrogen. Small pores are susceptible to condensation of water during FTS.

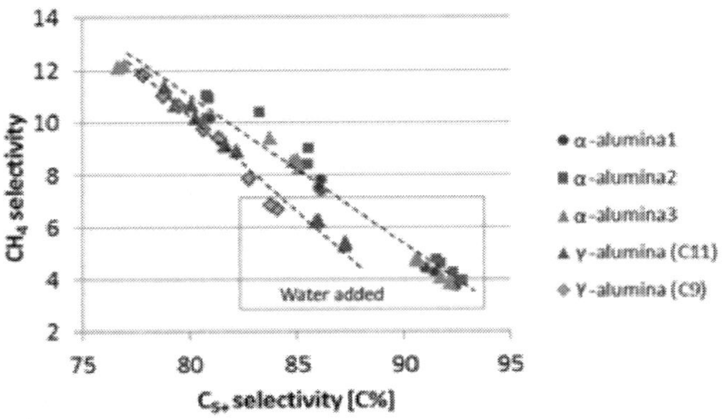

For the α-alumina supported catalysts, both activity and selectivity to higher hydrocarbons respond positively to water added or generated *in situ* by the reaction. A linear trend between formation of methane and C_{5+} products were found, but displaced to higher C_{5+} values compared to catalysts on γ-alumina; see Figure 1.

Figure 1. Methane selectivity plotted as function of C5+ selectivity for catalysts Cα1, Cα2, Cα3, Cγ (C11 and C9). See ref. 4 for notations.

Data points are for start and end of each test period. Five test periods were used: A: Synthesis gas with flow rate 250 ml/min. B: Synthesis gas with reduced space velocity to give an initial CO conversion of ca. 50% at 30 h time-on-stream (TOS). C: Keeping the synthesis gas flow-rate from period B and adding water vapor to give 21% water vapor pressure at the reactor inlet. D: Increasing the water vapor pressure to 35%. E: Returning to the conditions of period B.

Calculation of chain propagation probabilities (α_n) for α-alumina supported catalysts confirms that the first step, characterized by α_1, increases the most under higher water partial pressure. Moreover, α_1 is significantly higher for α- compared to all γ-alumina supports irrespective of pore sizes of the latter; see Figure 2.

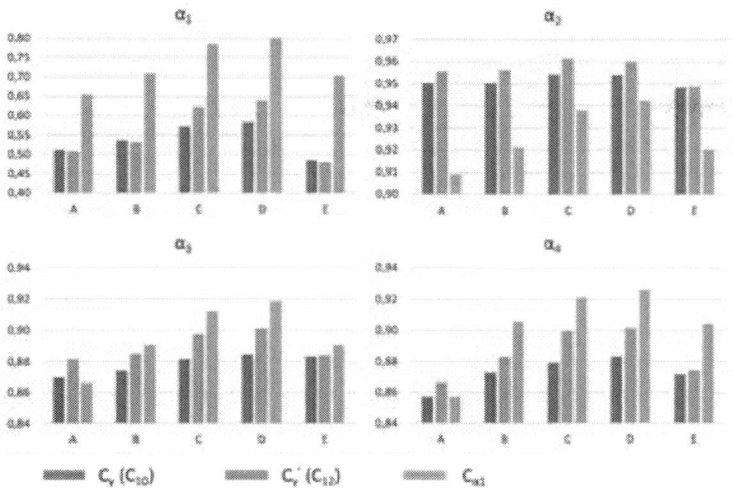

Figure 2. Chain propagation probabilities α_1, α_2, α_3 and α_4 in end of test periods A to E for two γ-alumina and one α-alumina supported catalysts. Catalyst notations in brackets refer to numbering in ref. 4; C10: average pore diameter 11.6 nm, C12: 18.3 nm.

These results, going from γ-alumina to α-alumina and increasing pore sizes, are ascribed to suppression of hydrogen coverage on the cobalt surface, linked to more regular cobalt crystallites, accompanied by enhanced water assisted generation of CH_x polymerization monomers. Surprisingly, the following α_2 probability is comparably low for α-alumina supports, although it increases significantly with water concentration. Linear correlations are found between each pair of parameters α_1, α_2, α_4 and S_{C5+}; giving support to a mechanistic model where all products are interlinked, including methane. Transients observed when water was added or removed from the system are ascribed to pore diffusion. Selectivities in these periods follow closely the general selectivity trends found for different process conditions.
<>3. Conclusion

4. Cobalt particle size increases with pore size of the support material, specifically γ-alumina, when deposited through incipient wetness impregnation, drying and calcination.

5. For broad or bimodal pore size distributions, cobalt preferably is deposited in the larger pores by the incipient wetness impregnation method.

6. Site-time yield is essentially constant for cobalt crystallites in the range 8-13 nm for supports with large variations in pore size and pore size distribution.

7. Broad pore size distribution in the catalyst causes a strong negative impact on catalyst activity when water vapor is added to the syngas feed.

8. Narrow pores are susceptible to condensation of water although direct experimental evidence is lacking.

9. The negative impact of water vapor on activity is strengthened for catalysts with narrow average pore size.

10. Well adapted pores size distribution, wide pores and sufficiently large cobalt crystallites causes a positive response on activity during addition of water vapor in moderate concentrations.

11. Water vapor pressure increases the selectivity to higher hydrocarbons with no apparent limit for the partial pressure.

12. C_{5+} selectivity increases with pore diameter of the catalyst.

13. There is a positive, but moderate, correlation between C_{5+} selectivity response to added water and pore diameter.

14. A concurrent response for C_{5+} selectivity and activity on added water implies a common mechanistic origin suggested to be water activation of CO which impacts the concentration of polymerization monomers.

15. The chain growth probabilities α_{3+} increase significantly as the pore diameter becomes wider. It follows that larger cobalt crystallites are more effective in CO activation.

16. Water increases the α_1 value significantly. This is attributed to suppressed hydrogen coverage on cobalt.

17. Increase in C_{5+} with water partial pressure is mainly due to higher α_1 and only to a minor extent to α_{3+}.

18. Deactivation by water reduces C_{5+} and increases α_1 indicating higher hydrogen coverage.

Unique properties of catalysts on α-alumina are:

- Very high α_1 values at all process conditions
- Low α_2 that respond positively to water vapor concentration
- Higher α_3 and α_4 probabilities that follow trend from γ-alumina supports.

<>4. References

[1] E. Rytter and A. Holmen, Consorted vinylene mechanism for cobalt Fischer-Tropsch synthesis encompassing water or hydroxyl assisted CO-activation. Topics Catal., 61 (2018) 1024-1034.

[2] E. Rytter and A. Holmen, Perspectives on the Effect of Water in Cobalt Fischer–Tropsch Synthesis. ACS Catal. 7(8) (2017) 5321.

[3] E. Rytter and A. Holmen, On the Support in Cobalt Fischer-Tropsch-Emphasis on Alumina and Aluminates. Catal. Today 275 (2016) 11-19.

[4] E. Rytter, Ø. Borg, N. E. Tsakoumis and A. Holmen, Water as key to activity and selectivity in Co Fischer-Tropsch synthesis; γ-alumina based structure-performance relationships, J. Catal., 365 (2018) 334-343.

545399 Deactivation of the Cobalt Fischer-Tropsch Catalyst - a Kinetic Study

Monday, June 3, 2019: 4:48 PM
Texas Ballroom EF (Grand Hyatt San Antonio)

Ljubisa Gavrilovic[1], Jan Brandin[2], Anders Holmen[1], Hilde Venvik[1], Rune Myrstad[3], Kumar Ranjan Rout[3], Erling Rytter[1], Magne Hillestad[1] and Edd Blekkan[1], (1)Chemical Engineering, Norwegian University of Science and Technology, Trondheim, Norway, (2)Department of Built Environment and Energy Technology, Linnæus University, Växjö, Sweden, (3)SINTEF Industry, Kinetics and Catalysis Research Group, Trondheim, Norway

12th Natural Gas Conversion Symposium

Deactivation of the Cobalt Fischer-Tropsch catalyst – a Kinetic Study

Ljubiša Gavrilović[a], Jan Brandin[b], Anders Holmen[a], Hilde J. Venvik[a], R. Myrstad[c], Kumar R. Rout[c], Erling Rytter[a], Magne Hillestad[a], Edd A. Blekkan[a,*]

[a]*Norwegian University of Science and Technology, Department of Chemical Engineering, Sem Sælands vei 4, 7491, Trondheim, Norway*

[b]*Linnæus University, Department of Built Environment and Energy Technology, 351 95 Växjö, Sweden*

[c]*SINTEF Materials and Chemistry, NO-7465 Trondheim, Norway*

Biomass to liquid (BTL) via gasification and integrated Fischer-Tropsch synthesis (FTS) is an attractive process for production of green diesel and jet fuel. The first step involves biomass gasification to produce syngas ($CO + H_2$)[1]. It was reported[2] that impurities in biomass feedstocks and partial gasification can lead to the contaminants (such as: tars, alkali, HCN, H_2S, etc.) in the produced syngas, which might cause catalyst deactivation and downstream problems. Therefore, series of cleaning steps and novel process design must be applied to achieve the required purity of the gas before the FTS reactor[3]. Alkali salts (mainly potassium) are the dominant salts in the fly ash composition after biomass gasification. Due to poor plant design or operational upsets in the cleaning section, these alkali salts might be present in the effluent gas composition in the produced syngas[3].

In order to do a good and realistic design of the BtL plant, it is necessary to have a kinetic model that captures major variations of reaction rates and selectivity subject to changes in process parameters. There are numerous kinetic models available in literature predicting reaction rates of the FT and there is a diversity of different model structures, due to different catalyst type, support, promoter, pore size etc.[4]. The product distribution from the FT is often described by the Anderson-Schulz-Flory (ASF) distribution, where the distribution is defined by a single parameter. In order to perform a FT reactor design with realistic evaluation of the products produced, there is a need for a way to quantify the product distribution. Hydrocarbon selectivity depends on local conditions like temperature, pressure, species concentrations (CO and H_2) and water produced. Therefore, a suitable growth model should include these variables. It is surprising that none of the models in literature have water partial pressure as a variable, which is found to have a significant positive impact on C_{5+} selectivity[5]. Recently, we have developed a chain growth model based on the data of Yang *et al.* [6](18 points), where effects of temperature, partial pressure of H_2O, H_2 and CO are taken into account. However, the model is not validated for a larger dataset covering the necessary range of variables, and also, the results from Yang et al.[6] do not include the effect of water and H_2/CO variation. The aim of the current work is to provide a detailed kinetic study, including effects of pressures (also water), and temperature. It is established that oxidation of bulk cobalt metal does not occur under typical FTS conditions, however small metal particle properties may differ considerably from those of bulk cobalt metal. Thus, a deactivation mechanism by the FT water byproduct might be thermodynamically favorable[7]. The water could also facilitate mobility of alkali species to the specific active sites on the Co catalyst[8]. A kinetic model encompassing catalyst deactivation by alkali will be developed.

A 20%Co/0.5%Re/γAl_2O_3 Fischer-Tropsch catalyst was poisoned by potassium salt (KNO_3) using the aerosol deposition technique, depositing up to 3500 ppm K as solid particles. Standard characterization techniques (H_2 Chemisorption, BET, TPR) showed no significant differences between poisoned samples and the reference catalyst. The Fischer-Tropsch activity, investigated at industrially realistic conditions (210 °C, H_2/CO = 2:1, 20 bar), showed a reduction in the site time yield (STY) with increasing potassium loading[8]. The selectivity towards heavier hydrocarbons (C_{5+}) was slightly increased, while the CH_4 selectivity was reduced with increasing potassium loading. In order to be able to predict the catalyst deactivation with potassium, the activity data were kinetically analyzed. The proposed simple (power rate law) kinetic model[9] developed for a supported cobalt catalyst is given in (1) :

$$r_{CO} = k^o {p_{H_2}}^{0.74} {p_{CO}}^{-0.24} \quad (1)$$

In expression (1), r_{CO} (ml/h/gcat) is the rate of CO consumption; p_{H_2}, p_{CO} (Pa) are the partial pressures of reactants and k^0 (ml/h/Pa$^{0.5}$/g$_{cat}$) is the kinetic constant. To be able to predict the catalyst deactivation with potassium, the expression (1) was modified with a term in the expression of the kinetic constant depending on the K loading:

$k = k^o f(K)$

where K is the potassium concentration on the catalyst. Three simple empirical equations[8] were selected to determine the effect of potassium on the catalyst activity:

Kinetic model	1) $k = k_0(1 - m * K)$	2) $k = k_0 exp(-m * K)$	3) $k = k_0/(1 + m * K)$

In this table K is the potassium impurity loading expressed in mg_K/kg_{cat}, and k^o is the kinetic constant of the reference sample. The value of the parameters m was estimated by means of a linear regression of CO conversion data[8]

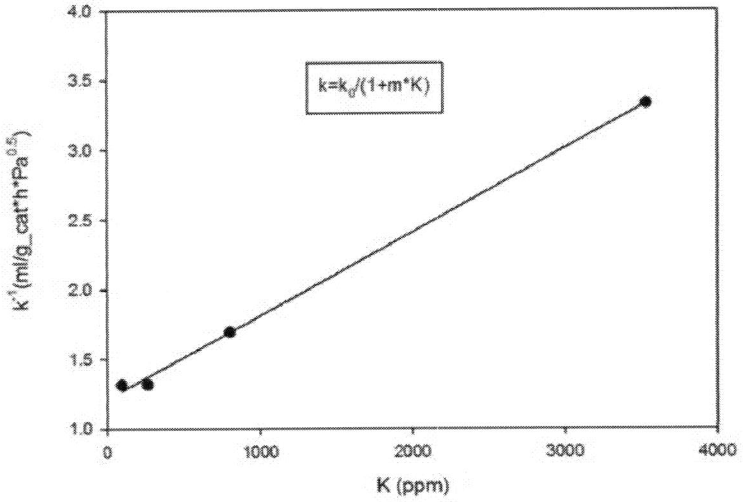

Figure 1. Linear regression of the deactivation rate equations

The results indicate that all the models fit the experimental results quite well, however the best fit was obtained with model (3), Figure 1. A hypothesis of selective K-poisoning of the Co FT catalyst is in the line with the strong effect of potassium on CO conversion. However, the number of data-points is low, so the apparent agreement does not constitute evidence of a mechanism, but the result can be used as a tool to predict the effect of potassium on the surface.

Further experiments are underway for a detailed kinetic study and kinetic model development including the deactivation behavior. The validated and fitted model can be implemented in a process simulation code thereby a realistic BtL plant can be designed in future.

[1] E. van Steen, M. Claeys, Chem. Eng. Technol. 31 (2008) 655–666.

[2] P.J. Woolcock, R.C. Brown, Biomass and Bioenergy 52 (2013) 54–84.

[3] H. Boerrigter, H.P. Calis, D.J. Slor, H. Bodenstaff, 2nd World Conf. Technol. Exhib. Biomass Energy, Ind. Clim. Prot. (2004) 10–14.

[4] M. Ostadi, E. Rytter, M. Hillestad, Chem. Eng. Res. Des. 114 (2016) 236–246.

[5] E. Rytter, A. Holmen, ACS Catal. 7 (2017) 5321–5328.

[6] J. Yang, S.B. Eiras, R. Myrstad, H.J. Venvik, P. Pfeifer, A. Holmen, in: Fischer-Tropsch Synth. Catal. Catal. Adv. Appl., 142 (2016) 223–242.

[7] C.E. Kliewer, S.L. Soled, G. Kiss, Catal. Today (2018) In Press, https://doi.org/10.1016/j.cattod.2018.05.021.

[8] L. Gavrilović, J. Brandin, A. Holmen, H.J. Venvik, R. Myrstad, E.A. Blekkan, Ind. Eng. Chem. Res. 57 (2018) 1935–1942.

[9] R. Zennaro, M. Tagliabue, C.H. Bartholomew, Catal. Today 58 (2000) 309–319.

545478 Conversion of Natural Gas to Energy through PEMFC to Apply at Commercial Application

Monday, June 3, 2019: 5:15 PM
Texas Ballroom D (Grand Hyatt San Antonio)

Abhi Soni[1], **Hiralkumar Morker**[1], **Naim Memon**[1], Vaishali Umrigar[1] and **Sopariwala Vishnu Manoj**[2], (1)Chemical Engineering, Sarvajanik College of Engineering & Technology, Surat, India, (2)Gujarat Technological University, Surat, India

Conversion of Natural Gas to Energy through PEMFC to apply at Commercial Application

Vaishali Umrigar, Abhi Soni*, Hiralkumar Morker, Naim Memon

Chemical Engineering Department,

Sarvajanik College of Engineering & Technology, Surat, India.

*Email:soni.abhi705@gmail.com

Due to depletion of fuel and the scarcity of crude in coming years, inclination towards alternative energy sources lead us to think for an alternative to the existing fuels' replacement. Also as many industrial processes, combustion processes releases excess amount of natural gas. This amount of natural gas is burned off or flared in the atmosphere. Much of such gas is being wasted which worth billions of dollars. Hence Natural gas can be utilized for the energy production and alternative to fuels' for automobile Industries. To fulfil this objective, the scope of the study targets the following project execution at application level.

1) Separation of Hydrogen from Natural Gas

2) Synthesis of Polymeric membrane

a. Synthesis of Nafion Membrane

b. Synthesis of alternative to Nafion membrane

c. Modification of existing Nafion synthesis process

3) Assemble the Polymeric electrolytic material Fuel Cell (PEMFC) for Electricity generation

4) Energy production trials and engine performance with Natural gas as a fuel

Each process from literature and well established mechanism has been explained in detail:

Step: 1

i. Removal of H_2S from Natural gas: - In natural gas two types of sulphur content are present such as H_2S and organic sulphur. H_2S is absorbed by zinc oxide. Organic sulphur is converted into inorganic sulphur (H_2S) by CoMoX (Cobalt Molybdenum Oxide) and absorbed. Many processes are available to remove Hydrogen sulphide.

ii. Reforming of gas to convert hydrocarbons into H_2 gas: - In reformer, natural gas is heated at high temperature (800-850 C) with steam to convert methane into hydrogen with CO & CO_2. Separation of H_2 in a pure form can be achieved by the following mechanism.

iii. Conversion of CO to CO_2:- To remove CO, it can convert into CO_2 by reacting it with steam at high temperature and where more yield of H_2 is achieved.

iv. Removal of CO_2:- After conversion of CO into CO_2 and less amount of H_2S, it can absorbed by Benfield solution (K_2CO_3) and form bicarbonate and bisulphide. Hence pure H_2 is ready for the use in for the process.

After this 4 step process hydrogen is formed and can be used in PEMFC for electricity generation.

Step: 2

In PEMFC, membrane is main component to transfer electrons for production of electricity. Now a day, Nafion is mainly used as PEM because of its better chemical stability, higher conductivity and oxidative stability. The most successful result of Nafion is its lifetime, which is four times higher than the crossed-linked polystyrene membranes about 50000 hours at 80 °C.

Nafion synthesis:

Nafion is synthesized in four steps process as shown below:

i. Reaction of tetrafluoroethylene with SO_3 to form cyclic sulfone

ii. Condensation reaction with Na_2CO_3 followed by the copolymerization of perfluoro vinyl ether sulfonyl fluoride with tetrafluoroethylene to afford an insoluble thermoplastic resin

iii. Hydrolysis of sulfonyl fluoride extruded film to form a perfluorosulfonic polymer

iv. Exchange of sodium counter ion with a proton in a suitable electrolyte

Due to serious disadvantages of high cost, low operating temperature (<80 °C), and high fuel crossover, various research groups have focused to develop some alternative polymer materials. Polymeric electrolytic materials (PEM) with low cost of synthesis and better performance under fuel cell operating conditions can be the alternative to overcome these drawbacks. These materials include sulfonatedpoly (phenylene oxide), polysulfone, sulfonated poly (ether-ether ketone), sulfonated poly (benzimidazole), phosphoric acid doped polybenzimidazole, and sulfonated polyimide. Polymeric materials such as sulfonated poly (arylene ether ether ketone) (PEEK) and polysulfones (PS) show good chemical and thermal stabilities under the real fuel cell conditions but they have the disadvantage of lower proton conductivities due to the weak acidity of the aryl sulfonic acid group than Nafion.

These are the alternatives of Nafion membrane but the cost of Nafion can also reduce by modification in process. The modification in process of Nafion is by Sol-gel method using SiO_2 which can increase characteristics and also the cost of production. Therefore, to find out alternative of Nafion is quite challenging for better optimization of PEMFC.

Objective will be satisfied by the modification in Nafion synthesis with economical process.

Step: 3

Once pure H_2 from natural gas is extracted and proper alternative for synthesis of modified nafion is identified with trial and error by doing experiments, the next target is assembling of this to frame the PEMFC as per the process available. It is formed with anode-cathode plate for the flow of O_2 and H_2 to produce heat and electricity.

The process deals again with feasibility check and experiments done with the installed set up. The small-lab scale set up will be commissioned at the college and the experiments will be performed based on the target objective. This project the group has targeted to present as their UG project as a part of graduation. It will be more focused on end application of this study.

The basic aim of this project is to generate electricity using PEMFC with natural gas as feed. Proton Exchange Membrane Fuel Cell (PEMFC) for is chosen due to its favourable characteristics. For efficient way of natural gas conversion to energy, PEMFC proposed has high efficiency of electricity production. Different operating parameters will be identified such as surface area, temperature, pressure, flow rate, concentration with a fruitful design of the system. The expected efficiency of above mentioned fuel cell is 45-60% when operated on optimum parameters.

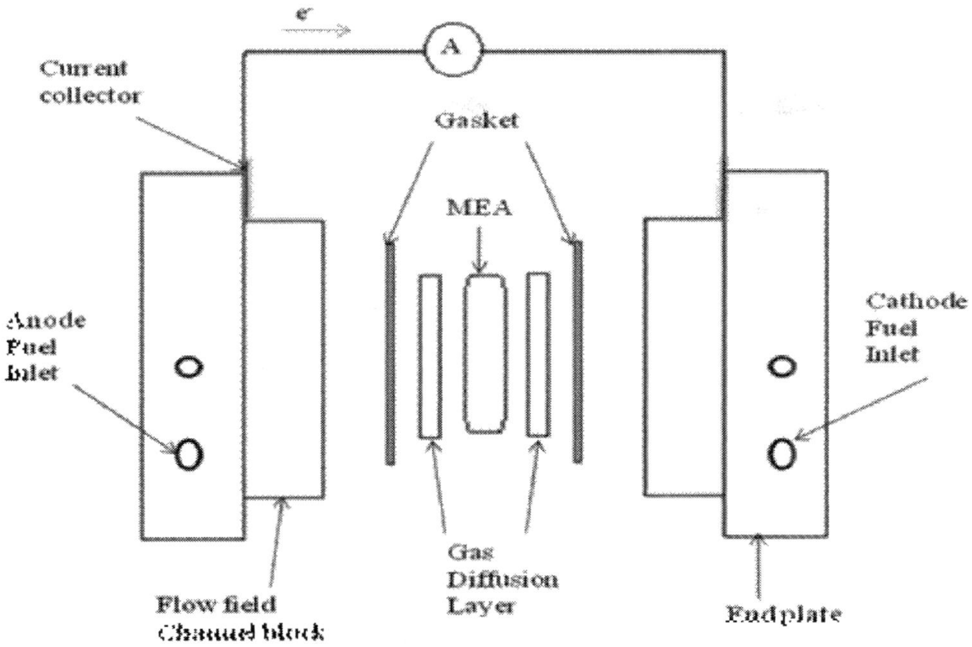

Figure 1 : PEMFC assembly

Step: 4

Applications need the stability and efficiency of the PEMFC to generate energy from Natural gas. Sources of PEM will be tried first to compare the efficiency. Then Fuel Cell (FC) will be assembled with a high capacity to replace fuel engine in automobile. Engine performance only can give final application to move beyond our explanation of this project. Target till assembly will be completed by December 2018. And trials and data will be generated by March 2019. The new abstract can further submitted with data inputs with energy production and PEMFC efficiency.

Keywords: Natural gas, PEM, PEMFC, Nafion, Feasibility

545481 Development of a Highly Efficient Ammonia Synthesis Using Base Metals in an Electric Field

Monday, June 3, 2019: 5:18 PM
Texas Ballroom D (Grand Hyatt San Antonio)

Kota Murakami, Yuta Tanaka, Ryuya Sakai, Tomohiro Yabe, Shuhei Ogo and Yasushi Sekine, Applied Chem., Waseda Univ., Tokyo, Japan

Ammonia is an important chemical feed stock which is mainly utilized as a resource of fertilizers, fibers and resins. These days, it is also anticipated as a hydrogen carrier owing to its high weight-based hydrogen content. The main ammonia synthesis process to date, Haber-Bosch process, requires high temperature (673-873 K) and high pressure (20-40 MPa) because of its thermodynamic and kinetic limitations.

Recently, our group reported the activation of ammonia synthesis using Ruthenium (Ru)-based catalyst by applying the electric field. It is revealed that the surface protonics in the electric field promote cleavage of nitrogen triple bond, which is the rate-determining step in ammonia synthesis. Theoretical calculations and *in-situ* IR measurements revealed that the N_2 dissociation proceed through N_2H as a specific intermediate in the electric field.

Regarding conventional heterogeneous catalysts, Ru-based heterogeneous catalyst exhibits higher turn-over-frequency (TOF) than other metals catalyst like Fe, Co, Mo and Ni. However, the ammonia synthesis mechanism in the electric field is different from conventional ones. Therefore, the activity trend would change by applying the electric field. Herein, we investigated the ammonia synthesis in the electric field over various active metals (Ru, Fe, Co, Ni, Pd, Pt). As a result, Fe-supported or Ni-supported catalysts exhibited higher TOF than Ru-supported catalyst in the electric field at low temperature region. Although the trend without the electric field was Ru>Fe>Co>Ni>Pd=Pt, the trend in the electric field was Fe>Ni>Ru>Co>Pd>Pt. N_2 dissociation energy, N_2H formation energy and N_2H cleavage energy over each metal was calculated using DFT method to identify the reasons for the specific activity order in the electric field. Results show that ammonia synthesis rate order in the electric field is determined by the N_2H formation energy over active metals.

545697 Analysis and Optimization of Gas to Liquid (GTL) Process Using Unconverted Syngas Recycle

Monday, June 3, 2019: 5:21 PM
Texas Ballroom D (Grand Hyatt San Antonio)

Saiedeh Arabi, Sandeep Patil and Brian Dennis, Mechanical & Aerospace Engineering, The University of Texas at Arlington, Arlington, TX

Gas to Liquid (GTL) process converts chemically natural gas to syngas and could be separated into different hydrocarbon fractions. GTL process comprised of three units: 1) Syngas production 2) Fischer-Tropsch synthesis (FTS) 3) Product separation. Partial oxidation of natural gas, steam methane reformer, and FTS are three main units which are simulated in this study. The developed simulation of auto thermal and Fischer-Tropsch reactors were carried out to get jet fuel from natural gas with different steam flow rates using Aspen HYSYS. The GTL pilot plant data was in reasonable agreement with simulation results. The recycling simulation of unconverted syngas was optimized. In this research, the effect of different conditions of feed stream such as: H_2O/CH_4, O_2/CH_4, and feed temperature on molar ratio of H_2/CO through the ATR were investigated. In addition, the different temperature and pressure operation conditions of syngas stream on syngas conversion and jet fuel selectivity were analyzed in fixed bed reactor based on cobalt catalyst. The results show that by increasing O_2/CH_4 and H_2O/CH_4 ratio the reduction and raising of H_2/CO ratio is gained respectively in GTL process. Also, the optimum operation condition of GTL process is 250 deg C.

545716 CFD Study of Low Temperature Fischer-Tropsch Reactor in Jet Fuel Application

Monday, June 3, 2019: 5:24 PM
Texas Ballroom D (Grand Hyatt San Antonio)

Saiedeh Arabi, Sandeep Patil and Brian Dennis, Mechanical & Aerospace Engineering, The University of Texas at Arlington, Arlington, TX

Fischer–Tropsch (FT) synthesis consists of hydrodynamic, chemical and thermal interactions. The optimal dimension of Fischer-Tropsch reactor is presented in this study. Drastic increasing temperature through FT reactor because of the high exothermic reactions nature, take place at the entrance region of the reactors. Temperature control is critical to the process of converting syngas to jet fuel. A steady-state computational fluid dynamics (CFD) simulation has been developed to investigate the effect of each tube size and coolant flow rate on moderate temperature profile. Increasing in tube size causes the thermal runaway because of weak heat transfer flux. The syngas consumption rate was determined by Langmuir kinetic model and the coolant heat flux were estimated to reduce deviations from the pilot plant data.

546770 Catalyst and Process Development for Reactions with Gas Recycle: Challenges and Opportunities

Monday, June 3, 2019: 5:27 PM
Texas Ballroom D (Grand Hyatt San Antonio)
Peter Kolb, R&D Solutions, hte GmbH, Heidelberg, Germany

Methanol is an extremely important chemical component, with a worldwide production capacity exceeding 100 Mtpa. It can be produced in a highly selective manner from synthesis gas, which in turn is available from a large variety of carbonaceous feedstocks. Methanol has a large range of applications. It is a starting material in the production of formaldehyde, acetic acid, dimethyl ether, or hydrocarbons via several MTX processes. It can also be used as a fuel or fuel additive and can also serve as an energy carrier in fuel cells. With these properties, methanol can play a pivotal role in an energy and chemical economy.

The industrial production of methanol from synthesis gas is a well-established process. Nevertheless, there is an ongoing need for process related research to test established or new catalyst formulations for the suitability with new feedstocks, and to find regions in the process parameter space with improved productivity or stability.

In this contribution, we present a study of a commercial methanol catalyst that was carried out under industrially relevant conditions (CO_2 ranging from 0 – 6 vol.-%, GHSV 3150 – 13000 h^{-1}, temperature 200 – 250 °C, pressure 40 bar) in a bench scale test system. The catalyst was tested both in once through mode and in recycle mode. In the latter case, a variable fraction of the reactor effluent was fed back to the reactor inlet and, together with fresh feed gas, passed over the catalyst bed again.

In once through mode, the CO_2 concentration, the space velocity, and the reaction temperature were varied. In the kinetically controlled regime (210-230 °C), the methanol formation rate was mainly determined by the reaction temperature. At higher temperatures (around 250 °C), the methanol formation rate final reached a broad maximum and then started to decrease, indicating operation at thermodynamic control. The formation rates of byproducts (methane, DME, higher alcohols) started to increase. However, the selectivity towards methanol remained on a high level (> 99%)

A special emphasis will be out on the results obtained in recycle mode. Operation in recycle mode dramatically increases the atom efficiency and the productivity of the process. Due to the back feeding of the condenser effluent to the reactor inlet, only a small fraction of the fed synthesis gas leaves the reactor system unconverted. On the other hand, the recycle operation increases the complexity of controlling the process parameters and keeping the entire reactor system in a productivity-maximized steady state. Overall, stable steady state operation was achieved both in once-through and recycle mode, with a maximum productivity of 1 g(MeOH)/(g(Cat)/h observed in recycle mode.

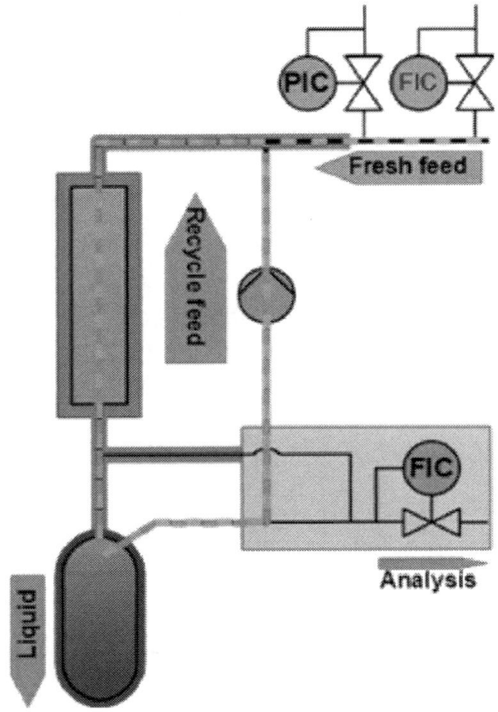

Figure 1. Once Through Mode

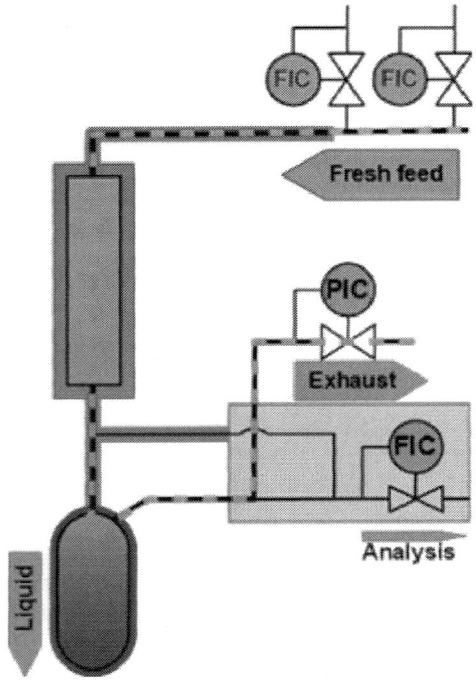

Figure 2. Recycle Mode

545625 Effect of Potassium on Highly Dispersed Iron Nanoparticles Support on Carbon for CO Hydrogenation

Monday, June 3, 2019: 5:42 PM
Texas Ballroom EF (Grand Hyatt San Antonio)

Joakim Tafjord, Department of Chemical Engineering, Norwegian University of Science and Technology, Trondheim, Norway

Effect of potassium on highly dispersed iron nanoparticles support on carbon for CO hydrogenation

*Joakim Tafjord[1], Rune Myrstad[2], Anders Holmen[1], Jia Yang[1]**

[1]Department of Chemical engineering, Norwegian University of Science and Technology, NO-7491, Trondheim, Norway

[2]SINTEF Materials and Chemistry, N-7463 Trondheim, Norway

** Corresponding author: jia.yang@ntnu.no*

1. Introduction

Fischer-Tropsch synthesis is a versatile process in terms of hydrocarbon production. Cobalt and iron are the most commonly employed catalysts in industry, where the former is used for long-chain paraffin production and the latter for wax formation as well as olefin production, depending on operation temperature. Iron also has the added benefit of water-gas shift (WGS) activity, utilizing the H_2O by-product to form additional H_2, thus enabling the use of H_2/CO lower than the stoichiometric number of the reaction. The role of the different iron phases present during reaction remains controversial, but the presence of iron carbides are clearly linked with high FTS activity. The bulk carbon from iron carbides might participate in a Mars-van Krevelen-like reaction, as suggested by both experimental work and DFT. [1,2]

In this work we employ an iron-carbon based catalyst (Fe-C) derived from pyrolysis of iron-containing polymers, with a narrow size distribution of iron particles. The effect of potassium promotion on CO hydrogenation is investigated using Steady-State Isotopic Transient Kinetic Analysis (SSITKA).

2. Experimental

The unpromoted catalyst was prepared by pyrolysis of iron-containing polymers, resulting in iron nanoparticles with a narrow size distribution in a carbon matrix. Potassium promotion was introduced by subsequent incipient wetness impregnation of unpromoted catalyst with an aqueous solution of $KHCO_3$. The sample was dried, then pyrolyzed in N_2 at 500°C, with a ramp rate of 2°/min and dwell for 4 hours.

The supported Fe nanoparticles were investigated by powder X-ray diffraction (XRD) (Bruker D8 Advance Da Vinci, Cu Kα, 20-80 2θ, V6 slit) and Transmission Electron Microscopy (200k eV, Tecnai). The apparent surface areas of the catalysts were measured by N_2-physisorption (Tristar-3020) applying the BET-equation, while Fe and K loadings were determined by ICP-MS (ELEMENT 2).

SSITKA experiments were carried out in a fixed-bed quartz reactor (44 mm i.d.). 150 mg of the catalyst (90-253 μm) was mixed with 150 mg of inert silicon carbide of the same fraction, to improve heat transfer in the reactor. The catalyst was reduced in H_2/Ar (10/10 mL/min) with a ramping rate of 3º min^{-1} to 350°C. After 2 hours of reduction, the temperature was reduced to 340°C, and syngas H_2/CO/Ar (15/3/7 mL/min) was introduced along with a pressure increase to 1.85 bar. The reactants and products were recorded with an online gas chromatograph (GC) equipped with a TCD and FID. After 12 hours on stream, an isotopic switch from $H_2/^{12}CO$/Ar to $H_2/^{13}CO$/Kr was performed, and the isotopic transient response was recorded with a Balmer QMG-422 quadrupole mass spectrometer (MS). A 16-loop storage was also used during isotopic switching, and subsequently analyzed with a GC-MS. The isotopic distribution of C_2-C_4 hydrocarbons was calculated by linear combination of the fragmentation patterns of the corresponding isotopic products. [3]

3. Results and discussion

The loading of the base catalyst (Fe-C_0.0) was 30 wt. %, with iron nanoparticles (Fe_3O_4 + Fe) well dispersed (average size of 3 nm) in a porous carbon matrix (350 m^2g^{-1}). The ICP-MS results confirmed that the desired potassium loading was achieved. These results are summarized in Table 1.

Table 1. Physical characterization

Sample	Iron phase	Fe particle size [nm]	Surface area [m^2g^{-1}]	Fe loading [wt. %]	K loading [wt. %]
Fe-C_0.0	Fe_3O_4 + Fe	3.05	353	30.0	0.0
Fe-C_0.2	Fe_3O_4 + Fe	-	350	31.5	0.2

The CO hydrogenation results show a relatively low selectivity towards methane given a H_2:CO ratio of 5.0 and 1.85 bar pressure, and also good selectivity towards lower olefins (25-27% of total carbon, 45-51% of total hydrocarbons) (Table 2.). The potassium promotion leads to slightly higher olefin selectivity, but also lowers overall activity. The CO_2 selectivity is also higher for the promoted sample, indicating higher WGS rates.

Table 2. Selectivity at 340ºC, H_2/CO ratio of 5 after reaching steady state (16 h), averaged over 8 analysis.

	X_{CO} [%]	CH_4 [%]	C_{2-4}p [%]	C_{2-4}o [%]	C_{5+} [%]	CO_2 [%]	C_2 o/p	C_3 o/p	C_4 o/p
Fe-C_0.0	7.4	16.0	7.9	24.9	8.8	44.5	1.9	5.7	6.5
Fe-C_0.2	6.2	15.7	7.5	26.9	5.2	47.1	2.2	6.4	6.5

The SSITKA technique is able to measure site coverage and residence time of the intermediates, and therefore it is very powerful technique to distinguish the contribution of the two on the overall reaction rate. The residence time and site coverages of CO, CH_x, C_2H_4, C_3H_6 and C_4H_8 intermediates are shown in Table 3.

Table 3. Residence times and site coverages of unlabeled species when switching from ^{12}CO to ^{13}CO. CH_4 and CO measurements were performed 4 times with MS, while C_2-C_4 results are from one GC-MS loop analysis. All residence times are corrected for the chromatographic effect of Ar, and hydrocarbons are corrected by $\frac{1}{2}\tau_{CO}$. A dispersion of 10% was assumed for both samples when calculating surface coverages. Standard errors for the last digit are listed in parenthesis where applicable.

	τ_{CO} [s]	τ_{CHx} [s]	τ_{C2o} [s]	τ_{C3o} [s]	τ_{C4o} [s]	θ_{CO}	θ_{CHx}	θ_{C2o}	θ_{C3o}	θ_{C4o}
Fe-C_0.0	1.5(3)	12.5(5)	87.0	77.3	90.7	0.040(7)	0.0043(2)	0.019	0.015	0.008
Fe-C_0.2	2.2(3)	11.8(3)	74.7	67.1	81.1	0.058(9)	0.0032(1)	0.014	0.011	0.006

The effect of potassium promotion has been debated; it has been suggested to act as an electronic promotor through donation of negative charge, or to cause blocking of active sites. The effect could be differentiated by comparing the effect of potassium on residence times and site coverages. The difference between the two samples show a greater relative difference in CH_x and olefin site coverage than residence time. These results suggest that electronic changes relating to surface adsorption are small, while the sites coverages are more severely reduced. The dispersion is important in the calculation of site coverage, and here we have assumed it to be equal for both catalyst. If the additional pyrolysis step performed on the promoted sample influences particle size, the comparison of site coverages falls short.

When performing a switch from ^{12}CO to ^{13}CO, the iron carbides will gradually increase the amount of interstitial ^{13}C over time. The process of exchanging one isotope of interstitial carbon with another is a much slower process than exchanging surface species. With a Mars-van Krevelen like mechanism, one would expect a lingering effect on the isotopic product distribution. Here we observe a slow depletion of isotopic products, as seen on the left in Figure 1 with C_3-olefins for the unpromoted catalyst, where particularly $^{12}C^{13}C_2H_6$ still have a relatively significant contribution until 4 minutes. In cobalt where the active phase is mainly metallic, one would expect the depletion to be much quicker. This is illustrated in Figure 1 with the Fe-C_0.0 to the left, and 20 wt.%Co 0.5 wt.%Re/ γ-Al_2O_3 from Ledesma et al. on the right. [3] From these results it would appear that a Mars-van Krevelen-like mechanism is present, but it isn't apparent to what degree it contributes to the overall activity.

At CO hydrogenation conditions, the effect of 0.2 wt. % potassium promotion is not very significant. To confirm the observed effect, higher loadings (0.6 wt. %, 1.0 wt. %) of potassium will be tested to observe greater differences on the effect of promotion. In addition, we plan to investigate the effect of the pyrolysis treatment with operando XRD and EXAFS to investigate the nature of the iron particles.

Figure 1. Isotopic distribution of C_3-olefin products for the Fe-C_0.0 and a 20 wt.%Co 0.5 wt.%Re/ γ-Al_2O_3 catalyst. [3]

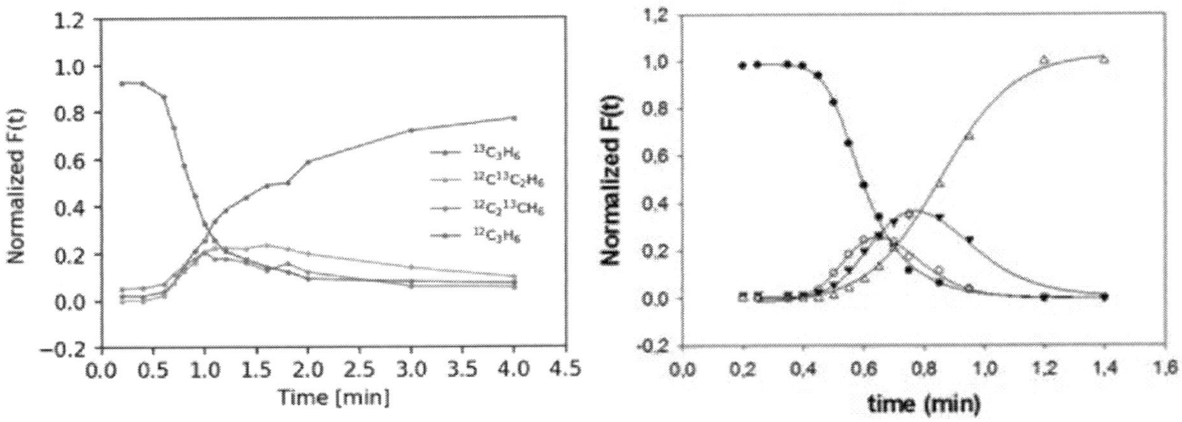

4. Conclusion

The addition of potassium lowers methane selectivity and enhances olefin selectivity slightly, which is reflected in the residence times and surface coverages, with the largest observed impact on the surface coverages. The slow decay of ^{12}C in the product distribution seem to indicate that a Mars-van Krevelen-like mechanism is present.

5. References

[1] Ordomsky V.V, Legras B, Cheng K, Paul S, Khodakov A.Y. Role of carbon atoms of supported iron carbides in Fischer-Tropsch Synthesis. **Catal. Sci. Technol.**, 2015; **5**, 1433-1437.

[2] Gracia J.M, Prinsloo F.F., Niemantsverdriet J.W. Mars-van Krevelen-like Mechanism of CO Hydrogenation on an Iron Carbide Surface. Catal. Lett. 2009; 133, 257-261

[3] Ledesma C, Yang J, Blekkan E.A, Holmen A, Chen D. Carbon Number Dependence of Reaction Mechanism and Kinetics in CO Hydrogenation on a Co-Based Catalyst. ACS Catalysis. 2016; 6(10):6674-6686.

545923 Kinetics for Cobalt Fischer-Tropsch Synthesis; A Multi-Dimensional Task

Monday, June 3, 2019: 5:45 PM
Texas Ballroom EF (Grand Hyatt San Antonio)

Magne Hillestad, **Erling Rytter** and Mohammad Ostadi, Chemical Engineering, Norwegian University of Science and Technology, Trondheim, Norway

Kinetics for Cobalt Fischer-Tropsch Synthesis; a multi-dimensional task

Magne Hillestad, [a] *Mohammad Ostadi,* [a] *and Erling Rytter* [a,b]

[a] *Department of Chemical Engineering, Norwegian University of Science and Technology (NTNU), N-7491 Trondheim*

[b] *SINTEF Industry, N-7465 Trondheim, Norway*

1. Introduction

In order to achieve an optimized and realistic design of a Gas-to-Liquid (GTL) or Biomass-to-liquid (BTL) plant based on cobalt Fischer-Tropsch synthesis (FTS), it is necessary to have available a kinetic model for FTS that correctly capture major variations in reaction rates and selectivities subject to changes in process parameters. Unfortunately, the kinetic responses vary to a large degree with the catalyst formulation, in particular the type of support material for cobalt crystallites. Water is a key component in the reactions, and the activity response changes from positive to negative depending on the support, even within one family of supports like γ-alumina.[1] Add to this experimental data with restricted variations in process parameters, data obtained in different reactor types, a multitude of approaches for formulating kinetics, lack of agreement on the FTS mechanism, and several parallel or interlinked reactions. The latter comprise formation of long chained paraffins, the associated chain-propagation probability, formation of olefins, and methane as an apparent separate product. The task becomes multi-dimensional with several possible outcomes.

Here we present comparison of thirty published kinetic rate models dating back to 1949, comprising twenty-four explicit rate expressions. Some of the models were used in optimization of a BtL plant. Further, the multiple-dimension of formulating a good and useful kinetic expression for FTS has prompted us to revisit both the experimental basis, formulation of useful and reasonable expressions based on understanding of the FT reaction,[2] and, provisionally, implementing new mechanistic based Langmuir-Hinshelwood formulations.[3] An important part of this effort is flexible handling of water responses and catalyst formulations.

2. Comparison of published kinetic expressions

Some of the reported kinetic models give a detailed description of formation rates of individual components, while most models only describe the overall consumption rate of CO. Twelve of them were analyzed based on different criteria, such as their behavior at high conversions, high water partial pressure, sensitivity to added water, selectivity to C_{5+} products, etc.[4] The rate models were implemented in a plug flow reactor model quite accurately representing both fixed-bed and microchannel reactors. The main objective was to see the kinetic effects with changing composition along the reactor. In order to predict product distribution, we proposed our own chain growth model and fitted parameters to published experimental data.[5] The chain growth model includes the effect of water and predicts the C_{5+} and methane selectivities quite well.

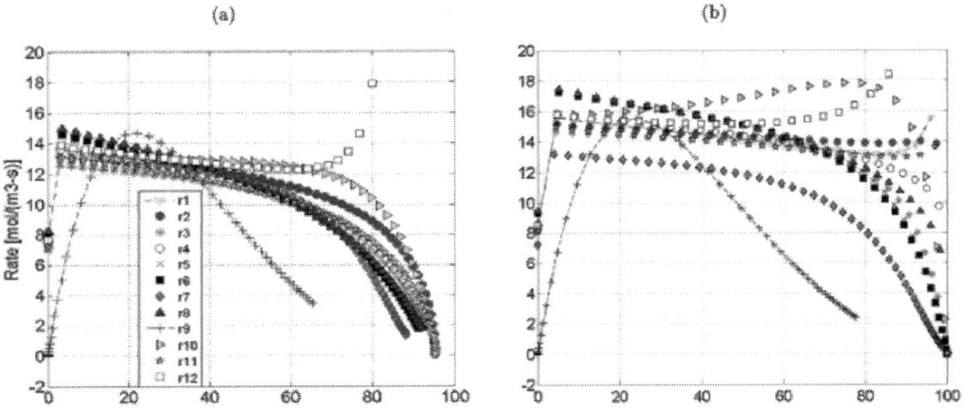

Figure 1. Activity as a function of CO conversion based on 12 different rate expressions; see ref. 4. a) Under-stoichiometric hydrogen feed; $H_2/CO=1.5$. b) Over-stoichiometric hydrogen feed; $H_2/CO=2.5$.

An example of the results is shown in Figure 1. The modelled rates span large variations depending on the rate expression with particular large discrepancies at high conversions. The spread is particularly large for feed gas with excess hydrogen. On the other hand, it is gratifying that the differences are not that large, except for one expression, for the core conversion range between 40 and 60% and with the industrially preffered under-stoichiometric feed. However, pushing the conversion to 70 or 75%, relevant for multichannel reactors, causes concern and necessitates careful selection of rate expression.

3. Optimizing a BTL plant depending on FT kinetics

This part of the study is concerned with optimization of a BtL plant. The basic configuration, shown in Figure 2, is based on biomass gasification in an entrained flow gasifier, supply of additional water from a high temperature electrolysis unit (SOEC), reverse water-gas-shift (WGS) reaction to convert CO_2 to CO, and multistage FTS reactors.[6] By adding the extra hydrogen, more than 90% carbon efficiency from biomass to products are achieved.

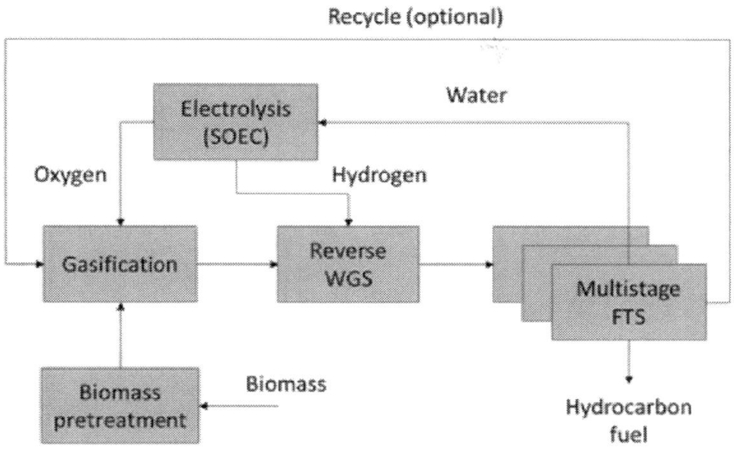

Figure 2. Block-flow diagram of BtL concept.

Several kinetic expressions from the study mentioned above are compared and used in process optimization of the plant. Fixing the sizes of the FT-reactors results in variations in total hydrocarbon productivity and the optimal H_2/CO feed ratio. Still, the main conclusion stands; it is profitable to add hydrogen from an external source, preferable made from renewable electricity, due to increase in carbon efficiency.

5. References

[1] E. Rytter, Ø. Borg, N. E. Tsakoumis and A. Holmen, Water as key to activity and selectivity in Co Fischer-Tropsch synthesis; γ-alumina based structure-performance relationships, J. Catal., 365 (2018) 334-343.

[2] E. A. Blekkan, M. Hillestad, E. Rytter, L. Gavrilovic and co-workers, work in progress.

[3] E. Rytter and A. Holmen, *Consorted Vinylene Mechanism for Cobalt Fischer-Tropsch Synthesis Encompassing Water or Hydroxyl Assisted CO-activation*, Topics Catal., 61 (2018) 1024-1034.

[4] M. Ostadi, E. Rytter and M. Hillestad, *Evaluation of kinetic models for Fischer-Tropsch cobalt catalysts in a plug flow reactor*, Chem. Eng. Res. Design, 114 (2016) 236-246.

[5] Yang, J., B, S., Myrstad, R., Venvik, H.J., Pfeifer, P., Holmen, A., 2016. *Fischer–Tropsch Synthesis on Co-based Catalysts in a Microchannel Reactor. Effect of Temperature and Pressure on Selectivity and Stability.* Catalysts and Catalysis: advances and Applications, vol. 223. CRC Press, pp. 259–266.

[6] M. Hillestad, M. Ostadi, G. d. Alamo Serrano, E. Rytter, B. Austbø, J. G. Pharoah, O. S. Burheim, *Increasing carbon efficiency of the biomass to liquid process with hydrogen from renewable power*, Fuel, 234 (2018) 1431-1451.

545624 Development of Highly Selective, Water Tolerant Cobalt Catalysts for Fischer-Tropsch Synthesis

Tuesday, June 4, 2019: 10:15 AM
Texas Ballroom EF (Grand Hyatt San Antonio)

Denzil Moodley[1], Thys Botha[2], Jana Potgieter[1], Michael Claeys[3] and Eric van Steen[4], (1)Fischer Tropsch & Syngas Conversion, Sasol Group Technology, Sasolburg, South Africa, (2)Technology & IP Management, Sasol Group Technology, Sasolburg, South Africa, (3)University of Cape Town, Cape Town, South Africa, (4)Department of Chemical Engineering, University of Cape Town, Cape Town, South Africa

Gas-to-Liquids (GTL) processes, employing Fischer-Tropsch technology, can be used to monetize remote sources of natural gas by converting the gas into readily transportable hydrocarbon liquid fuels. During times when the oil price is low, the flexibility of the process allows tuning to produce value added products like speciality waxes and high quality base-oil feedstock, which can have a significant product premium over the standard transportation fuels. Cobalt based Fischer-Tropsch synthesis (FTS) catalysts are an integral part of the Gas-To-Liquids (GTL) process. Due to the cost of both cobalt and noble metals, which are often used as promoters, extended catalyst life at high volumetric conversion rates and required minimum selectivity, is required to make the process economically feasible.

Operation of cobalt catalysts at high conversions is advantageous for the efficiency of the FTS process. These conditions leads to an increased reactor water partial pressure as well as a higher H2O:H2 exit partial pressure ratio. It is well known that several deactivation mechanisms such as oxidation, sintering of cobalt, formation of metal support compounds and loss of catalyst integrity are accelerated in presence of high water partial pressures. Thus, the development of more robust, water tolerant FTS catalysts with the desired product selectivity is an important objective.

In this contribution we showcase some of the work done at Sasol Group Technology to develop cobalt catalysts that are stable against high water partial pressures and that can give high selectivity in slurry phase reactors. Various strategies are discussed to tune robustness and performance, which include, (i) tailoring the cobalt particle size distribution, (ii) employing non-traditional hydrothermally resistant support materials such beta-SiC, and (iii) optimization of the physico-chemical properties of the support materials. The learnings from the work have cumulated in the development of a new cobalt-based GTL catalyst, highly selective to hard waxes and base-oil feedstock. The robustness of this catalyst towards water induced deactivation was studied with an in-situ magnetometer which allowed real time monitoring as a function of process conditions. Performance data of this catalyst in pilot scale slurry reactors will be demonstrated and this will be benchmarked against various state-of the art competitor systems.

545568 Co-Ni Supported on Reducible Oxides As Highly Active and Selective Fischer-Tropsch Catalysts

Tuesday, June 4, 2019: 10:39 AM
Texas Ballroom EF (Grand Hyatt San Antonio)

Carlos Hernández Mejía, Jessi E.S. van der Hoeven, Petra E. de Jongh and Krijn de Jong, Inorganic Chemistry and Catalysis, Utrecht University, Utrecht, Netherlands

The increasing worldwide demand for transportation fuels has stimulated the research towards efficient fuel production, particularly for liquid fuels since these possess a high energy density.[1] Fuels from processes such as natural gas to liquids (GTL) or coal to liquids (CTL) have emerged as an attractive alternative to the traditional fuels derived from crude oil. The GTL and CTL processes have several advantages like allowing for a feedstock diversification and generating ultra-clean fuels. A crucial step in GTL and CTL is the catalytic transformation of synthesis gas (a mixture of carbon monoxide and hydrogen) to hydrocarbons, also known as the Fischer-Tropsch (FT) synthesis. Cobalt-based catalysts are commonly employed in FT due to their high selectivity towards long-chain hydrocarbons (C_{5+}), low water-gas shift activity and stable catalytic performance.[2] A drawback however is that cobalt is a relatively expensive metal which in the last year has tripled in price,[3] harming the competitiveness of FT.

In this research, we investigated the effect of gradual cobalt substitution by nickel, a more abundant and affordable metal, on supported FT catalysts and the influence of the support's nature on the catalytic behaviour of the metals. For this, different cobalt to nickel molar ratios (x = Ni/(Ni+Co): 0, 0.25, 0.5, 0.75 or 1 mol/mol) supported on reducible (TiO_2 and Nb_2O_5) or non-reducible (a-Al_2O_3) oxides were studied. Several characterization techniques were employed to understand the effect of combining cobalt and nickel on the different supports. *Ex-situ* analysis of the reduced samples by transmission electron microscopy showed an increase in metal dispersion upon nickel addition independently of the support. EDX-mapping additionally showed an overlap of cobalt and nickel signals for the mixed compositions, indicating the formation of a cobalt-nickel alloy after the reduction treatment. H_2-chemisorption measurements of the reduced samples confirmed the increased metal dispersion upon nickel addition for the Al_2O_3-supported samples. In the case of the TiO_2- and Nb_2O_5-supported samples, the hydrogen uptake was lower than expected for their metal particle size and relatively constant for all metal compositions. The inhibition in hydrogen uptake is characteristic of reducible supports, known as Strong Metal-Support Interaction (SMSI).[4] During reductive conditions suboxides from the support might cover the metal nanoparticles, hindering hydrogen uptake. However, the metal-suboxide interface may give rise to promotional effects for FT.

The FT performance of the samples was evaluated at various reaction pressures and temperatures with an H_2 to CO ratio of 2 v/v. Figure 1 shows the metal-normalized activities, expressed as Metal-Time-Yield (MTY, $10^{-5}mol_{CO}\cdot g_{metal}^{-1}\cdot s^{-1}$), for TiO_2- and Nb_2O_5-supported catalysts as a function of the total pressure at 220 °C. At 1 bar, the Co-Ni samples displayed increased activities and C_{5+} selectivities, doubling the catalytic activity compared to the monometallic catalysts and reaching C_{5+} selectivities up to 77 % for the samples with 50 and 75 % Ni (x = 0.5 and 0.75). Contrastingly, Al_2O_3-supported catalysts had similar metal-normalized activities (~ $1\cdot10^{-5}mol_{CO}\cdot g_{metal}^{-1}\cdot s^{-1}$) and high selectivity towards methane for all Co-Ni compositions. Moreover, these catalysts deactivated severely throughout the experiment contrary to the more stable TiO_2- and Nb_2O_5-supported catalysts. Electron microscopy analysis of the spent catalysts revealed extensive metal particle growth for the Al_2O_3-supported catalysts containing nickel, explaining their fast deactivation, whereas only a slight growth was observed for the reducible supports. At these reaction conditions nickel readily forms volatile nickel carbonyl, leading to metal transport and particle growth. The use of reducible oxides appeared to have modified the reactivity of nickel in the alloy.

Increasing total pressure benefited the activity and selectivity of the samples with high cobalt content (x = 0.0, 0.25 and 0.5, Figure 1), particularly for the TiO_2-supported catalysts. Likewise, higher pressures improved their C_{5+} selectivity whereas a lesser increase was observed upon increasing nickel content. In contrast, the activity of the samples with a high nickel content (x = 0.75) diminished by increasing the total pressure.

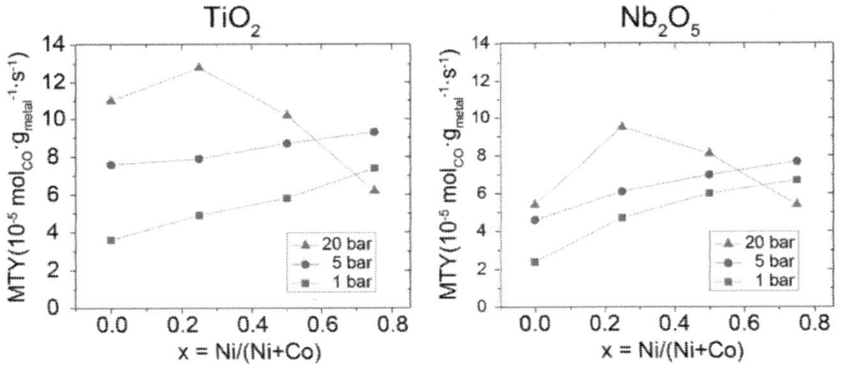

Figure 1. Metal-normalized catalytic activities against the different cobalt to nickel ratios supported on titania (left) or niobia (right). Reaction conditions: 1, 5 or 20 bar, 220 °C, H_2/CO = 2 v/v and CO conversions 5 – 35 %.

The catalytic behaviour at 20 bar, 220 °C, H_2/CO = 2 v/v and TOS = 100 h displayed similar trends for the TiO_2- and Nb_2O_5-supported catalysts. The samples with 25 % Ni had an activation period at the first hours of the experiment resulting eventually in the highest metal-normalized activities and turn-over-frequencies of the set of samples, moreover their C_{5+} selectivity was the same as the samples containing only cobalt (83 % for TiO_2 and 86 % for Nb_2O_5). Replacement of 50 %

cobalt by nickel (x = 0.5) showed similar activities as the cobalt-based catalysts (x = 0.0) however with a slightly lower C_{5+} selectivity. STEM-EDX was employed to characterize the catalysts after reaction, minor particle growth was observed for all samples. Additionally, the initial alloyed Co-Ni nanoparticles observed after reduction showed changes in their metal arrangement after catalysis (Figure 2) with some of the particles having a cobalt-enriched surface.

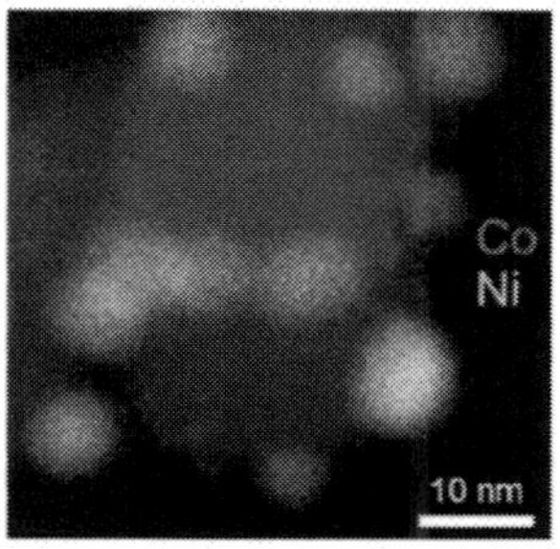

Figure 2. STEM-EDX mapping of Co-Ni (x = 0.75) supported on niobia after catalysis (20 bar, 220 °C, H_2/CO = 2 v/v and TOS = 100 h). Cobalt is shown in red and nickel in green.

The catalytic performance was additionally evaluated at different temperatures (220 – 260 °C) and 20 bar for the different Co-Ni compositions on both reducible supports. Apparent activation energies were derived from these results (Figure 3) and showed that at high nickel content (x = 0.75) the activation energy increases compared to the 100 % cobalt catalysts (x = 0.0) on both supports. A substitution of 50 % of the cobalt by nickel had the same activation energy as for x = 0.0. Finally, substitution of cobalt by 25 % nickel decreased the activation energy, indicating that up to a certain nickel concentration, it functioned as a promotor, which was also reflected on increased turnover frequencies. Interestingly, the selectivity was less affected at high temperatures when the nickel content was higher.

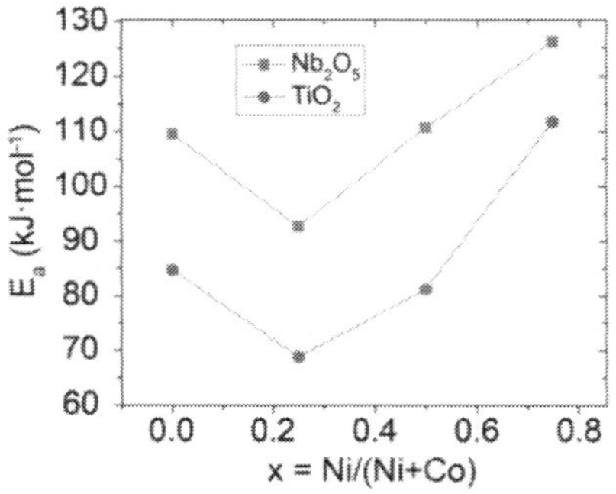

Figure 3. Apparent activation energies for the catalysts with different cobalt to nickel ratios supported on titania (red circles) or niobia (green squares). Reaction conditions: 20 bar, 220 - 260 °C, H_2/CO = 2 v/v.

In summary, we showed that the combination of cobalt and nickel supported on reducible oxides allow for 25 to 50 % cobalt substitution for nickel with increased FT activity and without sacrificing much C_{5+} selectivity. Furthermore, nickel-rich catalysts on reducible supports allowed for higher reaction temperatures while maintaining a good selectivity to C_{5+} products. Reducible oxides used as support material strongly modified the reactivity of Co-Ni alloys, opening new possibilities for more efficient and affordable FT catalysts.

1. De Jong, K. P. *et al.* The frontiers of energy. *Nat. Energy* **1,** (2016).

2. Dry, M. E. The Fischer–Tropsch process: 1950–2000. *Catal. Today* **71,** 227–241 (2002).

3. The Economist. Goblin metals. *Econ.* **426,** (2018).

4. Tauster, S. J., Fung, S. C. & Garten, R. L. Strong metal-support interactions. Group 8 noble metals supported on titanium dioxide. *J. Am. Chem. Soc.* **100,** 170–175 (1978).

545736 Improved Co/Ti Catalysts By Direct Reduction of Cobalt Nitrate and the Incorporation of Promoters

Tuesday, June 4, 2019: 11:03 AM
Texas Ballroom EF (Grand Hyatt San Antonio)

Mohammad Mehrbod[1], Michela Martinelli[2], Annabelle Martino[3], Burtron H. Davis[2], Donald Cronauer[4], A. Jeremy Kropf[5], Christopher Marshall[4] and **Gary Jacobs**[6], (1)Mechanical Engineering, University of Texas at San Antonio, San Antonio, TX, (2)Center for Applied Energy Research (CAER), University of Kentucky, Lexington, KY, (3)Chemical Engineering Program - Dept. of Biomedical Engineering, University of Texas at San Antonio, San Antonio, TX, (4)Argonne National Laboratory, Argonne, IL, (5)Chemical Sciences and Engineering Division, Argonne National Laboratory, Argonne, IL, (6)Chemical Engineering Program - Dept. of Biomedical Engineering / Dept. of Mechanical Engineering, University of Texas at San Antonio, San Antonio, TX

Fischer-Tropsch synthesis (FTS) is a collection of a heterogenously catalyzed reactions which converts synthesis gas (SynGas) to transportation fuels such as diesel and jet fuel. This reaction is also used to make premium lubricants and waxes. The most common metals which are used for FTS are iron (Fe), cobalt (Co) and ruthenium (Ru). The surfaces of these metal (or metal carbide, in the case of Fe) nanoparticles are active for FTS, but Fe has higher intrinsic water gas shift activity compared to Co and Ru. Thus, Co and Ru are often preferred for converting natural gas derived syngas, which has a higher H_2/CO ratio. Commercially, Co is more desirable, because it is considerably less expensive than Ru.

These metals are often supported on a metal oxide carrier to increase their reactive surface area. The most common supports are titania (TiO_2), alumina (Al_2O_3), and silica (SiO_2). For typical air calcined catalysts, Co has been found to have a stronger interaction with Al_2O_3 in comparison to TiO_2 or SiO_2. This interaction allows H_2-activated Co metal nanoparticles to be smaller in size compared to those supported on TiO_2 or SiO_2. Titania has a number of attractive characteristics including moderate surface area, mesoporosity, and stability. It is therefore not surprising that Shell® company incorporates TiO_2 into their FTS catalyst formulation.

Air calcination is a typical standard pretreatment method to convert cobalt nitrate to cobalt oxide and remove NO_X gases, as well as H_2O. However, Co oxide species in calcined Co/SiO_2 catalysts were found to have a weak interaction with SiO_2 support, and they were found to agglomerate into large clusters (e.g., 50+ nm) upon reduction. Previous studies revealed that direct reduction of $Co(NO_3)_2 \cdot xH_2O/SiO_2$, either unpromoted or accompanied by a reduction promoter such as platinum (Pt), ruthenium (Ru), rhenium (Re), or and silver (Ag), resulted in smaller cobalt metal nanoparticles than those produced from the reduction of the respective air calcined catalysts. Reduction of the cobalt oxides generated from cobalt nitrate decomposition was facilitated by the addition of promoter, leading to higher active site densities.

C_5+ selectivity and productivity in FTS reactor testing were improved as well.

Since using TiO_2 as a catalyst support is currently more practical than SiO_2 from an industrial standpoint, so this study revisits direct reduction of cobalt nitrate ($Co(NO_3)_2$), but in this case for unpromoted Co/TiO_2 as well as promoted catalysts incorporating reduction promoters such as Pt, Ru, Re, and Ag.

Catalyst preparation through the typical calcination method involves wet impregnation methods - e.g., incipient wetness impregnation (IWI) or slurry impregnation (SI) - of cobalt nitrate, followed by drying and air calcination in flowing air. Interestingly because of the vastly different degrees of interaction between different metal oxide supports and cobalt oxide species, activation of air calcined catalysts results in widely different average cobalt nanoparticle sizes.

In past studies, researchers have examined the effect of some promoters on Co/TiO_2. For example, the effects of boron (B), ruthenium (Ru) and rhenium were investigated in 2002. No effect of boron was observed on catalyst activity at lower space time but syngas conversion increased slightly at higher space time. While addition of ruthenium into Co/TiO_2 and CoB/TiO_2 enhanced both catalyst activity and selectivity, the study revealed that rhenium promoter exhibited the greatest effect on syngas conversion for Co/TiO_2 catalyst. Furthermore, the effect of water (in syngas) on the catalyst was investigated. While no significant change was observed for CO conversion at high space velocity (SV= 4~8 NL per g catalyst per h), the result showed that CO conversion was reduced at lower space velocity.

In 2007, the influence of support and promoter was investigated using temperature programmed reduction during EXAFS/XANES (TPR-EXAFS/XANES). The experiments illustrated that the nature of reduction of cobalt oxide and also demonstrated that the size of cobalt crystallites following activation depended on the strength of the interaction between the catalyst support and the cobalt oxide species prior to reduction, with silica offering a weak interaction and alumina providing a strong one. These experiments revealed a two step reduction process from Co_3O_4 to CoO and from CoO to Co^0. Adding platinum (Pt) played an important role in enhancing the reducibility of cobalt oxide species, in agreement with prior studies.

The most important properties of the catalyst considered in the current investigation are stability, selectivity and carbon monoxide conversion per gram of catalyst (conversion rate). Cobalt particle size may plays an important role in tuning the CO conversion rate; while studies shows that catalysts having particle sizes between 10 and 210 nm have almost identical turnover frequency (TOF), reported results for particle sizes smaller than 10 nm are not consistent. For instance, applying a deposition-precipitation method resulted in a Co particle size range of 3-5 nm, with results consistent with a linear trend between activity and cobalt surface area per gram of catalyst. However, other researchers reported that TOF and selectivity diminished on cobalt/carbon nanofiber (CNF) catalysts when the cobalt particle size was decreased from 16 nm to 2.6 nm. Thus, regarding prior results from the literature, the question remains as to whether changing from a conventional air calcination prior to activation to a direct cobalt nitrate reduction technique with the application of reduction promoters will result in smaller cobalt nanoparticles, and whether this change in cobalt metal cluster size on TiO_2 has an effect on activity and C_5+ selectivity.

Regarding catalyst stability, previous studies have illustrated that heavily loaded Co catalysts display greater stability against the catalyst oxidation (i.e., and cobalt-support compound formation) in Co/Al_2O_3. EXAFS/XANES experiments on catalysts exposed to FTS conditions containing water showed that higher loaded cobalt catalysts (e.g., 25% Co) with Pt promoter exhibited greater resistance toward oxidation and cobalt-support compound formation compared to lower loaded cobalt catalysts (e.g., 15%Co, and especially those containing Pt promoter that facilitated reduction of smaller CoO species). In 2016, Hughes et al. performed EXAFS/XANES experiments on Co/TiO_2 catalysts and reported that while catalysts with small average nanoparticle size (2-3 nm) prepared by CVD of cobalt underwent oxidation under conditions mimicking 50% conveersion, catalysts having larger cobalt clusters (nanoparticles ≥ 10nm) underwent net reduction at the same conditions. Thermodynamic studies suggest that small cobalt nanoparticles (cobalt particles ≤ 4 nm) may re-oxidize under FT reactions conditions, and that this may be exacerbated by the nature of the support.

In this study we assessed the effect of direct reduction of cobalt nitrate versus conventional calcination/reduction and incorporate reduction promoters to produce different extents of reduction and different cobalt crystallite sizes (i.e., Co-Co coordination in EXAFS) for different catalysts. Measurements include TPR-EXAFS/XANES, hydrogen chemisorption with pulse reoxidation, BET surface area and porosity measurements, and results of catalyst testing using a CSTR. CO conversion for the catalysts after H_2 activation and testing for FTS at steady state (process conditions: $T = 220^{o}C$, $P = 300$ psi, $H_2/CO = 2$ mol/mol, $SV = 3.8$ slph per g_{cat}) revealed the following trend: 0.5%Pt-12%Co/TiO_2 uncalcined ($X_{CO} = 56.4\%$) > 12%Co/TiO_2 uncalcined ($X_{CO} = 49\%$) > 0.5%Pt-12%Co/TiO_2 calcined (46.6%) > 12%Co/TiO_2 calcined ($X_{CO} = 29.8\%$). Methane selectivity was highest for the 12%Co/TiO_2 calcined catalyst (6.5%); the other catalysts were ~5%. TPR EXAFS/XANES results showed that while a typical two-step reduction occurs for calcined Co/TiO_2 catalysts, direct reduction of uncalcined catalysts proceeds through $Co(NO_3)_2$ decomposition to CoO_X, oxidation to a spinel resembling Co_3O_4, and then reduction of the spinel to more highly interacting CoO species (i.e., relative to the calcined case) prior to reduction to Co^0. Moreover, adding the metal promoter facilitates reduction, especially for the CoO to Co^0 step. EXAFS showed that the Co^0 species produced from direct reduction of cobalt nitrate had significantly lower Co-Co metal coordination numbers in the first shell (6.4 - 7.4 relative to 10.2 for the calcined Pt-containing catalyst). The results suggest that direct reduction of nitrate leads to smaller, more strongly interacting Co oxides, and that their reduction is facilitated by incorporation of a promoter. These catalysts outperformed their calcined counterparts in CSTR reactor tests.

545720 Optimizing Oxygenates Production in the Fischer Tropsch Reaction over Alkali-Promoted Co/Mn Catalysts

Tuesday, June 4, 2019: 11:27 AM
Texas Ballroom EF (Grand Hyatt San Antonio)

Norbert Kruse, Voiland School of Chemical Engineering and Bioengineering, Washington State University, Pullman, WA, Yizhi Xiang, Dave C. Swalm School of Chemical Engineering, Mississippi State University, Mississippi State, MS and Libor Kovarik, Pacific Northwest National Laboratory, Richland, WA

Introduction

Present efforts in Fischer Tropsch (FT) technology are devoted to increasing the selectivity of certain product classes and slates over others. While the traditional goal is to produce long-chain hydrocarbons for transport fuel applications, more recent research is focused on favoring either olefins or oxygenates production. As to the latter, long-chain terminal alcohols are important feedstocks for plasticizers, detergents and lubricants. The large-scale production of these compounds is based on homogeneous hydroformylation of C_n ($n \geq 3$) petroleum-derived 1-alkenes. One of the issues of this process is noble metal recovery; another is the limited regioselectivity dictated by the Markovnikov rule and predicting considerable product fractions to be branched rather than straight-chained. With this background, it is highly desirable to design a heterogeneous "one-pot" process based on CO hydrogenation according to Fischer-Tropsch (FT), in the absence of a target 1-alkene. Due to the polymerization-type hydrocarbon growth mechanism in FT, predominantly terminal alcohols with little chain branching are obtained. We made use of this strategy in the recent past[1-4] and provide here results demonstrating that the selectivity of heterogeneous CO hydrogenation can be tuned to long-chain terminal alcohols using alkali-promoted promoted CoMn catalysts[4]. In addition, we show that these catalysts are also active in long-chain aldehyde production. The systematic variation of the H_2/CO partial pressure ratios actually allows optimizing the various product classes without changing the overall Anderson-Schulz-Flory (ASF) chain lengthening probability. Interestingly, a reaction-induced change in the catalyst chemical composition is observed upon such partial pressure changes. High yields of chain-lengthened aldehydes under hydrogen-deficient conditions seems to be associated with the transformation of metallic Co into Co_2C.

Materials and Methods

Alkali-promoted CoMn catalysts were prepared using oxalate co-precipitation from metal nitrates in acetonic solution[1, 4]. Mixed-metal Co-Mn oxalate precursors are polymers with Metal Organic Framework (MOF) structure. Alkali precipitation was enforced by a solubility/entrainment effect since KNO_3 is soluble in water, but not in acetone. Oxalate presursors were subjected to temperature-programmed decomposition (TPDec) in the presence of hydrogen. Metal (Co) vs. metal oxide

(Mn_5O_8, MnO, K_2O) formation was evaluated from the relative amounts of gaseous CO and CO_2 released during the decomposition. Active catalysts contained nanosized metal particles and metal oxide phases acting as dispersant and promoter. Specific surface areas of catalysts were determined using the BET (Brunauer-Emmett-Teller) method. Values were found to be in the range between 35 and 80 m^2g^{-1} depending on the composition of the catalysts. Metallic surface areas were determined from H_2/D_2 dynamic exchange measurements and were usually in the range of 2-5 m^2/g.

Catalysts were characterized by in-situ X-Ray Diffraction (XRD) during the ongoing reaction. High resolution Transmission Electron Microscopy (HRTEM) and Electron Energy Loss Spectroscopy (EELS) were employed before and after reaction to receive information about catalysts' morphology and chemical phase composition.

High-pressure catalytic tests were performed in a fixed-bed reactor consisting of a quartz tubule (Φ_{inner} = 7 mm) inserted into a stainless steel housing. Oxalate precursors were diluted with SiC, followed by in situ thermal decomposition in H_2 at 370 °C for 1 h. Then the reactor was cooled to <100 °C before introducing syngas at a H_2/CO ratio of 1.5. The temperatures for the catalytic tests were approached using low heating rates of 1 °C min^{-1}. The CO conversion and product selectivities were measured by online by GCMS.

Results and Discussion

We show that the selectivity of the "one-pot" heterogeneous CO hydrogenation over alkali metal promoted CoMn catalysts can be tuned to receive various product classes, including paraffins, 1-alkenes, 1-alcohols and aldehydes. The relative atomic amounts of metal in these catalysts was systematically varied. Remarkably, low H_2/CO partial pressures (≤1.5) enabled oxygenate selectivities between 50 and 60 wt% (ex CO_2) could be obtained, thereof 70~97% (depending on the reaction temperature) as straight aldehydes. The best performance results in terms of oxygenates production were obtained by using small amounts of potassium as a promoter. Replacing K for Li shifted the product spectrum from mainly oxygenates to olefins.

Quite generally, while oxygenates production is favored by low temperatures (and high pressures) for thermodynamic reasons, high CO conversions can only be achieved at high temperatures. Thus, there is a selectivity-activity tradeoff that can only be mitigated by operating the synthesis at medium reaction temperatures. With this background, most of our studies were performed in the range between 220 and 240 °C where CO conversions up to 20% could be reached.

Strictly linear ASF distributions were obtained for C_{4+} products over 'CoMnK'. Major deviations from the ASF behavior were obtained for C_1-C_3 products: very low selectivities for C_1 (methanol+formaldehyde) and C_3 oxygenates were contrasted by high acetaldehyde selectivities.

Detailed kinetic studies were performed with $Co_4Mn_1K_{0.1}$ catalysts (indices indicating atomic ratios). Varying the H_2/CO partial pressure ratio had a significant effect on the catalytic performance. According to Figure 1, while aldehydes are clearly dominating at low H_2/CO ratios, they become less abundant when the H_2/CO ratio increases. The alcohol fraction is strongly increasing at the expense of aldehydes and reaches a maximum at $H_2/CO=5$. Paraffins increase from initially 20% and dominate for $H_2/CO>5$ while olefins starting from nearly 20% disappear at such high partial pressure ratios. Turning to the ASF chain lengthening characteristics, it is seen that a unique α value of 0.5 for the total C_n production (resulting from paraffins, terminal olefins and alcohols as well as aldehydes) is obtained independent of the H_2/CO pressure ratio. Because of the strong deviations of short-chain products from the general (linear) ASF behavior, only C_{4+} products are considered here.

An intriguing kinetic hysteresis effect was observed on cycling the partial pressure ratios from initially high ($H_2/CO= 9$) to low ($H_2/CO=0.5$) and back to high. Accordingly, the catalyst activity followed a clockwise hysteresis, i.e. overall low CO conversion when decreasing the H_2/CO ratio from high to low and increasingly higher CO conversion on the way back. Closed hysteresis loops were obtained when subjecting the catalyst to very high H_2/CO pressure ratios or pure H_2. Fascinatingly, chain-lengthened paraffins and alcohols showed counterclockwise and, respectively, clockwise hysteresis. Selectivity hysteresis was also observed, though less pronounced, for aldehydes, olefins and carbon dioxide. Note that a 60% aldehyde production at low H_2/CO pressure ratios was never before reported in the FT literature and makes the $Co_4Mn_1K_{0.1}$ formulation a unique catalyst.

To correlate the observed hysteresis with structural and chemical changes of the catalyst during the ongoing reaction, we performed *in-operando* XRD studies and found a Co fcc phase at high H_2/CO ratios (beginning of the loop) and a Co_2C phase at low such ratios. This Co_2C phase turned out to be rather reluctant when reducing it to metallic (hcp) cobalt in excess hydrogen. On the other hand, our results provide strong evidence that high aldehyde selectivities are favored by a Co_2C phase while high paraffin selectivities prefer a metallic (either fcc or hcp) Co phase.

HRTEM and EELS studies with a $Co_4Mn_1K_{0.1}$ catalyst provided further support for the formation of Co_2C after reaction with low H_2/CO ratios. Furthermore, a mixed-valence $Mn_2(II)Mn_3(IV)O_8$ phase was identified. We therefore advocate a synergistic interaction between Mn_5O_8 oxide and bulk Co_2C phases. Recent HRTEM studies focussed on the Mn-oxide phase and showed that part of the Mn_5O_8 underwent a reaction-induced decomposition forming highly dispersed MnO aggregates.

Conclusions

Potassium-promoted CoMn catalysts, in particular $Co_4Mn_1K_{0.1}$, show unprecedented catalytic performance in terms of straight-chain aldehyde production at low H_2/CO partial pressures. Catalysts undergo a reaction-induced restructuring and the bulk composition for optimal aldehyde formation is Co_2C/Mn_5O_8. Activity-selectivity hystereses are observed upon H_2/CO variation. High selectivities in oxygenates formation at low H_2/CO ratios may help design a 'heterogeneous hydroformylation' process in the absence of terminal olefins. To make this a viable option, further efforts are necessary to increase the activity of the catalysts.

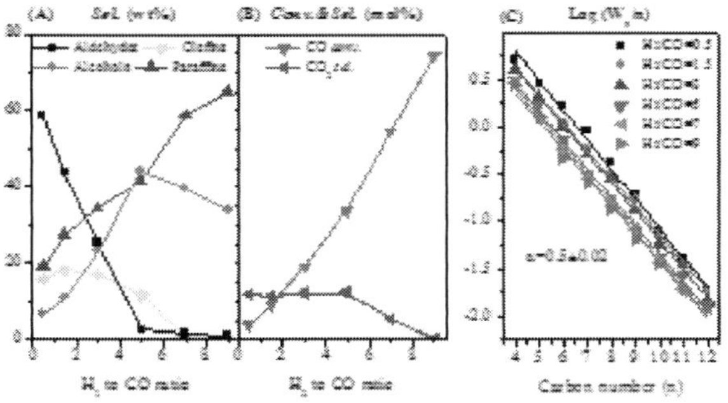

Figure 1. Effect of H_2 to CO ratio. (A) Products selectivity without CO_2, (B) CO conversion and CO_2 selectivity, (C) ASF behavior for total C_n of the same catalyst showing a chain lengthening probability independent of the H_2/CO pressure ratio. Reaction conditions: $Co_4Mn_1K_{0.1}$ catalyst 0.5 g, 220°C and 40 bar.

References

1. Xiang, Y.; Chitry, V.; Liddicoat, P.; Felfer, P.; Cairney, J.; Ringer, S.; Kruse, N., Long-Chain Terminal Alcohols through Catalytic CO Hydrogenation. *Journal of the American Chemical Society* **2013,** *135* (19), 7114-7117.

2. Xiang, Y. Z.; Barbosa, R.; Kruse, N., Higher Alcohols through CO Hydrogenation over CoCu Catalysts: Influence of Precursor Activation. *ACS Catalysis* **2014,** *4* (8), 2792-2800.

3. Xiang, Y. Z.; Barbosa, R.; Li, X. N.; Kruse, N., Ternary Cobalt-Copper-Niobium Catalysts for the Selective CO Hydrogenation to Higher Alcohols. *ACS Catalysis* **2015,** *5* (5), 2929-2934.

4. Xiang, Y.; Kruse, N., Tuning the catalytic CO hydrogenation to straight- and long-chain aldehydes/alcohols and olefins/paraffins. *Nature Communications* **2016,** *7*, 13058.

545758 MOF-Mediated Synthesis of Highly Active Fe- and Co-Based FTS Catalysts

Tuesday, June 4, 2019: 11:51 AM
Texas Ballroom EF (Grand Hyatt San Antonio)

Freek Kapteijn[1], Giusy Petty[1], Christian van der Sande[1], Patricia S Garcia Guerrero[1], Lide Oar-Arteta[1], Tim A. Wezendonk[1], Xiaohui Sun[1] and Jorge Gascon[2], (1)Catalysis Engineering, Department of Chemical Engineering, Delft University of Technology, Delft, Netherlands, (2)KAUST Catalysis Center, Advanced Catalytic Materials, King Abdullah University of Science and Technology, Thuwal, Saudi Arabia

Metal Organic Frameworks are well-defined crystalline porous hybrid structures with an excellent distribution of inorganic and organic building blocks. Many examples contain elements that are active in catalysis, including Fe and Co, well-known high and low temperature Fischer-Tropsch synthesis (FTS) catalysts, respectively.

Subjecting these specific MOFs to a thermal treatment in inert atmosphere, called MOF-mediated synthesis (MOFMS), causes decomposition, pyrolysis, reduction and particle growth phenomena resulting in a composite of nanosized oxidic or metallic particles covered by graphenic shells. The metal loadings may reach up to 60 wt.% [1]. Depending on the inorganic element the thermal decomposition, followed by in-situ XRD and TGA, can be described by a single- or a two-step solid conversion process, well captured by the Avrami-Erofeev nucleation models with as most relevant parameters heating rate and final temperature.

Fe-MOF precursors yield nanoparticles of Fe-suboxide encapsulated by a porous carbon shell. When exposed to the reducing FTS environment Fe-carbide is formed. Temperature and gas composition can be used to form the Hägg and/or epsilon carbide.

The Fe-loading (36–46 wt%) and nanoparticle size (3.6–6.8 nm) of the obtained Fe@C catalysts are directly related to the elementary composition and porosity of the initial MOFs. Furthermore, the carbonization leads to similar surface areas for the C matrix (S_{BET}between 570 and 670 m^2g^{-1}), whereas the pore width distribution is completely different for the various MOFs. Comparison of different Fe-MOF precursors shows that the high temperature FTS activity is proportional to the external particle surface. Promotion with alkali elements boosts the activity and increases the olefin/paraffin selectivity considerably. The catalytic performance @ 340 °C, stable >600 h, results in iron time yields (FTY) in the range of $1.9–4.6 \cdot 10^{-4} mol_{CO} g_{Fe}^{-1} s^{-1}$. In combination with the high Fe-loading these materials represent extreme highly active catalysts on a mass or volume basis [2-5].

This material, however, is not easy to pelletize, and these pellets disintegrate quickly. A procedure based on phase-inversion of a slurry of the MOF precursor and a selected polymer, followed by pyrolysis results in robust catalyst particles. Interestingly, the nitrogen stemming from the polymer already exerts a promotion influence on the selectivity of the catalyst.

A similar approach using a Co-MOF precursor (ZIF-67), yielded metallic Co nanoparticles covered with an impermeable graphene coating. This was nearly inactive for low temperature FTS. Adaptation of the synthesis procedure towards an oxidative treatment after impregnation with a silicon precursor and pyrolysis, yielded an extremely active LT-FTS catalyst consisting of highly loaded cobalt nanocomposites (~ 50 wt.% Co, ~12 nm Co), with cobalt oxide reducibility in the order of 80% and a good particle dispersion, that exhibit high activity, C_{5+}selectivity and stability in Fischer–Tropsch synthesis [6]. The structure of this material resembles that envisaged as optimal for this process, while the activity is about 30% of the maximum reachable for a pure Co catalyst of the optimal particle size [7].

References

1. Oar-Arteta, T. Wezendonk, X. H. Sun, F. Kapteijn, J. Gascon, *Materials Chemistry Frontiers* **2017**, *1*, 1709-1745.
2. P. Santos, T.A. Wezendonk, *et al.*, *Nat. Commun*. 6:6451, doi:10.1038/ncomms7451 (2015).
3. A. Wezendonk, X. H. Sun, A. I. Dugulan, A. J. F. van Hoof, E. J. M. Hensen, F. Kapteijn, J. Gascon, *Journal of Catalysis* **2018**, *362*, 106-117.
4. A. Wezendonk, Q. S. E. Warringa, V. P. Santos, A. Chojecki, M. Ruitenbeek, G. Meima, M. Makkee, F. Kapteijn, J. Gascon, *Faraday Discussions* **2017**, *197*, 225-242.
5. A. Wezendonk, V. P. Santos, M. A. Nasalevich, Q. S. E. Warringa, A. I. Dugulan, A. Chojecki, A. C. J. Koeken, M. Ruitenbeek, G. Meima, H.-U. Islam, G. Sankar, M. Makkee, F. Kapteijn, J. Gascon, *ACS Catalysis* **2016**, *6*, 3236-3247.
6. Sun, A. I. O. Suarez, M. Meijerink, T. van Deelen, S. Ould-Chikh, J. Zecevic, K. P. de Jong, F. Kapteijn, J. Gascon, *Nat Commun* **2017**, *8*:1680, doi:10.1038/s41467-017-01910-9
7. Guttel, T. Turek, *Energy Technology* **2016**, *4*, 44-54.

545351 Methanol Synthesis from $CO/CO_2/H_2$ and CO/H_2 over $Cu/ZnO/Al_2O_3$ and Cu/CeO_2 Catalysts

Tuesday, June 4, 2019: 1:30 PM
Texas Ballroom EF (Grand Hyatt San Antonio)

Leon G.A. van de Water[1], Sam Wilkinson[2], Brendon Miller[3], Mark. J Simmons[4], E. Hugh Stitt[5] and Michael J. Watson[1], (1)Johnson Matthey Technology Centre, Johnson Matthey, Billingham, United Kingdom, (2)Formulated Products, Process Systems Enterprise, LONDON, United Kingdom, (3)Jembec Consulting, Christchurch, New Zealand, (4)School of Chemical Engineering, University of Birmingham, Birmingham, United Kingdom, (5)Technology Centre, Johnson Matthey, Billingham, United Kingdom

1. Introduction

Methanol is an industrial chemical of high significance with an annual production of 62 million tons in 2014. Commercial production of methanol from carbon monoxide and hydrogen dates back nearly 100 years,[1] during which period significant improvements to the catalyst and process have been achieved. Methanol is currently produced from $CO/CO_2/H_2$ feed gas mixtures using a $Cu/ZnO/Al_2O_3$ catalyst which was developed by ICI in the 1960s.[1] The reaction mechanism and nature of the active site as a function of reaction conditions have been studied intensively and remain subject of active industrial and academic research. Based on numerous spectroscopic, isotope labelling and kinetic studies dedicated to study the mechanism and active site of this system, it has been established that CO_2 is the carbon source of methanol produced from $CO/CO_2/H_2$ feeds, where a linear correlation exists between Cu metal surface area and methanol synthesis activity. Whilst the presence of Cu is essential for the production of methanol, rates of reaction are enhanced significantly by the ZnO component in the catalyst. The correlation between Cu surface area and activity does not apply to all types of Cu catalyst however, and CO_2 is not the only possible carbon source: methanol formation over Cu/CeO_2 catalysts occurs via CO hydrogenation and involves a different active site and reaction mechanism.[2,3] In this contribution the generation of methanol synthesis active sites on different Cu-based catalysts is explored as a function of the feed gas composition, focusing on the interplay between metal, support and gas phase. Catalyst performance is evaluated under steady and non-steady state conditions using spatially resolved experimental techniques to assess catalyst performance in different catalyst bed sectors. The effect of reaction conditions on the active sites and prevailing reaction mechanism is studied by combining kinetic data with results from spectroscopic studies on catalysts measured under relevant reaction conditions.

2. Experimental

A commercial $CuO/ZnO/Al_2O_3$ (60/30/10 weight ratio) catalyst precursor and a 10 wt% CuO/CeO_2 mixed oxide, prepared by deposition precipitation of CuO onto a CeO_2 support,[3] were used throughout the study. Reaction studies were performed using a set of 6 parallel fixed-bed, down-flow micro-reactors. The oxidic $CuO/ZnO/Al_2O_3$ and CuO/CeO_2 catalyst precursors were reduced in dilute H_2 and subsequently exposed to relevant methanol synthesis conditions in the temperature range of 180 – 225 °C and pressures ranging from 10 to 35 bar. Gas compositions used were 0 - 3% CO / 0 – 3% CO_2 / 67% H_2 / N_2 for $Cu/ZnO/Al_2O_3$ and 2 or 33% CO / H_2 for Cu/CeO_2. Parallel difference testing [4,5] was used to obtain spatially resolved activity profiles throughout the catalyst beds. *In situ* DRIFTS spectra were recorded on a Thermo Nicolet 750 Magna FTIR, equipped with an environmental DRIFTS cell, operated at atmospheric pressure. Samples for XPS analysis were studied using a Thermo Fisher Escalab 250. The high-pressure gas cell was operated by introducing gas at ~3 bar under flow (~50 ml/minute at the exhaust). Cooling down of the samples was done under reactive gas flow, followed by evacuation of the cell and transfer of the sample to the analysis chamber.

3. Results and discussion

The exit gas composition measured during the first 40 minutes of exposure of the H_2/N_2 reduced $Cu/ZnO/Al_2O_3$ catalyst to $3\%CO/3\%CO_2/67\%H_2/N_2$ at 25 bar, 200 °C, reveals that the reverse water-gas shift (RWGS) reaction occurs from $t = 0$, whilst methanol synthesis is not seen until $t = 2.5$ min. Analysis of kinetic data obtained under the entire range of conditions indicates that the occurrence of RWGS during the first minutes of operation results in a build-up of O_s surface species. The detailed kinetic modeling study [5] revealed that a model including generation of bidentate surface carbonate species ($CO_3.2s$), formed upon reaction of CO_2 with O_s, gives the best fit of the experimental data. The effect of $CO:CO_2$ ratio in the feed on steady-state performance of $Cu/ZnO/Al_2O_3$ in different catalyst bed sectors is illustrated in Figure 1. The extent of the RWGS reaction is dependent on the CO_2 concentration in the feed, and RWGS equilibrium is reached at different stages along the length of the reactor, depending on the $CO:CO_2$ ratio. The results presented in the Figure illustrate that the methanol production is not affected by a change in $CO:CO_2$ ratio from 10 to 1.

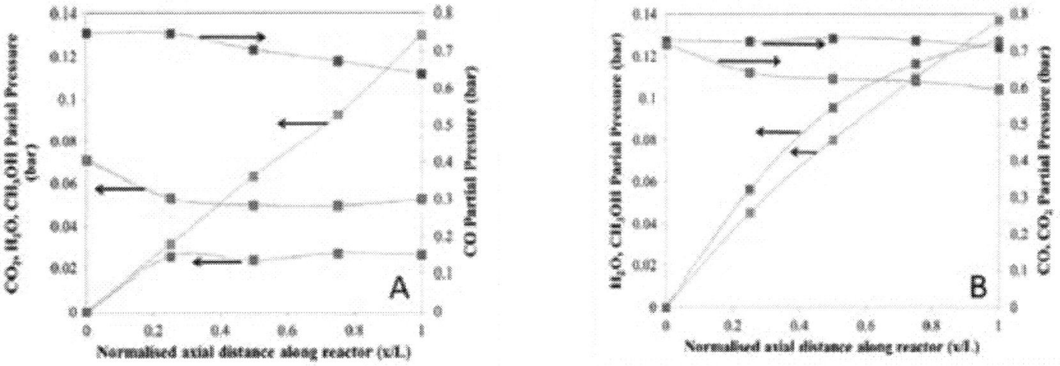

Figure 1. Steady state axial concentration plots for Cu/ZnO/Al$_2$O$_3$ under 3%CO/0.3%CO$_2$/67%H$_2$/N$_2$ (A) and 3%CO/3%CO$_2$/67%H$_2$/N$_2$ (B) conditions at 25 bar, 200 °C. Blue: CO; Red: CO$_2$; Green: CH$_3$OH; Purple: H$_2$O.

In the absence of CO_2 in the feed a different situation arises. Transient CO_2 and H_2O formation is observed during start-up, evenly distributed over the entire catalyst bed (Figure 2A). Methanol synthesis is observed after a short delay, at a 5 to 6 times lower rate than in the presence of CO_2. The methanol production rate decreases with time on stream, whilst the CO_2 and H_2O formation do not disappear entirely. These results suggest that the catalyst undergoes changes during exposure to CO/H_2. Subsequent exposure of the CO/H_2-treated $Cu/ZnO/Al_2O_3$ catalyst to 3%CO/3%CO_2/67%H_2/N_2 results in complete restoration of the activity as seen in the original activity test under $CO/CO_2/H_2$, highlighting the reversibility of the catalyst reduction process in CO/H_2.

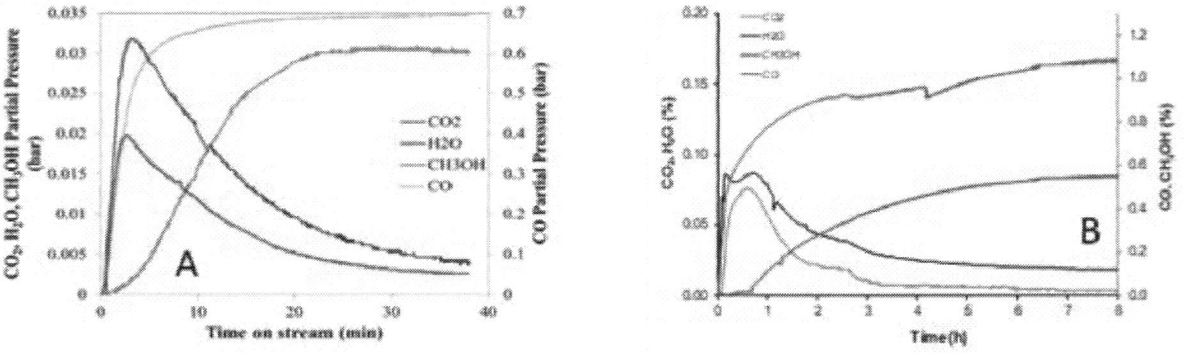

Figure 2. Catalyst performance for Cu/ZnO/Al$_2$O$_3$ during start-up under 3%CO/67%H$_2$/N$_2$ conditions at 25 bar, 200 °C (A), and for Cu/CeO$_2$ during start-up under 2%CO/H$_2$, 25 bar, 195 °C (B).

Studying the Cu/CeO_2 reference catalyst in CO/H_2 feeds revealed a number of similarities and key differences between these systems: *i)* Methanol synthesis was observed from CO/H_2 feeds, with transient CO_2 and H_2O formation during start-up (Figure 2B), in a similar vein to that seen for $Cu/ZnO/Al_2O_3$ in CO/H_2; *ii)* Methanol synthesis starts only after this CO_2 and H_2O formation; *iii)* Addition of CO_2 to the CO/H_2 feeds leads to irreversible catalyst deactivation due to build-up of surface carbonate and formate species in case of Cu/CeO_2. XPS and DRIFTS spectroscopy confirmed the presence of cationic Cu species, Cu^{d+}, and partially reduced ceria, CeO_{2-x}, under CO/H_2 reaction conditions.

4. Conclusions

Micro-kinetic modelling tools have increased the understanding of the methanol synthesis reaction over $Cu/ZnO/Al_2O_3$ under a wide range of reaction conditions. Generation of surface carbonate (CO_3.2s) species was found to be key to generation of catalytic activity from $CO/CO_2/H_2$ feeds. Methanol formation in the absence of CO_2 (i.e., from CO/H_2) occurs at a significantly reduced rate and involves a different mechanism and active site, involving the reducible ZnO component. Similar observations during catalyst start-up in CO/H_2 were made for the Cu/CeO_2 system, where cationic Cu species, Cu^{d+}, were found to be present under reaction conditions, the active state of the catalyst being Cu^{d+}/CeO_{2-x}. Similarly, a Cu^{d+}/ZnO_{1-x} active phase is likely to exist for the $Cu/ZnO/Al_2O_3$ catalyst in CO/H_2. Addition of CO_2 to the feed has a detrimental effect on Cu/CeO_2-based methanol synthesis, whereas a change from Cu^+/ZnO_{1-x} to a Cu^0/ZnO active phase occurs for $Cu/ZnO/Al_2O_3$, with a change from CO hydrogenation to CO_2 hydrogenation.

1. Twigg, M.V. Methanol Synthesis. In *Catalyst Handbook*, 2nd Edition ed.; Manson Publishing Ltd: London, **1989**; 441-468.

2. Nix, R.M.; Rayment, T.; Lambert, R.M.; Jennings, J.R.; Owen, G. *J. Catal.* **1987**, *106*, 216-234.

3. van de Water, L.G.A.; Wilkinson, S.K.; Smith, R.A.P.; Watson, M.J.. *J. Catal.* **2018,** *364*, 57-68.

4. Birtill, J.J. *Catal. Today* **2003,** *81* (4), 531-545.

5. Wilkinson, S.K.; van de Water, L.G.A.; Miller, B.; Simmons, M.J.H.; Stitt, E.H.; Watson, M.J. *J. Catal.* **2016,** *337*, 208-220.

545613 CO_2 Conversion to Meoh-DME Fuels:Catalytic and Technological Aspects

Wednesday, June 5, 2019: 2:48 PM
Texas Ballroom D (Grand Hyatt San Antonio)

Francesco Frusteri, Catia Cannilla, Fabio Costa, Aldo Mezzapica and Giuseppe Bonura, Energy, CNR-ITAE, Messina, Italy

CO_2 CONVERSION TO MEOH-DME FUELS:

CATALYTIC AND TECHNOLOGICAL ASPECTS

F. Frusteri, C. Cannilla, F. Costa, A. Mezzapica, G. Bonura*

CNR-ITAE, Via S. Lucia sopra Contesse 5, 98126 Messina, Italy

*Presenting author: Dr. Francesco Frusteri, PhD. Email: francesco.frusteri@itae.cnr.it

Abstract

One-step CO_2 hydrogenation reaction to dimethyl ether (DME) [1] on a hybrid system using different reactor bed configurations (physical mixing, dual-bed and mono-bed) will be discussed (see Fig. 1).

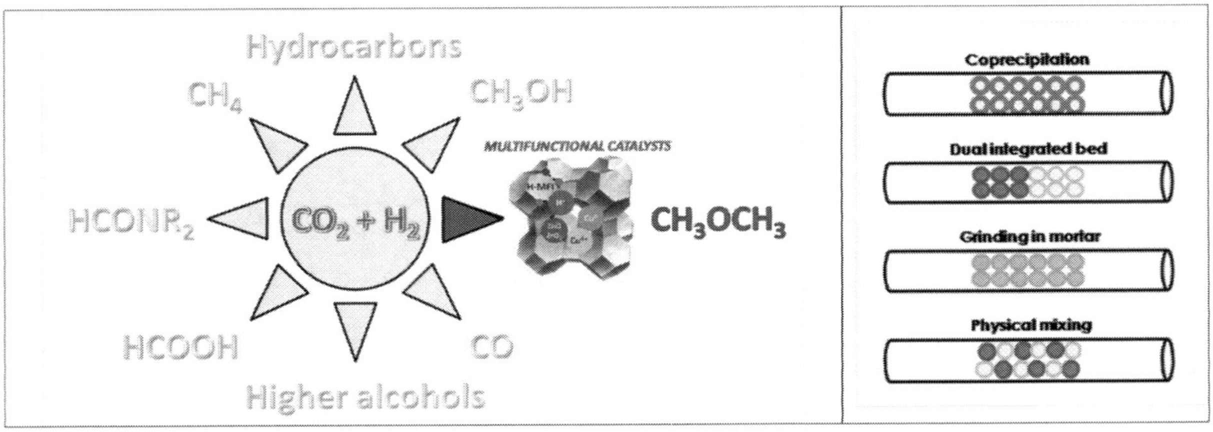

Fig. 1. Different reactor configurations adopted under CO2-to-DME hydrogenation conditions.

The influence of different materials and preparation methods to design a suitable hybrid Cu-ZnO-MeO_x–Zeolite catalytic system will be investigated with the aim to better define the way of achieving high DME productivity [2-4]. Catalytic results, in terms of CO_2 conversion and product distribution in the temperature range of 473-533 K at 3.0 MPa and CO_2/H_2 feed concentration of 1/3 v/v using a fixed bed reactor will be discussed to individuate the optimal reaction conditions ensuring both high CO_2 conversion per pass and high CH_3OH/DME selectivity [3,5-7]. Particular attention will be addressed to reaction mechanism with the aim to disclose the controlling step which prevents the achievement of high CO_2 conversion at temperature lower than 473 K. An effort will be also profuse to precisely ascertain, in the hybrid systems, the active sites for H_2 and CO_2 activation by correlating the results as a function of their nature and strength. Furthermore, the influence of catalyst structure and morphology on mass transferring phenomena of reaction intermediates will be investigated to highlight how these parameters affect DME/MeOH production.

The audience will have the opportunity to learn more about the feasibility of a process based on direct production of DME from CO_2 using hybrid catalytic systems. In particular, people working in this field should acquire new information to better address the research towards the development of optimized multi-component catalysts characterized by high CO_2 conversion and DME selectivity, without suffering from evident deactivation phenomena under reaction conditions.

Keywords: CO_2 hydrogenation; hybrid catalysts; DME synthesis.

Acknowledgements

Part of this work was realized within the bilateral agreement of Scientific and Technological Cooperation CNR-MTA/HAS (Hungarian Academy of Sciences).

References

[1]. M. Aresta, A. Dibenedetto, E. Quaranta, J. Catal. 343 (2016) 2–45.

[2]. E. Catizzone, G. Bonura, M. Migliori, F. Frusteri, G. Giordano, Molecules 23 (2018) 31-58.

[3]. F. Frusteri, M. Migliori, C. Cannilla, L. Frusteri, E. Catizzone, A. Aloise, G. Giordano, G. Bonura, J. CO2 Util. 18 (2017) 353-361.

[4]. F. Frusteri, G. Bonura, C. Cannilla, G. Drago Ferrante, A. Aloise, E. Catizzone, M. Migliori, G. Giordano, Appl. Catal., B 176 (2015) 522–531.

[5]. G. Bonura, M. Migliori, L. Frusteri, C. Cannilla, E. Catizzone, G. Giordano, F. Frusteri, J. CO2 Util. 24 (2018) 398-406.

[6]. M. Sánchez-Contador, A. Ateka, A.T. Aguayo, J. Bilbao, Fuel Process. Technol. 179 (2018) 258-268.

[7]. M. Gentzen, D.E. Doronkin, T.L. Sheppard, J.-D. Grunwaldt, J. Sauer, S. Behrens, Appl. Catal., A 562 (2018) 206-214.

545627 Carbon Dioxide Hydrogenation into Methanol with CuO-ZnO-ZrO_2 Catalyst Prepared By a Continuous Flow Microfluidic-Assisted Synthesis

Tuesday, June 4, 2019: 2:18 PM
Texas Ballroom EF (Grand Hyatt San Antonio)

L'hospital Valentin[1], Parkhomenko Ksenia[1], Christophe Serra[2] and **Anne Cecile Roger**[3], (1)UMR 7515, CNRS, Strasbourg, France, (2)ICS (UPR 22 CNRS), University of Strasbourg, Strasbourg, France, (3)ICPEES, Université de Strasbourg, Strasbourg, France

Carbon dioxide hydrogenation into methanol with CuO-ZnO-ZrO_2 catalyst prepared by a continuous flow microfluidic-assisted synthesis

Valentin L'hospital[1]**, Ksenia Parhomenko**[1]**, Christophe Serra**[2]**, Anne-Cécile Roger**[1]

[1] *ICPEES (UMR7515), Université de Strasbourg, 25 rue Becquerel, 67087 Strasbourg France*

[2] *ICS (UPR22-CNRS), 23 rue du Loess BP 84047 67034 Strasbourg Cedex 2, France*

E-mail :lhospital.valentin@gmail.com

Numerous measures to reduce anthropogenic greenhouse gas emissions and in particular CO_2 already exist such as its capture and storage as well as its conversion into valuable chemical compounds (methanol, methane, dimethyl ether and others). Currently, methanol is synthesized from syngaz (originally from oil or natural gas) but in this project, methanol is synthesized from pure CO_2 and green H_2, produced by water electrolysis with renewable energies. Methanol produced by CO_2 hydrogenation is considered as a platform molecule of great interest, for example for formaldehyde production, as a precursor in the plastic polymere formation (MTO, methanol to olefins) or it can be used in the hydrocarbons production (MTG, methanol to gasoline). This project is a promissing way of energy storage via CO_2 valorization and an important process for the overall World energy transition scenario.

During the methanol synthesis from CO_2, several reactions may take place: CO_2 hydrogenation (1), reverse water-gas shift (2) and possible CO hydrogenation (3). The majority of catalysts of this reaction are based on copper and the couple Cu/ZnO is considered as the base of the methanol synthesis catalyst, as it permits to increase the selectivity of methanol production versus competitive reactions[1]. Different supports have been studied, usually it is not inert, it has several roles in catalysis, like for example, increase of the Cu dispersion, creation of some active species for additional CO_2 adsorption. The use of ZrO_2 instead of alumina as catalyst support can improve the copper reductibility and so, increase the active copper phases and finally the catalyst sintering may be avoided[2].

Batch coprecipitation is a classical and easy way to prepare such catalytic materials but the main disadvantage of this method is the pH and concentration variations during the synthesis and variation of the composition during the coprecipitation and so, heterogeneity in the final catalyst. In this work, we propose to synthesize for the first time Cu-based catalytic materials by a continuous flow microfluidic-assisted method[3] and compare it to the classical way. During the microfluidic-assisted synthesis the precipitation zone is much smaller and the different variation of pH and concentration are greatly reduce. All the synthesis parameters are better controlled which may be beneficial in terms of kinetic control of the reactant mixing thus allowing improving the homogeneity of the final mixed oxide materials[4].

The metal oxides obtained (CZZ2) were characterized by various techniques and tested, after reducing treatment, as catalysts for CO_2 hydrogenation into methanol. The results obtained were compared to those of a catalyst of same composition prepared by classical co-precipitation in batch (CZZ1) and to those of an industrial copper-based catalyst (INDUS).

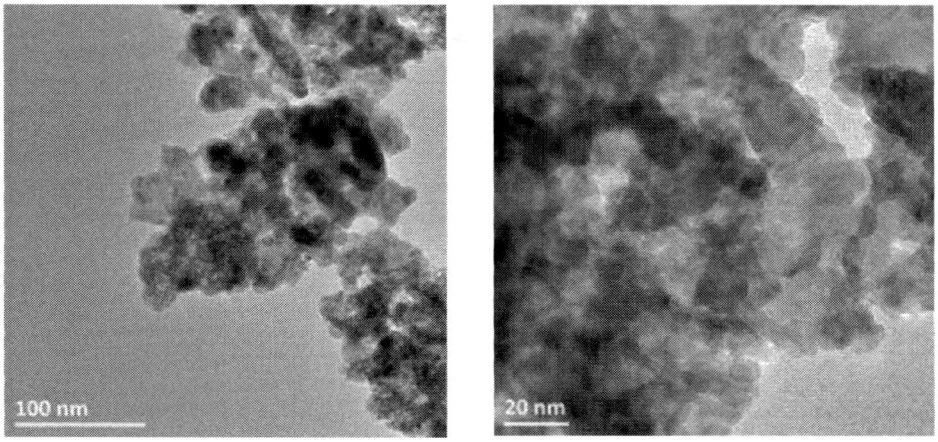

Figure 1: TEM pictures of (a) CZZ1(batch) and (b) CZZ2 (microfluidic)

Compared to the batch coprecipitation, the catalytic materials synthesized by a microfluidic system are more homogeneous (Figure 1) with much higher specific and metallic surfaces area (BET and TPD-N_2O, results not shown here).

The in depth study of the precursors generated by batch or by continuous microfluidic coprecipitation have been carried out. Single precursors (Cu, Zn or Zr carbonate/hydroxycarbonate), binary precursors (Cu-Zn, Cu-Zr or Zn-Zr carbonate/hydroxycarbonate and ternary precursors (Cu-Zn-Zr carbonate/hydroxycarbonate) have been prepared by both syntheses and have been characterised as prepared (XRD, TGA) and after calcination, under the oxide form (XRD, BET, microscopy, N_2O chemisorption, thermo-programmed reduction). It has been shown that no ternary species are formed in the precursors prepared by batch coprecipitation. On the contrary, a new ternary Cu-Zn-Zr precursor is obtained through continuous coprecipitation due to a better local control of parameters (lower composition gradient, pH). The formation of this ternary precursor is responsible for the better local homogeneity of the final material after calcination.

After synthesis and characterization, the catalytic materials were reduced under H_2 at 300°C and the CO_2 hydrogenation tests were carried out in a fixed bed stainless steel reactor at 240-300 °C, 50 bar, GHSV = 25 000 h^{-1} (STP) with H_2/CO_2 = 3.9[3, 5]. The gas phase is analyzed online with a μ-GC, and the liquid phase is condensed, collected and analyzed by GC at the end of the reaction. The results obtained at 280 °C are shown in Table 1 and Figure 2.

Table 1. Catalytic results at 280 °C, 50 bar, 25 000 h^{-1} (STP), H_2/CO_2=3.9

Catalyst	**H_2 conv.(%)**	**CO_2 conv.(%)**	**CH_3OH sel.(%)**	**CH_3OH productivity ($g \cdot kg_{cat}^{-1} \cdot h^{-1}$)**
INDUS	11.6	25.1	34	439
CZZ1 (batch)	7.4	17.9	36	725
CZZ2 (microfluidic)	9.9	21.4	33	1135

It can be seen that all the synthesized catalysts present lower conversions but better MeOH productivity compared to the INDUS catalyst.

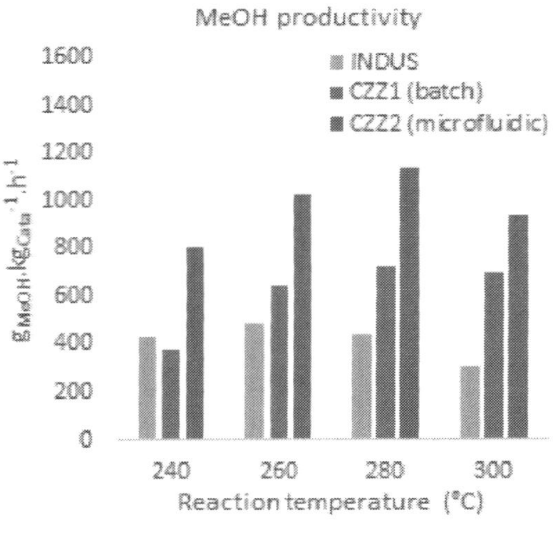

The microfluidic system (CZZ2) helps to increase the CO_2 and H_2 conversions and methanol productivity compared to batch (CZZ1).

We managed to control the physico-chemical characteristics of synthesized materials by tuning process parameters (the nature of vector fluids) which resulted in the improvement of the catalytic activity. The modification of other parameters (temperature of coprecipitation, pH precipitation and the residence time) have been

Figure 2: Methanol productivity from CO_2 hydrogenation

shown to enhance even more the homogeneity and thus the activity of these catalytic materials.

The possible production of dimethylether from CO_2 has been studied. The coupling of the most performant Cu-based catalyst with a catalyst with tunable acidity (zeolithe ZSM5) has been performed. The presence of the second acid catalyst clearly modifies the composition of the outlet flow. The selectivity to DME reaches almost 50% and the selectivity to CO is strongly decreased compared to "pure methanol catalyst". As a consequence, excellent dimethylether yields were obtained, up to 700 $g_{(DME+MeOH)}.kg_{cata}^{-1}.h^{-1}$ at 20 bar and 280 °C.

The whole study, synthesis, precursor characterization, catalytic activity for methanol synthesis and catalytic activity for DME synthesis, when combined to ZSM5, will be presented.

Funding: National Research Agency of France, project DIGAS ANR-14-CE05-0012

Acknowledgement: Prof Sébastien Paul and Dr Svetlana Heyte for performing the catalytic experiments on the RealCat platform (Lille, France).

References

[1] M. Behrens *et al.*, *J. of Inorg. and Gen. Chem.* 639, **2013**, 2683–2695

[2] C. Li *et al.*, *Appl. Catal. A Gen.*, 469, **2014**, 306–311

[3] L. Angelo *et al.*, *Catal. Today*, 270, **2015**, 59–67

[4] Y-T. Yang *et al.*, *Adv. Powder Technol.*, 26, **2015**, 156–162

[5] L. Angelo *et al.*, *Comptes Rendus Chim.*, 18, **2015**, 250–260

544653 Oxygenate Formation over K/Mo2c Catalysts - Role of Preparation and Promotion

Tuesday, June 4, 2019: 2:42 PM
Texas Ballroom EF (Grand Hyatt San Antonio)

Wijnand Marquart[1], David J. Morgan[2], Graham J. Hutchings[3], Michael Claeys[1] and **Nico Fischer**[1], (1)University of Cape Town, Cape Town, South Africa, (2)Carfidff University, Cardiff, United Kingdom, (3)School of Chemistry, Cardiff Catalysis Institute, Cardiff, United Kingdom

The Fischer-Tropsch (FT) process, producing long chained waxes and transportation fuels, is competing with fuels derived from crude oil, and its profitability is therefore dependent on the global oil price.[1] However, increasing the value of synthesized products could render the profitability of the FTS independent of fluctuations in the oil price (which are mostly due to global political trends rather than market demand). One way to achieve this, is to tailor the reactions selectivity through catalyst design and/or changes in reaction conditions to yield higher value compounds such as long chained oxygenates. Oxygenates are platform chemicals which can through further processing, be converted to alternatives for synthetic fuels or industrial chemicals and polymers.[2] Oxygenates are a typical by-product of the FT synthesis but due to selectivity limitations of the commonly employed catalysts, no commercial FT based process exists which produces oxygenates at a significant yield.[3]

Typically, transition metals such as Fe, Co, Rh and Ni are active for the FT synthesis. Based on reaction conditions employed, commercial Fe and Co based catalysts have been shown to produce between 6 and 12 C% oxygenates.[4] Rh has been demonstrated to have a higher oxygenate selectivity, but the associated high raw material cost becomes prohibitive for use as a commercial FT catalyst.[5] Catalysts other than the traditionally known FT active transition metals have also shown promising results. Transition metal carbides such as Mo_2C, were investigated under Fischer-Tropsch conditions. While the bare carbide produces mainly methane and other hydrocarbons, upon promotion with potassium the selectivity showed a significant shift towards oxygenates.[6]

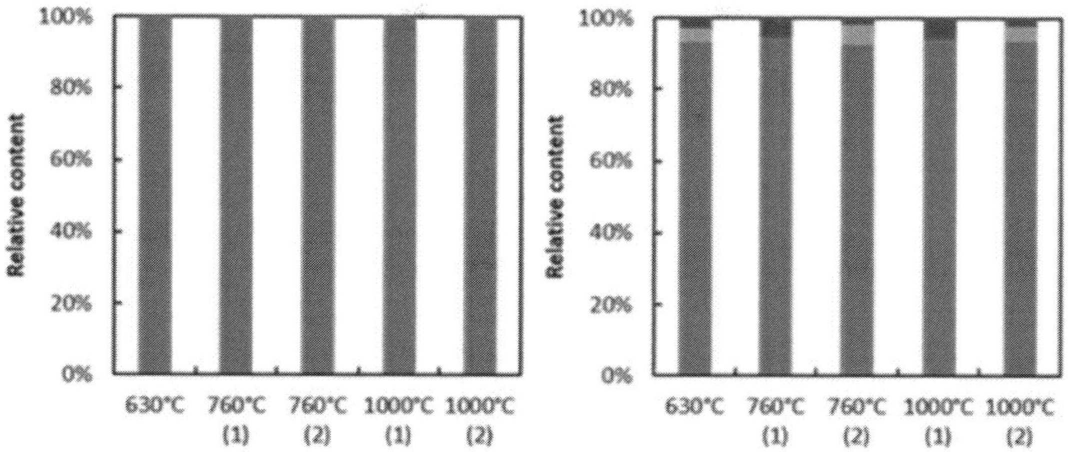

Figure 1: XPS analysis of the samples carburized at 630, 760 and 1000°C before (1) and after (2) the TPH treatment. Figure on the left displays the C 1s (red), O 1s (grey) and Mo 3d (blue) relative content on the surface of the samples. Figure on the right displays the graphitic (red), carbidic (grey), C-O (blue), C=O (green) and saturated graphitic (purple) carbon relative content on the surface of the samples (from [7]).

In the present study,[7] β-Mo_2C was obtained via carburization of MoO_3 in a mixture of methane and hydrogen at different temperatures. It could be confirmed that with increasing temperature the formation of carbon deposits is enhanced. Samples treated at 630-650°C did not show any evidence of carbon deposition in TEM or Raman spectroscopy and re-oxidized to MoO_2 upon exposure to air at room temperature. A mild passivation treatment in 1% O_2 in N_2 prevented the bulk re-oxidation. Samples carburized at 760 and 1000°C were apparently stabilized via a layer of carbon. It has to be noted though, that for all air-stable/stabilized samples, Raman spectroscopy evidenced the presence of surface MoO_3, suggesting that the carbon layer does not fully protect the Mo_2C surface but yields a passivated surface possibly through diffusion limitation.

Detailed XPS analysis of the materials support the trend of an increasing graphitic layer with increasing carburization temperature. However, samples synthesized at 630-650°C also displayed graphitic carbon depositions besides the expected carbidic one. Neither TEM nor Raman did detect this. With a a temperature programmed reduction of the samples carburized at 760 and 1000°C monitored by the methane concentration in the reactor outlet gas by online GC-TCD, reaction conditions were determined to remove the carbon layer. After treatment, the carbon speciation of the high temperature samples resembled that of the low temperature carburized sample.

Both the passivated and the carbon coated catalysts were activated in H_2 to remove the passivation/carbon layers before being exposed to syngas conversion conditions (T = 280°C, P = 33 bar, H_2/CO = 1). Compared to pristine β-Mo_2C (synthesized at 630°C, not passivated but directly exposed to reaction conditions) the CO conversion of the passivated and reduced and of the carbon bearing catalysts is significantly lower. Interestingly, at iso-conversion (adjusted through changes in space velocity and to a lesser extend reaction temperature), the passivated and reduced sample showed an eight times higher oxygenate content in the organic product while at the same time displaying a higher CO_2 selectivity (from 31 to 40 C%) and a lower methane selectivity (from 27 to 20 C%). The composition of the oxygenate fraction remained relatively constant at 74-79 C% MeOH. This change in selectivity is possibly linked to the introduced oxidic Mo species evidenced by Raman, supporting a different mechanistic pathway.

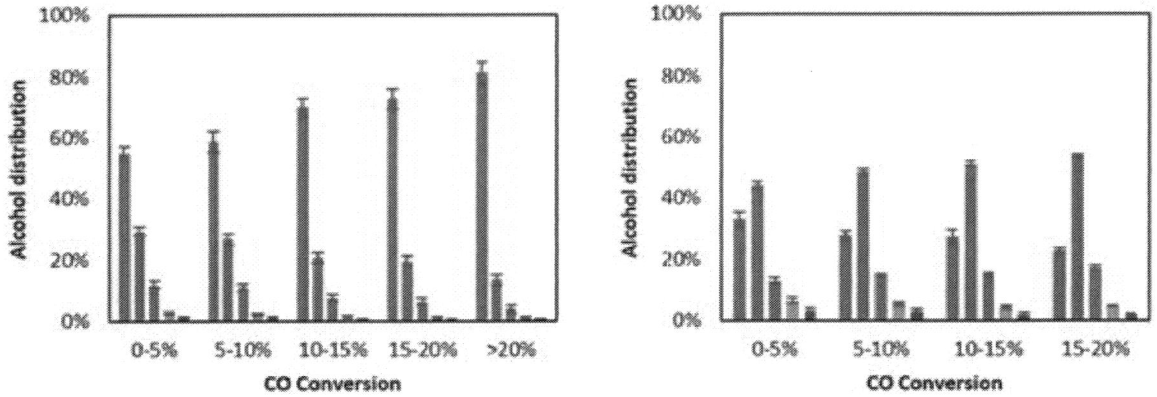

Figure 2: Alcohol distribution over the unpromoted samples (figure on the left) and the promoted samples (figure on the right) as a function of the CO conversion, with from left to right per series methanol (red, #1), ethanol (grey, #2), propanol (blue, #3), C4 alcohols (green, #4) and C5+ alcohols (purple, #5) (from [7]).

Passivated samples were subsequently promoted with up to 6.2 wt.-% potassium. Upon promotion the CO conversion of the catalyst decreased (apparent activation energies decreased from 122.4 kJ/mol for the unpromoted sample to 40.8 kJ/mol for the highest promotion level) in parallel to an increase in CO_2 and a decrease in CH_4 selectivity. The amount of potassium added did not influence the performance within the studied range (1.9 to 6.2 wt.-%). The oxygenate fraction in the organic product also increased upon promotion, either through a secondary formation route in the presence of potassium or rather through the suppression of secondary reaction of oxygenates forming hydrocarbons. The composition of the oxygenate fraction changes drastically. While still dominated by the alcohols, ethanol is now the most dominant product with significant amounts of aldehydes detected at lower conversion levels.

We have been able to demonstrate the mayor effect synthesis conditions have on the performance of a β-Mo_2C catalyst in the CO hydrogenation. In addition, our results suggest that amorphous surface oxide speciesinfluence not only on activity but also on selectivity. The oxygenate selectivity and composition could be altered via potassium promotion although the amount of potassium within the studied range had only a minor effect and the increase in oxygenate formation must be balanced with a decreased CO conversion, increased water gas shift activity and decreased methanation.

[1] Dry, M.E. Catalysis Today, 2002, 71, 227-241.

[2] Surisetty, V.R et al., Applied Catalysis A: General, 2011, 404, 1-11.

[3] Klimkiewicz, R. Chemistry Central Journal, 2014, 8, 77-85.

[4] Klerk, A. d. Energy & Environmental Science, 2011, 4, 1177-1205.

[5] Forzatti, P. et al., Catalysis Reviews - Science and Engineering, 1991, 33, 109-168.

[6] Woo, H.C. et al., Applied Catalysis, 1991, 75, 267-280.

[7] Marquart, W. et al., Catalysis Scinence and Technology, 2018, 8, 3806-3817.

545678 High Throughput Testing of Catalysts with Fast Deactivation for Methanol-to-Hydrocarbons (MTH)

Tuesday, June 4, 2019: 3:06 PM
Texas Ballroom EF (Grand Hyatt San Antonio)

Marius Kirchmann, Alfred Haas and Christoph Hauber, hte GmbH, Heidelberg, Germany

State-of-the-art catalyst SAPO-34 was tested for the reaction of Methanol-to-Hydrocarbons. In order to capture the rapidly changing product spectrum during deactivation, suitable test methods were developed at commercially relevant conditions. In order to follow the rapid deactivation of SAPO-34 in high time resolution, a fast spectroscopic method was used and evaluated by chemometrics.

1. Introduction

In this article we will focus on the Methanol-to-Hydrocarbons (MTH) reaction that has been thoroughly investigated with current state-of-the-art catalyst SAPO-34. This process represents a challenging application for parallel testing of catalysts which was met by appropriate test protocols and an advanced analytical setup to detect complex multi-component products in connection with a fully automated data evaluation.

2. Experimental

All experiments were carried out in 16-fold parallel fixed bed reactor systems. The unique process control guarantees constant pressure, flow, feed concentration and temperature (up to 570‹C) and allows sequential start-up of the reactors and product analysis. This means that catalysts with different deactivation time-scales can be compared at constant time-on-stream.

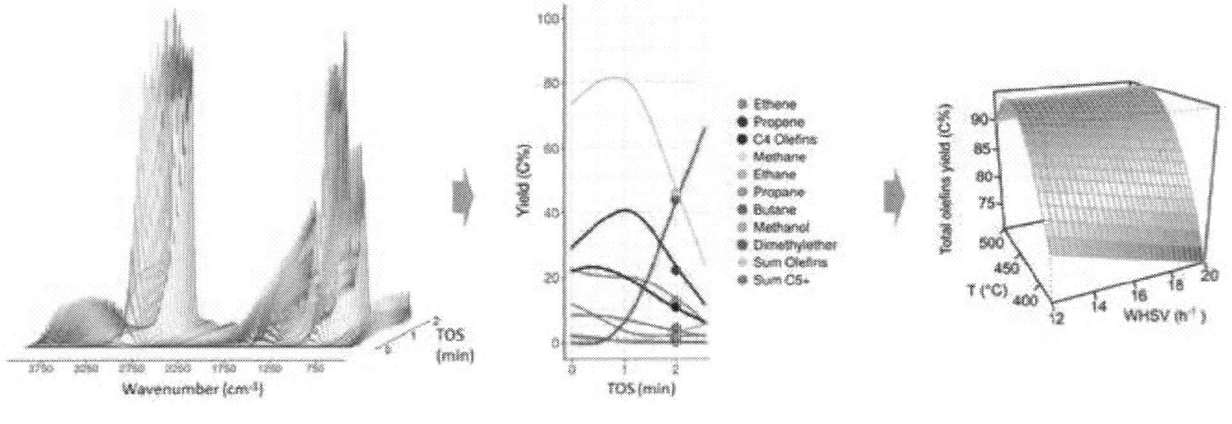

Figure 1. Data evaluation workflow from FTIR spectra (left) to chemometrical interpretation (middle) to regression of design of experiments DoE (right)

3. Results and discussion

The evolution and distribution of products on SAPO-34 was studied in a fixed-bed reactor at different severities ranging from 0.2-20 h^{-1}. Thereby, the timescale of deactivation at low severities smaller 1 h^{-1} is slow enough for online GC to give detailed information on changes of the product distribution and the reaction network. However, at higher and commercially relevant severities > 1 h^{-1} the timescale of deactivation becomes too fast to be adequately followed by online GC. In order to follow the rapidly changing reaction products, FTIR spectroscopy in the gas-phase was applied in measurement intervals down to 5 s. Multivariate analysis of the FTIR spectra and correlation to the online product analysis by GC was used for in-situ training of chemometrical models for all major products between C1-C5 . These models were used for the prediction of product selectivities from the FTIR spectra in high time resolution giving insights into the product formation during incubation, selectivation and deactivation period of the catalyst.

Partially burning-off the coke-on-catalyst indicated that the remaining coke-on-catalyst is linked to the product distribution and that a certain selectivation by coke is required to yield the maximum olefin selectivity. Finally, the temperature and WHSV were varied in a small design of experiments (DoE) showing the influence of these parameters on selectivity and conversion capacity.

4. Conclusions

Testing of catalysts with fast decay can be done in parallel fixed-beds by starting up reactors sequentially and comparing conversion and yields during TOS. More recently, catalysts can be tested at short contact times under high severity conditions using a fast analytical setup.

References

1. C. D. Chang, Catal. Rev.-Sci. Eng. **1983**, 25, 1.

2. U. Olsbye, , S. Svelle, M. Bjørgen, P. Beato, T. V. W. Janssens, F. Joensen, S. Bordiga, K. P. Lillerud, Angew. Chem. Int. Ed. **2012**, 51, 5810.

551437 Selective Conversion of Syngas to Chemicals and Fuels Enabled By Bifunctional OX-Zeo Catalysts

Wednesday, June 5, 2019: 10:30 AM
Texas Ballroom A (Grand Hyatt San Antonio)

Feng Jiao, Junhao Yang, Gen Li, Na Li, Yuxiang Chen, **Xiulian Pan** and Xinhe Bao,
Dalian Institute of Chemical Physics, Chinese Academy of Sciences, Dalian, China

The key of syngas chemistry lies in controlled C-C coupling but without polymerizing into long chain hydrocarbons and without forming methane and paraffins. We recently reported that coupling of partially reducible metal oxides and zeolites (OX-ZEO) makes it possible to catalyze direct conversion of syngas toward light olefins with the selectivity reaching 80%, surpassing the limit obtained over conventional Fischer-Tropsch based catalysts.[1] Such a bifunctional composite affords two types of active sites with complementary properties, turning syngas conversion into a tandem reaction, i.e. activating CO and H_2, generating intermediates over the partially reduced metal oxides, and C-C coupling within the pores of zeolites. [2-5] Thus, it is important to match the two components properly in terms of the compositions, structures and intimacy to achieve an optimum catalytic activity and selectivity. With this fundamental understanding, we show that syngas can be selectively converted to a variety of chemicals and fuels with a high selectivity, including ethylene, ethane, propane, gasoline range hydrocarbons, aromatics and even oxygenates.

1. Jiao, J. Li, X. Pan, J. Xiao, H. Li, H. Ma, M. Wei, Y. Pan, Z. Zhou, M. Li, S. Miao, J. Li, Y. Zhu, D. Xiao, T. He, J. Yang, F. Qi, Q. Fu, X. Bao. *Science* **351** (2016) 1065–1068.
2. Zhu, X. Pan, F. Jiao, J. Li, J. Yang, M. Ding, Y. Han, Z. Liu, X. Bao. *ACS Catal.* **7** (2017) 2800–2804.
3. P. de Jong, *Science* **351** (2016) 1030–1031.
4. Jiao, X. Pan, K. Gong, Y. Chen, G. Li, X. Bao, *Angew. Chem. Int. Ed.*57 (2018) 4692-4696.
5. Yang, X. Pan, F. Jiao, J. Li, X. Bao. *ChemComm* 53 (2017) 11146-11149.

551409 Intermediate Product Regulation in Tandem Syngas Conversion to Liquid Fuels

Wednesday, June 5, 2019: 11:09 AM
Texas Ballroom A (Grand Hyatt San Antonio)

Gonzalo Prieto, Max Planck Institute für Kohlenforschung, Muelheim an der Ruhr, Germany; ITQ Institute of Chemical Technology (CSIC-UPV), Valencia, Spain

The single-step production of liquid hydrocarbons from syngas (H_2+CO) via the tandem integration of Fischer-Trospch (FT) and hydrocracking (HC) reactions represents an attractive approach towards the intensification of gas-to-liquid (GTL) technologies for the valorization of unconventional and delocalized carbon resources into transportable liquid hydrocarbons, circumventing the wax handling required in conventional GTL processes.[1] However, the integration of two solid catalysts in a single reactor, to steer sequential reactions in tandem poses important challenges. The sharing of a common reaction medium generally means that each catalyst operates under individually suboptimal conditions. Besides, the dichotomy is often faced that close spatial proximity between the tandem catalysts – required for an effective transport of intermediate products between their active sites with minor derivation via undesired reaction pathways – prohibits individual adjustment of other key parameters such as the catalyst operation temperature. The FT/HC process is an excellent showcase for these limitations.

Operating the FT catalyst at the higher temperatures needed for HC, and the strong inhibition of the metal sites on HC catalysts by CO – which enhances secondary cracking and hydrogenolysis pathways – lead to high selectivities to gas (C_{4-}) hydrocarbons, which severely penalize the efficiency and economy of the tandem process. Our studies with model FT hydrocarbons on a Pt/ZSM-5 HC catalyst showed that, under the syngas atmosphere of the tandem process, α-olefin primary FT products can mitigate this undesired down-shift of the product distribution by curbing secondary cracking.[2] Moreover, we have found that the achievement of sub-micrometer effective transport distances in multimodally porous cobalt-based FT catalysts, enables the "channeling" of primary FT products onto the HC catalyst. In this manner, the two particulate catalysts can operate spatially distant – thus allowing a catalyst-specific temperature adjustment – albeit resembling the case of a nanoscale chemical intimacy between their active sites. The new catalyst and process configurations reconcile an effective wax depletion with high selectivities to middle distillates and minimum production of undesired gas hydrocarbons.[3]

References

[1] a) A. Martinez, G. Prieto, *Top. Catal.* 52 (2009) 75-90; b) Q. Zhang, K. Cheng, J. Kang, W. Deng, Y. Wang, *ChemSusChem* 7 (2014) 1251-64 c) S. Sartipi, M. Makkee, F. Kapteijn, J. Gascon, *Catal. Sci. Technol.* **4** (2014) 893-907; d) J. Li, Y. He, L. Tan, P. Zhang, X. Peng, A. Oruganti, G.Yang, H. Abe, Y. Wang, N. Tsubaki, *Nature Catal.* **1** (2018) 787–793.

[2] N. Duyckaerts, et al., *ACS Catal.* **6** (2016) 4229–4238.

[3] N. Duyckaerts, et al., *Angew. Chem. Int. Ed.* **56** (2017) 11480–11484.

545541 Bifunctional Catalysis for Direct Transformations of Syngas and Carbon Dioxide into Lower Olefins and Aromatics

Wednesday, June 5, 2019: 11:48 AM
Texas Ballroom A (Grand Hyatt San Antonio)
Xiaoliang Liu, Mengheng Wang, Jincan Kang, Kang Cheng, Qinghong Zhang and **Ye Wang**, Xiamen University, Xiamen, China

Bifunctional Catalysis for Direct Transformations of Syngas and Carbon Dioxide into Lower Olefins and Aromatics

Xiaoliang Liu, Mengheng Wang, Jincan Kang, Kang Cheng, Qinghong Zhang, Ye Wang*

State Key Laboratory of Physical Chemistry of Solid Surfaces, Collaborative Innovation Center of Chemistry for Energy Materials, College of Chemistry and Chemical Engineering, Xiamen University, Xiamen, 361005, China

**Corresponding author: wangye@xmu.edu.cn*

Syngas (CO/H_2), which can be produced from various carbon resources including natural/shale gas, coal and biomass, has attracted much attention as a key platform for the production of liquid fuels and chemicals.[1] At the same time, syngas derived from coal or biomass may contain a considerable concentration of CO_2, which has to be converted into useful products along with CO. More broadly, the hydrogenation of CO_2 by H_2 produced using renewable energy into value-added chemicals would offer a sustainable route for utilization of CO_2 as a renewable feedstock, contributing to low-carbon society. Lower (C_2-C_4) olefins and aromatics are key building-block chemicals. Many studies have been devoted to the conversion of syngas via Fischer-Tropsch (FT) synthesis into olefins (FTO), but the selectivity is limited by the Anderson-Schulz-Flory distribution, which shows a maximum C_2-C_4 selectivity of ~60%.[2,3] The hydrogenation of CO_2 to lower olefins is proposed to proceed via CO intermediate by the reverse water-gas shift (RWGS) reaction ($CO_2 + H_2$ ¨ $CO + H_2O$) followed by the FTO reaction over many catalysts. The selectivity of C_2-C_4 olefins is thus also limited to 60%. Aromatics are difficult to be obtained with a considerable selectivity via FT synthesis. The development of new routes for selective synthesis of C_2-C_4 olefins or aromatics from syngas and CO_2 is a highly attractive but challenging task.

Our idea is that the hydrogenation of CO into hydrocarbons involves a series of tandem elementary steps including CO activation and C-C coupling, and thus cannot be precisely controlled by one catalyst as in the conventional FT synthesis. The coupling of tandem steps using well-designed multifunctional catalysts containing components suitable to the respective steps would be a promising methodology. The use of one catalyst to integrate CO to C_1 intermediate (e.g., methanol) and methanol to olefins or aromatics (MTO or MTA) in tandem is a challenging task. Early studies using Cu-Zn-Al-O, the typical methanol-synthesis catalyst, in combination with the MTO or MTA catalyst led to the formation of C_2-C_4 paraffins and almost no olefins.[4]

Recently, we and a few other groups succeeded in integrating the activation of CO and the selective C-C coupling to lower olefins by designing highly selective bifunctional catalysts (Fig. 1).[5,6] We discovered that ZnO-ZrO_2 solid solution could catalyze the conversion of CO to CH_3OH/CH_3OCH_3 and was unique in keeping lower olefins from further hydrogenation. The selectivity of C_2-C_4 olefins reached 70-80% at CO conversion of 10-30% over a bifunctional catalyst composed of ZnO-ZrO_2 and zeolite SAPO-34 or SSZ-13.[6,7] Furthermore, the ZnO-ZrO_2/ZSM-5 catalyzed the direct conversion of syngas into aromatics with a selectivity of 80% at CO conversion of 20%.[8] The catalyst was very stable and no deactivation was observed in 1000 h.

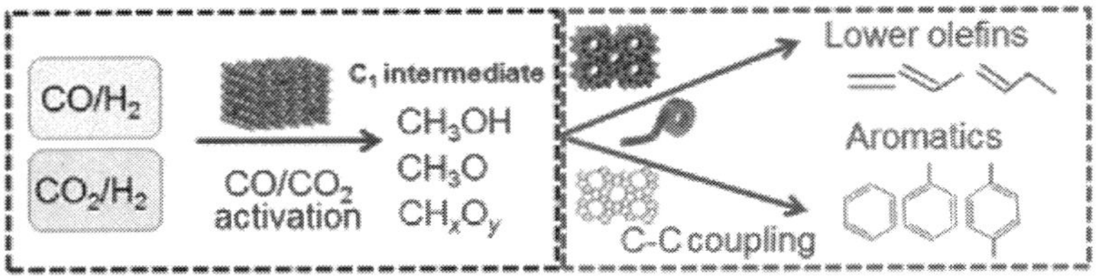

Fig. 1. Reaction coupling for direct production of lower olefins and aromatics from syngas or CO_2 by the integration of two active components.

We further found that bifunctional catalysts composed of Zn-Al-O or Zn-Ga-O binary oxide and zeolite SAPO-34 were highly selective for the conversion of syngas into C_2-C_4 olefins. Furthermore, these bifunctional catalysts were also very efficient for the hydrogenation of CO_2 into lower olefins. For example, the selectivity of C_2-C_4 olefins reached 86% at CO_2 conversion of 13% at 370 ºC.[9] The combination of the Zn-Al-O or Zn-Ga-O oxide with H-ZSM-5 could catalyze the hydrogenation of both CO and CO_2 to aromatics with selectivity higher than 70%.

Our studies on the effect of catalyst composition revealed that the catalyst with a Zn/Al or Zn/Ga molar ratio of 1/2 was the best for lower olefin or aromatic formation from either CO or CO_2. Characterizations using ESR suggested that the density of oxygen vacancies on the catalyst with Zn/Al or Zn/Ga ratio of 1/2 was the highest. The quantitative correlation suggests the oxygen vacancy site on binary metal oxides plays a crucial role in the activation and conversion of CO or CO_2 molecule (Fig. 2). We further investigated the activation of H_2 by H_2-D_2 exchange reaction. Our results indicated that the –Zn–O– domain on the Zn-Al-O oxide surface accounts for H_2 activation probably via a heterolytic dissociation mechanism (Fig. 2). We further confirmed that Brønsted acid sites in SAPO-34 or H-ZSM-5 are responsible for C-C coupling to form C_2-C_4 olefins or aromatics, but the density of Brønsted acid sites should be controlled because Brønsted acid sites also catalyze the hydrogenation of lower olefins to paraffins.

Our kinetic studies reveal that methanol and dimethyl ether are mainly formed from syngas and CO_2 on $ZnAl_2O_4$, which are transferred into SAPO-34 cages to be selectively converted to C_2-C_4 olefins or into H-ZSM-5 channels to be converted into aromatics owing to the shape-selective catalysis. In situ infrared spectroscopic studies and DFT calculations suggest that formate and methoxide species are key adsorbed species on $ZnAl_2O_4$ during the conversions of both CO and CO_2 in the presence of H_2 (Fig. 2).

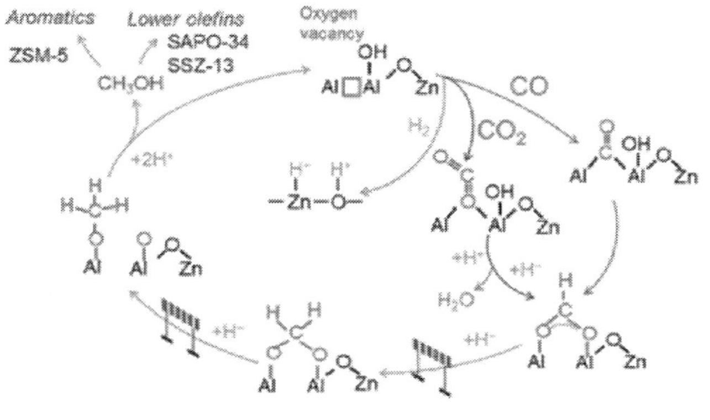

Fig. 2. Proposed reaction mechanism for the activation of CO, CO_2 and H_2, and the formation of lower olefins and aromatics via methanol intermediate over bifunctional catalysts.

References

1. K. Cheng, J. Kang, D. L. King, V. Subramanian, C. Zhou, Q. Zhang, Y. Wang, *Adv. Catal.* **2017**, *60*, 125-208.

2. H. M. T. Galvis, J. H. Bitter, C. B. Khare, M. Ruitenbeek, A. I. Dugulan, K. P. de Jong, *Science* **2012**, *335*, 835-838.

3. L. Zhong, E. Yu, Y. An, Y. Zhao, Y. Sun, Z. Li, T. Lin, Y. Lin, X. Qi, Y. Dai, L. Gu, J. Hu, S. Jin, H. Wang, *Nature* **2016**, *538*, 84-87.

4. K. Fujimoto, H. Saima, H. Tominaga, *J. Catal.* **1985**, *94*, 16-23.

5. F. Jiao, J. Li, X. Pan, J. Xiao, H. Li, H. Ma, M. Wei, Y. Pan, Z. Zhou, M. Li, S. Miao, J. Li, Y. Zhu, D. Xiao, T. He, F. Qi, Q. Fu, X. Bao, *Science* **2016**, *351*, 1065-1068.

6. K. Cheng, B. Gu, X. Liu, J. Kang, Q. Zhang, Y. Wang, *Angew. Chem. Int. Ed.* **2016**, *55*, 4725-4728.

7. X. Liu, W. Zhou, Y. Yang, K. Cheng, J. Kang, L. Zhang, G. Zhang, X. Min, Q.

Zhang, Y. Wang, *Chem. Sci.* **2018**, *9*, 4708-4718.

8. K. Cheng, W. Zhou, J. Kang, S. He, S. Shi, Q. Zhang, Y. Pan, W. Wen, Y. Wang, *Chem* **2017**, *3*, 334-347.

9. X. Liu, M. Wang, C. Zhou, W. Zhou, K. Cheng, J. Kang, Q. Zhang, W. Deng, Y. Wang, *Chem. Commun.* **2018**, *54*, 140-143.

544359 Transient Kinetics of the Fischer-Tropsch Reaction on Cobalt Catalysts

Wednesday, June 5, 2019: 10:30 AM
Texas Ballroom EF (Grand Hyatt San Antonio)

Wei Chen, Robert Pestman and Emiel Hensen, Schuit Institute of Catalysis, Department of Chemical Engineering and Chemistry,, Eindhoven University of Technology, Eindhoven, Netherlands

Transient Kinetics of the Fischer-Tropsch reaction on Cobalt Catalysts

Wei Chen, Robert Pestman, Emiel J.M. Hensen

Schuit Institute of Catalysis, Eindhoven University of Technology, 5600 MB Eindhoven, The Netherlands

Fischer-Tropsch (FT) synthesis continues to expand in importance as an effective rout to convert natural gas into liquid fuels and chemicals. Although the industrial application of the FT reaction preceded its scientific understanding, there remains a strong desire to understand the underlying reaction mechanism in order to develop better catalysts. In this context, the use of transient kinetics provides a unique method to furnish mechanistic information in a less ambiguous way than conventional steady-state kinetics. In this contribution, we provide novel insights into three topics of the kinetics and mechanism of cobalt-catalyzed FT reaction by combining transient kinetics with other techniques.

1) CO dissociation on Co/SiO_2 catalyst was probed by $^{12}C^{16}O/^{13}C^{18}O$ scrambling in the absence and presence of H_2 (Fig. 1). The initial scrambling rate without H_2 ($TOF = 1.6*10^{-2}\ s^{-1}$) is significantly higher than the rate of CO consumption under CO hydrogenation condition in steady state ($TOF = 3.0*10^{-3}\ s^{-1}$), meaning the surface contains sites sufficiently reactive to dissociate CO without the assistance of H atoms. Quantitative study reveals that only less than 20% of the cobalt metal surface, likely coordinatively unsaturated sites, is involved in CO scrambling. A minor influence of CO scrambling and CO residence time (by SSITKA) on H_2 partial pressure suggests that CO dissociation is only weakly affected by the presence of H_2. An additional proof of this conclusion was obtained by temperature-programmed in situ CO-IR studies on Co/SiO_2. With increasing temperature, the CO band shifts to higher wave number (from 2025 cm^{-1} to 2060 cm^{-1}) at approximately 150 ^oC. As confirmed by supplementary IR experiments and DFT calculations, this blue-shift of CO band was correlated to the lateral C-CO and O-CO interaction resulted from CO dissociation on cobalt surface. We observed that the onset temperature of The CO dissociation was independent of the H_2 pressure. This further underpins the conclusion that H_2 is not involved in CO dissociation on cobalt surface.

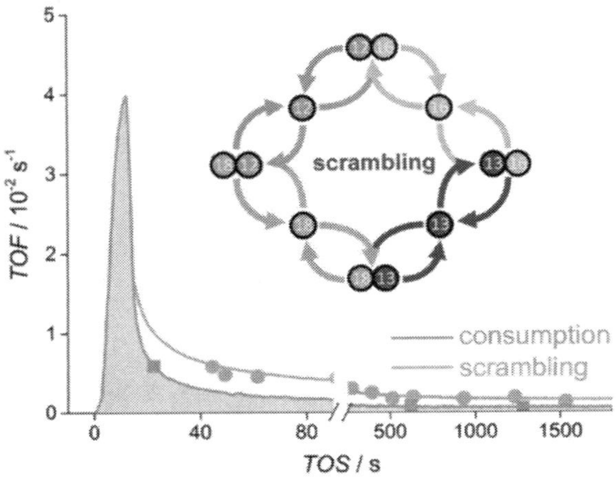

Fig.1. Dependence of CO scrambling (red) and consumption (blue) rate in the absence of H_2 on time-on-stream. Lines correspond to on-line MS measurement and points to GC-MS data. Condition: 220 °C, $p(^{12}C^{16}O) = p(^{13}C^{18}O) = 45$ mbar. The inset schematically presents the $^{12}C^{16}O/^{13}C^{18}O$ scrambling.

2) The mechanism of CO methanation at 260 °C was studied by using steady-state isotopic transient kinetic analysis (SSITKA) and backward and forward chemical transient kinetic analysis (CTKA) (Fig. 2) [2]. It was figured out that C and O hydrogenation are relatively slow compared to CO dissociation, and that the overall barrier for CO hydrogenation to CH_4 is lower than the CO adsorption energy. The combined transient kinetic data were used to fit an explicit micro-kinetic model for the methanation reaction. Micro-kinetics simulations based on the fitted parameters confirms that under methanation conditions the overall CO consumption rate is mainly controlled by C hydrogenation and to a smaller degree by O hydrogenation and CO dissociation. Under FT synthesis conditions, a model-based approach of SSITKA measurements involving hydrocarbon products containing up to five carbon atoms was employed to gain deeper insight into the relation between the kinetics and the composition of the surface adsorbed layer [3]. SSITKA data show that the rates of chain growth and chain decoupling are much higher than the rates of monomer formation and chain termination. An important corollary is that the fraction of free sites, which is mainly determined by CO pressure, has opposing effects on CO consumption rate and chain-growth probability. Lower CO pressure and more free sites leads to increased CO consumption rate but decreases chain-growth probability because of an increasing ratio of chain decoupling over chain growth (schematically presented in Fig. 3). The preferred FT condition involves high CO pressure in which chain growth probability is increased at the expense of the CO consumption rate. Degree of rate control analysis suggests that CO dissociation under FT condition controls the overall rate more significantly than that under methanation condition. The shift in rate-controlling steps relates to increased surface coverage at lower temperature.

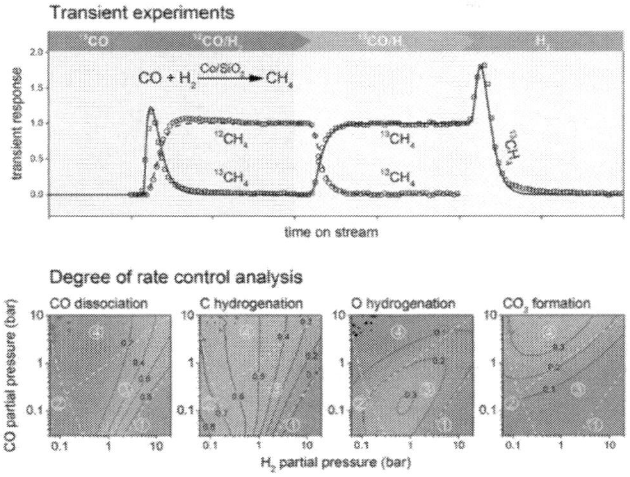

Fig.2. Combined transient experiments (top) and degree of rate control analysis based on model fitting (bottom). Condition: 260 °C, p(CO) =30 mbar, p(H_2) = 450 mbar.

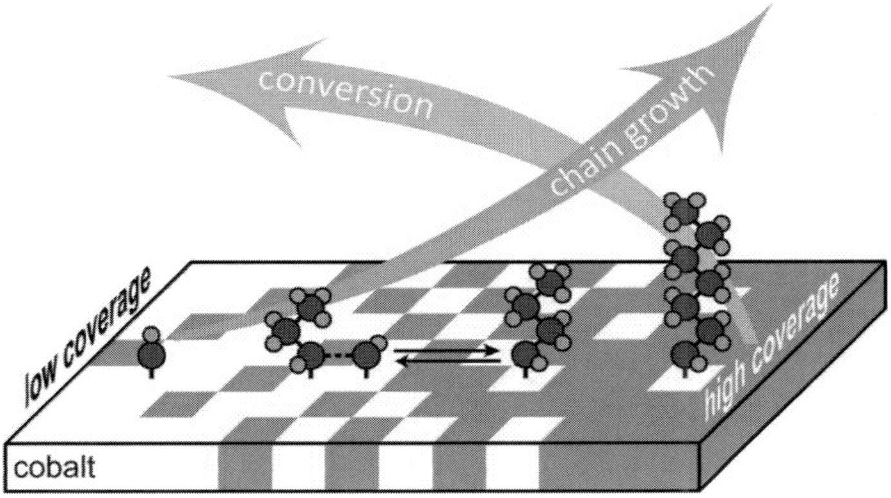

Fig.3. Schematic presentation of the influence of surface coverage on CO conversion and chain growth.

3) Effects of graphitic carbon deposits on FT performance is explained by a two-site model on the basis of the structure-sensitivity of the elementary steps [4]. Depositing graphitic carbon via the Boudouard reaction on cobalt catalyst strongly suppresses the non-ASF CH4, while the formation of higher hydrocarbons is much less affected. We provide the evidence that the amorphous carbon does not influence the FT reaction as it can be easily hydrogenated under reaction conditions, and that the graphitic carbon cannot be removed under the CO hydrogenation condition. This unreactive form of carbon is located on terrace sites and mainly decreases the CO conversion by limiting CH_4 formation. Despite nearly unchanged higher hydrocarbon yield, the presence of graphitic carbon enhances the chain-growth probability and strongly suppresses olefin hydrogenation. We demonstrate

the graphitic carbon slowly deposited on cobalt catalyst during the reaction influences the CO conversion and products distribution in a similar way to that for pre-deposited graphitic carbon. We present that the buildup of graphitic carbon by ^{13}CO facilitates the rate of C–C coupling in terms of the chain-growth probability during the $^{12}C_3H_6$ hydrogenation reaction, whose products follow an ASF-type product distribution of the FT reaction. We explain these results by a two-site model on the basis of structure sensitivity of the underlying reaction steps in the FT mechanism: step-edge sites catalyze CO dissociation and chain growth; carbon formed on step-edge sites is either involved in chain growth or migrate to terrace sites, where it is predominately hydrogenated to CH_4 (schematically presented in Fig. 4). Covering the terraces by graphitic carbon also increases the residence time of CH_x intermediates, therefor enhances the chain-growth probability. The occurrence of non-ASF is therefore attributed to the terrace sites that favor CH_4 formation. This two-site model is also demonstrated on silver-promoted Co/SiO_2 catalysts [5]. Characterization shows that mainly coordinatively unsaturated cobalt atoms are substituted at low silver content. Adding more silver results in replacement of both coordinatively unsaturated sites and planar sites, therefore affecting both structure-sensitive CO dissociation and structure-insensitive hydrogenation reactions. Transient experiments reveal that silver addition can accelerate oxygen-removal steps, compensating the negative effect on CO conversion. Our finding presents a possibility of improving the FT performance in terms of long-chain hydrocarbons production without much loss in activity toward CO conversion.

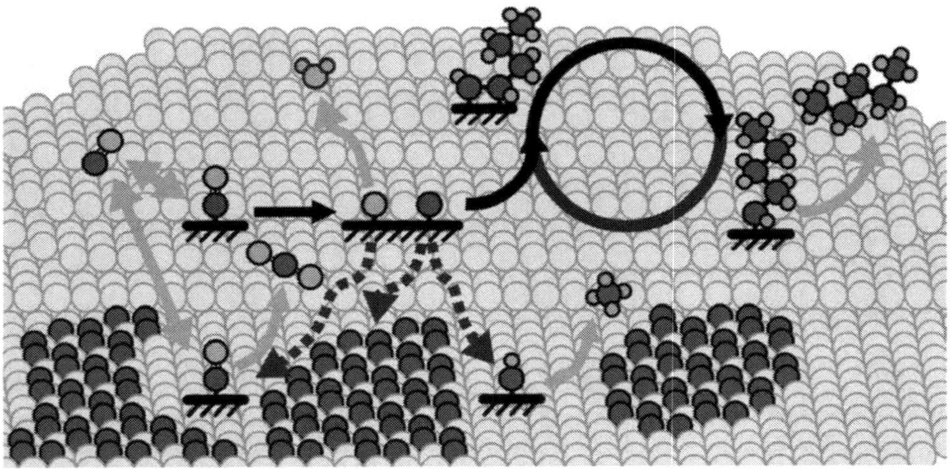

Fig.4. Schematic presentation of the effects of graphitic carbon on cobalt catalyst

These novel observations and the ensuing discussion will sharpen our thoughts on the extremely complex area of Fischer-Tropsch synthesis. We emphasize that the understanding in structure sensitivity of the elementary steps of the FT can help to design improved catalysts.

[1] W. Chen, B. Zijlstra, I.A.W. Filot, R. Pestman, E.J.M. Hensen, *ChemCatChem*, **2018**, *10* (1), 136–140.

[2] W. Chen, I.A.W. Filot, R. Pestman, E.J.M. Hensen, *ACS Catalysis*, **2017**. *7* (12), 8050–8060.

[3] W. Chen, R. Pestman, B. Zijlstra, I.A.W. Filot, E.J.M. Hensen, ACS Catalysis, 2017, *7* (12), 8061–8071.

[4] W. Chen, T.F. Kimpel, Y. Song, F.-K. Chiang, B. Zijlstra, R. Pestman, P. Wang, E.J.M. Hensen, *ACS Catalysis*, **2018,** *8* (2), 1580-1590.

[5] W. Chen, R. Pestman, F.-K. Chiang, E.J.M. Hensen, *Journal of Catalysis*, **2018**, *366*, 107-114.

551221 Some Enduring Mechanistic Puzzles in the Fischer-Tropsch Synthesis Revisited

Wednesday, June 5, 2019: 10:54 AM
Texas Ballroom EF (Grand Hyatt San Antonio)

David Hibbitts[1], Jianwei Liu[2], Hale Ay[2], Brett Loveless[2], Matthew Neurock[3] and Enrique Iglesia[4], (1)Department of Chemical Engineering, University of Florida, Gainesville, FL, (2)Chemical and Biomolecular Engineering, University of California, Berkeley, CA, (3)Department of Chemical Engineering and Materials Science, University of Minnesota, Minneapolis, MN, (4)Chemical and Biomolecular Engineering, University of California at Berkeley, Berkeley, CA

Density functional theory calculations and kinetic, spectroscopic, and isotopic experiments are combined here to address long-standing mechanistic puzzles in Fischer-Tropsch synthesis on Co and Ru-based catalysts: (i) the ability of low-index metal surfaces to activate CO at modest temperatures, high CO pressures, and thus high coverages of chemisorbed CO*; (ii) the increase in turnover rates and decrease in methane selectivity at higher water concentrations; (iii) the infrequent formation and rapid growth of hydrocarbon chains. Low-index surfaces exhibit higher reactivity as inferred from particle size studies despite the high barriers they exhibit for direct CO* activation because such routes are avoided in practice by more facile CO dissociation routes mediated by the addition of chemisorbed H-atoms before C–O cleavage to form *HCOH* species. These metal surfaces are covered by CO* and the densification of the CO* adlayer favors the dissociation of CO* because adlayer crowding preferentially destabilizes chemisorbed CO* relative to the *HCOH*-forming transition state as this reaction has a negative activation area. This H-assisted CO*-activation route, therefore, becomes facile at the moderate temperatures and large CO pressures typical of Fischer-Tropsch synthesis. Water, present at high concentrations in industrial reactors with large single-pass conversions, increases the rate of CO* activation by facilitating the H*-assisted CO* activation routes through proton-coupled electron transfer mechanisms for the addition of such H-atoms to chemisorbed CO. The rapid growth of few chains within dense CO adlayers reflect local disruptions of the CO adlayer around growing chains, which also facilitates CO* activation to form monomeric CH_x* species at sites vicinal to growing chains, thus avoiding high diffusional barriers of monomers through CO-saturated surfaces. These outstanding puzzles related to the ubiquitously observed effects of particle size, conversion, and chain growth patterns, therefore, are all resolved by a careful understanding of how CO* is activated on low-index surfaces through H*-assisted pathways that are facilitated by high CO* coverages, high H_2O concentrations, and CO*-adlayer disruption near growing hydrocarbon chains.

*** Formatting Issue - Best Available Paper/Graphic ***

546732 *CO and H_2 Adsorption on Co(11-20) - a Combined Experimental and Theoretical Investigation*

Wednesday, June 5, 2019: 11:18 AM
Texas Ballroom EF (Grand Hyatt San Antonio)

Hilde Venvik[1], Marie Døvre Strømsheim[2], Mari Helene Farstad[2,3], Ingeborg-Helene Svenum[4], Mehdi Mahmoodinia[2], Kees-Jan Weststrate[5] and Anne Borg[6], (1)Chemical Engineering, Norwegian University of Science and Technology, Trondheim, Norway, (2)Chemical Engineering, NTNU - Norwegian University of Science and Technology, Trondheim, Norway, (3)Jotun ASA, Sandefjord, Norway, (4)SINTEF Industry, Trondheim, Norway, (5)SynCat@DIFFER, Syngaschem BV, Eindhoven, Netherlands, (6)Physics, NTNU - Norwegian University of Science and Technology, Trondheim, Norway

Understanding the adsorption on and restructuring of cobalt (Co) surfaces is important towards controlling the phenomena occurring on Co-based Fischer-Tropsch catalysts. The surface of Co(11-20) is known to attain a (3x1) surface reconstruction under CO exposure [1], involving the anisotropic migration of Co atoms along [0001] to and from the step edges [2]. Co(11-20) was therefore chosen as a model system for the surface dynamics of a Fischer-Tropsch catalyst upon adsorption of CO and investigated with a combination of experimental and theoretical methods. The study is supplemented with investigations of hydrogen adsorption on different terminations of hcp cobalt to elucidate the nature of the Co(11-20) surface.

The CO-induced surface restructuring was studied with scanning tunnelling microscopy (STM), low energy electron diffraction (LEED), temperature programmed desorption (TPD) and density functional theory (DFT). The calculations were performed with the Vienna ab initio simulations package (VASP) [3]. Two theoretical model surfaces with either a missing row (MR) or an added row (AR) of Co atoms along [0001] were constructed for the CO/Co(11-20) system to represent alternatives for the (3x1) reconstruction. These were compared to (3x1) configurations of CO adsorbed on the unreconstructed surface (Figure 1). Transition state calculations with climbing image nudged elastic band (CI-NEB) [4] were performed to investigate the removal of a Co atom from the topmost layer and its mobility across the surface. In the case of (atomic) hydrogen, adsorption properties and coverage effects on Co(11-20) were studied by TPD and DFT and compared to results obtained for Co(0001) and Co(10-12).

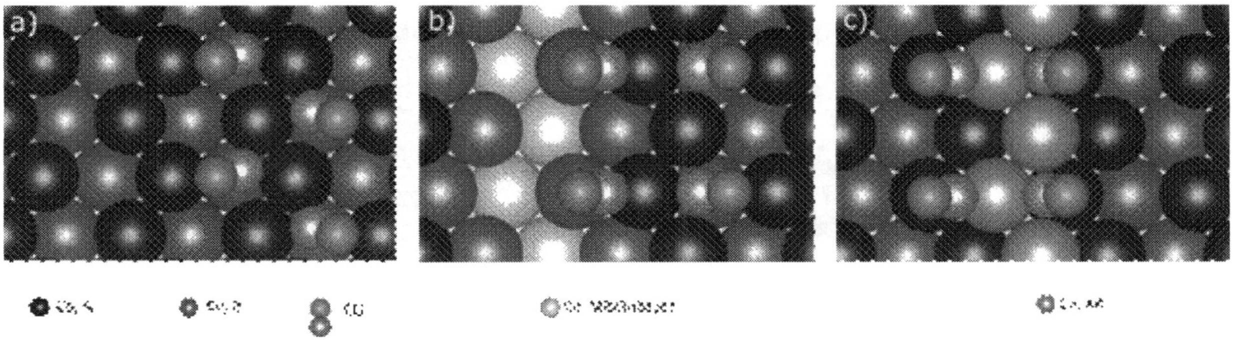

***** Formatting Issue - Best Available Paper/Graphic *****

Figure 1: the preferred adsorption geometries of (a) CO on unreconstructed (b) MR and (c) AR Co(11-20) surface.

It was inferred from LEED that, whereas the CO-induced reconstruction is facile at room temperature (RT), it may be kinetically hindered during CO exposure of clean Co(11-20) at low temperature. Previously, we have also reported that deposition of minor amounts of potassium (K) inhibits the rearrangement of CO atoms from the step edges, leading to a considerably slower and more disordered restructuring under CO [5]. Comparison of TPD data for the two cases shows significant differences in the range 200-350K, and somewhat higher temperature and coverage for the main desorption peak for the case where CO exposure was performed at RT before cooling in CO and subsequent TPD. The DFT calculations show that CO adsorption in coordination with the topmost layer of Co atoms is preferred, as this resulted in the most favourable adsorption energies irrespective of the Co(11-20) (3x1) model system configuration applied. The added row (AR) (3x1) reconstruction model yields the highest calculated adsorption energies for all coverages investigated so far, from 1 to 4 CO per (3x2) surface unit cell. Albeit small differences in adsorption energy, the unreconstructed surface remains the least stable. Moreover, surface energy values inferred increasing preference for the added row structure with increasing CO coverage. Jointly the computations and the experimental data suggest that if Co mobility is kinetically facile, the Co surface can rearrange to accommodate more and more strongly bound CO.

Co migration was investigated by comparing different initial and final configurations of a displaced single Co atom and an adsorbed CO molecule, and by obtaining the energy barrier for vacancy formation and displacement along the [0001] direction. The energy differences are not large, but imply that Co migration with CO attached is preferred over Co displacing alone, supporting the presumption that the migrating species is cobalt carbonyl. The relatively large barriers associated with the vacancy formation and migration further supports the notion of the restructuring being kinetically controlled.

From hydrogen TPD, a significantly lower desorption temperature was obtained for Co(11-20) than for Co(0001) and Co(10-12). In addition, a lower saturation coverage (H/nm^2) was found. The DFT calculations support these findings, in terms of both total and differential adsorption energies. However, there is a discrepancy in the maximum anticipated H coverage in DFT compared to TPD experiments, since our DFT calculations predict almost doubled saturation coverages for all three surfaces. Previous work by van Helden et al. [6] already explored this for H atoms on Co(0001). By experimentally comparing flat and stepped Co(0001) they could explain the difference in saturation coverage by kinetic effects; i.e. that pre-adsorbed hydrogen kinetically inhibits the adsorption as higher coverages are reached. The results support the notion that, although H_2 dissociation is facile on more open surfaces, atomic hydrogen binds stronger in (fcc) three-fold hollow sites. On Co(11-20), coordination to both top and second layer Co atoms is required for the preferred adsorption sites, and this weakens the binding. It still induces the same kinetic inhibition of H_2 dissociation.

In context of Fischer-Tropsch synthesis under industrially relevant conditions, the results obtained may point to cobalt restructuring as important for the activation of CO, and coverage effects as important for the activation of H_2. Defects such as step edges affect the adsorption of both CO and hydrogen.

References:

***** Formatting Issue - Best Available Paper/Graphic *****

[1] H. Papp, Surf. Sci. 149 (1985) 460–470.

[2] H.J. Venvik, A. Borg, C. Berg, Surf. Sci. 397 (1998) 322–332.

[3] G. Kresse, J. Hafner, Phys. Rev. B. 47 (1993) 558–561.

[4] G. Henkelman, B.P. Uberuaga, H. Jónsson, J. Chem. Phys. 113 (2000) 9901–9904.

[5] M. D. Strømsheim, I.-H. Svenum, M. H. Farstad, Z. Li, H. J. Venvik, Catal. Today 299 (2018) 37-46.

[6] P. van Helden, J.-A. van den Berg, and C. J. Weststrate, ACS Catal. 2 (2012) 1097–1107

545379 Mechanistic Understanding of Methanol Carbonylation: Interfacing Homogeneous and Heterogeneous Catalysis Via Carbon Supported Ir-La

Wednesday, June 5, 2019: 11:42 AM
Texas Ballroom EF (Grand Hyatt San Antonio)

Alyssa Hensley[1], Jianghao Zhang[2], Ilka Vinçon[3], Xavier Isidro Pereira Hernández[1], Diana Tranca[3], Gotthard Seifert[3], Jean-Sabin McEwen[1] and **Yong Wang**[4], (1)The Gene and Linda Voiland School of Chemical Engineering and Bioengineering, Washington State University, Pullman, WA, (2)Washington State University, Pullman, WA, (3)Theoretical Chemistry, Technische Universität Dresden, Dresden, Germany, (4)Institute for Integrated Catalysis, Pacific Northwest National Laboratory, Richland, WA

Mechanistic understanding of methanol carbonylation: Interfacing homogeneous and heterogeneous catalysis via carbon supported Ir-La

Alyssa J.R. Hensley [a], Jianghao Zhang [a], Ilka Vinçon [b], Xavier Pereira Hernandez [a], Diana Tranca [b], Gotthard Seifert [b,*], Jean-Sabin McEwen [a,c,d,e,*], Yong Wang [a,e,*]

[a] The Gene and Linda Voiland School of Chemical Engineering and Bioengineering, Washington State University, Pullman, WA 99164, United States

[b] Theoretical Chemistry, Technische Universität Dresden, Dresden 01062, Germany

[c] Department of Physics and Astronomy, Washington State University, Pullman, WA 99164, United States

[d] Department of Chemistry, Washington State University, Pullman, WA 99164, United States

[e] Institute for Integrated Catalysis, Pacific Northwest National Laboratory, Richland, WA 99352, United States

*Corresponding authors at: The Gene and Linda Voiland School of Chemical Engineering and Bioengineering, Washington State University, Pullman, WA 99164, United States (J.-S. McEwen and Y. Wang).

E-mail addresses: gotthard.seifert@chemie.tu-dresden.de (G. Seifert), js.mcewen@wsu.edu (J.-S. McEwen), yong.wang@pnnl.gov (Y. Wang)

1. Introduction

The carbonylation of methanol is one of the largest processes that is still being carried out under homogeneous catalysis. It allows to produce more than 6 million tonnes of acetic each year [1]. The energy requirements of the separation of catalyst and products, and the contamination associated to them, are considerably high for a process of this magnitude, therefore, finding and understanding the fundamentals of a heterogeneous catalysis homologous process is necessary and of high importance.

In this work, we report the synthesis, activity and characterization of a carbon-supported Ir-La catalyst which exhibited high activity and selectivity towards acetyls for methanol carbonylation, compared to the unpromoted carbon-supported Ir catalyst, as well as the homogenous Ir-Ru catalyst and a heterogeneous analog carbon-supported Ir-Ru catalyst [2]. The work focuses and answers two questions [3]: 1) what is the reaction mechanism and 2) what is the role La plays on the catalyst.

To provide information about these questions, X-ray photoelectron spectroscopy (XPS), high-angle annular dark-field scanning transmission electron microscopy (HAADF-STEM), in situ attenuated total reflectance (ATR) Fourier transform infrared (FTIR) spectroscopy, density functional theory (DFT) and ab initio molecular dynamics (AIMD) were used.

2. Results and Discussion

The activity measurements for the Ir-based catalysts mentioned above can be observed in Figure 1. The Ir-La/C catalyst is the only one that exhibits activity higher than the homogenous catalyst (close to 40% higher) and it exhibits around 3 times the activity of the unpromoted Ir/C catalyst. Selectivity to acetyls was > 99%. Even though Ir and Ru are used during the homogeneous process, a heterogeneous catalyst involving the same metals has about half the activity of the Ir-La/C catalyst. The reaction conditions are 240°C and 17 bar, a MeOH/MeI ratio of 70/30wt.% and a syngas molar composition of CO/H_2 = 4/1. Previous to reaction, the catalyst is activated by exposing it to a pretreatment with syngas of the same composition at the same reaction temperature for 1 h.

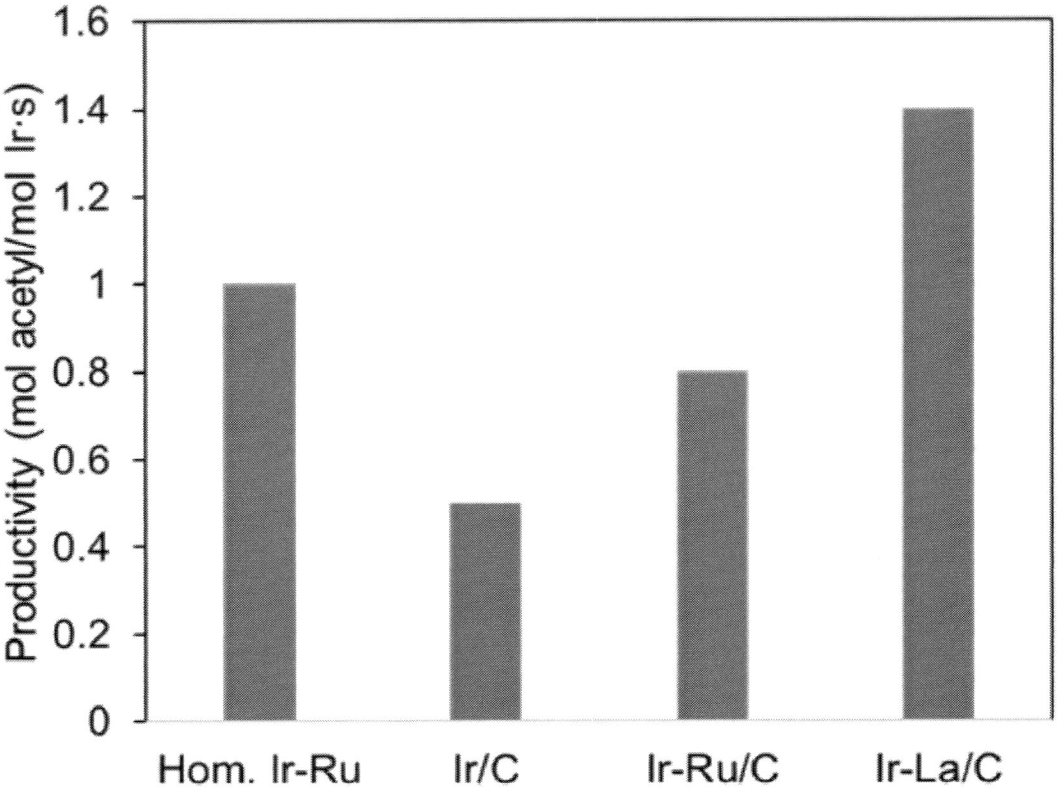

Figure 1. Activity results of Ir and M-promoted Ir catalysts supported on carbon compared to the homogeneous Ir-Ru catalyst. Reaction conditions: 17 bar, 240°C, MeOH/MeI: 70/30wt.%, syngas composition of $CO/H_2 = 4/1$.

To understand why the Ir-La/C catalyst exhibits good activity for the carbonylation of methanol, microscopic and spectroscopic characterization of the catalyst was performed. Figure 2 (left) shows that the catalyst has a very high dispersion and many of the particles contain 1-3 atoms, in which dimers are the predominating structure (50-60%). Dimers also show the pairing of atoms with high and low intensity, suggesting that Ir-La closely interact with each other. Intensity profiles indicate that the interatomic distances range between 2-3 Å. Furthermore, figure 2 (right) shows that exposing the catalyst to a syngas treatment changes the oxidation state of Ir from Ir^{3+} (that results from an inert treatment with He) to Ir^{+}. In all cases, La is always present as La^{3+}. CO-TPD experiments (not shown) indicated that the CO/Ir ratio is 1.9. This information suggests the heterogeneous Ir-La/C catalyst has a structure very similar to the homogeneous one, in which 2 CO molecules are bonded to an Ir^{+} metal center.

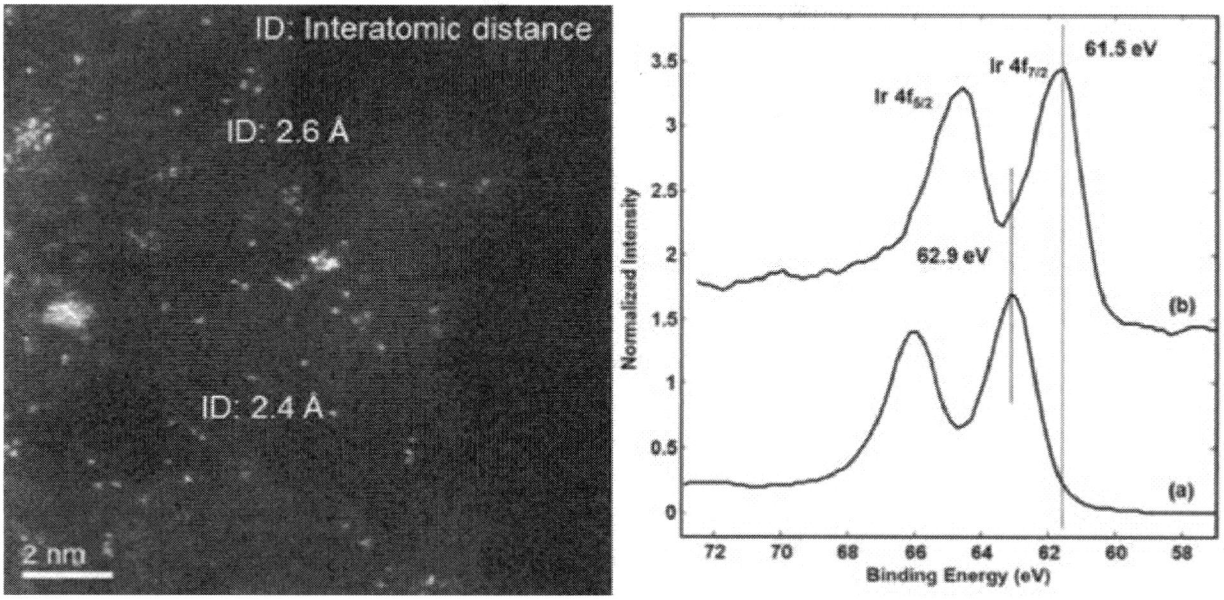

Figure 2. Left: HAADF-STEM image of the Ir-La/C catalyst. Right: XPS of the Ir-La/C catalyst after a) He pretreatment at 300°C and b) syngas pretreatment (CO/H_2 = 4/1) at 240°C after the He pretreatment at 300°C.

To provide more insights into the likely structure present in the catalyst, different configurations were studied by DFT. Three different configurations involving one trans and two cis were optimized and are presented in figure 3. The cis configurations were found to be ~ 65 kJ/mol more stable than the trans and of the two cis configurations, the open structure was found to be more stable by ~ 5 kJ/mol. ATR-FTIR spectra of the catalyst during activation conditions and the DFT calculated IR spectra of the CO stretching modes in the trans and open cis configurations of the Ir-La complex are shown in figure 4. These experimental and theoretical spectra suggest the cis structure is present when the catalyst is activated and the complexes are formed, which agrees well with its higher stability.

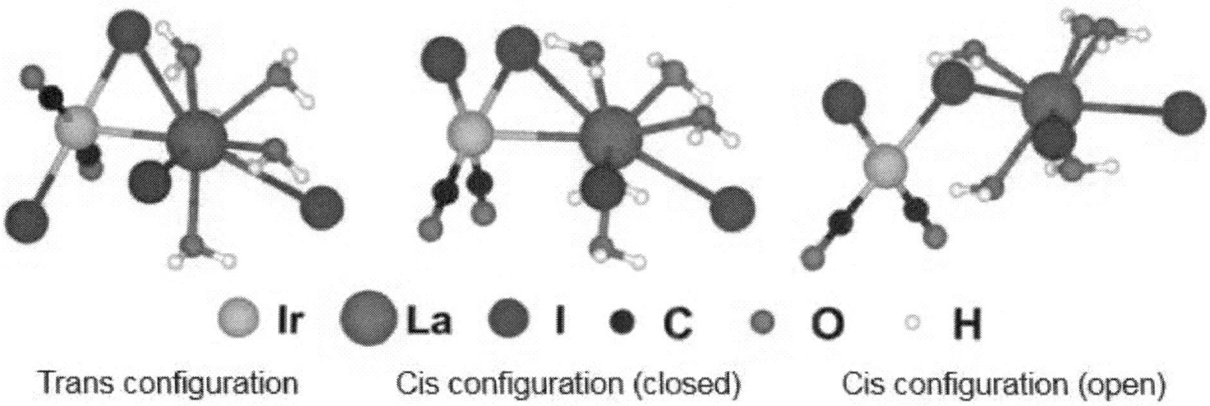

Figure 3. Optimized configurations for the trans, closed cis and open cis

configurations of the Ir-La complex with iodine ligands.

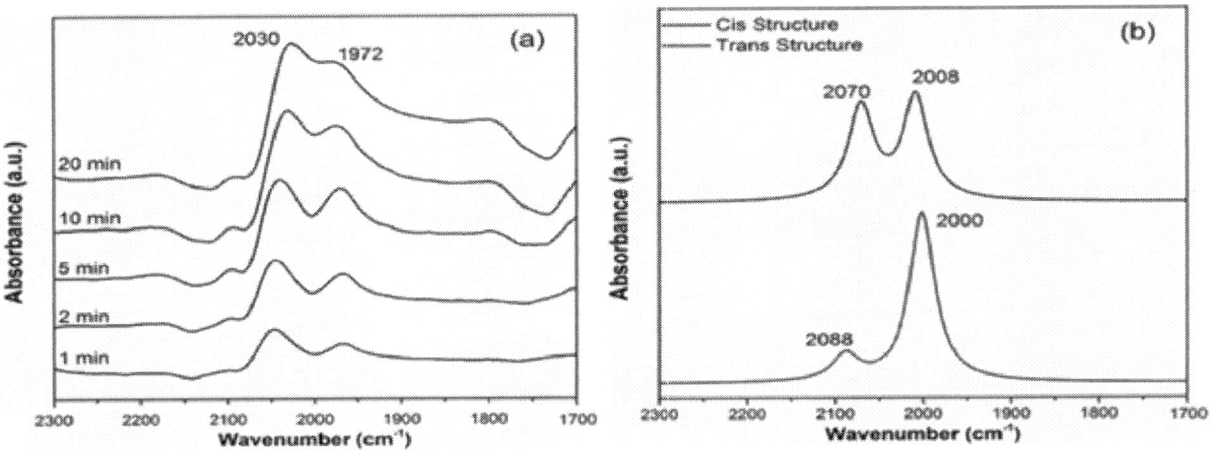

Figure 4. a) ATR-FTIR spectra of the Ir-La/C catalyst obtained during syngas pretreatment (CO/H_2 = 4/1) at 240°C and 3.75 bar, b) Spectra calculated for cis and trans configurations by density functional perturbation theory (DFPT).

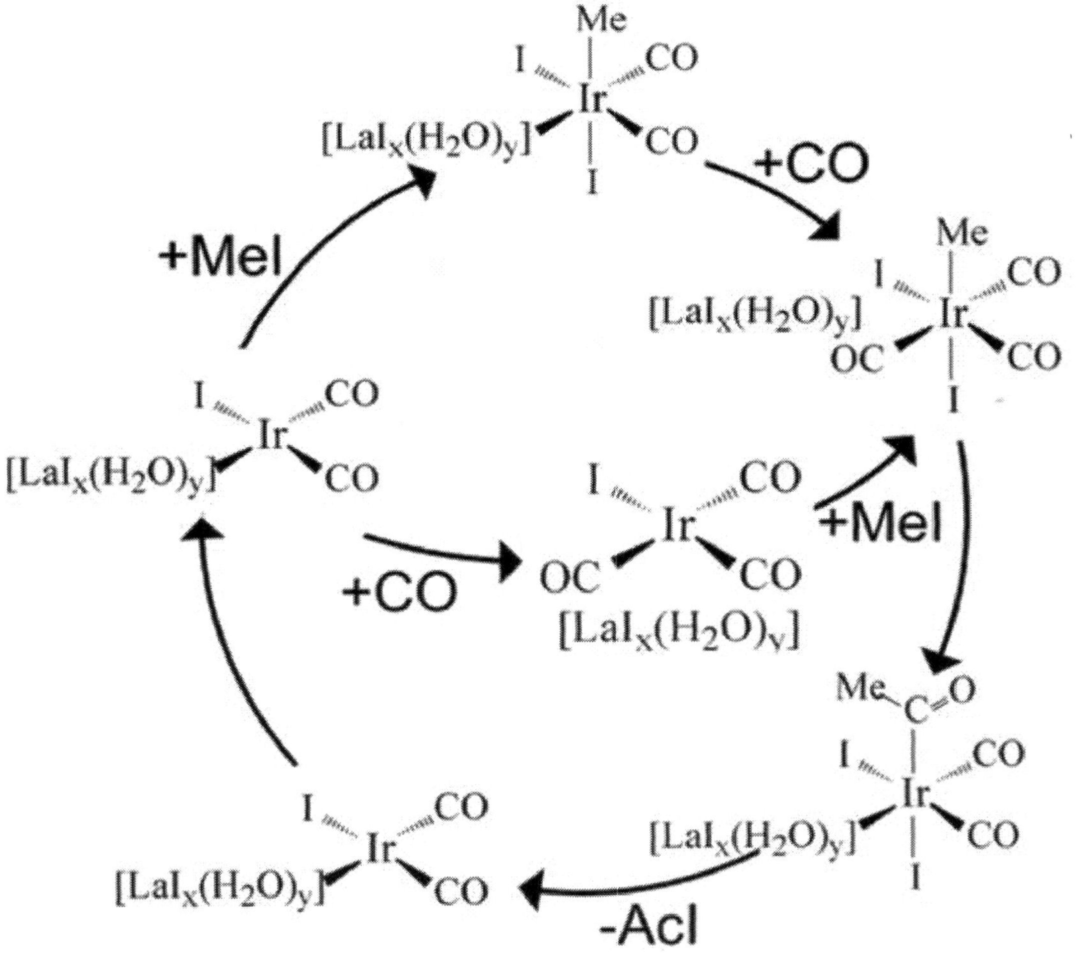

Figure 5. Reaction scheme for methanol carbonylation on the Ir-La heterogeneous catalyst complex with the La promoter acting as an iodine acceptor

Even though HAADF-STEM results would suggest the presence of the closed cis configuration due to the shorter bond between Ir and La, AIMD showed that at reaction conditions, this structure relaxes into the open cis structure. Therefore, the open cis configuration was used as a starting point in the reaction mechanism for the carbonylation of methanol obtained by DFT (figure 5). It was found that the catalyst follows a similar reaction mechanism to the homogenous analog. It starts with the oxidative addition of MeI to the Ir center to form a 6-fold coordinated structure, with a change in the oxidation state of Ir from Ir^{+} to Ir^{3+}. After the addition of MeI, La accepts an iodine from the Ir center which enables a third CO molecule to bind with the Ir center. The migratory insertion of CO into the Ir-Me bond is the next step, leading to the formation of the acetyl (Ac) group. Finally, a reductive elimination of AcI closes the cycle and restores the catalyst, in which the oxidation state of Ir changes from Ir^{3+} to Ir^{+}. It was also found that a possible variation exists in which first the third CO molecule binds to the Ir^{+} center before the oxidative addition of MeI, while the next steps remain the same. In both cases, La acts as an iodine acceptor.

Conclusions

We presented a carbon-supported Ir-La catalyst that exhibits high activity for the carbonylation of methanol. The catalyst exhibited around 40% higher activity than the homogeneous existing process and around 3 times the activity of the unpromoted carbon-supported Ir catalyst. Characterization of the catalyst showed the structure, reaction mechanism and promoter effect in the heterogeneous catalyst resembles that of the homogeneous catalyst with a very high dispersion of the metal. The present work has great implications on the possibility of using an industrially relevant heterogeneous catalyst that would not only lead to the reduction of energy requirements characteristic of homogeneous processes but also to the use of cheaper metals.

References

[1] A. Haynes, *Chapter 1 – Catalytic Methanol Carbonylation*, 1st ed., vol. 53, no. 10. Elsevier Inc., 2010.

[2] J. H. Kwak, R. Dagle, G. C. Tustin, J. R. Zoeller, L. F. Allard, and Y. Wang, "Molecular active sites in heterogeneous Ir-La/C-catalyzed carbonylation of methanol to acetates," *J. Phys. Chem. Lett.*, vol. 5, no. 3, pp. 566–572, 2014.

[3] A. J. R. Hensley, J. Zhang, I. Vinçon, X. I. Pereira Hernandez, D. Tranca, G. Seifert, J. Mcewen, and Y. Wang, "Mechanistic understanding of methanol carbonylation: Interfacing homogeneous and heterogeneous catalysis via carbon supported Ir-La," *J. Catal.*, vol. 361, pp. 414–422, 2018.

544664 Cobalt Nanocrystals As Model Catalysts for the Fischer-Tropsch Synthesis

Wednesday, June 5, 2019: 12:06 PM
Texas Ballroom EF (Grand Hyatt San Antonio)

Tom W. van Deelen and Krijn de Jong, Inorganic Chemistry and Catalysis, Utrecht University, Utrecht, Netherlands

Cobalt nanocrystals as model catalysts for the Fischer-Tropsch synthesis

Colloidal synthesis of metal nanocrystals (NC) potentially offers high control over their properties (size, shape or composition) and is therefore a promising approach to prepare well-defined model catalysts. For example, the controlled synthesis of cobalt (oxide) NC might enable new research linking the catalyst structure to its performance in the Fischer-Tropsch (FT) synthesis. However, the deposition of Co-NC onto supports and thereby their utilization as catalysts is still in its infancy. Here, we present the result of two studies to advance the application of Co-NC in supported FT catalysts.

We started by investigating an alternative to high-temperature treatments often applied to remove ligands that might otherwise block active sites and lower catalytic performance. Although such high-temperature treatments are effective, the harsh conditions can compromise the well-defined Co-NC.

Therefore, we synthesized Co-NC using a hot injection method and oxidized the NC at room temperature prior to their attachment to carbon nanotubes (CNT) as support. During this low-temperature oxidation, the as-synthesized ε-cobalt NC were oxidized to spherical polycrystalline CoO-NC, which decreased the magnetic interparticle interactions and facilitated their uniform distribution over the support (Figure 1A,B). On the other hand, the as-synthesized, non-oxidized ε-cobalt NC were more cubical and formed chains of particles, indicating magnetic interactions between the NC (Figure 1C). Consequently, direct attachment of Co-NC resulted in severe clustering of the NC (Figure 1D). Part of the CoO-NC/CNT sample was subsequently oxidized at 250 °C to evaluate the effect of a high-temperature oxidative treatment. Monocrystalline Co_3O_4-NC were formed during this treatment, which were partially embedded into the CNT support, and its ligands had been removed. *In situ* reduction also effectively removed the ligands from CoO-NC/CNT (TGA, data not shown), yielding two highly active catalysts with comparable performance to that of catalysts prepared by conventional synthesis techniques, such as impregnation of a precursor salt. However, more extensive particle growth was observed for the high-temperature treated sample, showing the adverse effects of severe oxidation. Therefore, low-temperature oxidation was found to be the preferred pre-treatment[1].

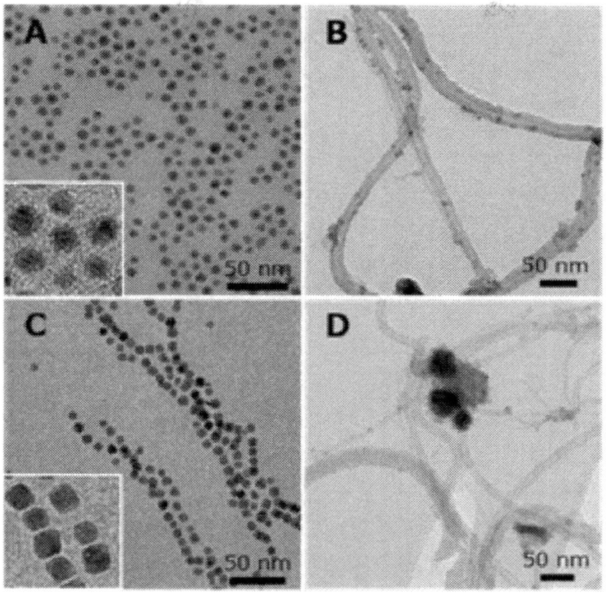

Figure 1 (cryo-)TEM results of the Co- and CoO-NC before and after attachment to the CNT support. Low-temperature oxidized CoO-NC (A) before and (B) after attachment to CNT. As-synthesized, non-oxidized Co-NC (C) before attachment to CNT, as imaged with cryo-TEM and (D) after attachment to CNT.

Having established a promising activation procedure, we focused on preparing model Co/TiO_2 and Co/SiO_2 catalysts using 3-12 nm CoO-NC. In particular on TiO_2, Co particle sizes below 10 nm are hard to obtain using conventional synthesis techniques, so colloidal techniques could offer a clear advantage in this case.

Co-NC were synthesized using a similar hot-injection method as before and their size was regulated between 3-12 nm by adjusting the temperature at which the precursor was injected. After low-temperature oxidation, uniform CoO-NC distributions were obtained for all NC sizes on TiO_2 as well as on SiO_2. The FT activity of the TiO_2-supported Co-NC of 6 and 12 nm was similar to that of Co/TiO_2 prepared by impregnation (turnover frequency ~0.07 s^{-1}), showing that relevant catalysts had been obtained. However, 3 nm Co-NC on TiO_2 were less active than anticipated. After FT, TiO_2-supported Co-NC of all sizes and 3 nm Co-NC on SiO_2 had grown to ~13 nm, while 6 and 9 nm Co-NC on SiO_2 had remained stable (Figure 2). Furthermore, during reduction up to 60 % of Co(-ions) were re-dispersed over TiO_2 against 15 % on SiO_2. The high precision in initial particle size enabled us to investigate the very different growth of Co-NC in Co/TiO_2 and Co/SiO_2 catalysts, showing that both Co-NC size and interaction with the support have a major influence on the Co-NC stability.

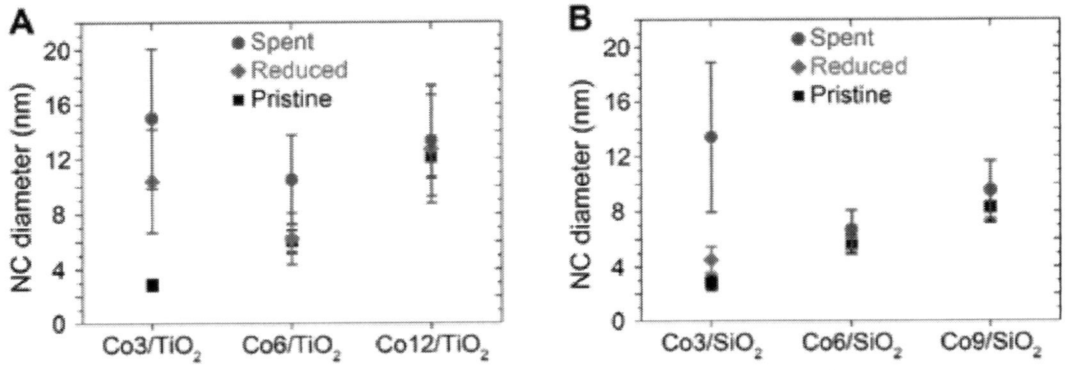

Figure 2 Average cobalt particle sizes of the catalysts in the pristine, reduced and spent state. Surface-volume-weighted mean sizes of the Co-NC supported on (A) TiO_2 and on (B) SiO_2 from TEM analysis. The bars give the standard deviation of the particle size distribution. The reduction was performed at 350 °C (TiO_2) or 500 °C (SiO_2) for 8 h with 1 °C·min^{-1} in 25 vol.% H_2 in He. FT was performed at 220 °C, 20 bar, 2 H_2/CO (V/V) for >100 h on stream.

Overall, the presented approach resulted in NC-based FT catalysts with high control over the Co-NC size on carbon and oxidic supports. We demonstrated their potential by investigating Co-NC growth and anticipate that these model catalysts will facilitate structure-performance studies.

1. T.W. van Deelen, H. Su, N.A.J.M. Sommerdijk and K.P. de Jong, Chem. Commun. 54 (2018) 2530–2533.

579749 Zeolite Catalysts as Enabling Materials for Carbon Management

Wednesday, June 5, 2019: 1:30 PM
Texas Ballroom A (Grand Hyatt San Antonio)

Ulrich Müller[1], S. Altwasser[1], Craig Cain-Borgman[2], A. Gordillo[1], A. Parvulescu[1] and L. Karwacki[1], (1)BASF SE, Ludwigshafen, Germany, (2)BASF Corporation, Iselin, NJ

Keynote Lecture – Session: Oxide-Zeolite Bifunctional Catalysts for Selective Conversion of Syngas

Title: Zeolite Catalysts as Enabling Materials for Carbon Management

Dr. Ulrich Müller

Dr. S. Altwasser, Craig Cain-Borgman, Dr. A. Gordillo, Dr. A. Parvulescu, Dr. L. Karwacki*

BASF SE, Ludwigshafen, Germany and *BASF Corporation, Iselin, NJ, USA

Heterogeneous catalysts are used in more than 85% of commercial chemical processes. By 2021, the overall catalyst market is estimated to be 31 billion US$, where it is estimated that 1 $ invest in chemical catalyst value leverages more than 100 times the value in manufactured products. Within BASF heterogeneous catalysts are broadly used and catalyst performance is of utmost importance for the profitability and sustainability of the respective value-added chains. In addition, BASF is also selling chemical catalysts to the automotive, chemical and refining industries where CAGRs of the strategic relevant markets are predicted at 3-5.6% (2015-2020).

Innovation in the chemical industry is

predominantly driven by catalyst research and development since the discovery and development of ammonia production in the Haber-Bosch process. Starting around 1970 zeolites at BASF were investigated and introduced as industrial chemical catalysts. The lecture will focus on manufacture, characterization and use of zeolites for chemical conversions. Beyond chemicals manufacture, a broader picture including carbon management and raw material change will be addressed. The contribution of zeolites in variable bifunctional metal-oxide/zeolite kits will be outlined. Recent contributions from BASF international academic network INCOE and impact of analytical tools like TEM-tomography and low dose TEM are introduced.

Biography

Ulrich Müller, Senior Vice President - Executive Expert Zeolite Catalysis at BASF SE

PhD from Johannes Gutenberg-University, Mainz, Germany, with K.K. Unger

BASF Zeolite Research since 1989

Catalyst and Process Innovation on cfc replacement, amines, olefin purification, active ingredients, HPPO propylene oxide, DeNOx

551323 Directly Producing Light Aromatics from Syngas over Modified Cezrox/ZSM-5 Bi-Functional Catalysts

Wednesday, June 5, 2019: 2:09 PM
Texas Ballroom A (Grand Hyatt San Antonio)
Hualong Xu, Chemistry, Fudan University, Shanghai, China

Directly Producing Light Aromatics from Syngas over Modified CeZrOx/ZSM-5 Bi-functional Catalysts

Zhen Huang, Shen Wang, Yinghong Yue, Weiming Hua, Wei Shen, Hualong Xu
The Department of Chemistry, Laboratory of Advanced Materials, Fudan University, Shanghai 200438, CHINA

Aromatics, especially light aromatics (BTX) play an important role in modern chemical industry, and are mainly produced from naphtha reforming and oil thermal cracking processes. Syngas can be converted into value-added hydrocarbons through a well-known Fischer-Tropsch synthesis (FTS) process, the products follow the Anderson-Schulz-Flory (ASF) distribution with little aromatics.[1,2] Although FTS catalysts combined with zeolites can produce aromatics, the selectivity is insufficient and generally limited by the operating conditions.[3,4] On the other hand, methanol can be produced from syngas, and then transferred into aromatics through methanol-to-aromatics (MTA) process catalysed by zeolites such as ZSM-5. However, the low selectivity to BTX and rapid deactivation of zeolites caused by coking have severely restricted the development of this process. Besides, the two-step process is not energy and cost efficient. The combination of these two steps could provide a strategy to highly efficient produce aromatics directly from syngas. Prior work has reported that aromatics could be obtained on a combined metal oxides and ZSM-5 catalyst with a moderate selectivity, implying that metal oxide can be adopted to activate CO and hydrogen molecules. But, development of a catalyst with improved activity and selectivity to light aromatics is highly desired for this process.

The OX-ZEO catalysts developed by Bao and co-workers using partially reduced metal oxides and zeolites exhibited an extraordinary activity and selectivity for direct conversion of syngas to olefins and aromatics.[5,6] The surface oxygen vacancy has been considered as active sites for C-O bond activation, where *C species formed can be afterwards hydrogenated into hydrocarbons as intermediates for the formation of olefins and aromatics on zeolites. Y. Wang and P. Zhang reported their works on modified ZSM-5 to enhance light aromatics, Zn-modified ZSM-5 catalyst decorated by Silicalite-1 shell, the selectivity to p-xylene can be significantly improved over modified ZSM-5.[7,8] Aromatics could be formed through the aromatization of intermediates diffusion from metal oxides, BTX mainly formed inside the micropores, $C9^+$ Aromatics are considered to produce on the acid sites in the pore-mouth region or on external surfaces of ZSM-5 through the alkylation of aromatic intermediates. So, the acidity modification of ZSM-5 on external surface is the key routes to enhance BTX selectivity.

Here in this report, we integrated Ceria-Zirconia solid solutions with modified ZSM-5 as a bi-functional catalyst for direct conversion of CO-rich syngas to aromatics. Ceria has been widely applied in heterogeneous catalysis due to its facile redox property and oxygen storage/release capacity (OSC), which is related to the oxygen vacancies and lattice oxygen mobility in ceria crystals. The oxygen vacancy could be improved by introducing various metal ions into ceria lattice.[9,10] In addition, ZSM-5 zeolite has been regarded as one of the most promising catalysts in the production of light aromatics due to the pore-size-initiated shape selectivity. Firstly, we synthesized a series of ceria-zirconia solid solutions (CZS) with different Ce content. The structure homogeneity of CZS in a wide range of Ce content can be attributed to the similar cation radius of Zr^{4+} (0.084 nm) to that of Ce^{4+} (0.097 nm). Surprisingly, aromatics can be obtained from CO-rich syngas (H_2/CO =1) with a high selectivity over these oxides which is integrated with HZSM-5 zeolite (Si/Al=40) by powder-mixing. However, when the Ce content in CZS samples increases from 20% to 100%, the selectivity of aromatics gradually decreases from 75.2% to 59.0% with a slight fluctuation of CO conversion around 5% at 380 °C, which indicates that the relatively low Ce content can favour the formation of aromatics. Moreover, the BTX components account for nearly half of the aromatics, making light aromatics as the main products. The experiment results indicated that the CO conversion increased constantly as reaction temperature increasing from 380 °C to 450 °C, while the aromatic selectivity decreased significantly from 75% to 45%. After optimizing reaction condition, the CO conversion reaches to 30% with selectivity of aromatics above 50%, the fraction of BTX in aromatics is about 23%.

In order to enhance the selectivity of BTX in the fraction of aromatics, modified ZSM-5 catalysts decorated by SiO_2 on external surface, combined with CZS have been also applied for enhancing BTX formation. The results indicated that CZS /ZSM-5-Si exhibited a high selectivity to aromatics, which is 55% with a CO conversion of 32% at 450 °C and 3.6 MPa, and the fraction of BTX reach about 41% among aromatics . After eliminating the strong acid sites on external surface of ZSM-5 by SiO2 decoration, the fractions of BTX in aromatics increase significantly from 25% to 41%. The light aromatic products (BTX) have increased due to the restrained production of heavier aromatics from alkylation of light aromatic intermediates.

Reference:

1. H. M. Torres Galvis and K. P. de Jong, ACS Catal., 2013, 3, 2130-2149.
2. C. D. Chang, W. H. Lang and A. J. Silvestri, J. Catal., 1979, 56, 268-273.
3. S. Sartipi, M. Makkee, F. Kapteijn and J. Gascon, Catal. Sci. Technol., 2014, 4, 893-907.
4. J. Plana-Pallejà, S. Abelló, C. Berrueco and D. Montané, Appl. Catal., A, 2016, 515, 126-135.
5. F. Jiao, J. Li, X. Pan, J. Xiao, H. Li, H. Ma, M. Wei, Y. Pan, Z. Zhou, M. Li, S. Miao, J. Li, Y. Zhu, D. Xiao, T. He, J. Yang, F. Qi, Q. Fu and X. Bao, Science, 2016, 351, 1065-1068.
6. J. Yang, X. Pan, F. Jiao, J. Li and X. Bao, Chem. Commun., 2017, 53, 11146-11149.
7. P. Zhang, L. Tan, G. Yang and N. Tsubaki, Chem. Sci., 2017, 8, 7941-7946.
8. K. Cheng, W. Zhou, J. Kang, S. He, S. Shi, Q. Zhang, Y. Pan, W. Wen, Y. Wang, Chem, 2017, 334-347?

9. Y. Wang, J. Posthuma de Boer, F. Kapteijn and M. Makkee, ChemCatChem, 2016, 8, 102-105.
10. Q. Li, Z. Huang, P. Guan, R. Su, Q. Cao, Y. Chao, W. Shen, J. Guo, H. Xu and R. Che, ACS Appl. Mater. Interfaces, 2017, 9, 16243-16251.

546762 Soldering Promoters for Iron and Cobalt Fischer-Tropsch Catalysts

Wednesday, June 5, 2019: 1:30 PM
Texas Ballroom EF (Grand Hyatt San Antonio)

Bang Gu[1], Vitaly Ordomsky[2], Mounib Bahri[3], Ovidiu Ersen[3], P.a. Chernavskii[4] and Andrei Khodakov[2], (1)Lille University of Science and Technology, Lille, France, (2)Univ. Lille, CNRS, Centrale Lille, ENSCL, Univ. Artois, UMR 8181 - UCCS - Unité de Catalyse et Chimie du Solide, F-59000 Lille, France, Villeneuve d'Ascq, France, (3)Strasbourg University, Strasbourg, France, (4)Moscow State University, Moscow, Russian Federation

Soldering promoters for iron and cobalt Fischer-Tropsch catalysts

Bang Gu[a], Vitaly V. Ordomsky[a], Mounib Bahri[b], Ovidiu Ersen[b], Petr A. Chernavskii[c] and Andrei Y. Khodakov[a*]

[a]*Univ. Lille, CNRS, Centrale Lille, ENSCL, Univ. Artois, UMR 8181 - UCCS - Unité de Catalyse et Chimie du Solide, F-59000 Lille, France*

[b]*IPCMS-UMR 7504 CNRS, Université de Strasbourg, 23 rue du Loess, BP 43-67034 Strasbourg Cedex 2, France*

[c]*Department of Chemistry, Moscow State University, 119992 Moscow, Russia*

Fischer-Tropsch (FT) synthesis is one of the most practicable ways to convert non-oil based feedstocks like natural gas, coal, biomass and shell gas to valuable fuels and chemicals [1]. Iron and cobalt catalysts have been widely used in FT synthesis. Many efforts have been dedicated to improve the activity and stability. Here, we found that metals usually used for soldering (Bi and Pb) had major positive effects on the catalytic performance and stability of iron and cobalt catalysts [2-3].

Iron catalysts have been widely used in high temperature FT synthesis for direct light olefin synthesis from syngas. We found that the promotion of iron catalysts with Bi and Pb results a remarkable increase (several times) in the FTO reaction rate and selectivity to the C_2-C_4 olefins (up to 55%-65%). The direct olefin synthesis from syngas occurs even at under atmospheric pressure. The presence of the promoters significantly slows down secondary hydrogenation of olefins and decreases the chain growth probability and selectivity to the C_{5+} hydrocarbons. A combination of characterization techniques uncovered notable migration of the promoting elements during the reaction and decoration of iron carbide nanoparticles with the promoters to form core-shell structures. The promoters seem to facilitate CO dissociation by removing O atoms from iron carbide.

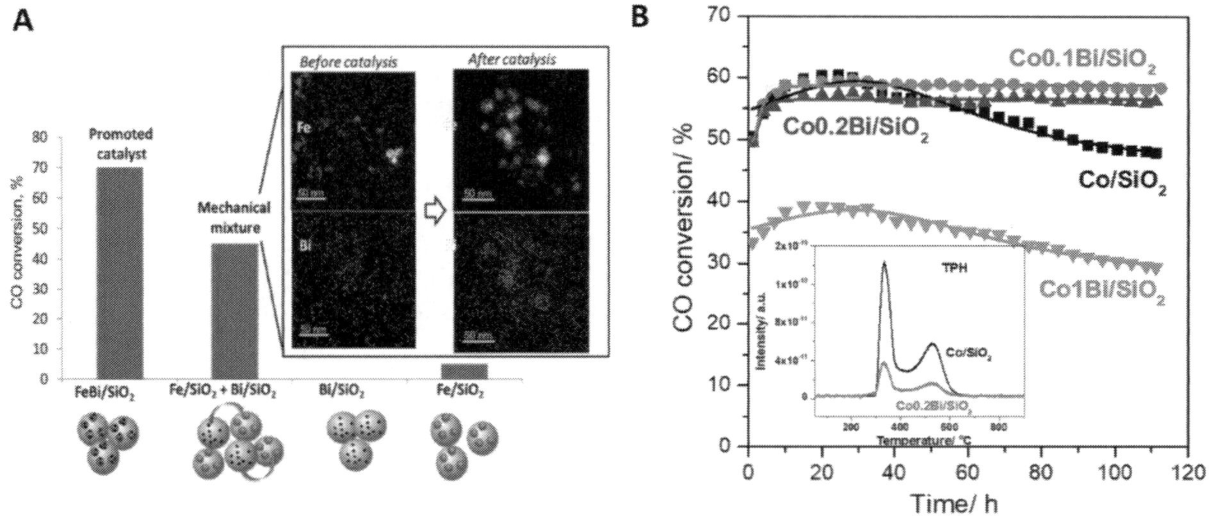

Figure 1. Promotion effect on activity and stability on iron(A)[2] and cobalt (B)catalysts in Fischer-Tropsch synthesis

Deactivation remains one of the main challenges of the cobalt-based catalysts in FT process. The major deactivation mechanisms of cobalt catalysts involve metal sintering and carbon deposition. Interestingly, the soldering promoter Bi also has an excellent effect of improving the stability of cobalt catalysts. Bismuth remarkably reduces both carbon deposition and metal sintering. At low content of the promoter, bismuth does not affect the selectivity of FT synthesis. Coupling of carbon oxidation with CO hydrogenation over the Bi-promoted catalysts reduces carbon deposition and results in the continuous catalyst self-regeneration under FT reaction conditions. Decoration of metal nanoparticles with Bi, which is highly mobile under the reaction conditions, protects them from sintering.

[1] A. Y. Khodakov, W. Chu, P. Fongarlond, Chemical Review, 2007, 107(5), 1692-1744

[2] Y. Luo, B. Gu, A.Y. Khodakov, et al., *ACS Catalysis*, 2017, 7(10), 6445-6452.

[3] B. Gu, V.V. Ordomsky, A.Y., Khodakov, et al., *Applied Catalysis B: Environment*, 2018, 234, 153-166.

545406 Promoted Cobalt Metal Catalysts Suitable for the Production of Chemicals and Fuels from Natural Gas

Wednesday, June 5, 2019: 1:54 PM
Texas Ballroom EF (Grand Hyatt San Antonio)

Jingxiu Xie[1], Pasi Paalanen[1], Tom W. van Deelen[1], Bert M. Weckhuysen[2], Manuel J. Louwerse[1] and Krijn de Jong[1], (1)Inorganic Chemistry and Catalysis, Utrecht University, Utrecht, Netherlands, (2)Inorganic Chemistry and Catalysis, Debye Institute for Nanomaterials Science, Utrecht University, Utrecht, Netherlands

Due to the strong surge of natural gas production, the feedstock for the chemical industry for the near future is expected to shift towards lighter hydrocarbons, in particular methane. The success of a Gas-to- Chemicals process via synthesis gas, a mixture of CO and H_2, depends on the ability of catalysts that suppress the formation of methane and of carbon dioxide. We designed a Co_1Mn_3-Na_2S catalyst, which is shown to have negligible Water-Gas-Shift activity and a hydrocarbon product spectrum, which deviates from the Anderson–Schulz–Flory distribution. At 240 °C and 1 bar, this catalyst showed a C_2-C_4 olefins selectivity of 54 %C with a C_2-C_4 olefin/paraffin ratio of 17. At more industrially relevant conditions of 240 °C, 10 bar, $H_2/CO = 2$, 18 – 30 % CO conversion, the catalytic performance of Co_1Mn_3-Na_2S was compared to other Co-based catalysts (Figure1). At 240 °C and 10 bar, this catalyst displayed still a selectivity towards chemicals of 30 %C and to fuels of 59 %C; the rest being methane and C_2-C_4 paraffins.

To obtain further mechanistic insights into the various catalytic systems, the detailed C product flow of 1-olefin and n-paraffin for each C number product was studied and Co_1Mn_3-Na_2S produced significantly more primary olefins than linear paraffins for each C containing hydrocarbon product. This suggests that β-H elimination is the dominant termination pathway for Co_1Mn_3-Na_2S and secondary hydrogenation of olefins was suppressed. Besides, the low fraction of 2-butenes in the C_4 hydrocarbon product spectrum of Co_1Mn_3-Na_2S implied the suppression of secondary isomerization of olefins. Tentatively the addition of Na_2S to Co_1Mn_3 was proposed to deactivate sites for secondary olefin hydrogenation and isomerization and for methanation, whereas the low "degree of alkalization" as compared with Na_2O is sufficiently low to inhibit the WGS reaction.

Detailed characterization of the spent catalyst using XRD and TEM revealed ~10 nm crystallites with hcp Co metal phase. Figure 2 shows the TEM images and particle size distribution of spent Co_1Mn_3-Na_2S after industrially relevant conditions (240 - 280 °C, 10 bar, and $H_2/CO = 2$), and STEM-EDX maps to differentiate Co and Mn. The Co particle size distribution from TEM revealed the average Co particle size to be 9.6 nm with a standard deviation of 4.4 nm, in agreement with the Co crystallite size of 9.2 ± 1.9 nm from XRD. The elemental maps of Co and Mn in Figure 2f confirmed that spent Co_1Mn_3-Na_2S consisted of Co nanoparticles well dispersed on MnO support. In summary, we propose a synergistic effect of Na plus S, which act as electronic promoters on the metallic Co surface, thus improving the selectivities towards lower olefins and fuels while largely reducing formation of CO_2 and CH_4.

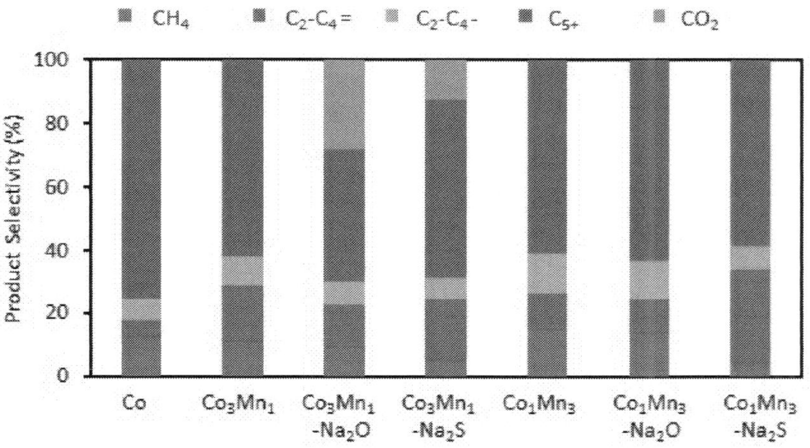

Figure 1. Product selectivities of Co catalysts at 240 °C, 10 bar, $H_2/CO = 2$, 18 – 30 % CO conversion and 100 hours time-on-stream.

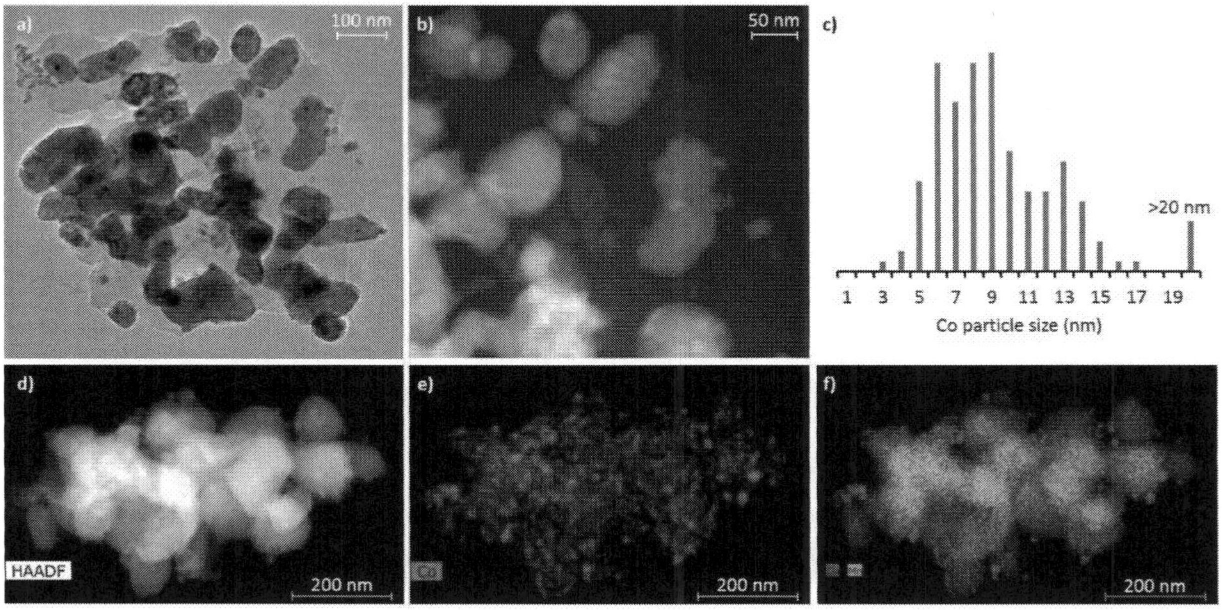

Figure 2. (a – b) TEM images, (c) Co particle size distribution and (d - f) STEM-EDX mapping of spent Co_1Mn_3-Na_2S after FT synthesis at 240 - 280 °C, 10 bar, H_2/CO = 2 and 100 h time-on-stream.

545476 Two-Dimensional Graphene-Assisted Formation of the Cylindrical Iron Carbide Nanocapsules for Fischer-Tropsch Synthesis

Wednesday, June 5, 2019: 2:18 PM
Texas Ballroom EF (Grand Hyatt San Antonio)
Feng Jiang, Chemical Engineering, Jiangnan University, Wuxi, China

Iron-based Fischer-Tropsch synthesis provides a promising alternative as a direct and non-petroleum route for producing olefins from syngas ($CO + H_2$). In general, the promoter is required to regulate the activity and olefins selectivity, and the major challenges come from the catalyst deactivation due to the carbon deposition and sintering at drastic reaction conditions (340 ^oC, high pressure). In this study, we show a totally different strategy for solving these problems. We find that the originally large spherical Fe_2O_3 particles of around 120 nm assisted by graphene without additional promoter can gradually evolve during the FTS reaction, forming much smaller cylindrical Hägg carbide nanocapsules of about 10 nm with single (510) crystal facet that displays excellent activity of nearly 40 times of the unmodified one, olefins selectivity of 76% (84% if methane is not considered) and high stability. In contrast, the original Fe_2O_3 catalyst and carbon-containing polymer material (e.g. PAA, PVP) modified Fe_2O_3 catalysts exhibit significantly lower selectivity to olefins, higher selectivity to methane and deactivate rapidly. Different from the above mentioned peculiar evolution of iron particles, the iron particles over these spent catalysts aggregate into larger ones having multiple crystal facets, e.g. χ-Fe_5C_2 (510), and χ-Fe_5C_2 (-202), and at the same time serious carbon deposition is observed, which mainly accounts for the catalyst deactivation. In the case of activated carbon and carbon nanofibers supported iron catalysts with initially small iron particles below 10 nm, irregular morphology and multiple crystal facets of iron carbides are also observed over the spent catalysts. The DFT calculations demonstrate that χ-Fe_5C_2 (510) crystal facet is more active and favorable for the formation of olefins than χ-Fe_5C_2 (-202) crystal facet, which is well in accordance with our experimental results. The presented methodology provides a new strategy to control the morphology and crystal facet of iron carbide during the catalytic reaction for solving the toughest problems in heterogeneous catalysis.

552245 Increased CO_2 hydrogenation to liquid products using Promoted Iron Catalysts

Wednesday, June 5, 2019: 2:42 PM
Texas Ballroom D (Grand Hyatt San Antonio)

Wilson D. Shafer, Asbury University, Wilmore, KY, Gary Jacobs, Chemical Engineering Program - Dept. of Biomedical Engineering / Dept. of Mechanical Engineering, University of Texas at San Antonio, San Antonio, TX, Uschi M. Graham, Topasol, Lexington, KY, Hussein H. Hamdeh, Wichita State University, Wichita, KS and Burtron H. Davis, Center for Applied Energy Research (CAER), University of Kentucky, Lexington, KY

The steady increase of carbon dioxide (CO_2) emissions into the atmosphere caused by human activities contributes to harmful global warming and climate change. The utilization of CO_2 as a feedstock for producing chemicals not only contributes to reducing global climate change caused by increasing CO_2 emissions, but also opens avenues for exploration of catalytic and industrial development opportunities. CO_2 hydrogenation to hydrocarbons is a modification of Fischer–Tropsch synthesis (FTS), where CO_2 is a reactant instead of carbon monoxide (CO). Hydrogenation of CO_2 is carried out with catalysts that have been demonstrated to be active for the FTS reaction. The effect of potassium (K), rubidium (Rb), and cesium (Cs) on the performance of precipitated iron-based catalysts was explored for carbon dioxide hydrogenation. Under similar reaction conditions, Cs and Rb promoted catalysts exhibited the highest initial CO_2 conversions to higher hydrocarbons. However, at steady state conversion, all three catalysts exhibited similar CO_2 conversions and selectivities. Results indicate a synergistic effect brought on by these promoters that, if balanced, could potentially yield superior CO_2 hydrogenation catalysts. With the aim of developing a better catalyst for CO_2 utilization, these iron catalysts initially reveal a turn in the right direction. However, the shortcomings indicate that carbon deposition could be the main problem. If coking could be mitigated, while still using the basicity strength of these alkali metal promoters, a much more effective, and cheaper, catalyst capable of converting CO_2 to higher hydrocarbons could be implemented.

Keywords: CO_2 hydrogenation; alkali promoter; iron-based catalysts; activity; product selectivity.

545691 Air Liquide Rethinks the Process for Transforming CO2 to Methanol

Wednesday, June 5, 2019: 2:51 PM
Texas Ballroom D (Grand Hyatt San Antonio)

Timm Schuhmann[1], Stéphane Haag[2], Holger Schlichting[1], Nga Thi Quynh Do[1], Martin Gorny[3], Veronika Gronemann[3], **Tobias Oelmann**[3] and Bryce A. Williams[3], (1)Air Liquide Forschung & Entwicklung, Frankfurt am Main, Germany, (2)1AIR LIQUIDE Forschung und Entwicklung GmbH, Frankfurt R&D Center, Frankfurt am Main, Germany, (3)Air Liquide Engineering & Construction, Frankfurt am Main, Germany

Air Liquide draws on a long history of expertise in methanol production including dedicated pilot plants at the Frankfurt Research and Technology Center (FRTC) in Germany, 18 commercial plants licensed with the Lurgi MegaMethanolTM technology and more than 60 methanol licenses for a total capacity of more than 47 million tonnes per year of methanol production. Air Liquide's methanol technology is well-known for its reliable performance and efficiency in the market for world scale capacities beginning at 5000 metric tons per day. One of the latest examples is the successful start-up of the NatGasoline plant in the US based on natural gas using a combined reforming front end to produce the syngas.

Air Liquide applies its experience to further improve the methanol synthesis technology and economics for unconventional feed sources such as stranded gas, unused syngas capacities and CO2 rich gases. In the development, the syngas generation is based upon different options from Air Liquide's proprietary technology portfolio including SMR, ATR, POX and off-gas pretreatment.
Resulting from this strong technology background are developments such as two-stage synthesis and minimized recycle ratios, being implemented from the lab directly into commercial plants.

One of the major trends in recent years is CO2 utilization by producing methanol with the addition of hydrogen from regenerative sources. The reduction of CO2 footprint as well as the valorization of CO2 rich gases are of increasing interest for many industries and methanol is a perfect fit for energy storage, for clean fuels and as a building block for producing high value chemicals. The production of methanol as an intermediate is an effective way to meet both environmental requirements and economic constraints.

Concepts for methanol synthesis based on CO2 and hydrogen exist but are not optimized for today's challenges such as smaller capacities, integration into other processes and fluctuating operation. This is why Air Liquide is developing a new solution for methanol synthesis from low-carbon hydrogen and CO2.

Air Liquide has embarked on a dedicated experimental program in 2017 to improve the know how on the production of methanol from CO2 rich gases and to tackle the technological details related to these.

In addition, as a founding member of the Hydrogen Council, Air Liquide brings a wide range of enabling technologies from the field of industrial gases including CO2 capture & purification, syngas & hydrogen production, hydrogen transport & storage. One option in the portfolio of Air Liquide Engineering & Construction that emphasizes carbon capture and utilization (CCU) is the production of methanol from CO2. Air Liquide is in the position to propose tailor made solutions for conversion of CO2 rich feedstocks to methanol for a wide range of capacities in an efficient way, combining lower investment costs with high productivity through process intensification to align with customers requirements.

The key points of these recent developments will be presented in this paper as listed here:

- Improved models for methanol synthesis based on CO2 rich feed gas to be able to offer better economics combined with robust and guaranteed performance.
- New philosophy for process design focussed on hydrogen efficiency at the expense of carbon efficiency.
- Extensive test campaigns in pilot plants to verify assumptions regarding hydrogen efficiency as a key parameter.
- Dedicated process design (methanol synthesis reactors) for small scale units

The application of CO2 to methanol technology must be seen in the context of current and future environmental regulations. Besides the regulatory framework, validation of dedicated business models for specific cases are mandatory to realize viable commercial applications. Often, the balance between fuel value, production of chemicals and the CO2 footprint have to be taken into consideration. The CO2 to methanol process can be implemented now with established technology offered by Air Liquide Engineering & Construction, complete with full commercial performance guarantees. Leveraging the recent developments a second generation CO2 to methanol process is expected to show better efficiency and substantially improved economics.

Air Liquide will provide information on how to integrate this process with the upstream hydrogen supply and the effect this has on the economics of methanol production.

545764 Behaviour of Cuznal and Cuznzr Ferrierite-Based Catalysts to Produce DME By CO_2 Hydrogenation

Wednesday, June 5, 2019: 2:54 PM
Texas Ballroom D (Grand Hyatt San Antonio)

Fabio Salomone[1], Giuseppe Bonura[2], Francesco Frusteri[2], Samir Bensaid[1] and Raffaele Pirone[1], (1)DISAT, Politecnico di Torino, Torino, Italy, (2)Energy, CNR-ITAE, Messina, Italy

Behaviour of CuZnAl and CuZnZr ferrierite-based catalysts to produce DME by CO_2 hydrogenation.

Fabio Salomone[a], Giuseppe Bonura[b], Francesco Frusteri[b], Samir Bensaid[a, *], Raffaele Pirone[a]

[a] *Department of Applied Science and Technology (DISAT), Politecnico di Torino, Corso Duca degli Abruzzi 24, 10129 Torino, Italy.*

[b] *ITAE - Istituto di Tecnologie Avanzate per l'Energia "Nicola Giordano", Consiglio Nazionale delle Ricerche, Salita Santa Lucia Sopra Contesse, 5 - 98126 Messina, Italy.*

* *Corresponding author. E-mail address:* samir.bensaid@polito.it

The direct hydrogenation of CO_2 into dimethyl-ether (DME) has been studied in presence of copper-based methanol (MeOH) selective catalysts and a commercial ferrierite-type zeolite [1 - 5]. These Cu-based systems are Cu-Zn-Al/ferrierite (CZA/FER) and Cu-Zn-Zr/ferrierite (CZZ/FER). Both hybrid catalysts, obtained through gel-oxalate coprecipitation (OX) and wet impregnation (WI), and physical mixed catalysts with 1:2 and 2:1 weight ratio [2] were explored to compare the activity and the selectivity of these different systems. MeOH-selective catalysts were prepared with a 6:3:1 atomic ratio.

All the samples were properly characterized with different physic-chemical techniques in order to determine the textural and the morphological nature of the surface of the catalysts. Firstly, physical adsorption measurements of N_2 were elaborated to assess the surface area and the porosity of the samples. Secondly, X-rays diffraction (XRD) patterns were analysed to identify the nature of the phases of the catalysts. In addition, FESEM and EDS techniques were carried out to investigate the morphology and the atomic composition of the surfaces. Eventually, temperature programmed reduction (TPR) measurements were performed for observing the surface and bulk reduction of metal oxides.

The experimental campaign was carried out in a fixed bed reactor using 250-500 μm catalyst particles. According to TPR measurements, the samples were activated by means of a reducing process at a temperature of 350 °C, at atmospheric pressure for 3 h. The reducing atmosphere was obtained with a H_2/N_2 molar ratio equal to 1:9. After the activation process, a degreening test was performed (2.5 MPa, 275 °C and 13.6 $Nl/kg_{cat}/h$) to stabilize the catalyst before the activity tests. This procedure was helpful in order to put in evidence if a fresh catalyst was affected by a strong and fast deactivation in reacting conditions [3]. This method highlighted that hybrid catalysts (i.e. CZZ/FER OX 1:2) seem to be very active for few hours, then their activity decreases up to 50 % before reaching a stable operating condition. On the contrary, commercial CZA/FER-based catalysts were not affected by fast deactivation. The diminishing of performances may be probably due to a rearrangement of the catalyst structures at high temperatures. For these reasons, both reduced and aged catalysts has also to be characterized in order to identify the causes of the deactivation and improve catalyst performance and stability.

The activity tests were carried out operating at 2.5 MPa with a temperature ranging from 200 °C to 300 °C with steps of 25 °C. A hydrogen (H_2), carbon dioxide (CO_2) and nitrogen (N_2) gas mixture with a $H_2/CO_2/N_2$ ratio of 3:1:1 was fed to the reactor. Besides, three different weighted hourly space velocity (WHSV) were investigated from 6.6 to 20.0 $Nl/kg_{cat}/h$. The results of the activity tests highlighted that hybrid catalysts are more selective towards methanol and dimethyl-ether also at higher temperatures, because the reverse water gas shift (WGS) reaction is more inhibited. Indeed, CZA/FER catalysts were less selective and produce non-negligible amounts of by-products (i.e. C_2 hydrocarbons) at temperatures higher than 275 °C. At low WHSV (6.6 $Nl/kg_{cat}/h$) the thermodynamic equilibrium was approached from 250 °C, but the cumulative MeOH/DME yield is lower than 80 $g/kg_{cat}/h$. Whereas, the cumulative yield increases at higher WHSV (20 $Nl/kg_{cat}/h$) reaching a cumulative yield of 256 $g/kg_{cat}/h$. The CO_2 conversion always rises as the reaction temperature grows up due to the endothermic reverse WGS reaction, which is favoured at higher temperatures. As illustrated in Fig. 1 (CZA/FER 1:2, 13.6 $Nl/kg_{cat}/h$), DME selectivity is extremely higher than CO selectivity at low CO_2 conversions, which implies a lower reaction temperature. In literature [3], hybrid CZZ/FER systems could reach extremely high DME yield of about 600 $g/kg_{cat}/h$, but the stability of these catalysts is a crucial aspect, which has to be improved.

These tests showed that hybrid CZZ/FER catalysts might to be improved exploring different synthesis methods in order to increase their activity at lower temperature and possibly also at lower pressures. However, at the same time, selectivity could be improved at higher temperature for increasing the yield through MeOH and DME. These goals could be achieved by trying to add promoters to the structure of these catalysts, or also changing the active phase of the catalysts [1].

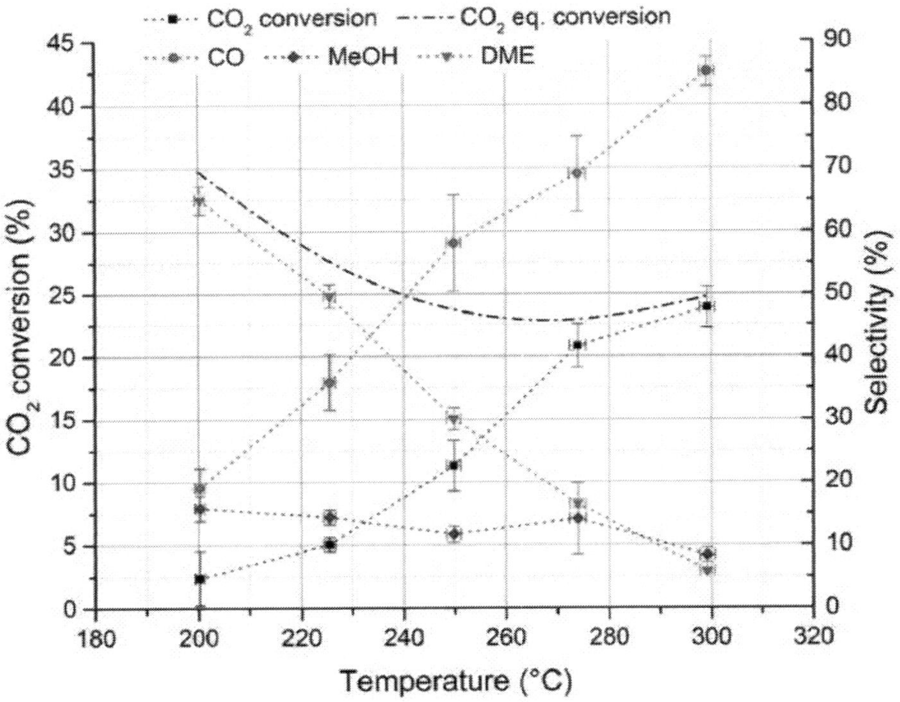

Figure 1 Activity tests on hybrid catalysts CZA/FER 1:2.

Conditions: $H_2/CO_2/N_2$ ratio of 3:1:1, 2.5 MPa, space velocity of 13.6 $Nl/kg_{cat}/h$.

Left axis: CO_2 conversion. *Right axis*: selectivity to CO, Methanol and DME.

References

[1] Andrea Álvarez et al., "Challenges in the Greener Production of Formates/Formic Acis, Methanol, and DME by Heterogeneously Catalyzed CO_2 hydrogenation Processes", Chemical Reviews, 117, 9804-9838, 2017.

[2] G. Bonura, C. Cannilla, L. Frusteri, A. Mezzapica, F. Frusteri, "DME production by CO_2 hydrogenation: Key factors affecting the behaviour of CuZnZr/ferrierite catalysts" Catalysis Today, 281, 337-344, 2017.

[3] F. Frusteri et al., "Direct CO_2-to-DME hydrogenation reaction: New evidences of a superior behaviour of FER-based hybrid systems to obtain high DME yield", Journal of CO_2 Utilization, 18, 353-361, 2017.

[4] G. Bonura et al., "Catalytic features of CuZnZr-zeolite hybrid systems for the direct CO_2-to-DME hydrogenation reaction", Catalysis Today, 277, 48-54, 2016.

[5] G. Bonura, "Acidity control of zolite functionality on activity and stability of hybrid catalysts during DME production via CO hydrogenation", Journal of CO

Utilization, 24, 398-406, 2018.

544796 Direct CO_2 Hydrogenation into Liquid Fuels and Chemicals with High Selectivity

Wednesday, June 5, 2019: 2:57 PM
Texas Ballroom EF (Grand Hyatt San Antonio)
Peng Gao Sr., **Hui Wang** and Yuhan Sun, Shanghai Advanced Research Institute, Chinese Academy of Sciences, Shanghai, China

As an economical, nontoxic, renewable and abundant carbon source, chemical utilization of CO_2 to value-added products is very attractive because it can not only alleviate the global warming caused by increasing atmospheric CO_2 concentration, but also offer a solution to replace dwindling fossil fuels. Thus, much attention has been paid to the CO_2 conversion to various C1 feedstock. Due to the extreme inertness of CO_2 and a high C–C coupling barrier, it is still a great challenge to synthesize C_{2+} products directly from CO_2, such as olefins, gasoline and aromatics.

We have developed a bifunctional catalyst contained partially reducible metal oxides (In_2O_3) and HZSM-5, which exhibited excellent performance for the direct production of C_5~C_{11} hydrocarbons from CO_2 hydrogenation with high selectivity. The C_{5+} (hydrocarbons with five or more carbons) selectivity in hydrocarbons distribution reached up to 78.6% with only 1% for CH_4 selectivity at a CO_2 conversion of 13.1%. There was no obvious catalyst deactivation over 150 h, and much better performance for CO_2 hydrogenation to C_{5+} hydrocarbons was observed using pellet catalyst with internal gas recycling. Such results thus suggest a promising potential for its industrial application.

With the above oxide/zeolite bifunctional catalyst for CO_2 hydrogenation to hydrocarbons, the hydrocarbon distribution can be easily tuned through the shape selectivity of the zeolite. When SAPO-34 was chosen as the active phase for C–C coupling, the selectivity to lower olefins reached 80% in hydrocarbons with only 4% of CH_4 at more than 35% CO_2 conversion.

We also fabricated a highly efficient multifunctional catalyst composed of a spinel structural oxide with hierarchical nanocrystalline HZSM-5 zeolites for selective transformation of CO_2 into aromatics. This catalyst displays high aromatic yield (>18%) with high CO_2 conversion (~35%), and the total selectivity of CH_4 and CO has been successfully suppressed to less 20%.

545148 Catalytic Hydrogenation of CO_2 to Isoparaffins over Fe-Based Multifunctional Catalysts

Wednesday, June 5, 2019: 3:00 PM
Texas Ballroom EF (Grand Hyatt San Antonio)

Jian Wei, Ruwei Yao, Jian Sun and **Qingjie Ge**, Dalian National Laboratory for Clean Energy, Dalian Institute of Chemical Physics, Chinese Academy of Sciences, Dalian, China

The unceasingly rising concentration of CO_2 in the atmosphere, mainly resulting from fossil fuel combustion, is confidently thought to have induced global warming and other climate change phenomena.[1] Therefore, numerous solutions have been proposed to develop efficient CO_2 capture and utilization systems in order to reduce CO_2 emissions. Actually, CO_2 is also a cheap (or even negative-value) and abundant carbon source to manufacture valuable fuels and commodity chemicals in view of the development of a resource-efficient and low-carbon economy.[2,3] It would be a sustainable way of reducing CO_2 emissions, replacing fossil fuels and facilitating the storage of renewable energy, if H_2 used for converting CO_2 was supplied by water electrolysis with renewable electricity.[4,5] Current researches into the catalytic hydrogenation of CO_2 mainly focus on the production of low-carbon chemicals,[6] such as CO, methane, formic acid, methanol, ethanol and lower olefins.[7-23] Due to the weak capability of carbon chain growth during CO_2 hydrogenation,[24,25] relatively, the selective hydrogenation of CO_2 into long-chain hydrocarbons is the least studied and characterized process.[26-33]

CO_2 hydrogenation to long-chain hydrocarbons can be catalyzed through modified Fischer-Tropsch synthesis (FTS) route or methanol-mediated route to promote the hydrocarbon chain growth.[5] Very recently, our team[26]and Gao et al.[27] have both reported the direct conversion of CO_2 into gasoline-range (C_5–C_{11}) hydrocarbons with high selectivity and a very low CH_4 production over composite catalysts. However, the yields to isoparaffins are not high due to poor matching among reverse water-gas shift (RWGS), C-C coupling and isomerization. To reduce the emission of pollutants in the automobile exhausted gas, the olefin and aromatic contents in gasoline fraction were strictly restricted by law. By contrast, environmentally benign and high-octane isoparaffins are ideal gasoline components.[34-36] Meanwhile, isobutane is clean hydrocarbon fuel components of liquefied petroleum gas (LPG), and also used to synthesize high-octane isooctane via alkylation. Despite the great importance of selective synthesis of isoparaffins via CO_2 hydrogenation, developing a high-efficient catalyst is still challenging.

The zeolite component in a multifunctional catalyst is a key factor to precisely control hydrocarbon selectivity for above reactions.[37] Except HZSM-5, other types of zeolites possessing different framework topology, such as HMCM-22 and HBeta, have shown unusual catalytic characteristics in the skeletal isomerization reactions of hydrocarbons, while they undergo quick deactivation as a result of coke deposition.[38,39] To date, however, there is little known about the catalytic properties and deactivation behaviors of HMCM-22 and HBeta in CO_2 hydrogenation reaction.

Herein, we develop a high-efficient multifunctional catalyst system for synthesizing middle (C_4–C_{11}) isoparaffins from CO_2 hydrogenation (Table 1). The effects of zeolite pore structure and acidity on the product distribution and coke formation were investigated over Na–Fe_3O_4/Zeolite composite catalysts containing different topology but similar SiO_2/Al_2O_3 ratios of zeolites: HMCM-22, HBeta and HZSM-5. The product distributions over HMCM-22 and HBeta were strikingly different from that obtained over HZSM-5 and displayed very high initial selectivities to isoparaffins (Figure 1). In addition, the zeolite deactivation behavior and nature of coke formation have been studied in detail, which provide important information to take into account for the regeneration of zeolites in this study.

Studies on the reaction schemes of CO_2 hydrogenation to isoparaffins over Na-Fe_3O_4/HMCM-22 catalysts show that CO_2 and H_2 are activated on Na–Fe_3O_4 to generate olefins as intermediates, and their oligomerization and isomerization are preferentially performed on HMCM-22 or HBeta. The high yield to isoparaffins was derived owing to well matching of three tandem reactions comprising reverse water-gas shift, C-C coupling and isomerization. Unique pore structure and appropriate Brønsted acid properties of HMCM-22 effectively suppressed aromatization, whilst promoting isomerization (Figure 2). Coke deposition inside the micropores of HMCM-22 causes the decline of isoparaffin yield without changing the total yield of heavy hydrocarbons. Both the physico-chemical properties and catalytic performances of the catalysts could still keep their original levels after several reaction-regeneration cycles, indicating a promising potential application in the future commercialization process of CO_2 hydrogenation.

Table 1. Reaction performance for CO_2 hydrogenation. [a]

Catalysts	CO_2 conv. (%)	CO sel. (%)	Hydrocarbon distribution (C-mol %) C_1	$C_{2\text{-}3}$	N-C_{4+} [b]	Isoparaffins	C_{ole}/C_p [c]	STY [d] ($mg_{iso}\ g_{cat}^{-1}\ h^{-1}$)
Na–Fe_3O_4/HMCM-22	25.9	17.1	8	10	25	57	0.08	102
Na–Fe_3O_4/HBeta	25.8	17.4	9	11	21	59	0.02	105

$Na–Fe_3O_4$/HZSM-5	25.6	17.3	6	16	44	34	0.02	60
$Na–Fe_3O_4$	25.1	17.7	8	26	60	6	5.11	12

[a] Reaction conditions: H_2/CO_2 = 2, 320 °C, 3 MPa, 4,000 ml h^{-1}, time on stream of 90 min.

[b] N-C_{4+}: C_{4+} products except for isoparaffins. [c] C_{ole}/C_p is the molar ratio of all olefins to all paraffins with n > 1. [d] STY: Space time yield of isoparaffins.

Figure 1. The detailed hydrocarbon product distribution obtained over different catalysts.

a, $Na–Fe_3O_4$/HMCM-22. **b**, $Na–Fe_3O_4$/HBeta. **c**, $Na–Fe_3O_4$/HZSM-5. **d**, $Na–Fe_3O_4$. Reaction conditions as in Table 1.

Figure 2. Reaction scheme of isoparaffin synthesis and coke formation during CO_2 hydrogenation over $Na–Fe_3O_4$/HMCM-22 catalyst.

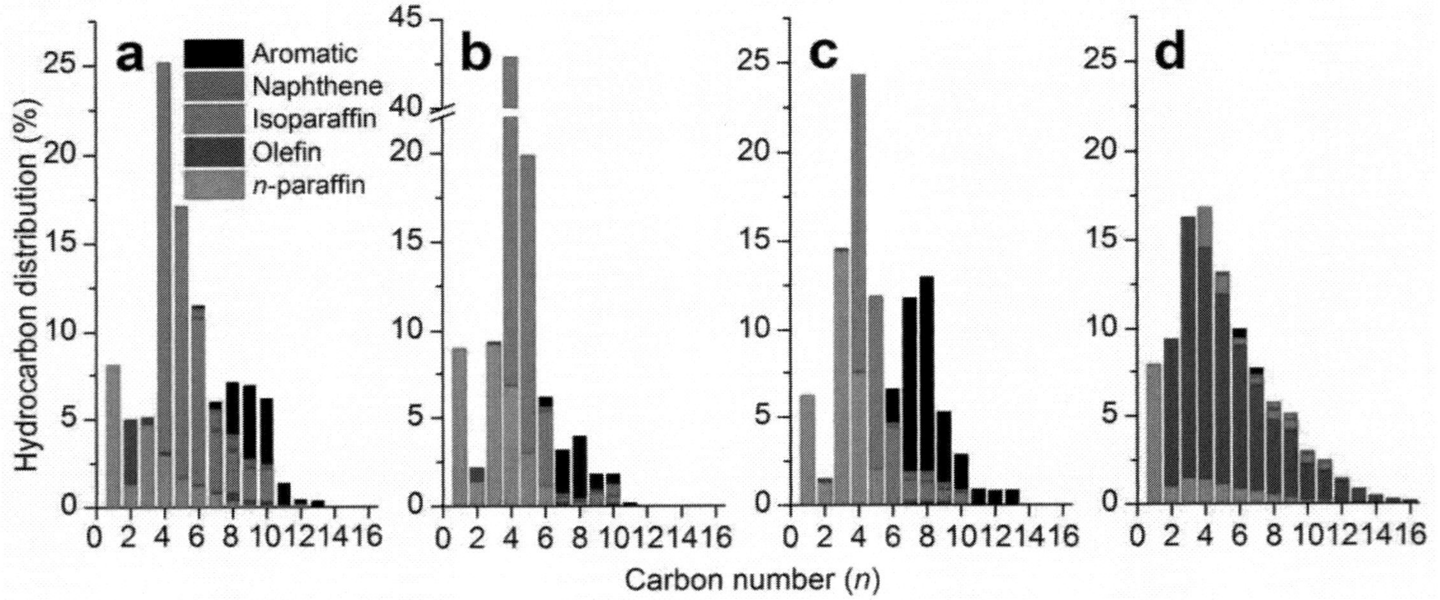

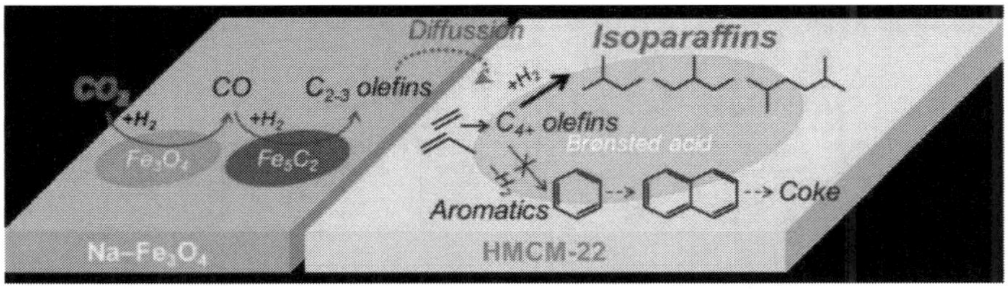
Diffussion
Isoparaffins
CO_2
CO
$C_{2\text{-}3}$ olefins
$+H_2$
$+H_2$
Fe_3O_4
Fe_5C_2
C_{4+} olefins
Brønsted acid
Aromatics
Coke
$Na\text{–}Fe_3O_4$
HMCM-22

544652 Zeolite a Catalysts Reduced By $NaBH_4$ for Hydroformylation of 1-Hexene

Wednesday, June 5, 2019: 3:03 PM
Texas Ballroom EF (Grand Hyatt San Antonio)

Yuan Fang, Guohui Yang, Yoshiharu Yoneyama and Noritatsu Tsubaki, Department of Applied Chemistry, School of Engineering, University of Toyama, Toyama, Japan

Zeolite A catalysts reduced by $NaBH_4$ for hydroformylation of 1-hexene

*Yuan Fang, Guohui Yang, Yoshiharu Yoneyama, Noritatsu Tsubaki**

Department of Applied Chemistry, School of Engineering, University of Toyama, Toyama, 930-8555, Japan

**Corresponding author: tsubaki@eng.u-toyama.ac.jp*

A series of cobalt catalysts have been prepared by impregnation and stirring process using $NaBH_4$ as the reducing agent over the 3A, 4A and 5A zeolites as supports. The preparation process employed $Co(NO_3)_2 \cdot 6H_2O$ as Co resource and $NaBH_4$ as in situ reductant to directly prepare Co supported zeolite catalysts through two different preparation ways. In the hydroformylation of 1-hexene, the major by-product is iso-heptanal that has branch and larger thermodynamic radius than the desired product of normal heptanal. Therefore, we aim to utilize the confined space afforded by zeolite cell cage to restrict the formation of iso-heptanal, at the same time facilitating the generation of normal heptanal. With zeolite as the cage of active catalysts, we can effectively construct a confined reaction space to direct the hydroformylation reaction route and control the products distribution. The in-situ reduction method here significantly improved the conversion of 1-hexene to 92.2% for 4A zeolite supported catalyst and normal/iso ratio of aldehyde to 1.93 for 3A zeolite supported catalyst during the reaction.

It is indicated that the preparation way for $NaBH_4$/XA-15%Co (X=3, 4, 5) catalysts can effectively depress the formation of internal hexane, at the same time facilitating the formation of normal heptanal. Above all, this developed method combined the two advantages of shape selectivity of zeolite supports and the remarkable reducing effect of $NaBH_4$.

544947 *Directly Converting Carbon Dioxide to Linear Î ±-Olefins on Bio-Promoted Catalysts*

Wednesday, June 5, 2019: 3:06 PM
Texas Ballroom EF (Grand Hyatt San Antonio)

Lisheng Guo[1,2], Jian Sun[2], Qingjie Ge[2] and Noritatsu Tsubaki[1], (1)Department of Applied Chemistry, School of Engineering, University of Toyama, Toyama, Japan, (2)Dalian National Laboratory for Clean Energy, Dalian Institute of Chemical Physics, Chinese Academy of Sciences, Dalian, China

Directly converting carbon dioxide to linear α-olefins on bio-promoted catalysts

Lisheng Guo[1, 2], Jian Sun[2], Qingjie Ge[2], Noritatsu Tsubaki[1]**

[1]Department of Applied Chemistry, School of Engineering, University of Toyama, Gofuku 3190, Toyama 930-8555, Japan

[2]Dalian National Laboratory for Clean Energy, Dalian Institute of Chemical Physics, Chinese Academy of Sciences, Dalian 116023, China

Corresponding authors: sunj@dicp.ac.cn; tsubaki@eng.u-toyama.ac.jp

Higher olefins and in particular linear α-olefins (LAOs), which contain a terminal carbon-carbon double bond, are important and expensive industrial feedstocks for producing highly-value-added chemicals, such as lubricants, detergents, and polyolefins. However, no related reports concerning LAO production from CO_2 have been reported to our knowledge, in spite of the fact that CO_2 conversion to heavy hydrocarbons has been demonstrated in recent years.

Recently, we design an enzyme-like integrated catalyst comprising iron carbides and alkali promoters from calcined corncob (CC) ash, accomplishing direct conversion of CO_2 hydrogenation into LAO with selectivity in hydrocarbons of higher than 40% and total olefin selectivity in hydrocarbons of 72%. Contrary to the introduction of potassium (K) from a chemical promoter, K derived from biopromoters (CC) shows stronger migration ability than chemical promoters during the CO_2 hydrogenation process. According to the results from ^{57}Fe Mössbauer spectra and X-ray photoelectron spectroscopy, we learn that these surface enriched K ions could promote the formation of carbides (Fe_5C_2) and further enhance the activity of C-C coupling. Besides, the surface-enriched alkali K from CC ash suppresses the hydrogenation of alkenes and leads to more formation of higher alkenes based on a propylene pulse experiment. The synergetic effect from enhanced C-C coupling and subdued hydrogenation ability led to a high selectivity of linear α-olefins. Thus, this design of biopromoters-modified catalyst provides a new strategy for selective hydrogenation of CO_2 into high value-added products, and sheds a light on the rational design of highly efficient catalysts referenced from biocatalysis in nature.

545548 Investigation of the Active Phases in Alkali Metal Promoted Molybdenum Sulfide Catalysts for the Synthesis of Methanethiol from Syngas and Hydrogen Sulfide

Wednesday, June 5, 2019: 3:09 PM
Texas Ballroom EF (Grand Hyatt San Antonio)

Miao Yu[1], Nikolay Kosinov[1], **Emiel Hensen**[2] and **Wei Chen**[2], (1)Eindhoven University of Technology, Eindhoven, Netherlands, (2)Schuit Institute of Catalysis, Department of Chemical Engineering and Chemistry,, Eindhoven University of Technology, Eindhoven, Netherlands

1. Introduction

Methanethiol (CH_3SH) is an important industrial raw material widely used in the synthesis of valuable organosulfur compounds such as pesticides, pharmaceuticals and petrochemicals [1]. A state-of-art production method for CH_3SH is the thiolation of methanol (CH_3OH) over transition metal sulfide catalysts [2]. Even if the production of CH_3SH is an efficient process involving high CH_3OH conversion and CH_3SH yield, this route needs a multi-step pathway for the synthesis of CH_3SH, which results in high production cost. Considering that CH_3OH is generated from syngas (CO and H_2), Olin et al. [3] proposed to synthesize CH_3SH directly from H_2S-containing syngas. The benchmark catalysts for this reaction are alkali metal promoted molybdenum sulfide (MoS_2) supported on silica or alumina [4].

Most of the earlier works have been focused on the optimization of the reaction conditions, while the active phase itself started to attract attention only recently. Cordova et al. proposed that the active phase related to the direct synthesis of methanethiol from $CO/H_2/H_2S$ using K_2MoO_4/Al_2O_3 and K_2MoS_4/Al_2O_3 catalysts precursors, consists of layered 1T-MoS_2 crystallites in which potassium ions are intercalated between the layers [4]. This K_xMoS_2 phase was characterized by X-ray photoelectron spectroscopy (XPS) and the amount of this phase was correlated with the CH_3SH productivity.

It should be noted, however, that 1T-MoS_2 has been reported to be unstable compared to 2H-MoS_2 phase: it easily transforms to the 2H phase at high temperature [5]. The correlation between the amount of 1T phase and CH_3SH productivity was solely based on XPS result before the reaction. Therefore, aim of this work was to study the stability of 1T-MoS_2 under reaction conditions to examine whether the 1T- K_xMoS_2 phase is the actual active component for CH_3SH synthesis.

2. Experimental

Potassium-promoted molybdenum sulfide catalysts were prepared by co-impregnation of a 75-125 μm sieve fraction of crushed γ-Al_2O_3 extrudates (V_p=0.6 cm^3/g) with an aqueous solution of suitable metal salts, ammonium heptamolybdate ($(NH_4)_6Mo_7O_{24}\cdot 4H_2O$), Sigma Aldrich) and potassium nitrate (KNO_3, Sigma Aldrich). After impregnation, the catalyst precursors were dried at 110°C for 24 h and subsequently calcined at 450°C for 2 h in flowing air for 2 h. The molybdenum loading was 8 wt.%, whereas the loading of potassium was varied to achieve the atomic ratio of K to Mo as 0.5, 1 and 2, which was confirmed by inductively coupled plasma atomic emission spectroscopy (ICP-AES). The catalysts are denoted as K_xMo/Al_2O_3, in which x is the atomic ratio of K to Mo. Besides, the catalysts were also characterized by transmission electron microscopy (TEM), X-ray photoelectron spectroscopy (XPS) and operando extended X-ray adsorption fine structure (EXAFS).

The activation of catalyst precursors and CH_3SH synthesis were performed in a fixed-bed reactor. Prior to catalytic test, the precursors were in-situ sulfided with 10 vol% H_2S/H_2 under atmospheric pressure and 350°C for 2 h. The same procedure was also applied for "quasi" in situ XPS and TEM measurements. After activation, the feed gas was changed to mixed $CO/H_2/H_2S$ (molar ratio equals to 1/2/1) with a total flow rate of 60 mL/min and 3 mL/min N_2 as internal standard. The catalytic performance was measured under 10 bar from 280°C to 320°C, the outlet product composition was analyzed by an online compact gas chromatograph (GC).

3. Results

The XPS spectra of Mo 3d core level of K_0Mo/Al_2O_3 and K_2Mo/Al_2O_3, which were in-situ sulfided at atmospheric pressure, are shown in Fig. 1a,. These spectra can be deconvoluted into four doublets: Mo^{VI}, Mo^{V}, Mo^{IV} and K-Mo^{IV}. The doublet of Mo^{IV} at 228.9 eV and 232.0 eV corresponds to Mo in pristine 2H-MoS_2, while the doublet of K-Mo^{IV} at 228.0 eV and 231.1 eV corresponds to Mo in 1T-MoS_2 [4]. Obviously, K-Mo^{IV} exists only in the spectrum of K_2Mo/Al_2O_3, which means that the addition of K in MoS_2 can induce its phase conversion from 2H to 1T. By comparing the area of doublet Mo^{VI} and KMo^{VI}, the ratio of 1T to 2H phase in K_2Mo/Al_2O_3 is about 1.2.

The sulfided samples were also analyzed by EXAFS at the Mo K-edge, and the Fourier transforms of K_0Mo/Al_2O_3 and K_2Mo/Al_2O_3 are shown in Fig. 1b. In the spectrum of K_0Mo/Al_2O_3, there are two main peaks corresponding to the nearest Mo-S (2.40 Å) and Mo-Mo (3.17 Å) bonds, respectively. On the other hand, there is one extra scatterer at 2.77 Å, which is assigned to a distorted shorter Mo-Mo bond. Previous studies showed that the structure of MoS_2 can be distorted when it changes from 2H-phase to 1T-phase [6], so the shorter Mo-Mo bond peak in EXAFS spectrum can be assigned to the 1T-phase in K_2Mo/Al_2O_3.

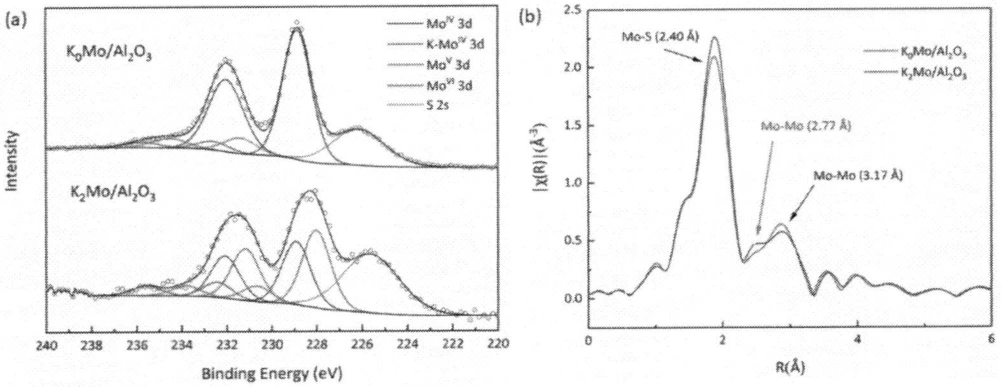

Fig. 1. (a) Mo 3d XP spectra of K_0Mo/Al_2O_3 and K_2Mo/Al_2O_3 (b) Fourier transform of Mo K-edge EXAFS spectra of K_0Mo/Al_2O_3 and K_2Mo/Al_2O_3

The catalytic performance of all four catalysts at 320°C is shown in Fig. 2a. It is clear that the addition of K can both increase the CO conversion and CH_3SH selectivity. Meanwhile, the XPS spectra of the K_2Mo/Al_2O_3 after 2 h, 5 h, 10 h and 20 h under reaction conditions (10 bar and 350°C) were also measured, and the ratios of 1T to 2H phase were calculated and shown together with CO conversion and CH_3SH selectivity in Fig. 2b.

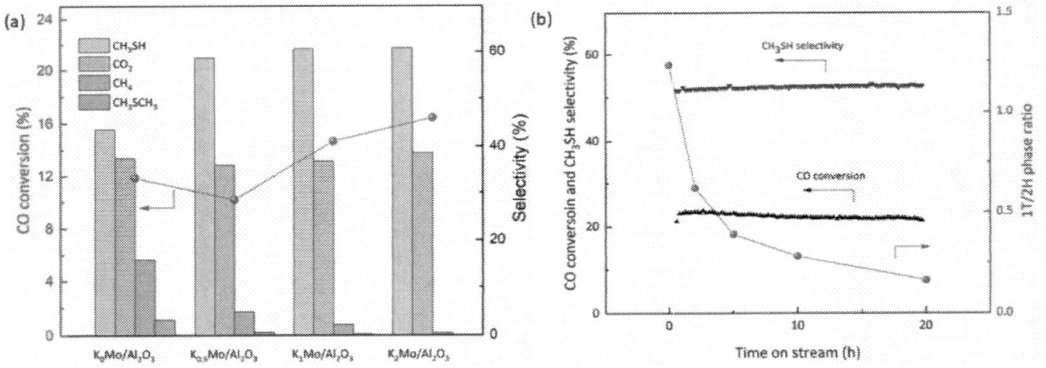

Fig. 2. (a) CO conversion and product selectivity of all four catalysts under 320°C (b) Catalytic performance and 1T/2H ratio of K_2Mo/Al_2O_3 in 20h

4. Discussion

The results from XPS and EXAFS clearly show that the addition of K in MoS_2 drives the conversion of 2H phase to 1T phase. Additionally, the ratios of 1T/2H calculated from XPS and EXAFS (by coordination numbers for different scattering paths) are very similar, which makes XPS a reliable method to identify the amount of 1T-MoS_2 in this system. The 1T/2H ratio values (Fig. 2b) show that 1T phase converts to 2H phase during the reaction very quickly. Since the reaction temperature is the same as that for the sulfidation, high pressure should be responsible for the 1T to 2H transition. However, even though the ratio of 1T/2H decreases dramatically, the catalytic performance was quite stable for 20 h, suggesting that the structure of 1T-MoS_2 is not an all-important aspect of K-promoted MoS_2 for CH_3SH synthesis.

5. Conclusions

In our work, K-promoted MoS_2 catalysts were prepared. The existence of 1T-MoS_2 was confirmed by XPS and EXAFS. However, 1T-MoS_2 quickly converted to 2H-MoS_2 during the CH_3SH synthesis reaction because of its instability at high pressure. Therefore, we argue that under CH_3SH reaction conditions the 2H-MoS2 phase is the real active phase.

6. References

[1] J. Sauer, W. Boeck, L. von Hippel, W. Burkhardt, S. Rautenberg, D. Arntz, W. Hofen, US patent 5852219, 1998.

[2] Mashkin, V.; Kudenkov, V.; Mashkina, A. *Ind. Eng. Chem. Res.* 1995, 34, 2964–2970.

[3] J. Olin, B. Buchholz, B. Loev, R. Goshorn, US patent 3070632, 1962.

[4] Cordova A, Blanchard P, Lancelot C, et al. *ACS Catalysis*, 2015, 5(5): 2966-2981.

[5] Eda G, Yamaguchi H, Voiry D, et al. *Nano letters*, 2011, 11(12): 5111-5116.

[6] Liu Q, Li X, He Q, et al. *Small*, 2015, 11(41): 5556-5564.

545760 Impact of Catalyst Deactivation on a Carbon Dioxide Methanation "Power-to-Gas" Process: Deactivation Kinetic Study and Reactor Modelling

Wednesday, June 5, 2019: 3:12 PM
Texas Ballroom EF (Grand Hyatt San Antonio)

Eduard Alexandru Morosanu[1], Albin Chaise[2], Alain Bengaouer[2] and Samir Bensaid[1], (1)DISAT, Politecnico di Torino, Torino, Italy, (2)LITEN, CEA, Grenoble, France

<>Impact of catalyst deactivation on a carbon dioxide methanation “Power-to-Gas” process: deactivation kinetic study and reactor modelling <>Eduard Alexandru Morosanu[a,b], Albin Chaise[b], Alain Bengaouer[b], Samir Bensaid[a,*]

[a] Department of Applied Science and Technology (DISAT), Politecnico di Torino, Corso Duca degli Abruzzi 24, 10129 Torino, Italy.

[b] Laboratory of Heat-Exchanger Reactor, Univ. Grenoble Alpes, CEA, LITEN, DTBH, Grenoble F-38000, France

* *Corresponding author. E-mail address:* samir.bensaid@polito.it

Power to synthetic natural gas (SNG) is a candidate technology to be used for long term storage of renewables (wind and solar). In the past years this technologie has received much interest from the scientific community and a consistent number of demonstrator plants have been built over the years. One of the aspects that need investigation is the ability of the system to maintain the productivity and quality of the gas in time. Decrease in the conversion rate has been observed and it is generally attributed to the loss of the catalyst activity. Some studies are present in scientific literature on the phenomena causing catalyst deactivation (i.e. sintering). However, few publications linking the decrease in kinetics with the reactor performance evolution can be found for the SNG process. In this work we've performed an experimental campaign on a commercial catalyst to determine the deactivation kinetics for the CO_2 methanation system. The deactivation law was then used in transient model to evaluate the performance in time of the reactors.

A set of 100h long tests at different temperatures were performed (example reported in Figure 1), with final restoring of the original temperature to determine the true deactivation effect.

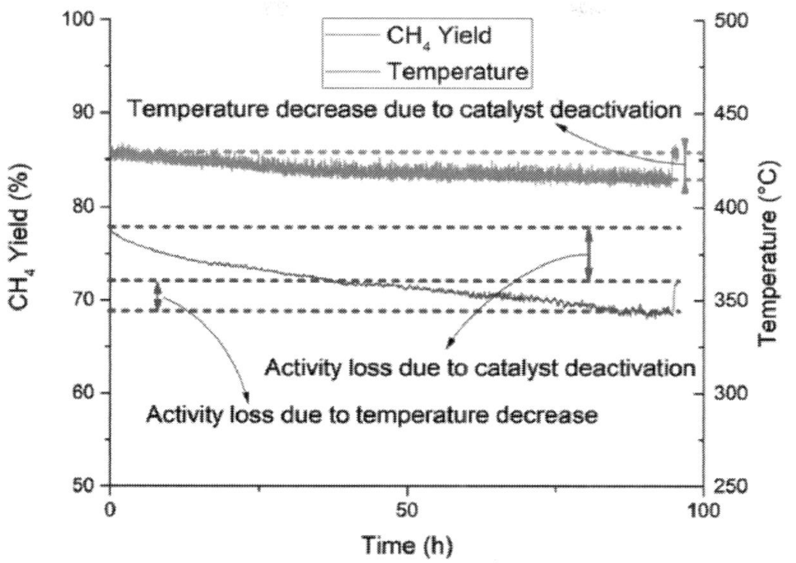

Figure 1: Example of long-term test results.

Inlet: 60% H_2, 16% CO_2 and 20% N_2; pressure: 5 bar; temperature: 425 °C; duration: ~ 100 h.

Sintering was determined to be the deactivation mechanism (BET surface reduction and Nickel particle diameter increase). In literature sintering processes are modelled using first or second order general power low expression (GPLE). The experimental data was best fitted by a second order GPLE. Adiabatic and cooled reactors with different geometries were simulated using the deactivation law in a model developed in COMSOL. The adiabatic reactor exhibited a high decrease in time of the performance due to the high temperature reached. The cooled reactor showed a relevant decrease during the first hours caused by the exothermal high temperature peak. Afterwards the peak diminished in intensity and broadened. This caused the reduction of the deactivation kinetics and stabilization of the reactor performance.

546390 CO Hydrogenation over Ni Catalysts Supported on MAl_2O_4@Al (M=Mn, Mg, Zn)

Wednesday, June 5, 2019: 3:15 PM
Texas Ballroom EF (Grand Hyatt San Antonio)

Thien An Le[1], Ji Eun Kim[1], Jong Kyu Kang[1] and **Eun Duck Park**[1,2], (1)Department of Energy Systems Research, Ajou University, Suwon, Korea, Republic of (South), (2)Department of Chemical Engineering, Ajou University, Suwon, Korea, Republic of (South)

Much attention has been paid to develop non-fossil fuel energy sources to reduce the CO_2 emission and create a sustainable energy system for the future. Although methane is a well-known fossil fuel, it can also be synthesizd from CO and CO_2 through hydrogenation which is called Sabatier reaction. Methane is a versatile energy source that can be used directly for heat and electricity generation. It can be fueled to the compressed natural gas vehicles and be a viable option for renewable energy storage. Ni-based catalysts have been widely used in the commercial process to carry out the Sabatier reaction because of their relative fair activity, low cost, and high availability compared with the others. However, the easy deactivation at high temperatures caused by the metal sintering and coke deposition is still the big challenge for Ni catalyst. Furthermore, a highly exothermic nature of Sabatier reaction limits the complete conversion of reactants at high temperatures and causes the hot spots throughout the catalyst bed. Therefore, the development of catalysts active at low temperatures with high thermal conductivity is highly required for this reaction.

Until now, many kinds of supports combined with the novel preparation methods have been investigated to fabricate the Ni-based catalysts owning the high methanation catalytic activity at low temperatures with the high resistance of sintering and coking. Spinel aluminates (MAl_2O_4) have gained much interest as supports in catalysis because of their inherent properties including high thermal stability. The unique core-shell spinels MAl_2O_4@Al (M=Mn, Mg, Zn), which are obtained by a simple hydrothermal surface oxidation (HTSO) of Al metal particles in an aqueous solution of heterometal ions at elevated temperature (120-200 ^oC), are used as the supports for Ni catalysts for CO methanation. Various techniques including N_2 physisorption, H_2 chemisorption, H_2-TPR, chemisorption, XRD, ICP-OES, in-situ FTIRS, STEM-EDX, and TEM are employed to characterize the catalysts.

The combination of the core-shell microstructures and spinel-type makes the outstanding catalytic performance of the Ni supported by MAl_2O_4@Al (except for $ZnAl_2O_4$@Al) for CO methanation compared to commercial γ-Al_2O_3 at low reaction temperatures due to the improvement of the dispersion of NiO nanoparticles, as well as the increase of the quantity of the reduced active Ni species. $MnAl_2O_4$@Al is considered as the best support for Ni-based catalysts for CO methanation. The good stability of Ni/MAl_2O_4@Al catalysts are also confirmed.

546667 Mechanistic Insights in the Methanol-to-Hydrocarbon Process Using Advanced and Combined Operando Spectroscopy

Wednesday, June 5, 2019: 3:00 PM
Texas Ballroom D (Grand Hyatt San Antonio)

Sophie H. van Vreeswijk, Joris G. Goetze and Bert M. Weckhuysen, Inorganic Chemistry and Catalysis, Utrecht University, Utrecht, Netherlands

The Methanol-to-Hydrocarbon (MTH) process is currently of great interest as the feedstocks for the production of methanol can be obtained from practically any carbon source. Thereby, an alternative route for the production of more renewable plastics and fuels can be obtained. The MTH process, in which methanol is converted to hydrocarbons, can be catalyzed by zeolites. The amount of Brønsted acid sites, the pore size and structure and the cage size and structure of the zeolites determine which actual products are formed.[1,2] The most accepted mechanism of the formation of hydrocarbons within the structures is the hydrocarbon pool mechanism or the dual cycle mechanism, in which aromatic hydrocarbons and olefins function as intermediate products in the zeolite cages. Insights in the formation of these intermediates during the conversion of methanol are required to define the actual mechanism and to gain more knowledge about the tunability of the process within the frameworks of the zeolites. Additionally, the formation of large hydrocarbon species as polyaromatic compounds can block the pores as they occupy the whole cavity opening or large coke species can accumulate on the outside of the zeolites by which the accessibility of the pores and cavities is hindered. In the recent years, our group developed multiple operando spectroscopic techniques to analyze the formation of different species within the cages of the zeolites during the MTH reaction to, in the end, follow the active and deactivating period of the process.

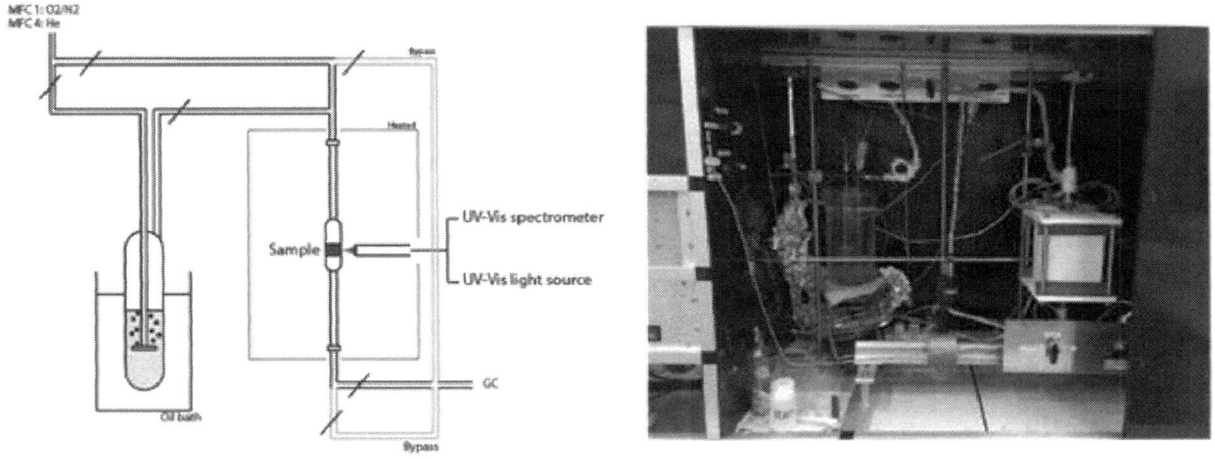

Figure 1: Schematic representation and a picture of the operando UV-vis set-up. A high temperature UV-Vis probe is used to obtain information about the intermediate species formed during the MTH reaction. Multiple UV-vis probes can be used to

record the UV-Vis spectra at different positions along the catalyst bed to follow the so called "cigar burn" principle.

Charged and neutral aromatic intermediates absorb visible and ultraviolet light. Therefore, operando UV-vis spectroscopy (UV-vis spectroscopy combined with on-line product analysis) provides us of a way to follow the formation of the aromatics formed within the hydrocarbon pool. Figure 1 represents the schematic of the operando UV-vis set-up used for our group's research. This technique, combined with chemical analysis of carbonaceous deposits on the zeolites, were used to obtain information about the mechanism of zeolites with different topologies. It was found that the exact size and shape of the cages of a zeolite is significantly influencing the mechanism, deactivation and product distribution of the MTH reaction. To obtain as much as possible information about the evolution of the UV-vis spectra and thereby the formation of the aromatic compounds within the structure, multivariate analysis can be used. With this technique, it was found that, besides the significant influence of the nature of the zeolite, especially for zeolites with 2D pore networks, the reaction temperature also has an important effect on the number and role of certain active and deactivating hydrocarbon pool species.[1]

Introduction of alkaline earth metals can drastically influence the reaction mechanism and thereby the products formed.[2] However, the way of introducing these metals is very important to the final contribution to the mechanism.[3,4] With operando spectroscopy, the differences between both the formed products and reactivity as well as the differences between the reaction intermediates can be analyzed and combined to enable the rationalization of the relation of these intermediates and products.

Additionally, operando spectroscopy can provide information about the location of the deactivating carbon deposition on the zeolites and enables us to follow the so called "cigar burn" principle.[1,3] Regeneration studies combined with operando spectroscopy can provide information about the nature of the coke species within the pores of the zeolites.

The accumulation of hydrocarbon species in the zeolite framework result in the expansion of the lattice. With operando X-ray diffraction this lattice expansion can be correlated to the activity and selectivity of the zeolites in the MTH process. In our lab the operando XRD set-up and an operando UV-Vis set-up are combined to be able to correlate possible lattice expansion to actual hydrocarbon species. The exact lattice expansion and parameters can be determined using Rietveld refinement. These experiments were done for three small pore 8-ring zeolite frameworks: CHA, DDR and LEV. All three structures showed lattice expansion. The origin of this lattice expansion was correlated to different hydrocarbon species and is illustrated in Figure 2.[5]

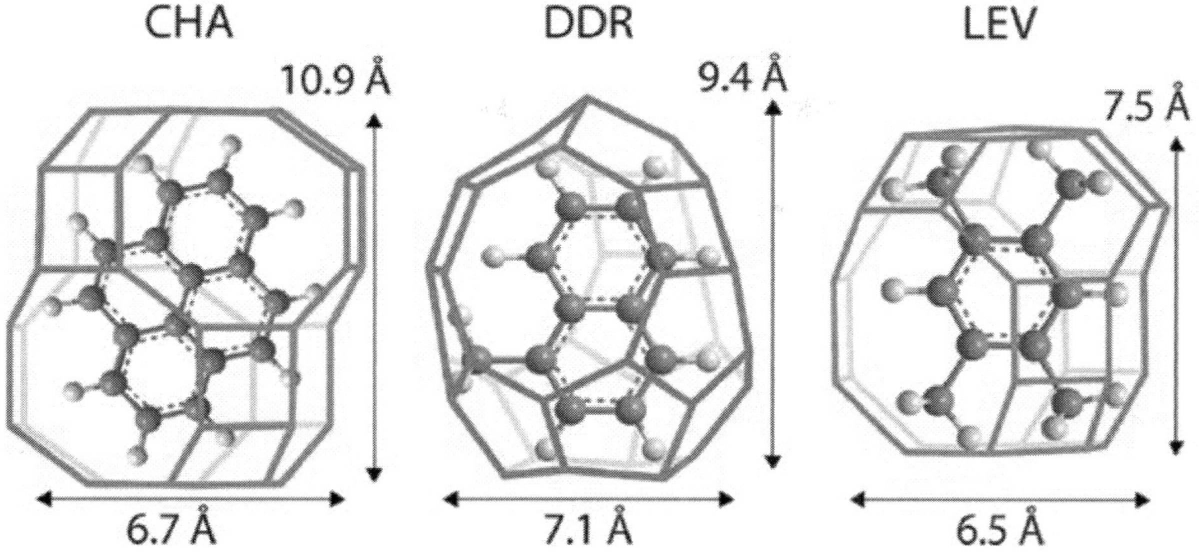

Figure 2: The three zeolite frameworks and the species causing the lattice expansion during MTO observed with operando XRD experiments.[5]

To conclude, in the work to be presented we will include the operando spectroscopic results of the MTH process on different zeolite frameworks with and without incorporation of hetero atoms to be, in the end, able to contribute to the unraveling of the complicated and promising MTH process.

Reference list:

(1) Goetze, J.; Meirer, F.; Yarulina, I.; Gascon, J.; Kapteijn, F.; Ruiz-Martínez, J.; Weckhuysen, B. M. Insights into the Activity and Deactivation of the Methanol-to-Olefins Process over Different Small-Pore Zeolites As Studied with Operando UV-Vis Spectroscopy. *ACS Catal.* **2017**, *7*, 4033–4046.

(2) Yarulina, I.; Chowdhury, A. D.; Meirer, F.; Weckhuysen, B. M.; Gascon, J. Recent Trends and Fundamental Insights in the Methanol-to-Hydrocarbons Process. *Nat. Catal.* **2018**, *1*, 398–411.

(3) Goetze, J.; Weckhuysen, B. M. Spatiotemporal Coke Formation over Zeolite ZSM-5 during the Methanol-to-Olefins Process as Studied with: Operando UV-Vis Spectroscopy: A Comparison between H-ZSM-5 and Mg-ZSM-5. *Catal. Sci. Technol.* **2018**, *8*, 1632–1644.

(4) Yarulina, I.; De Wispelaere, K.; Bailleul, S.; Goetze, J.; Radersma, M.; Abou-Hamad, E.; Vollmer, I.; Goesten, M.; Mezari, B.; Hensen, E. J. M.; et al. Structure–performance Descriptors and the Role of Lewis Acidity in the Methanol-to-Propylene Process. *Nat. Chem.* **2018**, *10*, 804–812.

(5) Goetze, J.; Yarulina, I.; Gascon, J.; Kapteijn, F.; Weckhuysen, B. M. Revealing Lattice Expansion of Small-Pore Zeolite Catalysts during the Methanol-to-Olefins Process Using Combined Operando X-Ray Diffraction and UV-Vis

Spectroscopy. *ACS Catal.* **2018**, *8*, 2060–2070.

545703 Kinetic Parameters Optimization of Fischer-Tropsch Process for Oil Production and Experimental Validation

Wednesday, June 5, 2019: 3:06 PM
Texas Ballroom D (Grand Hyatt San Antonio)

Saiedeh Arabi, Sandeep Patil, Ashkan Akbariyeh, Siddarth Chintamani and Brian Dennis, Mechanical & Aerospace Engineering, The University of Texas at Arlington, Arlington, TX

Fischer-Tropsch (FT) process involves a mixture of predominantly CO and H_2, obtained from combustion of natural gas which is converted to a multicomponent mixture of hydrocarbon. FT process has been setup in our lab for the reaction of Syngas in tube filled with catalyst. Some direct measurements and some indirect calculated values for quantities of chemical species are obtained from conducted experiments.

As the first step, simple model using chemical reaction rates equations was developed and constructed a system of differential equations. Chemical equilibrium equation incorporating reaction rate was included in system for each chemical reaction. A four-factor design was proposed to study the effect of reaction temperature, H_2/CO, oil selectivity, and syngas conversion. A mathematical model for flow conditions inside the FT reactor was developed to simulate the chemical reaction process. Values for some of the reaction rates are the uncertainties of the model and the aim of the work is to find optimum values and validate with experimental results.

The solution of the system gives us concentration of all chemical species as a function of time. Final values of the concentrations and convert them to appropriate quantities to compare with experimental data. An objective function is defined as errors between measured and simulated values. Objective function value close to zero indicates a good fit. Optimization problem was setup using MATLAB with the defined objective function to find an optimum set of values for the uncertain reaction rates.

Developed mathematical model with optimized values were evaluated with results obtained for Oil production with temperature. It shows close agreement with experimental data.

545891 Evaluating Cu/BEA Catalyst Performance with Techno-Economic Analysis to Develop a Market-Responsive Biorefinery Concept Around the Conversion of Methanol to High-Octane Hydrocarbons

Wednesday, June 5, 2019: 3:09 PM
Texas Ballroom D (Grand Hyatt San Antonio)

Daniel Ruddy[1], Jesse E. Hensley[1], Connor Nash[1], Eric Tan[2], Earl Christensen[3], Carrie A. Farberow[4] and Joshua A. Schaidle[5], (1)National Bioenergy Center, National Renewable Energy Laboratory, Golden, CO, (2)National Renewable Energy Laboratory, Golden, CO, (3)Transportation Technologies, National Renewable Energy Laboratory, Golden, CO, (4)Department of Chemical and Biological Engineering, University of Wisconsin-Madison, Madison, CO, (5)National Bioenergy Center, National Renewable Energy Laboratory, Golden, MI

A recent advancement in the conversion of methanol to hydrocarbon fuels is the development of a Cu-modified beta zeolite catalyst (Cu/BEA) that exhibits a higher productivity and yield than the parent H-form BEA catalyst in the production of branched C_4-C_8 hydrocarbons. This Cu/BEA catalyst enables process improvements in the High-Octane Gasoline (HOG) pathway, which operates under lower severity conditions than traditional methanol-to-fuels processes, offering a lower coke formation rate, higher selectivity to C_{5+}gasoline-range hydrocarbons, and subsequently, improved process economics based on a recent process model and techno-economic analysis (TEA). Here, a market-responsive biorefinery concept around methanol and the HOG pathway is presented, where the major C_4 products, isobutane and isobutene, serve as versatile intermediates that can either be recycled to increase the yield of the HOG product or directed to jet-fuel range hydrocarbons. Based on simulated recycle experiments with ^{13}C-labeled isobutane, the Cu/BEA catalyst achieved increased yield to the HOG products during DME homologation by activating the co-fed isobutane and incorporating it into the chain growth pathway, as evidenced by the presence of isotopically-labeled ^{13}C in C_{5+} products. Olefin coupling of an isobutene-rich mixture was also investigated, and the fuel properties of the resulting synthetic kerosene product were within the specifications for a typical jet-fuel. These experimental results were incorporated into an updated Aspen-based process model and TEA to quantify their impact on a market-responsive biorefinery concept. The improvements offered by C_4 recycling with the Cu/BEA catalyst result in a decrease in the methanol-to-fuel synthesis cost from 101¢ per gallon of gasoline equivalent (GGE) with the baseline BEA catalyst to 72¢/GGE, due in part to a significantly increased HOG yield from 40 to 55 GGE/dry-ton biomass. Expanding the process model to direct C_4 products to a jet-range hydrocarbon product demonstrated a yield of 42 GGE/dry-ton HOG with 12 GGE/dry-ton distillate, resulting in a minor increase in the fuel synthesis cost to 77¢/GGE versus the HOG-only case. With continued development, these advancements in fuel yield and reductions in cost position a market-responsive biorefinery around the HOG pathway to be competitive against mature fuel-synthesis technologies from a variety of carbon sources through methanol.

549387 Catalytic Options for Methane Valorization *Catalyst Performance Evaluation Using Parallel Fixed Bed Reactor Systems and Data Driven Catalyst Development*

Wednesday, June 5, 2019: 3:12 PM
Texas Ballroom D (Grand Hyatt San Antonio)

Roel Moonen[1], Amy Triesscheijn[1], Erik-Jan Ras[2] and Carlos Ortega[1], (1)Avantium, Amsterdam, Netherlands, (2)Catalysis, Avantium, Amsterdam, Netherlands

The abundance of shale gas has increased the interest in methane valorization via catalytic conversion. Methane can be used as a feedstock to produce valuable chemicals, mostly by first producing syngas via reforming, followed by one or more catalytic conversion steps. Well-known examples are the methanol-to-olefins (MTO), methanol-to-gasoline (MTG) and the Fischer-Tropsch (FT) processes. Other routes for the valorization of shale gas exist, e.g. the catalytic oxidative coupling of methane (OCM) to produce ethylene.

For the development of new catalyst formulations for these processes, it is crucial to perform well-defined catalyst performance tests. The catalyst tests should provide information on the intrinsic activity of the materials, which facilitates improvement developments between generations of catalysts.

Avantium has used its parallel miniature fixed bed catalyst-testing platform (Flowrence) in many areas of catalysis. This platform has been successfully used to develop and optimize a large variety of catalytic processes, ranging from refining applications, gas-to-liquids, chemicals and biomass conversion. This miniaturized technology has demonstrated to achieve scalable results [1], even in the presence of highly exothermic reactions, which could hinder catalytic performance. In addition, the our high throughput technology (HTT) allows us to simultaneously evaluate up to 64 catalyst / process conditions , reducing significantly the time required to screen new materials or evaluate process parameters.

In this talk, we will show the design considerations taken to evaluate materials in a miniaturized parallel system, under kinetically control and near ideal conditions. These design considerations include, among others, precise and accurate control of the feed flow rate using glass chips and our proprietary active liquid distributors based on microfluidics technology, improved reactor-to-reactor pressure control by means of an active pressure controller, which can reduce the pressure difference in parallel reactors well below 5% of the operational pressure and accurate temperature control that guarantee quasi-isothermal operation.

Several applications will be used to explain the capabilities and design considerations. These include Fischer-Tropsch, OCM and MTO. Each of these applications have challenges to achieve proper control of the test conditions. For the MTO process, temperature and pressure were shown to have a critical effect. In addition, the fast deactivating behavior of SAPO-34 materials used in MTO represented an analytical challenge that was overcome by the use of a 16-loop sampling valve coupled with a gas chromatograph.

References:

[1] R Moonen, J Alles, E Ras, C Harvey, JA Moulijn, Chemical Engineering & Technology 40 (11), 2025-2034

[2] EJ Ras, S Gomez-Quero, Topics in Catalysis 57 (17-20), 1392-1399

[3] JC van der Waal, EJ Ras, CM Lok, R Moonen, N van der Puil, Modern Applications of High Throughput R&D in Heterogeneous Catalysis, 197

551132 One-Pass Tunable Formation of Chemicals or Super-Clean Fuels Via Syngas Conversion: Precise Design of Multiple Functional Catalysts

Thursday, June 6, 2019: 10:30 AM
Texas Ballroom A (Grand Hyatt San Antonio)

Noritatsu Tsubaki, Department of Applied Chemistry, School of Engineering, University of Toyama, Toyama, Japan

One-pass tunable formation of chemicals or super-clean fuels via syngas conversion: precise design of multiple functional catalysts

Noritatsu Tsubaki

Department of Applied Chemistry, School of Engineering, University of Toyama, Gofuku 3190, Toyama 930-8555, Japan

Corresponding authors: tsubaki@eng.u-toyama.ac.jp

The ongoing depletion of finite petroleum resources and emphasis on sustainable development compels humanity to seek new routes to energy supplement. In various transformation processes, catalytic conversion of syngas to chemicals or super-clean fuels is a promising and powerful approach. Thereinto, the formation of syngas produced from nonpetroleum sources such as natural gas, coal and biomass, can be viewed as a green process. Hence, directly converting syngas to ideal target products has become the focus of research, especially achieving directional transformation. Given the product distribution from syngas conversion occurring on Fischer-Tropsch catalysts (i.e. iron or/and cobalt, etc.), is limited by the Anderson-Schulz-Flory (ASF) distribution, precise control of chain propagation process or regulation of chain length is a huge challenge. Single-type active site does not well meet the requirements for catalytic performance, and then multifunctional catalysts (oxide-zeolite) containing several types of active sites are developed to overcome these problems. Using the synergistic catalysis between multiple active sites to achieve an excellent performance, can preferably meet product diversity and selectivity.

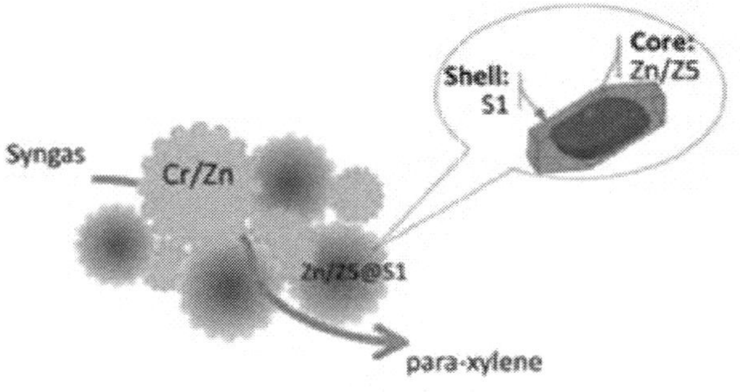

Figure 1 Illustration of the one-pass selective conversion of syngas to para-xylene over the designed hybrid catalyst Cr/ZnOx¨CZn/Z5@S1.

Part I: Based on the above considerations, we fabricate a promising multifunctional catalyst, named Cr/ZnOx-Zn/Z5@S1, to effectively realize the one-pass conversion of syngas to para-xylene, as shown in Figure 1. This hybrid catalyst exhibited enhanced activity on syngas conversion (CO conversion of 55.0%), good stability of catalyst lifetime and considerable selectivity of para-xylene (27.6% in the total products and 77.3% in xylene). The characterization and catalytic performance evaluation revealed that the well-designed Zn/Z5@S1 zeolite capsule catalyst, as a vital part of this Cr/Zn-Zn/Z5@S1 hybrid catalyst, substantially contributed to its extreme performance for the para-xylene one-pass precise synthesis from syngas. The concerted combination of two components in this hybrid catalyst can effectively depress the formation of unwanted by-products and facilitate the oriented synthesis of para-xylene from syngas with unprecedented efficiency at the same time (1).

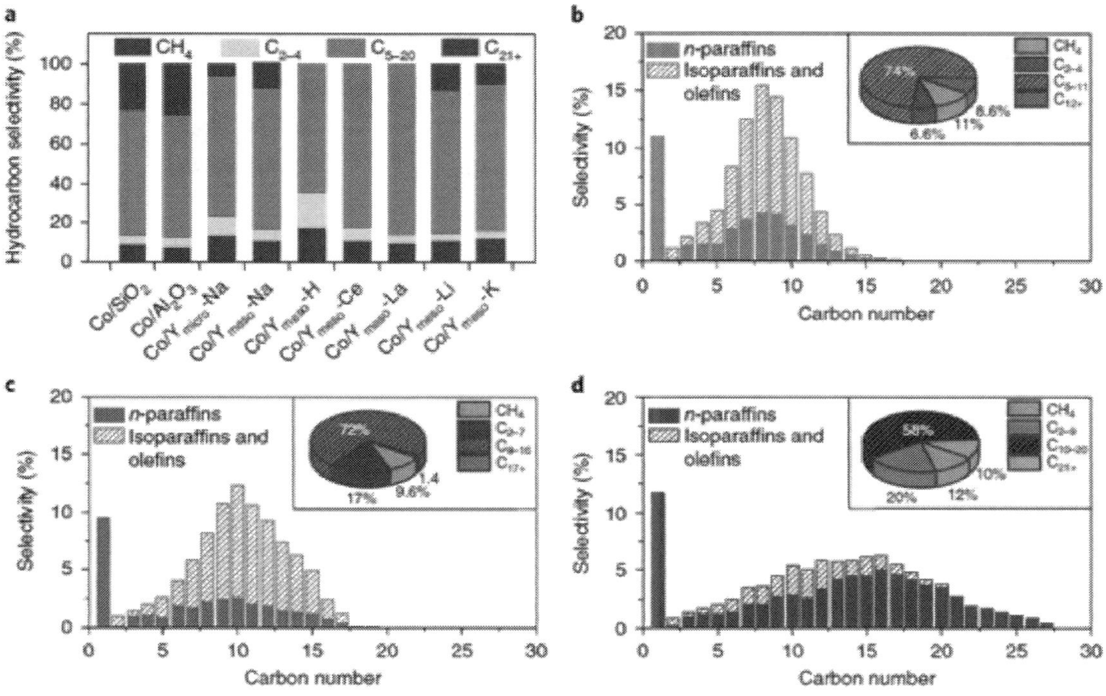

Figure 2 a) Fischer-Tropsch performance over the catalysts with conventional supports or Ymeso zeolites. (b-d) Detailed product distribution of Ymeso catalysts: Co/Ymeso-Ce (b), Co/Ymeso-La (c) and Co/Ymeso-K (d).

Part II: In general, liquid hydrocarbon fuels possess higher energy content compared with other fuels such as dimethyl ether, hydrogen, and so on. Moreover, super-clean Fischer-Tropsch-based fuels with a very low content of sulfur and aromatic compounds are more environmentally friendly than petroleum-based fuels. However, to control the products of Fischer-Tropsch synthesis distributing in a specified range (such as gasoline C_{5-11}, jet fuel C_{8-16} or diesel fuel C_{10-20}) is highly challenging. In order to break the limitation of ASF distribution, we present an integrated catalytic process for the direct conversion of syngas (CO/H) into different types of liquid fuels without subsequent hydrorefining post-treatments of Fischer-Tropsch hydrocarbons. As for this integrated catalytic process, outstanding selectivities for gasoline, jet fuel and diesel fuel as high as 74, 72 and 58% are achieved, respectively, by only using mesoporous Y-type zeolites in combination with cobalt nanoparticles (Figure 2). The porous and acidic properties of Y_{meso} zeolites play two key roles in tuning the product distribution, thus the types of liquid fuels can be readily tuned by controlling the porosity and acid properties of the zeolites (2). A new Fischer-Tropsch products distribution model is also proposed to replace the classic ASF model.

References

(1) P. Zhang, L. Tan, G. Yang, N. Tsubaki, Chemical Science, Vol.8, 7941-7946, 2017

(2) J. Li, Y. He, L. Tan, P. Zhang, X. Peng, A. Oruganti, G. Yang, H. Abe, Y. Wang, N. Tsubaki, Nature Catalysis, Vol.1, 787-793, 2018

545679 Bifunctional Catalysts for Selective Conversion of Syngas beyond Fischer-Tropsch Synthesis

Thursday, June 6, 2019: 11:09 AM
Texas Ballroom A (Grand Hyatt San Antonio)

J. Lennart Weber[1], Daniel Martinez[2], Carmen Martos[2], Arturo Vizcaino[2], **Petra E. de Jongh**[1] and Krijn de Jong[1], (1)Inorganic Chemistry and Catalysis, Utrecht University, Utrecht, Netherlands, (2)University Rey Juan Carlos, Madrid, Spain

Conversion of Synthesis Gas to Aromatics using Bifunctional Catalysis

J. Lennart Weber[1], Daniel Martinez[2], Carmen Martos[2], Arturo Vizcaino[2], Petra E. de Jongh[1], Krijn P. de Jong[1] [1] Inorganic Chemistry and Catalysis, Debye Institute for Nanomaterials Science, Utrecht University, The Netherlands.

[2] Department of Energy and Chemical Technology, University Rey Juan Carlos, Campus de Móstoles, Spain.

The production of materials such as plastics, as well as platform chemicals such as olefins and aromatics is mainly based on processing of crude oil. In order to reduce the carbon footprint alternative pathways are needed. Synthesis gas (a mixture of carbon monoxide and hydrogen) can be derived from carbon containing feedstocks such as natural gas, coal and biomass and therefore offers a potentially more sustainable pathway for the production of chemicals and fuels. The Fischer-Tropsch synthesis allows to convert synthesis gas to hydrocarbons with a wide range of chain length varying from methane to heavy waxes with more than 30 carbon atoms. These Fischer-Tropsch products can be further processed to diesel, gasoline, olefins or waxes with very high purity. Alternatively, synthesis gas can be converted to methanol, which can be further processed to olefins and other hydrocarbons or dimethyl ether.

In the past years the conversion of synthesis gas to chemicals using bifunctional catalysts has received great attention. The combination of a metal oxide and a zeolite into a bifunctional catalyst allows to convert synthesis gas to short olefins such as ethylene [1] or aromatics [2] with selectivities as high as 80% with oxygenates as intermediates. The catalyst bed design plays an important role for the product distribution and activity of these bifunctional catalysts [3]. Alternatively, synthesis gas can be converted into C_2-C_4 olefins with high selectivity of 63% using the Fischer-Tropsch to olefins (FTO) process. Adding sodium and sulfur promoters to an iron (-carbide) based catalyst suppresses methane formation and increases activity compared to an unpromoted catalyst [4,5]. Such an FTO catalyst can be combined with an H-ZSM-5 zeolite to convert the olefins into aromatics at ~400 C and 1 bar pressure [6]. To reach more industrially relevant conditions, a medium temperature and high pressure process is required. Here, we show a process to convert synthesis gas to aromatics operating at temperatures between 250 °C and

300 °C and 20 bar pressure forming olefins in the C_4-C_8 range as intermediates report the impact of temperature and zeolite content in the catalyst on the formation and distribution of aromatic products.

We prepared a bulk iron catalyst (surface area 40 m^2/g, pore volume 0.3 mL/g) by precipitation from a 0.1 mol/L iron nitrate solution with ammonia solution. After washing, drying at 120 °C and calcination at 300 °C in static air, the resulting iron oxide was impregnated with a potassium nitrate solution to achieve a mass ratio of iron to potassium of 100Fe:3K (denoted as FeK). This potassium promoted bulk iron catalyst was combined with an H-ZSM-5 zeolite (Si:Al = 15 at/at) in a stacked bed configuration with ratios of zeolite to FeK between 0.5 m/m and 4 m/m and tested in a 16-channel high throughput fixed bed unit (Flowrence, Avantium) under medium alpha conditions (alpha of 0.67 at 250 °C and alpha of 0.60 at 300 °C). After reduction at 350 °C for 2 h in 30 % H_2 in N_2 at 1 bar, synthesis gas was introduced at temperatures between 250 °C and 300 °C at 20 bar pressure. Here, we show the influence of temperature and zeolite to FeK ratio in stacked bed mode on the formation of aromatics and distribution within the aromatic fraction.

The initial activity normalized to the mass of iron (iron time yield, FTY) was found to be around 5.7 $10^{-5}mol_{CO}\ g_{Fe}^{-1}\ s^{-1}$ for all experiments performed at 300 °C and decreased to a stable FTY of 2.2 $10^{-5}mol_{CO}\ g_{Fe}^{-1}\ s^{-1}$ after 40 h on stream, whereas the experiments performed at 250 °C stabilized at an activity of 0.9 $10^{-5}mol_{CO}\ g_{Fe}^{-1}\ s^{-1}$ and at 275 °C at an FTY of 1.1 $10^{-5}mol_{CO}\ g_{Fe}^{-1}\ s^{-1}$. Further, we found that the selectivity to C_6-C_{10} aromatics increased with increasing reaction temperature operating at a zeolite to FeK ratio of 2 m/m. In stacked bed mode, selectivity to C_6-C_{10} aromatics of 2.3 $\%_C$ at 250 °C, 2.8 $\%_C$ at 275 °C and 4.1 $\%_C$ at 300 °C were observed. Furthermore, a shift to heavier aromatic products was noticed with on average 7.9 carbon atoms at 250 °C to 8.3 carbon atoms at 300 °C within the aromatics distribution. The potassium promoted bulk iron catalyst without zeolite showed a selectivity to C_2-C_3 olefins of 21 $\%_C$ and to C_4-C_8 olefins of 30 $\%_C$ at 300 °C and no aromatics were formed (see Figure - top). With increasing the zeolite to FeK ratio between 0.5 m/m and 1.5 m/m C_4-C_8 olefins selectivity decreased from initially 30 $\%_C$ to 18-21 $\%_C$, whereas the selectivity to C_2-C_3 olefins remained constant between 20 $\%_C$ and 21 $\%_C$. C_6-C_{10} aromatics were formed with 4-6 $\%_C$ selectivity (87 % xylenes in C_8 aromatics) and an average carbon number of 8.1 at at a zeolite to FeK ratio of 1 m/m. Increasing the zeolite to FeK ratio to 2 m/m led to the decrease in C_2-C_3 olefins selectivity to 7 $\%_C$ and the formation of heavier aromatics with an average carbon number of 8.3 at. A further increase in zeolite to FeK ratio to 2.5 m/m led to the formation of C_6-C_8 aromatics with 19 $\%_C$ selectivity, whereas the selectivity towards both C_2-C_3 olefins and C_4-C_8 olefins decreased. At a high zeolite to FeK ratio of 4 m/m we observed a selectivity to C_6-C_{10} aromatics of 21 $\%_C$ and almost no olefins were formed. These results show that C_8-aromatics were formed preferentially from longer C_4-C_8 olefins when a low amount of zeolite was present (zeolite to FeK ratio between 0.5 m/m and 1.5 m/m). Increasing the amount of zeolite led to alkylation of light aromatics such as toluene with ethylene as well as oligomerization of C_2-C_3 olefins towards C_4-C_8 olefins, resulting in lower C_2-C_3 olefins selectivity and increased selectivity to aromatics with higher average carbon number. Operating with zeolite to FeK ratio of 4 m/m the catalytic data suggested the aromatization of shorter olefins to form aromatics. In contrast to a high temperature and low pressure process the aromatization followed a hydrogen transfer pathway, forming paraffins as side product (data not shown).

We show that combining a potassium promoted bulk iron catalyst with an H-ZSM-5 zeolite in a stacked-bed enabled a medium temperature route to convert synthesis gas to olefins and aromatics with adjustable product composition depending on reaction temperature and catalyst composition. The CO conversion of this medium temperature process is scalable by changing space velocity (CO conversion as high as 87 % and 62 % aromatics in the liquid C_{5+} fraction, detailed data will be shown at the conference). This demonstrates an advantage over the processes using oxygenates intermediates to form aromatics, which seem to be limited to CO conversions of around 20 %. Furthermore, this medium temperature process shows a long lifetime of the catalysts compared to the high temperature process operating at 400 °C. These findings allow to design targeted bifunctional catalyst systems for various applications.

[1] F. Jiao, X. Pan, K. Gong, Y. Chen, G. Li, X. Bao, Angew. Chemie Int. Ed. 57 (2018) 4692–4696.

[2] K. Cheng, W. Zhou, J. Kang, S. He, S. Shi, Q. Zhang, Y. Pan, W. Wen, Y. Wang, Chem 3 (2017) 334–347.

[3] W. Zhou, J. Kang, K. Cheng, S. He, J. Shi, C. Zhou, Q. Zhang, J. Chen, L. Peng, M. Chen, Y. Wang, Angew. Chemie Internatioanl Ed. (2018) 10.1002/anie.201807113.

[4] H.M. Torres Galvis, J.H. Bitter, C.B. Khare, M. Ruitenbeek, A.I. Dugulan, K.P. De Jong, Science 335 (2012) 835–838.

[5] M. Oschatz, T.W. van Deelen, J.L. Weber, W.S. Lamme, G. Wang, B. Goderis, O. Verkinderen, A.I. Dugulan, K.P. de Jong, Catal. Sci. Technol. 6 (2016) 8464–8473.

[6] J.L. Weber, I. Dugulan, P.E. de Jongh, K.P. de Jong, ChemCatChem 10 (2018) 1107–1112.

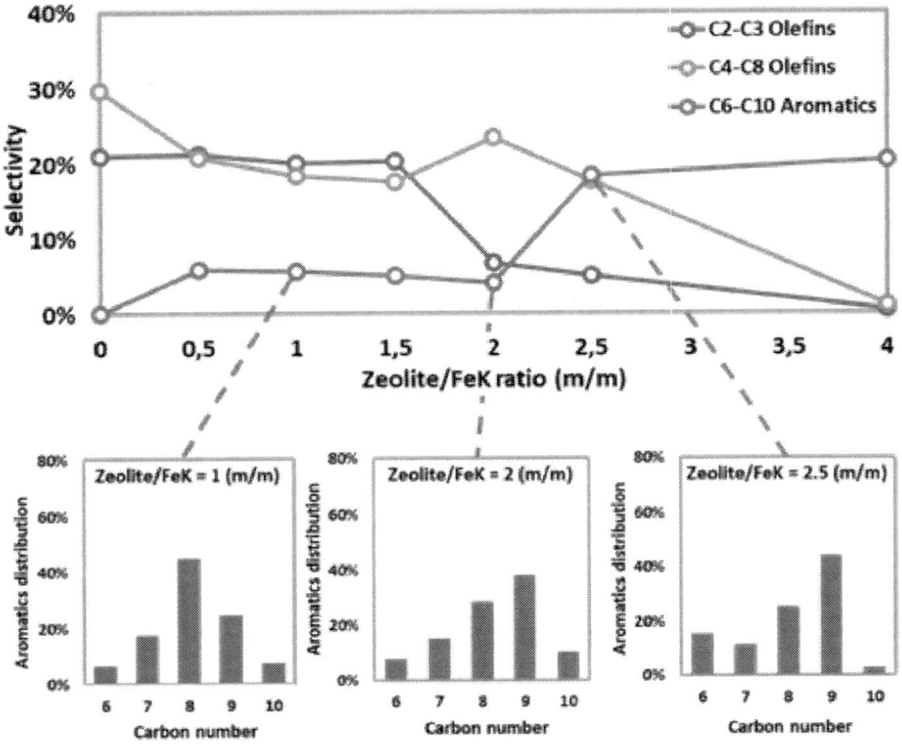

Top: Selectivity of hydrocarbons (CO_2 free) to C_2-C_3 olefins, C_4-C_8 Olefins and C_6-C_{10} aromatics as function of zeolite/FeK mass ratio in a stacked bed configuration at 300 °C.
Bottom: Distribution of aromatics for Zeolite/FeK mass ratio of 1, 2 and 2.5, respectively.
Reaction conditions: 300 °C, 20 bar, CO:H_2=1 (v/v), TOS: 6-12 h

551289 Direct Conversion of Syngas to Light Olefins: Construction of Reaction Network and Design of Oxide-Zeolite Bifunctional Catalysts

Thursday, June 6, 2019: 11:48 AM
Texas Ballroom A (Grand Hyatt San Antonio)

Yangdong Wang[1], Junjie Su[1], Haibo Zhou[1], Wenqian Jiao[1], Su Liu[1], Dong Wang[1], Weimin Yang[1], Zaiku Xie[2] and Mingyuan He[3], (1)State Key Laboratory of Green Chemical Engineering and Industrial Catalysis, Shanghai Research Institute of Petrochemical Technology, SINOPEC Corp., Shanghai, China, (2)China Petrochemical Corporation, Beijing, China, (3)Shanghai Key Laboratory of Green Chemistry and Chemical Processes, School of Chemistry and Molecular Engineering, East China Normal University, Shanghai, China

Abstract: The direct synthesis of light olefins from syngas over a bifunctional catalyst containing oxide and zeolite has proven to be a promising strategy. Nevertheless, unclear reaction network hinder the further enhancement of catalytic performance, such as increasing CO conversion and olefin/praffin ratio. We herein synthesis a series of bifunctional catalysts containing oxides (InZr, ZnAl, ZnCr, MoS_2, etc.) and zeolites (SAPO or AlPO zeolites with CHA, AEI, ERI, AFN topologies). A complete reaction network is proposed via DFT calculations and model reactions, including CO activation, intermediate diffusion, C-C coupling, hydrogenation, etc. We demonstrate that methanol is the key intermediate to be transformed into light olefins in zeolites and inhibiting excessive hydrogenation is an effective strategy to achieve higher performance. In addtion, we also find that zeolite is the key component for determining olefin/paraffin ratio of hydrocarbon products. By optimizing the particle size, the density and strength of acid sites and the framework structure of zeolites, we design a novel bifunctional catalyst showing superior olefin/paraffin ratio and high stability at mild reaction conditions. According to the reaction network, we also propose an integration manner of the bifunctional catalysts which can be used in commercial production. These findings will greatly boost the industrialization of lower olefins synthesis from syngas.

References:

1. Haibo Zhou, *et al. Ind. Eng. Chem. Res.*, 2018, 57, 6815-6820.
2. (2) Junjie Su, *et al. ChemCatChem,* 2018, 10, 1536-1541.

552134 Conversion of synthesis gas to light olefins: impact of hydrogenation activity on process selectivity over Cr-Zn and Cu-Zn with SAPO-34

Thursday, June 6, 2019: 12:12 PM
Texas Ballroom A (Grand Hyatt San Antonio)

Alexey Kirilin[1], Joseph DeWilde[2], Adam Chojecki[1], Vera Santos[1], Kinga Scieranka[1] and Andrzej Malek[2], (1)Dow Benelux B.V., Hoek, Netherlands, (2)The Dow Chemical Company, Midland, MI

Abstract for the 12th Natural Gas Conversion Symposium. Invited lecture

Conversion of synthesis gas to light olefins: impact of hydrogenation activity on process selectivity over Cr-Zn and Cu-Zn with SAPO-34

Alexey Kirilin[1]*, Joseph DeWilde[2], Adam Chojecki[1], Vera Santos[1], Kinga Scieranka[1] and Andrzej Malek[2]

[1] Dow Benelux B.V., Herbert H. Dowweg 5, building 443 (BBB), 4252 NM, Hoek, Netherlands; [2] The Dow Chemical Company, Midland, building 1776, MI 48674, USA.

**akirilin@dow.com*

1. Introduction

Light olefins (C2-C3) are industrially vital feedstocks for production of plastics, functional materials, and as platform chemicals for production of other derivatives. Short chain hydrocarbons can be produced from synthesis gas by using a hybrid catalyst combining Cu-based methanol synthesis and SAPO-34 [1]. Recently it was shown that direct conversion of synthesis gas to light olefins with high selectivity is also possible via the so-called Ox-Zeo process by combining $ZnCrO_x$ catalyst with SAPO-34 [2].

In the present study, we aim to compare two bifunctional catalytic mixtures containing a Cu-Zn (low temperature methanol catalyst) [3] or Cr-Zn (high temperature methanol catalyst) [3] in combination with a zeolite component SAPO-34 for the direct conversion of synthesis gas to olefins. The focus of this study is the distribution of products and kinetics of product formation [4].

2. Methods

Cu-based methanol catalyst "HiFuel R120" (Johnson Matthey, sold by Alfa Aesar) and Cr-Zn catalyst (prepared by co-precipitation [5]) were used. SAPO-34 was synthesized according to the literature procedures [6]. Materials were crushed and sieved to 60-80 mesh. Catalyst were characterized by N_2 physisorption, XRF, XRD, TPR and SEM techniques. Hybrid catalysts were prepared by mixing Cu-Zn or Cr-Zn with SAPO-34. Catalytic tests were performed in a tubular stainless steel fixed-bed microreactors (i.d. 3 mm) at 370-410°C, 20 bar, GHSV = 1200 h^{-1} and different H_2/CO ratios. Online GC analysis of components (N_2, H_2, He, CO, CO_2, C1-C5 alkanes and olefins) was performed periodically. Mass balance in all experiments was 95-105% based on carbon.

3. Results and discussion

For Cu-Zn/SAPO-34 only saturated hydrocarbons (C1-C5) and carbon dioxide (due to water-gas-shift reaction) were observed while for the Cr-Zn/SAPO-34 system olefins were also present among the reaction products (Figure 1). The C3/C2 ratio is consistently higher for Cr-Zn/SAPO-34 systems compared to Cu-Zn/SAPO-34 systems at the same conversion level for all measured conditions (Figure 1). The difference in C3/C2 ratios demonstrates the role of the relative hydrogenation behavior in affecting SAPO-34 product distributions. We propose that C3/C2 ratio is predominantly controlled by the relative rates of olefin cycle propagation and the rate of cycle termination by the formation of paraffins on the Cr-Zn catalyst – loading additional SAPO-34 at low conversions increases the relative contribution of olefin cycle propagation on the reactor effluent [4].

A kinetic model was built for the Cr-Zn/SAPO-34 system to better understand and describe the observed patterns in product distribution [4]. Model parameter fits were performed using standard Bayesian techniques employing the Athena Visual Studio software package (M. Caracotsios and W.E Stewart, v14.2).

The parity plots for conversion, combined paraffin carbon yield, and combined olefin carbon yield show considerable scatter around perfect model prediction. Considering these kinetic model shortcomings, improving the quantitative description of this reaction network, potentially necessitating SAPO-34 transport restriction analysis, will be a useful avenue of future investigation to predict C3/C2 product distributions as a function of process conditions.

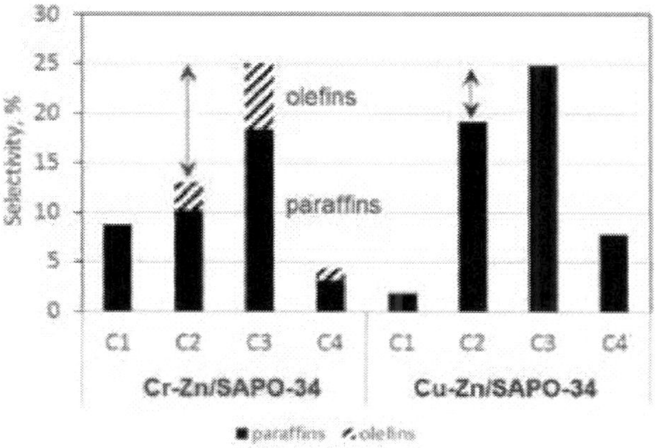

Figure 1. Product composition (C mol%) in conversion of synthesis gas over bifunctional catalysts. Conversion of syngas 40-50%. CO_2 selectivity was 45-50%.

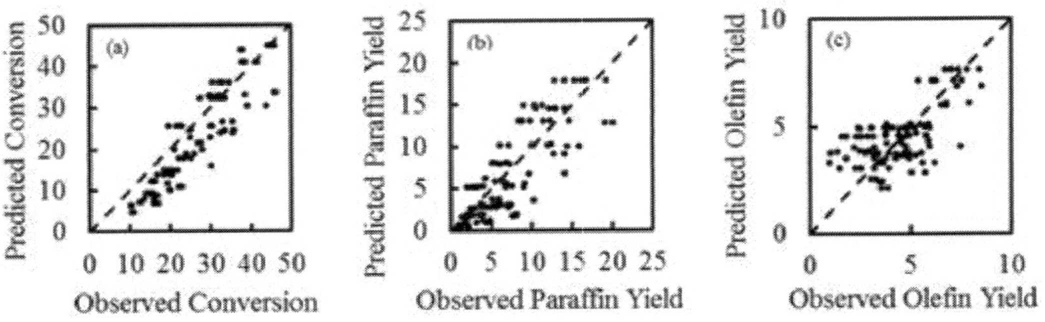

Figure 2. Parity plots for (a) conversion (%), (b) combined paraffin (ethane, propane, and butane) carbon yield (%), and (c) combined olefin (ethylene, propylene, and butene) carbon yield (%) for the kinetic model for Cr-Zn/SAPO-34 system data.

4. Conclusions

The choice of methanol synthesis catalyst alters product selectivity and the C3/C2 product ratios. The relative rates of propagation of olefin methylation and cracking (olefin cycle) in SAPO-34 and olefin hydrogenation on the methanol synthesis catalyst dictates the observed C3/C2 ratio. A kinetic model for the hybrid system is proposed to describe the observed selectivity patterns. We identify the balance of methanol synthesis and olefin hydrogenation rates on the mixed-metal-oxide catalysts as a potentially strong factor to control the product distribution in synthesis gas-to-olefin/paraffins technologies.

References

[1] D. Nieskens, A. Ciftci, P. Groenendijk, M. Wielemaker, A. Malek, Ind. Eng. Chem. Res. 56 (2017), 2722-2732.

[2] F. Jiao, J. Li, X. Pan, J. Xiao, H. Li, H. Ma, M. Wei, Y. Pan, Z. Zhou, M. Li, S. Miao, J. Li, Y. Zhu, D. Xiao, T. He, J. Yang, F. Qi, Q. Fu, X. Bao, Science 351 (2016), 1065-1068.

[3] J.B. Hansen, P.E. Højlund Nielsen, In Handbook of Heterogeneous Catalysis, Wiley-VCH Verlag GmbH & Co. KGaA: 2008. pp. 2920-2949.

[4] A.V. Kirilin, J.F. Dewilde, V. Santos, A. Chojecki, K. Scieranka, A. Malek, Ind. Eng. Chem. Res. (2017), 56 (45), pp 13392–13401.

[5] Collins, U.S. Patent 3,850,850A, 1974.

[6] B.M. Lok, C.A. Messina, R.L. Patton, R.T. Gajek, T.R. Cannan, E.M. Flanigen, U.S. Patent 4,440,871A, 1984.

Keywords

Olefins, syngas, mechanism, kinetic model

545543 The Effects of Intraparticle Diffusion Phenomena on Dimethyl Ether Direct Synthesis

Thursday, June 6, 2019: 10:30 AM
Texas Ballroom EF (Grand Hyatt San Antonio)
Simone Guffanti, Politecnico di Milano, Milano, Italy, Carlo Giorgio Visconti, Department of Energy, Politecnico di Milano, Milan, Italy and **Gianpiero Groppi**, Politecnico di Milano, Dipartimento di Energia, Via La Masa, Milan, Italy

Introduction

Dimethyl ether (DME) is widely used as propellant and it is considered a valid alternative fuel. It can be employed as alternative to diesel fuel, having higher cetane number (55-60) and less pollutants emission, or as alternative to liquefied petroleum gas (LPG), because of the similar physical properties (-24°C boiling point; liquefied at 5 bar pressure at ambient temperature). Traditionally DME is produced from syngas in a two-step process consisting in methanol synthesis and dehydration (indirect synthesis) [1]. To improve the process efficiency, many efforts have been made in the last decades to develop the one-step direct synthesis, combining methanol ($Cu/ZnO/Al_2O_3$) and dehydration catalyst (zeolites or γ-Al_2O_3). The two catalysts can be either mechanically mixed as different pellets in a random fixed bed or intimately coupled by producing hybrid pellets [2][3]. Core-shell pellet configuration has also been proposed [4]. The intraparticle diffusion phenomena have a strong influence on the reactants and products profiles inside the pellets, affecting the reaction kinetics and the catalyst efficiency. As a consequence the thermal behavior and the performances of the reactor are also affected, especially under industrial relevant conditions. However, these effects have not been deeply investigated in literature.

Methods

In order to compare the different configurations (bed of mixed pellets vs. bed of hybrid pellets vs. core-shell pellets), two different kind of simulation have been performed. First, single pellet 1D models for each configuration have been used in order to evaluate the effects of diffusion in the different cases analyzed considering same boundary conditions. The models account the i-species concentration gradients inside the catalyst pellet assuming isothermal conditions. A reaction scheme including methanol (MeOH) synthesis from CO_2, Reverse Water Gas Shift (RWGS) and MeOH dehydration has been adopted with kinetics taken from the literature [5][6][7]. Then, in order to evaluate the effect of diffusion on industrially relevant conditions, a heterogeneous 2D model of a single tube of a multi tubular fixed bed technical reactor for the direct DME synthesis has been used. The models consist of i-species mass, energy and momentum balances for the gas-phase, coupled with i-species mass and energy balances for the catalyst phases (one solid phase for the hybrid and core-shell pellets, two solid phases for the mixed pellets) accounting for concentration gradients (1D) in isothermal pellets. An externally cooled multi tubular reactor of 6 m long and 1 inch internal diameter tubes, filled with 4.8 mm diameter spherical catalyst pellets, has been considered in the simulations. The model equations have been implemented in gPROMS® commercial software for the numerical resolution of the boundary value problem.

Results

The simulations have been performed considering different feed compositions (consisting of H_2, CO, CO_2, CH_4). The results show that the performances of the hybrid pellet configuration in terms of conversion and DME yield are higher compared to the mechanical mixture. So, as consequence of the exothermicity of the reaction involved, the reactor loaded with hybrid pellets has a more difficult thermal control, that is shown by the higher hot-spot temperature, and this can represent an issue considering the possibility of catalyst deactivation. These results can be explained considering that, in the hybrid catalyst pellet, the DME synthesis reaction consumes the methanol produced by the MeOH synthesis, while WGS removes H_2O produced by both MeOH and DME syntheses. This results in a synergistic effect on the conversion rate with respect to the mechanical mixture, where, due to intraparticle diffusion limitation, the catalyst efficiencies decrease both in the MeOH catalyst pellets, because of the equilibrium approach, and in the DME catalyst pellets due to combined effect the lower methanol concentration and the higher water content.

However, the coupling of the two catalysts in hybrid pellets can cause the deactivation of the support, due to the bad interaction between the two catalytic functions [8], with a consequent decrease in performances. The core-shell configuration can be a good compromise in order to guarantee an advantageous contact of the two catalyst functions while avoiding unwanted deactivation effects. The core-shell configuration is at present under analysis.

References

- [1] Azizi, M. Rezaeimanesh, T. Tohidian, M.R. Rahimpour, Chem. Eng. Process. 82 (**2014**) 150-172
- [2] -H. Kim, M.J. Park, S.J. Kim, O.-S. Joo, K.-D. Jung, Appl. Catal., A 264 (**2004**) 37-41
- [3] Catizzone, A. Aloise, M. Migliori, G. Giordano, Appl. Catal., A 502 (**2015**) 215-220
- [4] Yang, N. Tsubaki, J. Shamoto, Y. Yoneyama, Y. Zhang, J. Am. Chem. Soc., 132 (23) (**2010**) 8129–8136
- [5] M. Vanden Bussche, G.F. Froment, J. Catal. 161 (**1996**) 1-10
- [6] Bercic, J. Levec, Ind. Chem. Eng. Res. 31 (**1992**) 1035-1040
- [7] L. Ng, D. Chadwick, B.A. Toseland, Chem. Eng. Sci. 54 (**1999**) 3587-3592
- [8] Garcìa-Trenco, A. Martìnez, Cat. Today, 227 (**2014**) 144–153

Acknowledgements This contribution has received funding from the European Union's Horizon 2020 research and innovation programme under grant agreement No 727600.

545037 Designing a Hierarchical Nanosheet ZSM-35 Zeolite to Realize More Efficient Ethanol Synthesis from Dimethyl Ether and Syngas

Thursday, June 6, 2019: 10:54 AM
Texas Ballroom EF (Grand Hyatt San Antonio)

Xiaobo Feng, Guohui Yang, Yoshiharu Yoneyama and Noritatsu Tsubaki,
Department of Applied Chemistry, School of Engineering, University of Toyama, TOYAMA, Japan

Designing a hierarchical nanosheet ZSM-35 zeolite to realize more efficient ethanol synthesis from dimethyl ether and syngas

Xiaobo Feng, Guohui Yang, Yoshiharu Yoneyama, Noritatsu Tsubaki*

Department of Applied Chemistry, School of Engineering, University of Toyama, Gofuku 3190, Toyama 930-8555, Japan

**Corresponding author: (+81)76-445-6848,* tsubaki@eng.u-toyama.ac.jp

As a clean and alternative fuel, ethanol (EtOH) has attracted increasing attentions due to the shortages of fossil energy and worsening environment. In this work, a dual-catalyst bed reactor packed with the combination of hierarchical nanosheet HZSM-35 (Hi-NZ35) zeolite and CuZnAl catalyst was proposed to realize more efficient ethanol synthesis from dimethyl ether (DME) and syngas ($CO+H_2$). The nanosheet ZSM-35 (NZ35) zeolite was synthesized using hydrothermal synthesis and the CuZnAl catalyst was prepared by co-precipitation method. Moreover, a series of Hi-$NZ35_x$ zeolites were obtained from NZ35 zeolite by further treatment with varied NaOH aqueous solution using hydrothermal process ("x" means the NaOH solution concentration of 0.2-0.6 M). It was found that Hi-$NZ35_{0.4}$ zeolite was more effective to obtain hierarchical porosity with mesopore volume up to 0.131 $cm^3\ g^{-1}$. For single DME carbonylation reaction, the NZ35 zeolite exhibited superior catalytic activity (32.2%) and stability compared with conventional ZSM-35 (CZ35) zeolite. Furthermore, improved catalytic activity (42.0%) was observed on Hi-$NZ35_{0.4}$ zeolite owing to its abundant mesoporous structure. This result revealed that the hierarchical porosity of zeolite could effectively promote the catalytic performance of zeolite for DME carbonylation reaction. For the ethanol synthesis using the optimized catalysts combination of Hi-$NZ35_{0.4}$ zeolite and CuZnAl catalyst, the DME conversion was 47.2% and the ethanol productivity reached 840.2 mmol $kg^{-1}\ h^{-1}$, much better than that of the combination of the CZ35 zeolite with CuZnAl catalyst. This new alternative ethanol synthetic approach from syngas and DME will reduce the dependence on petroleum resources and cut the emission of the greenhouse gas.

545623 Adoption of Highly Conductive Open Cell Foams As an Effective Solution to Enhance the Heat Transfer Performances of Packed-Bed Fischer-Tropsch Reactors

Thursday, June 6, 2019: 11:18 AM
Texas Ballroom EF (Grand Hyatt San Antonio)

Laura Fratalocchi[1], Carlo Giorgio Visconti[2], Luca Lietti[2], Gianpiero Groppi[3] and Enrico Tronconi[4], (1)Energy, Politecnico di Milano, Milano, Italy, (2)Department of Energy, Politecnico di Milano, Milan, Italy, (3)Politecnico di Milano, Dipartimento di Energia, Via La Masa, Milan, Italy, (4)energy, politecnico di milano, milano, Italy

In the last decade, the interest in the Fischer-Tropsch synthesis (FTS) in compact reactors has been considerably renewed in view of exploiting both associated and remote natural gas fields and biomass. In view of the highly exothermic nature of the reaction, the heat removal is a key issue for the development of an intensified reactor [1]. The use of conventional packed-bed reactors (PBRs) is severely limited by heat removal causing a series of constraints which limit the catalyst performances. The poor heat transport properties in PBRs lead to non-isothermal operation of the reactor. This negatively affects the catalyst selectivity and, in the worst case, leads to the onset of thermal runaway [2].

In line with these premises, the efficient thermal management in the modular reactor technology suitable for the FTS on a compact scale (< 3000 BPD) is a key issue [2]. In this regard, Politecnico di Milano recently proposed the adoption of honeycomb catalysts with high-conductivity supports within externally cooled multitubular fixed-bed reactors [3]. In this way, the heat transfer is strongly enhanced because the primary radial heat exchange mechanism is changed from convection to conduction within the thermally connected solid matrix of the honeycomb monolith [3].

In this work, we propose to enhance the heat transport properties of a FTS packed bed reactor through the adoption of highly conductive open-cell metal foams packed with catalyst pellets [4]. These materials were recently proposed as a strategy to intensify heat transfer in strongly endo- and exo-thermic processes [5], but were never used for the FTS. They exploit the same conductive heat transfer mechanism of the monolithic substrates but, in addition, they have the advantage of enabling radial mixing within their structure, thus enhancing flow uniformity [5]. Our strength is also the adoption of "packed" open-cell foams, which can overcome the inherently limited catalyst inventory of washcoated structured reactors.

The open-cell Al-foam (40 PPI and $e_{foam} \approx 0.906$) with a cylindrical shape (d_{foam}= 2.78 cm and l_{foam}= 4 cm) is provided by ERG AEROSPACE. One axial through hole of 0.32 cm diameter is located at the centerline for the insertion of the stainless steel thermowell (1/8" O.D.), protecting a sliding J-type thermocouple.

Once the foam is loaded in the tubular reactor and the thermowell is positioned in the foam, the system is packed as follows. Initially, 5 g of α-Al_2O_3 pellets (d_{pellet}=

300 μm) are poured into the foam, thus forming a 1 cm deep layer. Then, 7.2 g of $Co/Pt/Al_2O_3$ catalyst diluted with a very small amount of α-Al_2O_3 (catalyst:α-Al_2O_3 = 6:1 w/w) with the same particle size are poured into the foam, thus forming a catalyst layer of 1.89 cm. Eventually, 5 g of α-Al_2O_3 pellets (d_{pellet}= 300 μm) are packed into the foam so to fill the last 1 cm of the structure. The resulting average catalyst volumetric density is 0.63 g/cm^3, calculated as the ratio of the catalyst mass (7.2 g) to the reactor volume occupied by the catalyst bed (11.4 cm^3).

In the case of the packed-bed reactor, 7.2 g of $Co/Pt/Al_2O_3$ catalyst are again randomly packed in the same tube used for the foam, but diluting with a large amount of α-Al_2O_3 (catalyst: α-Al_2O_3 = 1:1.7 w/w) pellets with the same particle size (d_{pellet}= 300 μm), so to form a catalyst bed with length of 4 cm, equal to the length of the foam (≈4 cm). This leads to a catalyst volumetric density near 0.29 g/cm3, i.e. almost half of that corresponding to the packed-foam reactor. Accordingly in the packed bed, at the same CO conversion level, the volumetric heat duty (kW/m^3) generated by the reaction is less than in the case of the packed foam.

The performances of a highly active $Co/Pt/Al_2O_3$ catalyst packed into the foam metallic structure have been assessed at industrially relevant process conditions (190-240 °C, 25 bar, H_2/CO= 2 mol/mol, 6410 $cm^3(STP)/h/g_{cat}$) for more than 750 h and compared with those obtained in a conventional randomly packed fixed-bed reactor [4]. The structured reactor reaches outstanding performances (CO conversions >67.5%) with a remarkable temperature control. The volumetric heat duty (Q) calculated in the experiment with the packed-foam reactor increases with temperature, being a function of the CO conversion. It starts from 80.6 kW/m^3 obtained at 180 °C, up to a remarkable value of 1360.4 kW/m^3 at 240 °C. To our knowledge, this is the first time in the scientific literature that such high values are reported for the FTS reaction in a lab-scale apparatus.

Limited T-gradients along the catalyst bed are obtained at all the temperatures investigated, thus resulting in very small $\Delta T_{cat} = T_{max} - T_{min}$ measured along the catalyst bed, even if in the presence of high volumetric heat duty. In this regard, the ΔT_{cat} is negligible at 180 °C and 190 °C, in line with the very low catalyst activity. Also in the 195–205 °C T-range, it is small ($\Delta T_{cat} \approx 1$ °C) although the CO conversion level increases up to 22%. Increasing the reaction temperature up to 210 °C and then to 215 °C, the T-gradients are only slightly affected, with a ΔT_{cat} of 2 and 3 °C, respectively. The CO conversion values at these temperatures are 28% and 33%, respectively. At 220 °C and 225 °C, with significant CO conversions of 44% and 50%, the ΔT_{cat} is still small and equals 4 °C and 4.5 °C, respectively. The most significant effect on the ΔT_{cat} is obtained at 230 °C and 240 °C when the CO conversion reaches the highest values (55% and 67.5%). Accordingly, the ΔT_{cat} becomes 5 °C and 6 °C, respectively [4].

Noteworthy, although the catalyst was tested for several hours (≈800 h), which is unusual for lab-scale runs, and frequently varying the process conditions, it was found to be very stable with Time on Stream (T.o.S.). In fact, the CO conversion measured by replicating the standard conditions (200 °C, 25 bar, H_2/CO= 2 mol/mol, 6410 $cm^3(STP)/h/g_{cat}$) at different T.o.S. (200, 520 and 760 h) was always around 16%. This is particularly interesting since the catalyst worked under severe conditions (i.e. high P_{H2O}/P_{H2} in particular in the second half of the catalyst bed [6]) for over 200 h. We believe that this is a further indication of the excellent heat transfer properties of the packed foam reactor, which prevent the deactivation of the catalyst by avoiding strong temperature gradients in the catalyst bed.

The results herein reported prove the excellent ability of the "highly conductive packed-foam reactor" concept to manage the strong exothermicity of the FTS reaction [4]. In contrast, when the same experiment was carried out over the same $Co/Pt/Al_2O_3$ catalyst just randomly packed in the reactor, an abrupt increase of the catalyst temperature occurred already at low temperature (duties <244 kW/m^3 with CO conversions <12.1%) eventually leading to thermal runaway.

Our data confirm that, thanks to the adoption of the conductive foams, the mean temperature inside the reactor can be much better controlled providing new operating windows, which are not accessible using the conventional packed-bed reactor technology. The conductive packed-foams enable in fact running the FT reaction under severe conditions (i.e. high CO conversion and large heat duty) with an intensified temperature control. Indeed, in a crucial comparative experiment the conventional packed-bed reactor, although operated under milder conditions (i.e. the catalyst density was halved with respect to the foam), experienced thermal runaway already at very low temperatures and CO conversions, i.e. with limited release of reaction heat. These results are a direct indication that the heat exchange is significantly enhanced thanks to the structured conductive substrate of the foam.

Furthermore, the "packed-foam" configuration also represents an innovative solution to increase the catalyst inventory in structured tubular reactors, since the catalyst load which can be packed in the open-cell foam is much greater than the amount which can be loaded by washcoating the same foam. In this way, the productivity per reactor volume can be boosted. In addition, "packing" the foam means overcoming all the problems linked to the coating process, to the catalyst loading and unloading in the reactor, and to the replacement of the spent catalyst [4].

References:

[1] C. G. Visconti, E. Tronconi, G. Groppi, L. Lietti, M. Iovane, S. Rossini, R. Zennaro, Chem. Eng. J. 171 (2011) 1294-1307

[2] C. G. Visconti, E. Tronconi, L. Lietti, G. Groppi, P. Forzatti, C. Cristiani, R. Zennaro, S. Rossini, Appl. Catal. A-Gen. 370 (2009) 93-101

[3] M. Iovane, R. Zennaro, P. Forzatti, G. Groppi, L. Lietti, E. Tronconi, C. G. Visconti, S. Rossini, E. Mignone, Pat. Appl. WO/2010/130399

[4] L. Fratalocchi, C.G. Visconti, L. Lietti, G. Groppi, E. Tronconi, Chem. Eng. J. 349 (2018) 829-837

[5] G. Groppi. E. Tronconi, C.G. Visconti, A. Tasso, R. Zennaro, Pat. Appl. WO/2015/033266

[6] L. Fratalocchi, C.G. Visconti, L. Lietti, G. Groppi, E. Tronconi, E. Roccaro, R. Zennaro Catal. Sci. Technol., 6 (2016) 6431-6440

Acknowledgment:

This project has received funding from the European Research Council (694910/INTENT).

551452 Fischer-Tropsch Synthesis on Freeze-Cast Hybrid-Backbone Meso-Macroporous Micromonolith Catalysts: Reconciling High Pore Transport Rates with Improved Reaction Heat Management

Thursday, June 6, 2019: 11:42 AM
Texas Ballroom EF (Grand Hyatt San Antonio)

Kai Jeske[1], Jonglack Kim[1], Valentina Nese[1], Jochen Joos[2], Nicolas Duyckaerts[1], Norbert Pfänder[3] and Gonzalo Prieto[4], (1)Heterogeneous Catalysis, Max-Planck Institut für Kohlenforschung, Mülheim an der Ruhr, Germany, (2)Institut für Angewandte Materialien, Karlsruhe Institute of Technology (KIT), Karlsruhe, Germany, (3)Heterogeneous Reactions, Max-Planck Institut für chemische Energiekonversion, Mülheim an der Ruhr, Germany, (4)ITQ Institute of Chemical Technology (CSIC-UPV), Valencia, Spain

Fischer-Tropsch synthesis on freeze-cast hybrid-backbone meso-macroporous micromonolith catalysts: reconciling high pore transport rates with improved reaction heat management

Kai Jeske,[a] Jonglack Kim,[a] Valentina Nese,[a] Jochen Joos,[b] Nicolas Duyckaerts,[a] Norbert Pfänder,[c] and Gonzalo Prieto[a,*]

[a] Max-Planck-Institut für Kohlenforschung, Kaiser-Wilhelm-Platz 1, 45470 Mülheim an der Ruhr (Germany), [b] Institute for Applied Materials (IAM-WET), Karlsruhe Institute of Technology (KIT), Adenauerring 20b, 76131, Karlsruhe (Germany), [c] Max-Planck-Institut für Chem. Energiekonversion, Stiftstraße 34-36, 45470 Mülheim an der Ruhr (Germany)

* Email: prieto@mpi-muelheim.mpg.de

Background

Pore mass transport phenomena in the cobalt-catalyzed Fischer-Tropsch (FT) synthesis are of kinetic significance for the extent of secondary processing of primary 1-olefin products and thus for the overall reaction selectivity under industrially relevant conditions.[1] This has driven attention to catalysts with large, hierarchically organized porosities, particularly if high selectivity to C_{5+} olefins is targeted.[2] Highly porous catalysts are, however, intrinsically thermally insulating, which poses a challenge for managing the significant heat of reaction, and the prevention of hot-spots that can have a deleterious impact on selectivity. Common strategies to improve heat dissipation, such as deposition of the catalytically active species within metallic open foams or meshes,[3] lead to low volumetric catalyst loadings. Albeit barely explored in the field of catalysis, the so-called freeze-casting synthesis route for macroporous materials is compositionally flexible, and thus provides a means for the assembly of hierarchically porous composite catalysts uniting high mass transport rates with thermally conductive skeletons.[4]

Aim

Here we report the synthesis of micromonolith bodies with a hybrid backbone composed of carbon nanotubes (CNTs) and mesoporous alumina and their application as scaffolds for Co-based FT catalysts.[5] A honeycomb-shaped and axially oriented macroporous architecture, which closely resembles the structure of conventional macroscopic monolithic catalyst bodies at a micrometer scale, enhances pore mass transport, while the mesoporous ZrO_x/Al_2O_3 oxide backbone component is an excellent support for cobalt nanoparticles. Percolating CNT skeletons contribute low thermal resistance pathways and thus improve the spatial dissipation of heat within the catalyst body, as reflected in the selectivity obtained to liquid olefins at different operation temperatures.

Methods

Hierarchically porous monolithic scaffolds with Al_2O_3/CNT (50–50 wt%) hybrid backbones and reference all-alumina backbones were synthesized by unidirectional freeze casting using zirconium acetate as ice growth modulator (Fig. 1a-d). Impregnation yielded supported Ru-promoted Co catalysts which were tested in a fixed-bed reactor setup for Fischer-Tropsch synthesis at 463-503 K, WHSV ~ 1.44 g_{CO} g_{Co}^{-1} h^{-1}, P = 20 bar, and H_2/CO = 2.0. Catalysts were characterized by N_2-sorption, H_2-chemisorption, XRD, SEM, HAADF-STEM/EDX and quantitative X-ray tomography.

Results and discussion

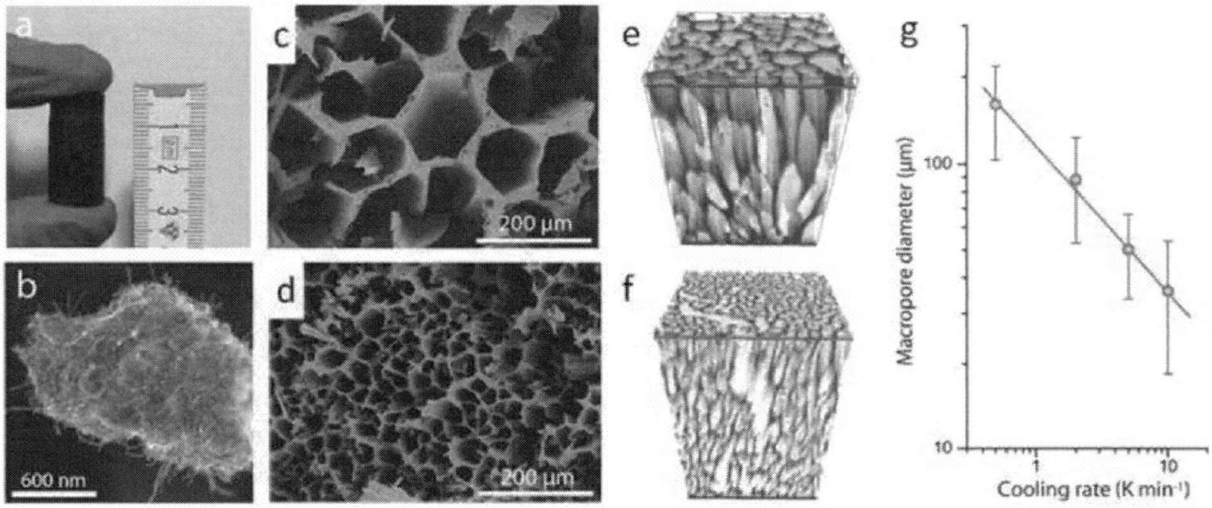

Figure 1: a) Picture of a CNT-$ZrAlO_x$ monolith body after freeze-casting and annealing. b) High-resolution SEM micrograph showing the structure of the channel walls in a CNT-$ZrAlO_x$ monolith. c,d) Cross-sectional SEM micrographs perpendicular to the monolith axial axis, and e,f) surface-rendered 3D reconstructed X-ray tomograms for CNT-$ZrAlO_x$ micromonoliths freeze cast at cooling rates of -0.5 K min^{-1} (c,e) and -10 K min^{-1} (d,f). g) Relationship between the volume-averaged macropore diameter and the cooling rate applied during freeze casting for CNT-$ZrAlO_x$ micromonoliths. Error bars correspond to the standard deviation of the average.

The reconstructed tomograms for a representative sub-region of monoliths cast at -0.5 and -10 K min^{-1} were analysed using a local thickness algorithm to determine the dependency of macropore diameter and macropore wall thickness on the cooling rate (Fig. 1e,f). The macropore diameter along the z-axis was found to be a function of the cooling rate (σ) according to the power law, $PD=k\cdot \sigma^n$, with n being 0.50 (Fig. 1g), suggesting an increasingly higher supercooling at the freezing interface, and higher ice nucleation rates during casting at increasingly higher cooling rates.

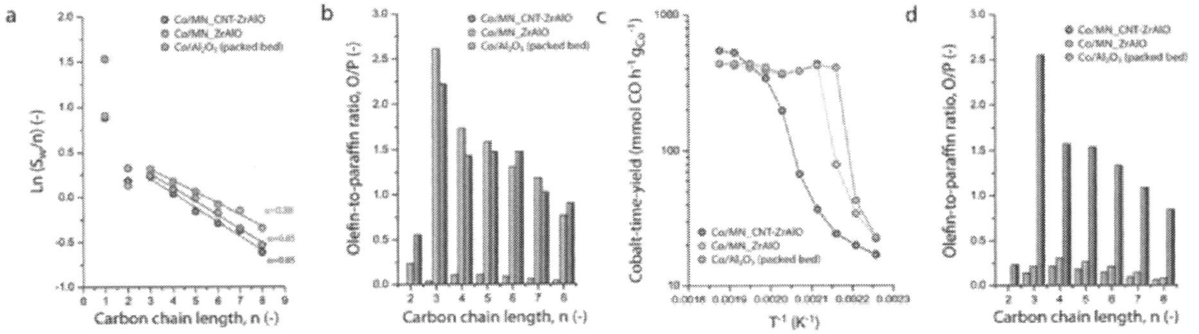

Figure 2: a) Anderson-Schulz-Flory hydrocarbon chain-length distribution plots, and b) olefin-to-paraffin molar ratio for hydrocarbon products in the carbon chain length range from C_2 to C_8, for different Co-based Fischer-Tropsch catalysts at 463 K. c) Evolution of the metal-specific Fischer-Tropsch reaction rate (cobalt-time yield) with the reaction temperature for different Co-based Fischer-Tropsch catalysts. d) Olefin-to-paraffin molar ratio for hydrocarbon products in the carbon chain length range from C_2 to C_8, for different Co-based Fischer-Tropsch catalysts at 483 K. WHSV=1.44 g_{CO} g_{Co}^{-1} h^{-1}, P=20 bar, H_2/CO=1.0.

The incorporation of CNTs into the skeleton led to a 2.5-fold improvement in the effective thermal conductivity in the composite bodies. The cobalt-loaded structured micromonoliths proved to be effective catalysts for the FT synthesis (Fig. 2). Under industrially relevant reaction conditions, higher rates of evacuation of primary reaction products from the metal active sites through the directional macropore system led to enhanced selectivities to liquid (C_{5+}) α-olefin products compared to conventional, microgranulate catalysts based on unimodally mesoporous γ-Al_2O_3 support materials (Fig. 2b). Moreover, the higher effective thermal conductivity associated with the hybrid CNT-$ZrAlO_x$ micromonolith backbone inhibits the development of hotspots within the catalyst body and thus a more gradual increase in the CO conversion rate on increasing the reactor outer wall temperature (Fig. 2c). This prevented undesired secondary hydrogenation of primary olefin products and methanation in a wider range of operation temperatures (Fig. 2d). These findings illustrate the significance of a dual compositional and structural design of multimodally porous bodies by directional freeze casting to produce effective solid catalysts for intensified syngas conversion processes.

References

[1] E. Iglesia, *Appl. Catal. A-Gen*, **1997**, *161*, 59-78.

[2] N. Duyckaerts, M. Bartsch, I. T. Trotus, N. Pfander, A. Lorke, F. Schuth and G. Prieto, *Angew. Chem. Int. Edit.*, **2017**, *56*, 11480-11484.

[3] a) L. Fratalocchi, C. G. Visconti, G. Groppi, L. Lietti and E. Tronconi, *Chem. Eng. J.*, **2018**, *349*, 829-837. b) M. S. Challiwala, B. A. Wilhite, M. M. Ghouri and N. O. Elbashir, *AIChE Journal*, **2018**, *64*, 1723-1731.

[4] K. Michaela, A. Idris, G. Christian, V. Céline and D. Sylvain, *Adv. Eng. Mater.*, **2012**, *14*, 1123-1127.

[5] J. Kim, V. Nese, J. Joos, K. Jeske, N. Duyckaerts, N. Pfänder and G. Prieto, *J. Mater. Chem. A*, **2018**, *44*, 21978-21989.

547889 Challenges and Opportunities in Direct Conversion of Methane to Higher Value Products

Monday, June 3, 2019: 10:15 AM
Texas Ballroom A (Grand Hyatt San Antonio)
Benjamin A. McCool and **Randall Meyer**, ExxonMobil, Annandale, NJ

Industrial and academic researchers have long sought opportunities for direct conversion of methane to higher value products. This talk will present an overview of R&D efforts, both industrial and academic, over the past several years. We will discuss the fundamental challenges associated with methane activation and attempts to solve them through myriad approaches. Notable topics to be covered include: bio-inspired heterogeneous catalysis, electrocatalysis, oxidative routes and process intensification through membrane reactors and plasma based conversion. We will discuss the challenges and opportunities associated with each of these areas. As we seek to convert methane to higher value products we also seek to deliver those products with a lower carbon footprint vs. incumbent technologies. To that end we will discuss energy intensity and carbon footprint associated with today's commercial gas conversion technologies and contrast with various approaches being pursued by the scientific community. Finally, we will present a perspective on the key areas for future work in direct methane conversion needed to turn promising leads into viable commercial scale processes.

544992 Microkinetic Modeling of Non-Oxidative Methane Coupling to Ethylene over Iron/Silica Catalyst

Monday, June 3, 2019: 10:39 AM
Texas Ballroom A (Grand Hyatt San Antonio)

Hilal Ezgi Toraman[1], Konstantinos Alexopoulos[1] and Dionisios G. Vlachos[2], (1)Delaware Energy Institute, Department of Chemical and Biomolecular Engineering, University of Delaware, Newark, DE, (2)Chemical and Biomolecular Engineering, University of Delaware, Newark, DE

The production of large amounts of shale gas, lower overall costs, and sustainability concerns have generated interest for the production of chemicals directly from methane. Currently, methane is converted into fuels and chemicals through syngas, an energy intensive and large capital investment process. A recent study has reported that isolated iron atoms embedded in a silica matrix ($Fe@SiO_2$) enables direct conversion of methane to ethylene under non-oxidative conditions at high temperatures (1223 – 1363 K). It has been suggested that methane conversion is initiated by catalytic generation of methyl radicals and hydrogen followed by subsequent gas-phase reactions.[1] Although the concept was demonstrated experimentally, fundamental understanding of the reactions is missing. In this study, we model this process by combining detailed catalytic and gas-phase reaction mechanisms. The latter includes up to 600 species and 10,000 reactions. For the former, we investigate the C-H activation of methane and C-C coupling reactions using density functional theory (DFT) calculations. The combined catalytic and gas-phase reaction mechanisms allow not only to capture the trends in terms of conversion and selectivity but also allow to assess the predictive capability of the model. Microkinetic simulations show that ethylene is the only primary product among the major species including acetylene and higher molecular weight aromatics. Overall, the present study exemplifies the interplay between catalytic and gas phase reactions that controls selectivity.

1. Guo, X.; Fang, G.; Li, G.; Ma, H.; Fan, H.; Yu, L.; Ma, C.; Wu, X.; Deng, D.; Wei, M.et al. Direct, nonoxidative conversion of methane to ethylene, aromatics, and hydrogen. Science 2014, 344 (6184), 616.

545581 Electrochemical Oxidation of Methane to Olefins in HF/HSO_3f

Monday, June 3, 2019: 11:03 AM
Texas Ballroom A (Grand Hyatt San Antonio)

Yuxin Fang[1], Christopher G. Arges[1], John C. Flake[1] and **Craig Plaisance**[2], (1)Cain Department of Chemical Engineering, Louisiana State University, Baton Rouge, LA, (2)Department of Chemical Engineering, Louisiana State University, Baton Rouge, LA

Advances in unconventional natural gas extraction have driven a renewed interest in the direct oxidation of methane to olefins. In 1968 Olah et al.[1]reported direct oxidation of methane using strong chemical oxidants in the presence superacids. The superacid is thought to facilitate the protonation activation of methane to trimethylcarbonium ion (methanium), however this reaction requires elevated temperatures (at 140 C). Later research by Jasinski et al.[2]reported electrochemical oxidation of propane and butane with Sb(V) in superacids; however, the electrochemical oxidation of methane and ethane yielded no measurable products.

In this work, we demonstrate the direct electrochemical oxidation of methane to olefins in super acid (HF/HSO_3F) electrolytes. Linear sweep voltammetry experiments with acids of different acidity suggest only strong acids are capable of the activation of methane for oxidative coupling. Products include ethylene and a variety of liquid-phase olefins at relatively high Faradaic efficiencies. We will discuss the efficiency of the direct-electrochemical methane to olefins process and opportunities to selectively form high value olefins.

Reference:

1. Olah, G. A.; Schlosberg, R. H., *J. Am. Chem. Soc.* **1968,***90*, 2726-2727.
2. Brilmyer, G.; Jasinski, R., *J. Electrochem. Soc.* **1982,***129*, 1950-1954.

545230 Synergy of Thin Catalytic Layers and DBD Plasma for Coupling of Methane

Monday, June 3, 2019: 11:27 AM
Texas Ballroom A (Grand Hyatt San Antonio)

Nuria Garcia-Moncada and Leon Lefferts, Catalytic Processes and Materials (CPM) - TNW Faculty, University of Twente, Enschede, Netherlands

The conversion of methane into fuels and value-added chemicals shows a promising alternative for steam cracking of petroleum derivatives. Non-oxidative coupling of methane in a non-equilibrium plasma, generated in a dielectric barrier discharge (DBD) reactor, allows activation of methane in absence of oxygen at mild temperature, avoiding thereby formation of CO and CO_2 in benefit of hydrocarbons (mainly alkanes and olefins). In this work, the DBD plasma for coupling of methane has been studied at ambient conditions assisted by a Pd/Al_2O_3 catalyst.

Combining catalysis and plasma is rather complex as introduction of a packed bed in a DBD plasma strongly influences the electrical field in the plasma, changing the properties of the plasma and the chemistry taking place in the plasma. Possible effects may include formation of microdischarges, changes in the discharge type, changes in the catalyst surface area, hot spot formation, changes in the catalytic surface properties and chemisorption of reactants. Besides, the morphology of the particles as well as the inter-particle space have a remarkable influence in the plasma features, as plasma is especially generated at the contact point between particles. Therefore, in this work, the catalyst was introduced as a thin layer on the internal wall in a DBD plasma reactor, minimizing the influence of the catalyst on the plasma. In addition, by this configuration with thin layers, the reactant volume occupied by the gas phase to create the plasma is very similar for the empty DBD reactor and the catalyst-assisted DBD reactor, making comparative tests possible.

Different amounts of catalysts have been tested up to loading of 15 mg. The thickness of catalyst layers are 20 μm or less to ensure the absence of any internal diffusional limitation. The catalytic wall reactors have been tested for methane coupling and compared to the non-catalytic DBD reactor. Important differences in product distribution are achieved in presence of the catalyst, *i.e.* more saturated hydrocarbons are formed and this effect increases with the amount of catalyst. In absence of catalyst, the selectivity to acetylene is *ca.* 20 %, whereas a loading below 6 mg of catalyst results in ethane selectivity of 20 % without any formation of acetylene. Also, the amount of carbonaceous deposits is strongly decreased by the presence of the catalyst. Indeed, the carbonaceous deposits are reduced up to an 80 % when 15 mg of catalyst are loaded on the reactor wall, compared to the empty DBD reactor in the same operation conditions. On the other hand, the methane conversion is determined by the plasma power and is not influenced by the amount of catalyst. For instance, the voltage input has been constant during all the experiments with different catalyst loadings providing the same plasma power (*ca.* 6.5 W). Hence, considering the same operation conditions, the amount of catalyst is only modifying the product distribution but not the plasma power or the methane conversion.

Remarkably, only very minor correction of the power source was needed in catalytic wall reactors to keep the plasma power the same as the non-catalytic reactor. This correction is significantly smaller compared to the correction that would be needed in a fixed bed reactor.

In short, the DBD plasma successfully activates the methane molecules at ambient conditions and the presence of the catalyst allows shifting the product distribution towards saturated hydrocarbons and less carbonaceous deposits.

546867 Strategies for Valorization of Methane: A Computational Perspective

Monday, June 3, 2019: 11:51 AM
Texas Ballroom A (Grand Hyatt San Antonio)

Ambarish Kulkarni, University of California, Davis, Davis, CA

Selective partial oxidation of methane to methanol has long been one of the holy grails in catalysis. Currently, methane is transformed into methanol through the two-step syngas process that necessitates high temperatures and centralized production. Despite over a century of research an industrially viable process for direct conversion to methanol does not exist. Beginning with a universal descriptor model [1] that describes C-H activation energies for a range of catalysts (with special attention to cation-exchanged zeolites)[2-4], we will present a simple kinetic model to explain the selectivity-conversion tradeoff that hampers continuous partial oxidation of methane to methanol.[5] We show that the selectivity of methane to methanol in a direct, continuous process can be fully described by the methane conversion, temperature, and the catalyst-independent difference in methane and methanol activation free energies, ΔG^a (dictated by the relative reactivity of the C-H bonds in methane and methanol). Using this model, we successfully rationalize a large number of experimental studies from the diverse fields of heterogeneous, homogeneous, biological and gas-phase catalysis. Based on this understanding, we suggest several design strategies for increasing methanol yields under the constraint of constant ΔG^a. These strategies include (1) "collectors," materials with strong methanol adsorption potential that can help to lower the partial pressure of methanol in the gas phase, (2) aqueous reaction conditions, and/or (3) diffusion-limited systems.

1. *ACS Catal.*, **2018**, *8*, pp 6894–6907
2. *ACS Catal.*, **2016**, *6*, 10, pp 6531–6536
3. *ACS Catal.*, **2016**, *6*, 6, pp 3760–3766
4. *Catal. Sci. Technol.*, **2018**, 8, pp 114-123
5. *ACS Catal.*, **2018**, *8*, pp 6894–6907

546678 Direct Methane Conversion to Olefins and Aromatics Under Non-Oxidative Condition

Monday, June 3, 2019: 1:30 PM
Texas Ballroom D (Grand Hyatt San Antonio)

Xinhe Bao, Jianqi Hao, Guangzong Fang, Xiaoguang Guo, Pierre Schwach and Xiulian Pan, Dalian Institute of Chemical Physics, Chinese Academy of Sciences, Dalian, China

With the discovery of new large methane reserves (e.g. shale gas and methane hydrate), the direct conversion of methane into high value-added fuels and chemicals such as light olefins and aromatics, has ignited interest worldwide. [1] Significant progress has been made in the development of several catalytic processes, such as oxidative coupling of methane (OCM),[2,3] methane dehydroaromatization (MDA)[5-10]. We previously reported that iron single sites embedded within the lattice of silica are active in catalyzing methane conversion to olefins, aromatics and hydrogen (MTOAH) under non-oxidative conditions.[11] Characterization showed that the reaction involves catalytic activation of methane generating methyl radicals, which subsequently go through gas phase C-C coupling forming ethylene, aromatics and hydrogen as the products.[11] DFT calculations suggested that the H· free radicals may also play an important role in the selective reaction pathway of the MTOAH reaction.[11] Therefore, in this study, we further investigate the role of methyl radicals and hydrogen radicals. Furthermore, the strong endothermal reaction causes a steep temperature gradient in the catalyst bed along the radial direction. In addition, the gas phase reaction is also affected by the space of the reactor due to the lifetime of radicals. Thus, we explore the type of reactors containing catalysts directly on their walls, by coating a layer of catalysts onto the inner wall of the reactors including quartz and SiC-based reactors.

A catalytically active layer of around 15-20 μm thick SiO_2 containing the iron single sites with a loading in a range of 0.16-0.34 wt.% was coated on the inner wall of a quartz reactor with an inner diameter (id) of 14 mm, outer diameter (od) of 16 mm, and length of 100 mm. It gives a methane conversion of 7% at 1273 K and the conversion increases to 17.5% at 1323 K and a flow rate of 60 ml/min. The conversion and product selectivities can be tuned by changing the temperatures and flow rate. Interestingly, by providing hydrogen free radicals, generated by thermal decomposition of tetrahydronaphthalene (THN) or benzene, the MTOAH reaction is enhanced and methane conversion increases to 25.5% without changing much the products' distribution. The involvement of hydrogen in the activation of methane is verified by the isotope experiments.

The same strategy can be scaled up to a catalytic reactor with id = 15.8 mm, od = 20 mm, and isothermal length as long as 700 mm. At a flow rate of 3800 ml/min of $50\%H_2/45\%CH_4/5\%N_2$ and 1463 K, methane conversion is 23.6%, which can be increased beyond 30% with a C_2 selectivity of 66.8% and and BTX 22.2% without coke formation in the presence of hydrogen radicals.

In order to meet the requirement of applications, we also develop a new type of reactor based on SiC containing iron sites in the reactor wall with a loading of 0.37wt%. Methane conversion reaches about 30% at 1423 K with the feeding of 50%H_2/45%CH_4/5%N_2 at a flow rate of 1700 ml/min. A 700 h test showed a rather good stability.

References:

(1) Schwach, P.; Pan, X.; Bao, X., *Chem. Rev.* **2017,** 117 (13), 8497-8520.

(2) Lunsford, J. H., *Angew. Chem. Int. Ed.* **1995,** 34 (9), 970-980.

(3) Wang, P.; Zhao, G.; Wang, Y.; Lu, Y., *Science Advances* **2017,** 3 (6).

(4) Keller, G. E.; Bhasin, M. M., *J. Catal.* **1982,** 73 (1), 9-19.

(5) Gao, J.; Zheng, Y.; Jehng, J.-M.; Tang, Y.; Wachs, I. E.; Podkolzin, S. G., *Science* **2015,** 348 (6235), 686-690.

(6) Lezcano-González, I.; Oord, R.; Rovezzi, M.; Glatzel, P.; Botchway, S. W.; Weckhuysen, B. M.; Beale, A. M., *Angew. Chem. Int. Ed.* **2016,** 55 (17), 5215-5219.

(7) Kosinov, N.; Coumans, F. J. A. G.; Uslamin, E.; Kapteijn, F.; Hensen, E. J. M., *Angew. Chem. Int. Ed.* **2016,** 55 (48), 15086-15090.

(8) Morejudo, S. H.; Zanón, R.; Escolástico, S.; Yuste-Tirados, I.; Malerød-Fjeld, H.; Vestre, P. K.; Coors, W. G.; Martínez, A.; Norby, T.; Serra, J. M.; Kjølseth, C., *Science* **2016,** 353 (6299), 563-566.

(9) Wang, L. S.; Tao, L. X.; Xie, M. S.; Xu, G. F.; Huang, J. S.; Xu, Y. D., *Catal. Lett.* **1993,** 21 (1-2), 35-41.

(10) Liu, H. M.; Bao, X. H.; Xu, Y. D., *J. Catal.* **2006,** 239 (2), 441-450.

(11) Guo, X.; Fang, G.; Li, G.; et al., *Science* **2014,** 344 (6184), 616-619.

545475 Catalysis of Liquid-Metal Indium for Direct Dehydrogenative Conversion of Methane into Higher Hydrocarbons

Monday, June 3, 2019: 2:09 PM
Texas Ballroom D (Grand Hyatt San Antonio)

Yuta Nishikawa[1], Ayumi Nakaya[1], Hitoshi Ogihara[2], Yuki Ohtsuka[3], Akira Nakayama[3], Jun-ya Hasegawa[3], Shoji Iguchi[1] and Ichiro Yamanaka[1], (1)Department of Chemical Science and Engineering, School of Materials and Chemical Technology, Tokyo Institute of Technology, Tokyo, Japan, (2)Department of Applied Chemistry, Graduate School of Science and Engineering, Saitama University, Saitama, Japan, (3)Institute for Catalysis, Hokkaido University, Sapporo, Japan

Catalysis of Liquid-Metal Indium for Direct Dehydrogenative Conversion of Methane into Higher Hydrocarbons

Yuta Nishikawa,[a] Ayumi Nakaya,[a] Hitoshi Ogihara,[b] Yuki Ohtsuka,[c] Akira Nakayama,[c] Jun-ya Hasegawa,[c] Shoji Iguchi,[a] Ichiro Yamanaka[a,]*

[a]*Department of Chemical Science and Engineering, School of Materials and Chemical Technology, Tokyo Institute of Technology,2-12-1-S1-16, Ookayama, Meguro-ku, Tokyo, 152-8550, Japan*

[b]*Department of Applied Chemistry, Graduate School of Science and Engineering, Saitama University, 255, shimo-ookubo, Sakura-ku, Saitama, 338-8570, Japan*

[c]*Institute for Catalysis, Hokkaido University, Kita 21, Nishi 10, Kita-ku, Sapporo, 001-0021, Japan*

**Corresponding author: +81-3-5734-2144, E-mail address: yamanaka.i.aa@m.titech.ac.jp*

Abstract: Dehydrogenative conversion of methane (DCM) is an attractive reaction. We already reported that liquid-metal indium supported on silica (In/SiO_2) was an effective catalyst for DCM. Characterization studies and kinetic studies indicated that liquid-metal indium catalyzed cleavage of a C-H bond of methane and coupling methane to ethane selectively. In addition, conversion of ethane and temperature-programmed reaction in ethane (TPR-C_2H_6) suggested that indium did not activate both a C-H bond and a C-C bond of ethane. Indium has the unique catalysis to activate only a C-H bond of methane, therefore, selectivity to hydrocarbons was still high despite high temperature.

Keywords: Methane conversion, Liquid metal catalysis, Indium catalysis.

1. Introduction

Methane (CH_4) is the main component of natural gas, which is an abundant energy sources in the earth. Most of CH_4 is currently used to generate electric power and heat because it is difficult to convert CH_4 to valuable compounds. Steam reforming of CH_4 to syngas ($CO + H_2$) and following catalytic reactions are only realized as industrial processes for utilization of CH_4. These processes need large amount of energy and costs because of mult-step reaction. Therefore, direct conversion of methane to chemicals and liquid fuels is a promising way to utilize natural gas.

In dehydrogenative conversion of CH_4, molybdenum/zeolite (Mo/zeolite) catalysts have been extensively studied and tested to overcome coke deposition[1,2]. Recently, a few new DCM catalysts were reported. For example, Guo et al. reported the Fe/SiO_2 catalyst (single Fe sites in a silica matrix) to convert CH_4 to ethylene, benzene and naphthalene at >1223 K[3]. We also reported the In/SiO_2 catalyst to convert CH_4 to ethylene, propylene and benzene[4]. CH_4 conversion was 4.8% and selectivity to hydrocarbons was 75 % by the In/SiO_2 catalyst at 1173 K. In this work, we study detailed reaction mechanisms for the DCM reaction on the In/SiO_2 catalyst.

2. Experimental

The In/SiO_2 catalyst was prepared by a conventional impregnation method. Indium nitrate hydrate was dissolved in deionized water and CARiACT Q-3 (SiO_2 support) was added to the solution. The mixture was dried up at 393 K. This catalyst precursor was calcined at 773 K in air and was reduced with H_2 at 873 K. The DCM reaction tests were conducted using a fixed-bed quartz reactor (I.D. 12 mm) with In/SiO_2 (100 mg) and CH_4 (1 atm, 10 mL min^{-1}) was flowed. Hydrocarbons and H_2 were analyzed by gas chromatographs or an online mass spectrometer. In the cases of conversion of C_2H_6, 5% C_2H_6/Ar (1 atm, 30 mL min^{-1}) or 0.5% C_2H_6/H_2 (1 atm, 20 mL min^{-1}) was flowed. A Ni/SiO_2 catalyst as a reference one was prepared from nickel nitrate hexahydrate using the same method.

3. Results and discussion

Figure 1 (a)-(d) show the profiles of temperature-programmed reaction (TPR) from 323 K to 1173 K with 4 K min^{-1} monitored by a mass spectrometer. In the TPR using CH_4 on In/SiO_2 (a), formation of C H

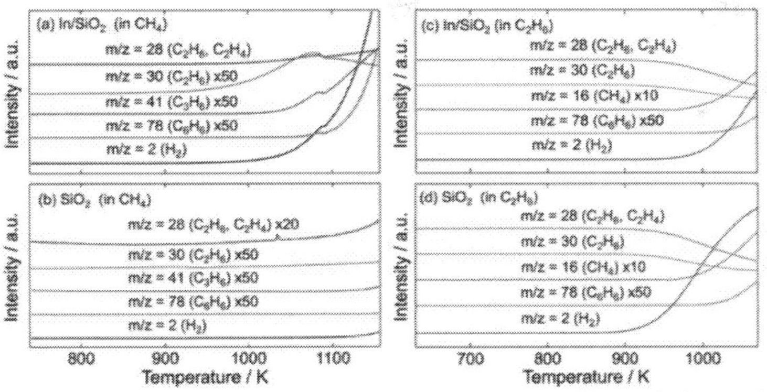

Figure 1. Mass spectra of temperature-programmed reaction in CH_4 over (a) SiO_2 and (b) In/SiO_2 and in C_2H_6 by (c) SiO_2 and (d) In/SiO_2 with 4 K min^{-1}.

2 6

(m/z = 30) was observed from 800 K. On the other hand, no formation of C_2H_6 on SiO_2 (b).It was clear that indium catalyzed CH_4 activation and propylene and benzene were successively produced following the C_2H_6 formation. In the TPR using C_2H_6 on In/SiO_2 (c), significant products were not detected at 800 K. Formations of H_2 (m/z = 2) and CH_4 (m/z =16) were observed over 950 K, corresponding to conversion of C_2H_6 (m/z = 28, 30). In the case of SiO_2 (d), similar profiles were observed except for hydrogen production which was slightly suppressed by In/SiO_2. Figure 1 indicated that indium could activate a C-H bond of CH_4, however, a C-H bond of C_2H_6 cannot.

Catalyst	C_2H_6 conv. / %	Selectivity / %		
		CH_4	C_2H_4	Coke
Ni/SiO_2	91	67	8	25
In/SiO_2	58	18	76	6
None	61	21	67	12

Table 1. C_2H_6 conversion in H_2 over Ni/SiO_2, In/SiO_2 and no catalyst at 1073 K.

To obtain more information of indium catalysis, conversion of C_2H_6 in H_2 was conducted. As references, results of no catalyst and Ni/SiO_2 catalyst were shown in Table 1. Ni/SiO_2 catalyst converted C_2H_6 into CH_4 and coke, therefore, a C-C bond of C_2H_6 was cleaved. On the other hand, dehydrogenation

of C_2H_6 to C_2H_4 was mainly proceeded on In/SiO_2.

The selectivities on In/SiO_2 were very similar to that of no catalyst. This indicated that distributions were thermodynamically decided. Indium cannot activate C-H and C-C bonds of C_2H_6, therefore, carbon deposition was suppressed. In the presentation, DFT-calculation results will be discussed.

4. Conclusions

Results of TPR-CH_4 and TPR-C_2H_6 on In/SiO_2 indicated that indium could activate only a C-H bond of CH_4 but not a C-H bond of C_2H_6. Additionally, indium cannot cleave a C-C bond of C_2H_6 from the results of conversion of C_2H_6. These unique catalysis of In/SiO_2 for CH_4 and C_2H_6 results in higher selectivity to hydrocarbons in the DCM reaction.

References

1. L. Wang, L. Tao, M. Xie, G. Xu, J. Huang, Y. Xu, Catal. Lett., 21 (1993) 35.

2. S. H. Morejudo, R. Zanon, S. Escolastico, I. Yuste-Tirados, H. Malerod-Fjeld, P. K. Vestre, W. G. Coors, A. Martinez, T. Norby, J. M. Serra, C. Kj_lseth, Science, 353 (2016) 563.

3. X. Guo, G. Fang, G. Li, H. Ma, H. Fan, L. Yu, C. Ma, X. Wu, D. Deng, M. Wei, D. Tan, R. Si, S. Zhang, J. Li, L. Sun, Z. Tang, X. Pan, X. Bao, Science, 344 (2014) 616.

4. Y. Nishikawa, H. Ogihara, I. Yamanaka, ChemistrySelect, 2 (2017) 4572.

545434 Methane Dehydroaromatization Is Mediated By Surface Carbon

Monday, June 3, 2019: 2:33 PM
Texas Ballroom D (Grand Hyatt San Antonio)

Nikolay Kosinov, Eindhoven University of Technology, Eindhoven, Netherlands and **Emiel Hensen**, Schuit Institute of Catalysis, Department of Chemical Engineering and Chemistry,, Eindhoven University of Technology, Eindhoven, Netherlands

Methane dehydroaromatization is mediated by surface carbon

Nikolay Kosinov, Emiel J.M. Hensen

Inorganic Materials Chemistry, Eindhoven University of Technology, The Netherlands

Non-oxidative dehydroaromatization of methane (MDA) is a promising catalytic process for direct valorization of natural gas to liquid hydrocarbons. The application of this reaction in practical technology is hindered by lack of understanding about the mechanism and nature of the active sites in benchmark zeolite-based Mo/ZSM-5 catalysts, which precludes the solution of problems such as rapid catalyst deactivation. Here by applying spectroscopy and microscopy, we demonstrate that the active centers in Mo/ZSM-5 are partially reduced, single-atom Mo sites stabilized by the zeolite framework. Further, by combining a pulse reaction technique with isotope labeling studies, we demonstrate that MDA is governed by a hydrocarbon pool mechanism, in which benzene is derived from secondary reactions of confined polyaromatic carbon species with the initial products of methane activation. Although this mechanism shares some similarities with the well-understood chemistry of methanol-to-hydrocarbons process, MDA was found to proceed via a radical pathway and not the carbocation one.

Important challenges in gaining insight into the mechanistic aspects of MDA were the high reaction temperature at which the reaction takes place and its transient nature, which involves rapid activation and deactivation stages when the fresh Mo/ZSM-5 catalyst is exposed to a methane feed. These factors complicate operando spectroscopy and kinetic investigations. A novel aspect of this work is the increased temporal resolution obtained by periodically pulsing small amounts of the reactant over the catalyst (Fig. 1) and the use of recovered samples at different stages of the reaction for spectroscopy (XPS, MAS NMR, EPR, etc.) characterization and operando XAS spectroscopy performed under exact reaction conditions.

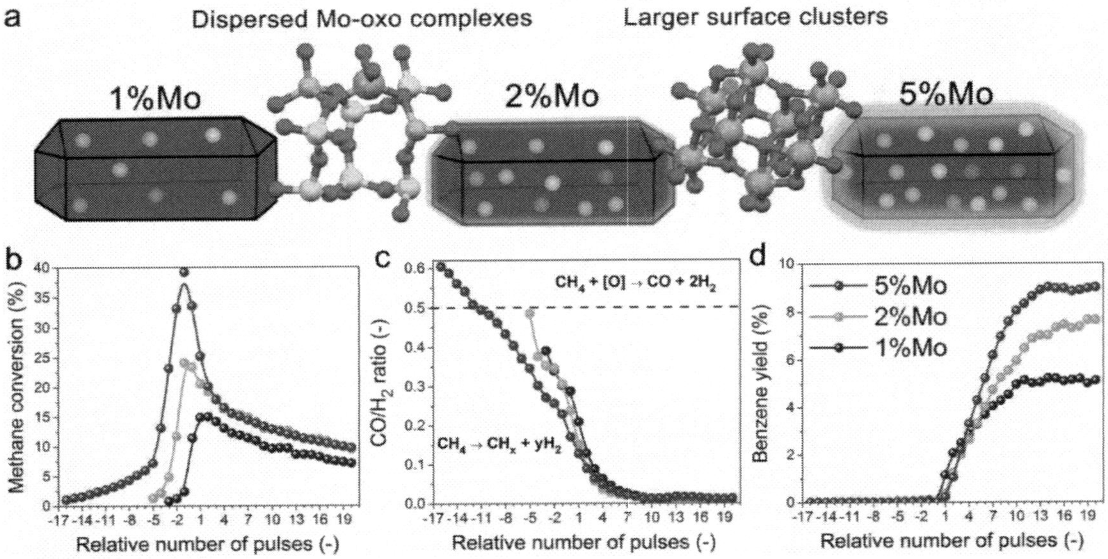

Figure 1. Schematic representation of catalyst samples applied in this study, featuring exclusively atomically dispersed Mo(VI) species stabilized in the zeolite pores for the 1%Mo sample and a mixture of the dispersed species and larger Mo-oxo clusters on the external surfaces of 2%Mo and 5%Mo catalysts (a). Per-pulse profiles of methane conversion (b), CO/H2 ratio (c), and benzene yield (d), recorded while pulsing 5 mL methane every 200 sec with 30 mL/min flow of Ar carrier.

References

- N. Kosinov, A. S. G. Wijpkema, E. Uslamin, R. Rohling, F. J. A. G. Coumans, B. Mezari, A. Parastaev, A. S. Poryvaev, M. V. Fedin, E. A. Pidko and E. J. M. Hensen, *Angew. Chemie Int. Ed.*, 2018, **57**, 1016–1020.
- N. Kosinov, E. A. Uslamin, F. J. A. G. Coumans, A. S. G. Wijpkema, R. Rohling and E. J. M. Hensen, *ACS Catal.*, 2018, **8**, 8459-8467.

E-mail: n.a.kosinov@tue.nl

546026 On the Dynamic Nature of Mo Sites for Methane Dehydroaromatization

Monday, June 3, 2019: 2:57 PM
Texas Ballroom D (Grand Hyatt San Antonio)

Freek Kapteijn[1], Ina Vollmer[1], Bart van der Linden[1], Samy Ould-Chikh[2], Irina Yarulina[1,2], Alma I. Olivos Suarez[1] and Jorge Gascon[2,3], (1)Catalysis Engineering, Department of Chemical Engineering, Delft University of Technology, Delft, Netherlands, (2)KAUST Catalysis Center, Advanced Catalytic Materials, King Abdullah University of Science and Technology, Thuwal, Saudi Arabia, (3)KAUST Catalysis Center (KCC), Advanced Catalytic Materials, King Abdullah University of Science and Technology, Thuwal, Saudi Arabia

Abstract: The mechanism of methane activation on Mo/HZSM-5 is poorly described, despite the great interest in methane dehydroaromatization (MDA) to replace oil refineries for producing aromatics. It is difficult to assess the exact nature of the active site due to fast coking. By pre-carburizing Mo/HZSM-5 with carbon monoxide, the active site for MDA can be isolated and studied without the formation of coke. This strategy helped us examine how methane is activated on the catalytic site by carrying out MDA using isotopically labelled methane ($^{13}CH_4$). We show that carbon originating from the pre-formed carbide is incorporated into the main products of the reaction, ethylene and benzene, demonstrating the dynamic nature of these active sites.

Keywords: Methane dehydroaromatization, Mo/HZSM-5, labelling experiments, molybdenum carbide

1. Introduction

For methane dehydroaromation (MDA), a great deal of research has focused on the most promising systems Mo/HZSM-5 and Mo/HMCM-22.[1, 2] The active phase for this catalyst, reduced Mo, was previously found to form in an initial period where no desired products are formed yet.[4-9] Several phases, MoC, Mo_2C, coke modified Mo_2C, Mo_2C on outside surface and reduced oxides in the pores of the zeolite, any kind of Mo^{6+} or partially reduced Mo^{6+} as $MoO_{(3\ x)}$ are proposed as the active phase for this reaction, but no clarity on the exact nature has been achieved so far.[3] Observing the active site formation in the initial phase of MDA is difficult because of the simultaneous fast coking.

2. Experimental

Catalysts were based on a commercial HZSM-5 zeolite having Si/Al = 13 (HZ-13) and were prepared with 1, 2 and 5 wt.% loading (*x*) of Mo denoted as *x*MoHZ-*13* using $(NH_4)_6Mo_7O_{24}$ as a precursor.

3. Results and discussion

To understand the conditions necessary to prepare the active Mo site by CO carburization, CO consumption and simultaneous CO_2 production were monitored by mass spectrometry (MS) during Temperature Programmed Carburization (TPC) with $C^{18}O$. Figure 1 presents the typical profile of CO consumption and CO_2 production during CO carburization. It is evident from this figure that much less ^{18}O evolves as $CO^{18}O$ or $C^{18}O_2$ than what is consumed as $C^{18}O$, presenting proof that oxygen is part of the active Mo site. This supports previous claims that the active Mo site is in an oxycarbidic phase.[5, 6] Measuring *operando* X-ray absorption near edge structure (XANES), the reduction of a sample during catalyst activation under methane was compared with that of samples during carburization in CO. These experiments confirmed that after 1 h at 780 C, CO reduces Mo to the same extent as achieved with CH_4, while reduction remains incomplete at 600 C or 700 C. The TPC and *operando* XANES experiments confirm that CO carburization can be used to produce an active Mo phase that is equivalent to the one forming during the activation period of the reaction with CH_4. The active site prepared this way is free of coke and can be freely accessed by probe molecules for infrared (IR) measurements. The active site was characterized by CO IR and ^{13}C NMR uncovering three types of active sites with a Mo oxidation state between 4+ and 6+. The absence of coke surrounding the active site also allows probing the interaction of CH_4 solely with the active Mo-carbide phase isolating this interaction from interactions with undefined (hydro-)carbonaceous species surrounding the active site, that would be present after carburization with CH_4. To understand how methane is activated on the reduced Mo species, we performed a series of pulsing experiments using labelled methane, $^{13}CH_4$.

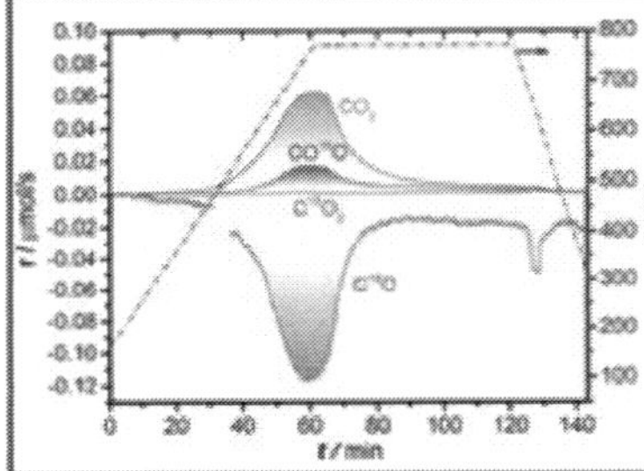

Figure 1. $C^{18}O$ consumption and simultaneous CO_2, $CO^{18}O$ and $C^{18}O_2$ production in TPC of 2MoHZ-13 with 30 ml/min, 2.5% $C^{18}O$ in He. Temperature was increased to 780 C at a rate of 10 C/min (right axis).

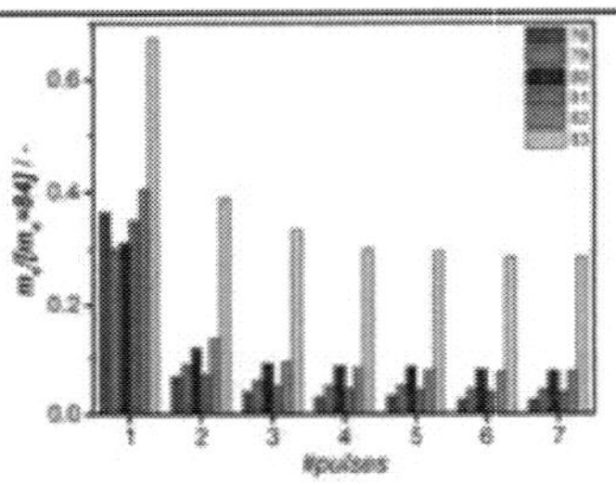

Figure 2. Evolution of area under each pulse for masses 78 to 83 typical for the fragmentation of benzene normalized by $m_e = 84$. under consecutive pulsing of 223 µmol $^{13}CH_4$ to 300 mg 2MoHZ-13 catalyst carburized at 780 ¡C with 30 ml/min 2.5% ^{12}CO in He.

Prior to this pulsing, the catalyst was carburized using ^{12}CO, forming ^{12}C based Mo-carbidic or -oxycarbidic species. This way, it was possible to track the incorporation of ^{12}C from the catalytic Mo site into the products. Firstly, masses 84 to 78 arising from fragmentation of labelled benzene, $^{13}C_6H_6$ as well as benzene where some ^{12}C is incorporated, were recorded on the MS. Figure 2 shows the abundance of the masses normalized by the one with highest abundance, $m/z = 84$. The ratio 83/84 is most informative in assessing the incorporation of ^{12}C into the observed benzene, because $m/z = 83$ is the most abundant mass for $^{12}C^{13}C_5H_6$ and should lead to a higher 83/84 ratio than for the control experiment where $m/z = 83$ only represents the $^{13}C_6H_5$ fragment. When using ^{12}C for carburization and ^{13}C for methane pulsing (Figure 2), the ratio of 83/84 reaches a value of 0.67 for the first pulse and decreases to 0.28 over the next 8 methane pulses. This value of 0.28 is the constant fragmentation ratio in the control experiment. The higher value of 83/84 during the initial pulses can clearly be attributed to the formation of $^{12}C^{13}C_5H_6$. Similarly development of mixed carbon isotopes was shown in ethylene, showing the incorporation of the carbide carbon in the product species.

4. Conclusions

We demonstrate that CO carburization is a powerful approach to isolate the formation of the active site in Mo/HZSM-5 for the non-oxidative methane conversion without producing undesired catalyst coking. This strategy allowed us to characterize the active site in more detail and let us distinguish among three different kind of active sties. We were also able to study, at the molecular level, the activation of methane on the Mo catalytic active site. We show that the catalytic Mo site actively takes part in the reaction rather than acting as an adsorption site to lower the activation barrier of CH_4. Rapid exchange reactions with the dynamic Mo-site results in the incorporation of carbidic carbon into the products ethylene and benzene. It provides a good starting point for finding the precise molecular structure of the reduced Mo formed as it demonstrates that the carbon at the metal site in this structure is easily replaced by another carbon from methane, pointing at a dynamic active site. Further, we provide proof that oxygen is present at the active site.

References

1. J. J. Spivey and G. Hutchings, *Chemical Society Reviews*, 2014, **43**, 792-803.

2. S. Ma, X. Guo, L. Zhao, S. Scott and X. Bao, *Journal of Energy Chemistry*, 2013, **22**, 1-20.

3. Y. Xu and L. Lin, *Applied Catalysis A: General*, 1999, **188**, 53-67.

4. H. Jiang, L. Wang, W. Cui and Y. Xu, *Catalysis Letters*, 1999, **57**, 95-102.

5. I. Lezcano-Gonz‡lez, R. Oord, M. Rovezzi, P. Glatzel, S. W. Botchway, B. M. Weckhuysen and A. M. Beale, *Angewandte Chemie International Edition*, 2016, **55**, 5215-5219.

6. F. Solymosi, A. Szške and J. CserŽnyi, *Catalysis Letters*, 1996, **39**, 157-161.

7. D. Wang, J. Lunsford and M. Rosynek, *Topics in Catalysis*, 1996, **3**, 289-297.

8. S. Liu, L. Wang, R. Ohnishi and M. Lchikawa, *Kinetics and catalysis*, 2000, **41**, 132-144.

9. W. Ding, S. Li, G. D Meitzner and E. Iglesia, *The Journal of Physical Chemistry B*, 2001, **105**, 506-513.

10. J. Dang, G. Zhang, L. Wang, K. Chou and P. C. Pistorius, *Journal of the American Ceramic Society*, 2016, **99**, 819-824.

11. B. M. Weckhuysen, D. Wang, M. P. Rosynek and J. H. Lunsford, *Journal of Catalysis*, 1998, **175**, 347-351.

12. B. M. Weckhuysen, D. Wang, M. P. Rosynek and J. H. Lunsford, *Journal of Catalysis*, 1998, **175**, 338-346.

545614 Direct Methane Conversion to Value-Added Products over Fe-FER Catalysts

Monday, June 3, 2019: 3:21 PM
Texas Ballroom D (Grand Hyatt San Antonio)

Guangyu Zhao[1], Michael Stockenhuber[1], Eric M. Kennedy[1], Adesina Adesoji A.[2], Meng-Jung Li[1] and **Matthew Drewery**[3], (1)Chemical Engineering, The University of Newcastle, Newcastle, Australia, (2)ATODATECH LLC, Pasadena, CA, (3)University of Newcastle, Callaghan, Australia

Direct methane conversion to value-added products over Fe-FER catalysts

G. Zhao[1], E. Kennedy[1], M. Stockenhube[1], Adesoji A. Adesina[2], Meng-Jung Li[1]*

1. Chemical Engineering, School of Engineering, The University of Newcastle, Callaghan, NSW, 2308

2. Atodatech LLC, 23586 Calabasas Road, CA 91302

Keywords: Direct methane conversion, active sites, incipient wetness impregnation, solid state ion exchange <>1. Introduction

Direct oxidation of methane to value-added products, such as methanol and formaldehyde, has received extensive attention in the field of chemical engineering and catalysis [1]. However, current direct methane oxidation processes at high temperature (normally 600¡ãC¨C900¡ãC) suffer from poor selectivity of C_1-oxygenates [2]. Conversely, a relatively high selectivity to desired products was observed at low temperature (lower than 200¡ãC), but these routes are not regarded as truly catalytic processes because an additional extraction step is necessary to obtain the desired products [3]. Therefore, the current study on methane oxidation under moderate temperatures (350¡ãC) has been conducted, aiming to synthesize methanol, formaldehyde and dimethyl ether (DME) over heterogeneous catalysts with the need for an extract step [4].

In this study, methane oxidation over Fe-FER catalysts prepared by incipient wetness impregnation (IWI) and solid state ion exchange (SSIE) techniques at 350¡ãC was investigated. The catalysts were characterised with in situ IR spectroscopy, XAS, XPS and microreactor studies. The aims of the research were to determine the effect of preparation methods on transition metal site structure and in turn methane conversion to value products generation, and infer the involved mechanisms of the reactions. <>2. Results and discussion

TEM micrographs of IWI and SSIE prepared Fe-FER catalysts are compared in Fig. 1. Large iron oxide aggregates are observed in both samples, and the size of the aggregates is generally larger in IWI catalyst than in SSIE catalyst.

The EDS mapping images of two catalysts are presented in Fig. 2. It is shown that particles over 100 nm were detected in IWI Fe-FER catalyst, and the particles size in SSIE sample are much smaller, which indicated better metal dispersion of SSIE sample.

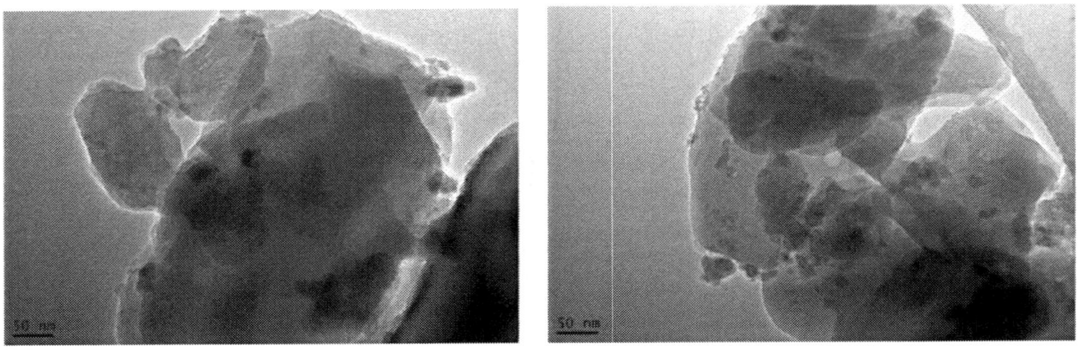

a b

Fig. 1 TEM micrographs of (a) IWI and (b) SSIE Fe-FER

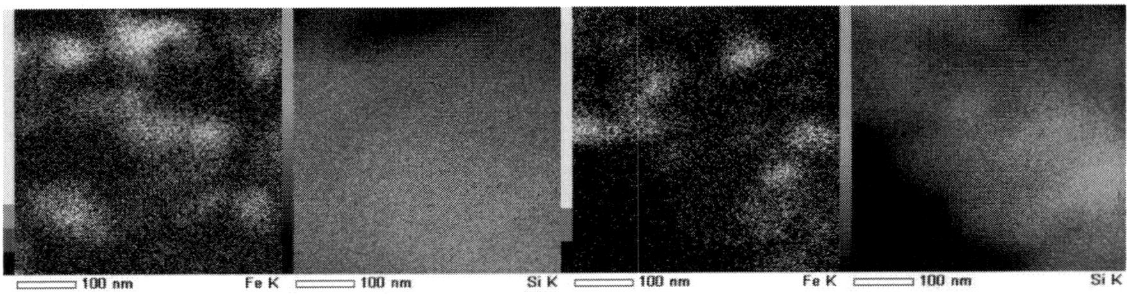

a b

Fig. 2 EDS mapping images of (a) IWI and (b) SSIE Fe-FER

Fig. 3 compares the H_2-TPR profiles of IWI and SSIE catalysts that were activated in a stream of air with or without extra N_2O pre-treatment (at 250¡ãC). The results show that the N_2O pre-treatment leads to a new reduction peak at around 230¡ãC, and the area of the peak observed correlates with the number of active sites for the N_2O decomposition [5]. In comparison to IWI Fe-FER, this peak is larger for SSIE Fe-FER.

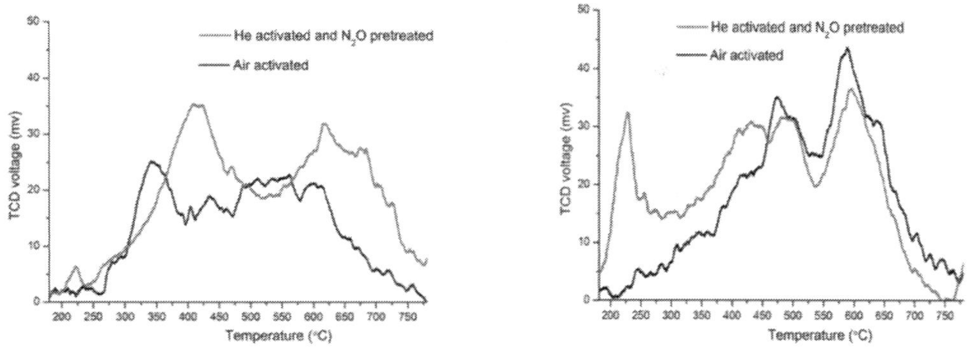

a b

Fig. 3 TPR profiles of (a) IWI and (b) SSIE catalysts

The activity test results show that the SSIE Fe-FER obtained higher methane conversion, N_2O decomposition and more amount of methanol, formaldehyde and DME, as shown in Table 1.

Table 1. Activity rest results of IWI and SSIE prepared Fe-FER catalysts

	Conversion or decomposition (%)		Concentration (μmol)		
	methane conversion	N O decomposition	methanol	formaldehyde	DME
IWI	2.0	13.7	13.4	3.8	11.0
SSIE	2.8	20.6	24.7	6.4	25.6

<>3. Conclusion

The experimental results demonstrated that methane can be converted to methanol, formaldehyde and DME over Fe-FER catalysts using N_2O as the oxidant. Additionally, in comparison to catalyst prepared by IWI technique, the Fe-FER catalyst prepared by SSIE method displayed better iron dispersion on zeolites, more active sites for N_2O decomposition, and consequently provided more significant performance, in terms of methane conversion, N_2O decomposition, and the amount of value-added products. In this paper, details of the site structure and its correlation with activity are discussed. <>4. Reference

1. Z. C. Wang, N. Dietl, R. Kretschmer, J. B. Ma, T.Weiske, M. Schlangen, H. Schwarz, *Angew. Chem. Int. Ed. Engl.* **2012**, 51, 3703-37077.

2. R.G. Herman, Q. Sun, C. Shi, C. Shi, K. Klier, C.B Wang, H. Hu, I. E. Wachs, M. M. Bhasin, *Cataly. Today* **1997**, 37, 1¨C14.

3. E.V. Starokon, M. V. Parfenov, L.V.Pirutko, S.I. Abornev, G.I. Panov, *J. Phys. Chem. C* **2011**, 115, 2155-2161.

4. B. Michalkiewwicz, *Appl. Catal. A. Gen.* **2004**, 277, 147-153

5. K. J¨ªša, J. Nov¨¢kov¨¢, M. Schwarze, A. Vondrov¨¢, S. Sklen¨¢k, Z. Sobalik, *J. Cataly.* 2009, 262, 27¨C34.

545057 Control of Kinetic and Diffusive Length-Scales during Absorptive Hydrogen Removal for Enhanced Aromatics Yield in Methane Dehydroaromatization on Mo/H-ZSM-5 Catalysts

Monday, June 3, 2019: 4:00 PM
Texas Ballroom D (Grand Hyatt San Antonio)

Neil Razdan, Anurag Kumar and Aditya Bhan, Chemical Engineering and Materials Science, University of Minnesota, Minneapolis, MN

Control of kinetic and diffusive length-scales during absorptive hydrogen removal for enhanced aromatics yield in methane dehydroaromatization on Mo/H-ZSM-5 catalysts

Neil Razdan[1], Anurag Kumar[1], and Aditya Bhan[1,*]

[1] *University of Minnesota, Minneapolis MN 55455 (USA)*

[*] abhan@umn.edu

Introduction

Direct, non-oxidative conversion of methane to benzene requires high reaction temperatures (950 – 1023 K) and is limited by thermodynamic constraints ($6CH_4 \rightarrow C_6H_6 + 9H_2$) to ~10% equilibrium conversion. Mo/H-ZSM-5 catalysts have been shown to convert methane to benzene at greater than 70% selectivity with forward rates unaffected by hydrogen pressure [1]. We demonstrate interpellet physical mixtures and staged packed-bed configurations of pre-carburized MoC_x/H-ZSM-5 catalyst and Zr, a known H_2 absorbent, lift equilibrium limitations, thereby increasing single-pass conversion to near the kinetic limit (~14% conversion to benzene), as dictated by forward synthesis rates. A reaction-transport model rigorously accounts for catalyst-absorbent proximity effects through consideration of kinetic, diffusive, and convective length scales. Simulation results demonstrate introduction of zirconium metal enhances net methane pyrolysis rates throughout the catalyst bed and accurately predict benzene yield and effluent H_2 partial pressure for various reactor configurations.

Materials and Methods

NH_4-ZSM-5 (Si/Al = 11.5, CBV 2314) was converted to H-ZSM-5 by heating in dry air (0.67 $cm^3\ s^{-1}$) from room temperature to 773 K at 0.0165 K s^{-1} and holding for ~36 h to decompose NH_4^+ to H^+ and $NH_{3(g)}$. Mo/H-ZSM-5 formulations were prepared from intimate physical mixtures of MoO_3 and H-ZSM-5 powders ($Mo:Al_f$ ~0.25) heated from room temperature to 623 K at 0.0167 K s^{-1}, held at this temperature for

15 h in dry air (0.67 cm^3 s^{-1}), and finally heated to 973 K at 0.0167 K s^{-1} and held for 10 h to promote migration of MoO_x into the zeolite pores.

MoC_x/H-ZSM-5 catalysts were prepared by treating pelletized and sieved (mesh 40-80) Mo/H-ZSM-5 samples in 0.27 cm^3 s^{-1} of a 90%/10% CH_4/Ar mixture for ~15.5 ks at 973 K. Upon carburization, MoO_x moieties convert to zeolite-encapsulated MoC_x nanoparticles of 0.5-2 nm and ~10 Mo atoms with concomitant oxygen removal [2, 3].

Reactor configurations of MoC_x/H-ZSM-5 with pelletized Zr particles of the same size (mesh 40-80) were prepared in a glove-bag under inert (helium) atmosphere to prevent oxidation of the carburized catalyst.

Results and Discussion

CH_4 dehydroaromatization on MoC_x/H-ZSM-5 with and without Zr

Methane dehydroaromatization rates were measured employing: (i) a MoC_x/H-ZSM-5 catalyst bed, (ii) an interpellet mixture of MoC_x/H-ZSM-5 and Zr, and (iii) MoC_x/H-ZSM-5 with a succeeding Zr bed. Interpellet physical mixtures showed single-pass conversion as high as ~28%, as depicted in Fig. 1(a). The identity and distribution of products is unaltered upon addition of zirconium, suggesting that the bifunctional reaction pathways of methane pyrolysis were unperturbed by hydrogen removal. Interpellet formulations were partially regenerated by treatment in helium at 973 K, removing all absorbed H_2 and yielding above-equilibrium methane conversion in successive regeneration cycles.

Fig. 1(b) shows benzene yield as a function of time-on-stream (TOS) for all three reactor configurations. Addition of Zr, in both case (ii) and (iii), enhances benzene yield ~2x. The increase in methane conversion and aromatic production arises as a consequence of continuous hydrogen scavenging circumventing equilibrium constraints, as is evident in comparison of benzene yield to thermodynamic and kinetic limits shown in Fig. 1(b). Benzene synthesis rates decrease with time-on-stream owing to carbonaceous deposits that limit accessibility of residual Brønsted acid sites. With addition of Zr, the rate of deactivation is accelerated by transient saturation of hydrogen-removal sites.

Kinetic-Transport Model

We present a kinetic-transport model inclusive to all reactor configurations. Differential mole balance in the catalyst bed yields a non-dimensional equation

$$(1)\quad \frac{1}{Pe}\frac{d^2y_j}{dx^2} - \frac{dy_j}{dx} - \nu_j Da\, y_{CH_4}(1-\eta) - \delta_{H_2,j} Da_{Zr}\, y_{H_2} = 0$$

The last term in Eq. (1) is only included in case (ii). Case (iii) requires consideration of a second differential mole balance for the Zr bed placed downstream of Mo/H-ZSM-5

$$(2)\quad \frac{1}{Pe_{Zr}}\frac{d^2y_j}{dx^2} - \frac{dy_j}{dx} - \delta_{H_2,j}Da_{Zr}y_{H_2} = 0$$

Approach to equilibrium for benzene production, η, is calculated from gas-phase partial pressures and equilibrium constant, K_{eq}

$$(3)\quad \eta = \frac{P_{C_6H_6}^{\frac{1}{6}} P_{H_2}^{\frac{3}{2}}}{P_{CH_4} K_{eq}}$$

Non-dimensionalization of design equations naturally gives rise to dimensionless parameters

$$Da = \frac{kL}{u_s} \qquad Da_{Zr} = \frac{k_{Zr}L_{Zr}}{u_s} \qquad Pe = \frac{u_sL}{D_{eff}} \qquad Pe_{Zr} = \frac{u_sL_{Zr}}{D_{eff}}$$

Da and Pe^{-1} reflect ratios of the characteristic convective length-scale to kinetic and diffusive length-scales. Pe and Pe_{Zr} are Péclet numbers in the catalyst and Zr bed, both taken to be unity per numerical fit. Da = 0.16 is the Damköhler number for the pseudo 1st-order forward benzene synthesis rate, calculated using measured values taken from previous investigations [1]. H_2 scavenging by Zr is rapid and precludes kinetic measurement in current reactor configurations. $Da_{Zr} > 10^4$ give essentially invariant axial methane conversion and H_2 pressure profiles and are in quantitative agreement with net synthesis rates and outlet η, as shown in Fig. 2.

In case (iii), Danckwerts boundary conditions are applied at both bounds of the catalyst and absorbent beds.

$$(4)\quad \frac{1}{Pe}\frac{dy_j}{dx}\bigg|_{x=0^+} = y_j(0^+) - y_j(0^-) \qquad (5)\quad y_j(1^-) = y_j(1^+)$$

$$(6)\quad \frac{1}{Pe}\frac{dy_j}{dx}\bigg|_{x=1^-} = \frac{1}{Pe_{Zr}}\frac{dy_j}{dx}\bigg|_{x=1^+} \qquad (7)\quad \frac{1}{Pe_{Zr}}\frac{dy_j}{dx}\bigg|_{x=1+\frac{L_{Zr}}{L}} = 0$$

Eqs. (4) and (7), "closed" Danckwerts boundary conditions, account for discontinuous change in dispersion length-scales between entrance and reaction sections and absorptive and exit sections, respectively. Eqs. (5) and (6), "open" Danckwerts boundary conditions, demand continuity of mass and diffusive flux between catalyst and absorbent beds. Case (i) and (ii) simply require "closed" Danckwerts boundary conditions at each end of the single reactor bed.

Proximity and length-scale effects during polyfunctional catalysis

The impact of catalyst-absorbent proximity is evident in comparison of benzene yield in cases (ii) and (iii), as shown in Fig. 1(b). Physical interpellet mixtures of MoC_x/H-ZSM-5 and Zr show the largest maximum benzene yield, but deactivate most rapidly.

Intimacy of product formation and scavenging functionalities in interpellet mixtures disproportionately promotes formation of higher aromatics which congest zeolite pores and sterically hinder dehydrocyclization steps. Separation of catalytic sites from H_2 absorptive function mitigates deactivation and increases cumulative benzene yield from 3.03 to 3.78 mol C_6H_6/mol Mo at 8.7 ks TOS from case (ii) to case (iii).

Profiles for benzene yield and approach to equilibrium along the catalyst bed predicted by the kinetic-transport model are shown in Fig. 2. Non-zero benzene yield and η values at the reactor inlet arise from "closed" Danckwerts boundary conditions which demand discontinuous change in species concentration from $x = 0^-$ to $x = 0^+$ as dispersion considerations become relevant in the finite reaction section. Predicted values of $\eta \rightarrow 0$ at reactor outlet in case (iii) result from rapid H formation of ZrH_x such that $P_{H2} \rightarrow 0$ at $x = 1$. Balance of convective and diffusive length scales, reflected in $Pe \sim 1$, permit response of H_2 pressure profiles throughout the catalyst bed. In contrast, in ideal PFR reactors, dominance of convective mass transport (i.e. $Pe \rightarrow \infty$) precludes bed-scale changes in axial profiles by successive product removal.

Significance

Introduction of a continuous H_2 scavenging functionality in interpellet and staged-bed configurations of MoC_x/H-ZSM-5 and Zr lifts thermodynamic equilibrium constraints for non-oxidative methane conversion, enhancing aromatic synthesis rates and cumulative product yields without changes to product distribution. Kinetic-transport models and dimensionless parameters presented herein quantitatively capture effects of hydrogen absorption and provide a framework to investigate performance of various reactor configurations through the formalism of Da and Pe which capture interplay between kinetic, diffusive, and convective length-scales.

Notation

y_j	dimensionless partial pressure of component j, normalized by 1 bar
x	dimensionless length along reactor bed
u_s	superficial linear velocity [m/s]
L	length catalyst bed [m]
L_{Zr}	length of absorbent bed [m]
D_{eff}	effective diffusivity [$m^2\ s^{-1}$]
k	pseudo 1st-order rate constant for benzene production reaction [s^{-1}]
k_{Zr}	pseudo 1st-order rate constant for H_2 absorption [s^{-1}]

Greek letters

v_j	stoichiometric coefficient of component j in benzene production reaction
$d_{H2,j}$	Kronecker delta, permitting absorption of only H_2 by Zr
η	approach to equilibrium

References

[1] Bedard, J., Hong, D-Y., Bhan, A., *J. Catal.*, 306, 58-65 (2013).

[2] Kumar, A., Song, K., Liu, L., Han, Y., Bhan, A. *Angew. Chem. Int. Ed.*, (2018). *In Review*

[3] Ding, W., Li, S., Meitzner, G.D., Iglesia, E., *J. Phys. Chem. B*, 105, 506-513 (2001).

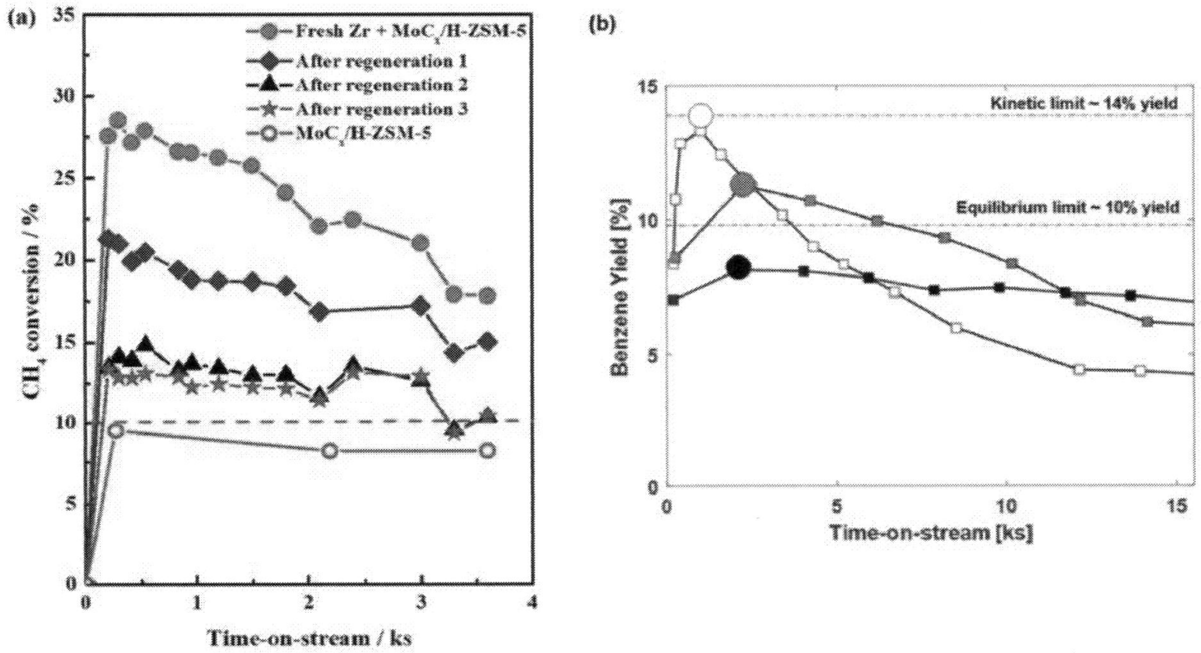

Figure 1: (a) CH_4 conversion vs TOS for case (i) and case (ii) before (fresh) and after regeneration in He flow. Regenerations 1, 2, and 3 of interpellet mixtures by flushing in He flow at 973 K for 61.2 ks, 84.6 ks, and 34.2 ks respectively. Dashed red line indicates ~10% equilibrium conversion. (b) Measured benzene yield vs TOS for cases (i), (ii), and (iii) in black, white, and gray-filled squares, respectively. Large circles are predictions of the kinetic-transport model in the absence of deactivation.

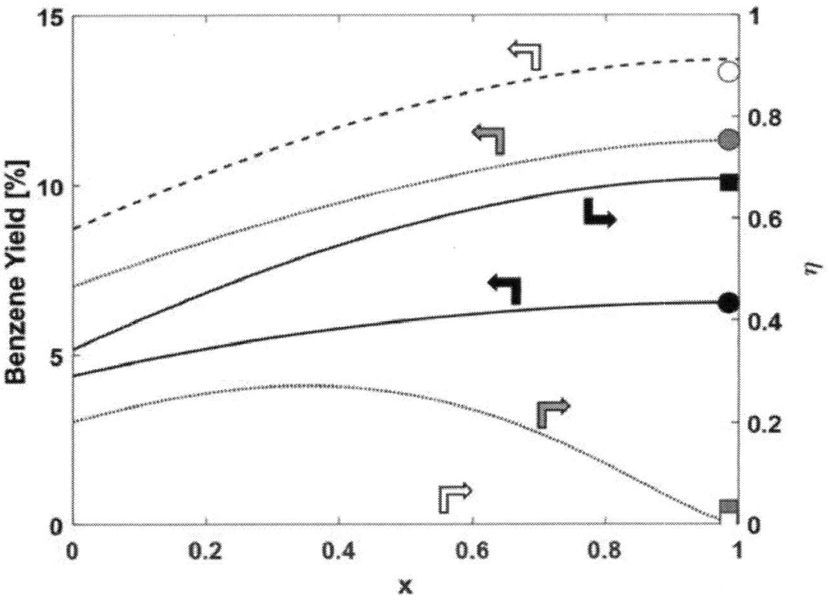

Figure 2: Axial profiles of benzene yield and approach to equilibrium, as predicted by the kinetic-transport model, for cases (i), (ii), and (iii), shown in solid, dashed, and dotted lines, respectively. Measured benzene yield (approach to equilibrium) for cases (i), (ii), and (iii) shown in black, white, and gray-filled circles (squares), respectively. Simulations predict $\eta \rightarrow 0$ throughout the bed for interpellet mixtures.

545298 A Stable Mo@HZSM-5 Catalyst for Direct Conversion of Methane to Aromatics

Monday, June 3, 2019: 4:24 PM
Texas Ballroom D (Grand Hyatt San Antonio)

Yi Liu[1], Tianyun Wang[2], Franklin (Feng) Tao[3] and Yi Zhang[2], (1)State Key Laboratory of Organic–Inorganic Composites, Beijing University of Chemical Technology, Beijing, China, (2)Beijing University of Chemical Technology, Beijing, China, (3)Departments of Chemical Engineering and Chemistry, The University of Kansas, Lawrence, KS

A Stable Mo@HZSM-5 Catalyst for Direct Conversion of Methane to Aromatics

Yi Liu, [1] Tianyun Wang, [1] Franklin Feng Tao,[2] Yi Zhang[1*]

[1]State Key Laboratory of Organic-Inorganic Composites, Beijing University of Chemical Technology, 15 Beisanhuan East Road, Beijing 100029, PR China.

[2]Department of Chemical and Petroleum Engineering, University of Kansas, KS 66045, USA

Email: yizhang@mail.buct.edu.cn

Introduction

Methane (CH_4), the principal component of natural gas and shale gas, is considered to be an alternative feedstock for the production of commodity chemicals and fuels currently produced from crude oil. Thus, there are strong drivers for direct methane conversion. One of such pathways, non-oxidative methane dehydroaromatization (MDA), is particularly promising for the direct conversion of methane into aromatics using catalysts with Mo nanostructures supported on shape-selective zeolites such as Mo/HZSM-5, the most adequate catalysts. However, from the beginning in 1993 (1), application of this technology is still limited by rapid catalyst deactivation (severe carbon deposition) (2). It is proposed that the state and location of Mo species in zeolite matrix are crucial in determining the catalytic performance of MDA, especially for the formation of carbonaceous deposits, and the ensuing catalyst deactivation. Herein, we report the imbedment of Mo sites into ZSM-5 crystal using an in situ synthesis strategy via hydrothermal method. Thus obtained Mo@HZSM-5 catalyst with imbedded structure, highly dispersed stabilizing Mo species and confined reaction environment, showed a remarkably stable catalytic performance for MDA, and no any deactivation was observed during a 150 h test. Meanwhile, via careful characterization of the active sites, the unique and simple structure of this catalyst would contribute to discover the truth of MDA catalyst.

Materials and Methods

The precursor catalyst MoO3/SiO2 was prepared by incipient wetness impregnation using commercial silica support with 6.7 nm pore diameter. The molybdenum-imbedded microporous zeolite catalyst (Mo@HZSM-5) with nominal 5 wt% molybdenum loading was then obtained by using MoO3/SiO2 as the only Si source for the crystallization of HZSM-5; meanwhile the Al species added to the synthetic solution were incorporated into the zeolite framework, according to our previously reported method (3). By comparison, a conventional zeolite-supported 5 wt% Mo/HZSM-5 catalyst was also prepared by incipient wetness impregnation using HZSM-5 support. MDA reactions were conducted with 0.3 g of catalyst in a fixed bed, tubular quartz reactor (8 mm id) at atmospheric pressure and 973 K.

Results and Discussion

Figure 1 shows that the reaction performance of Mo@HZSM-5 was extremely stable, and no deactivation was observed during a 150-hour test. Methane conversion remained at ~12% throughout this long run. Meanwhile, selectivities to benzene (69.3%), toluene (13.1%) and naphthalene (17.2%) were constant with TOS, and the total selectivity to these products remained >99%. In contrast, CH4 conversion decreased sharply with TOS and dropped to less than 6% after 10 h because of the rapid carbon deposits on the catalyst. On the basis of HR-TEM, SRPES, NMR, in-situ XANES and EXAFS observations before and after reaction, the MoOC(O) active phase were formed and maintained the initial structure as the reaction proceeded in 150 h-consistent with the stability in catalytic performance. The suitable interaction between Mo and Bronsted acid sites and the encapsulating system prevent reduction of highly active molybdenum oxycarbide (MoOC(O)) species to less active molybdenum carbide (Mo2C) large particles, resulting in stable activity of such catalysts. Moreover, the modified BAS nature and shaped channels of zeolite suppress the formation of the PAHs and preferentially produce the single ring aromatics and naphthalene, contributing to the stability of MDA catalyst. Extensive characterization work will be discussed in this presentation together with more recent results on a much improved catalyst.

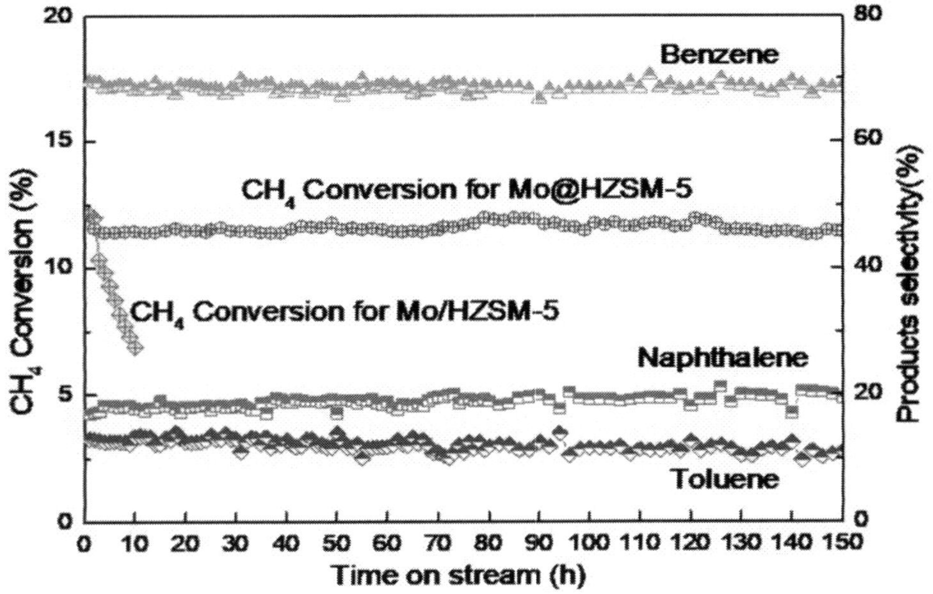

Figure 1. Long-term stability test of Mo@HZSM-5 catalyst at atmospheric pressure, 973 K and 2000 mL (CH_4)/gcat h for MDA reaction.

References

1. L. Wang, L. Tao, M. Xie, G. Xu, J. Huang and Y. Xu, Catal. Lett. 1993, 21, 35-41.
2. Y. Liu, D. Li, T. Wang, Y. Liu, T. Xu, and Y. Zhang, ACS Catal. 2016, 6, 5366−5370
3. J. Y. Liu, J. F. Chen, Y. Zhang, Catal. Sci. Technol. 2013, 3, 2559-2564.

546705 Activity and Selectivity of Oxidative Coupling of Methane on Doped La_2O_3 Catalysts: A Density Functional Theory Study

Monday, June 3, 2019: 5:12 PM
Texas Ballroom EF (Grand Hyatt San Antonio)
Zhiqiang Wang and **Xueqing Gong**, East China University of Science and Technology, Shanghai, China

The oxidative coupling of methane (OCM) on rare earth metal oxides has been studied by periodic density functional theory (DFT) calculations. Detailed energetic, structural and electronic properties of the related reactions have been determined. The calculated activation energies of breaking the first C-H bond in methane on the La_2O_3(001) and CeO_2(111) surfaces are compared, which suggest the high selectivity and low activity on the La_2O_3(001) surface and the opposite on the CeO_2(111) surface for the OCM reaction. It has been found that the methyl radical is the key species, and the activity of the OCM reaction can be increased by modifying the catalyst from the perspectives of structural and electronic properties for tuning the formation of the methyl radical. The La_2O_3(210) surface has low-coordinated oxygen and it is beneficial for breaking the first C-H bond in methane; and with doped Sr and Ce, the surface can maintain the activity to dissociate first C-H bond while favor the occurrence of methyl radicals for further coupling. This work illustrates that the $SrCeLa_2O_3$ catalyst can be highly active and selective for catalytic OCM reactions.

544795 Bifunctional Catalysis Involving Surface Coupling for Low Temperature Oxidative Coupling of Methane

Monday, June 3, 2019: 5:15 PM
Texas Ballroom EF (Grand Hyatt San Antonio)
Zhinian Li, Lei He, Shenliang Wang, Shihui Zou, Liping Xiao and Jie Fan,
Department of Chemistry, Zhejiang University, Hangzhou, China

Bifunctional catalysis involving surface coupling for low temperature oxidative coupling of methane

Zhinian Li, Lei He, Shenliang Wang, Shihui Zou*, Liping Xiao, Jie Fan*

Since large reserves of natural gas have been explored and mature technologies of producing methane have been developed,[1,2] converting methane to chemicals of higher value, such as methanol[3], light olefins[4-6] and aromatics[7,8] is of both scientific and strong economic interest. Currently, the industrial-scale conversion of methane adopts an indirect route, which is energy and capital intensive. The oxidative coupling of methane (OCM) as a direct conversion of methane to C2 hydrocarbons due to its low energy consumption and simple process has been gotten widespread attention since it was firstly reported by Keller and Bhasin in 1982.[9]

Up to date, several catalyst systems such as Li/MgO, rare-earth oxides, $Mn/Na_2WO_4/SiO_2$ and perovskite composite oxides have been investigated for this reaction. Among these catalysts, La_2O_3, a model of rare earth oxides has been widely studied due to its low ignition temperature of methane and long-term stability. However, the C2 selectivity is poor due to inefficient coupling at gas phase.[10-12] $Mn/Na_2WO_4/SiO_2$ discovered by Li group in 1992 with highest C2 yield (23.9%) and stability is taken as the most promising catalyst for industrialization.[13,14] One drawback is that it requires as high as above 800 ¡ãC to work. Although close to 40 years extensive research and development has been devoted to this reaction[15], it has never met the lowest requirement for commercialization. It is very challenging to gain activity and selectivity at the same time.

All these catalysts involved 68 elements have been developed for OCM reaction in the past 40 years, but it haven't made much progress to its mechanism. It is generally accepted that OCM follows a ¡°heterogeneous-homogeneous¡± mechanism, which involves CH_4 initiated on the surface of catalyst to generate methyl radicals[16], and CH_3 radicals couple to form C_2H_6 and C_2H_4 in the gas phase[17-20]. The homogeneous coupling of methyl radical in gas phase usually need relatively high temperature and was uncontrollable. Besides, the resulting C2 product tend to further oxidize to unwanted CO and CO_2. So, there is a limit of 25% to the yield of C2-hydrocarbons in catalytic OCM due to its homogeneous gas-phase coupling. Arutyunov et. al proposed the noticeable exceeding of this limit can be obtained only in the case if catalyst plays significant role not only in heterogeneous generation of methyl radicals but as well in their subsequent transformations.[21]

Since the contribution from heterogeneous catalysis was restricted to the activation of methane to produce methyl radicals while the subsequent homogeneously coupling of methyl radicals was out of control, which consequently led to limited selectivity to C2 species (ethane and ethylene). To overcome this challenge, we first experimentally confirmed the possibility of coupling of methyl radicals on a Na-W/SiO_2 catalyst surface at low reaction temperature (700 ^oC) by VUV photoionization mass spectroscopy (SVUV-PIMS). Thereafter, we propose to optimize the OCM reaction based on the design of bifunctional catalysts that provide two types of active sites (methane activation site to generate methyl radicals and methyl radical coupling sites to produce ethane and ethylene). Further regulation of the proximity of these two active sites lead to a synergic OCM catalytic system that is capable of achieving high C2 yield at low reaction temperature (reaction temperature <600 ^oC, C2 yield > 10%).

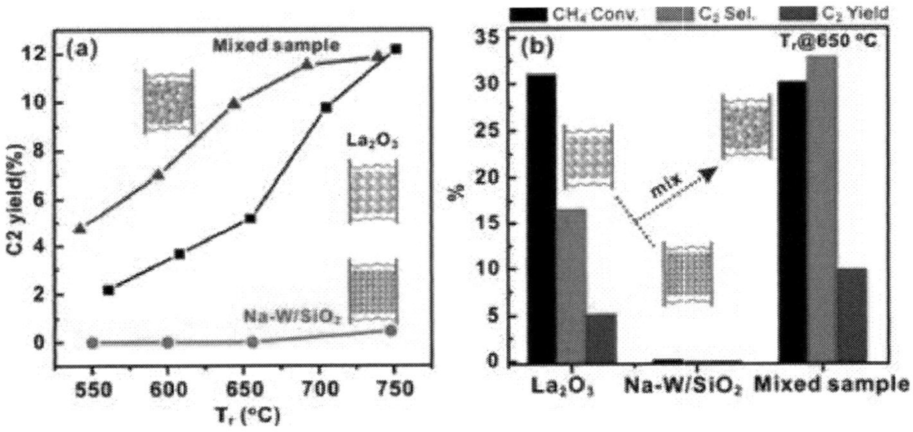

Figure 1. Mixture effect of temperature on catalytic performance. (a) C2 yield of La_2O_3 nanoparticles, Na-W/SiO_2 and their mixed sample from 550 - 750 ^oC. (b) The conversion, selectivity and C2 yield of La_2O_3 nanoparticles, Na-W/SiO_2 and their mixed sample at 650 ^oC. Reaction conditions: catalyst weight, 0.10 g for La_2O_3 nanoparticles and 0.10 g for Na-W/SiO_2; CH_4¡ÃO_2¡ÃN_2 = 3¡Ã1¡Ã2.6, total flow rate 66 mL min^{-1}.

We found that the C2 yield significantly increased as the C2 selectivity enhanced after mixed La_2O_3 with Na-W/SiO_2 below as shown in Figure 1. At lower temperature (< 750 ^oC), the C2 yield on mixed sample is much higher than La_2O_3 or Na-W/SiO_2 catalyst. The specific comparison of CH_4 conversion and C2 selectivity among mixed samples, pure La_2O_3 and Na-W/SiO_2 nanoparticles at 650 ^oC was shown in Figure 1(b). The substantial increase of C2 selectivity instead of CH_4 conversion led to much higher C2 yield on mixed sample than only on La_2O_3 or Na-

W/SiO_2 particles. In addition, there is almost no contribution to methane activation for Na-W/SiO_2 catalyst at 650 °C. Therefore, the addition of Na-W/SiO_2 is beneficial for C2 selectivity, which can offer sites for surface coupling and prevent C2 species from over-oxidization.

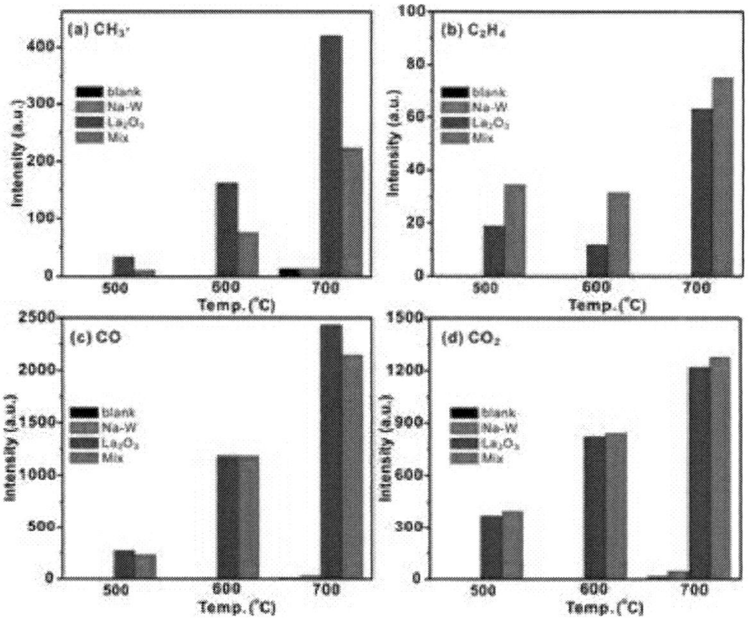

Figure 2. Integrated SVUV-PIMS peak intensity of gas-phase methyl radicals, ethylene and carbon oxide and carbon dioxide during OCM reaction catalyzed by various catalysts at 600 °C.

Reaction conditions: CH_4 (42 SCCM), O_2 (14 SCCM), pressure (2 torr). Mixed sample: 70 mg La_2O_3 with 70 mg Na-W/SiO_2; The intensity is normalized by the amount of La_2O_3.

As shown in Figure 2, the enhanced C2 selectivity of mixed sample was substantiated by SVUV-PIMS experiment. In Figure 2(a), no detected methyl radical on Na-W/SiO_2 even at 700 °C demonstrated Na-W/SiO_2 is too inert to activate methane, which is consistent with the experimental results. Compared to La_2O_3, more ethylene with less methyl radical was produced on mixed sample while almost the same CO and CO_2 was generated, which further demonstrated that Na-W/SiO_2 could facilitate to turn methyl radical to C2. In other word, the coupling of methyl radical partly was taking place on Na-W/SiO_2 surface.

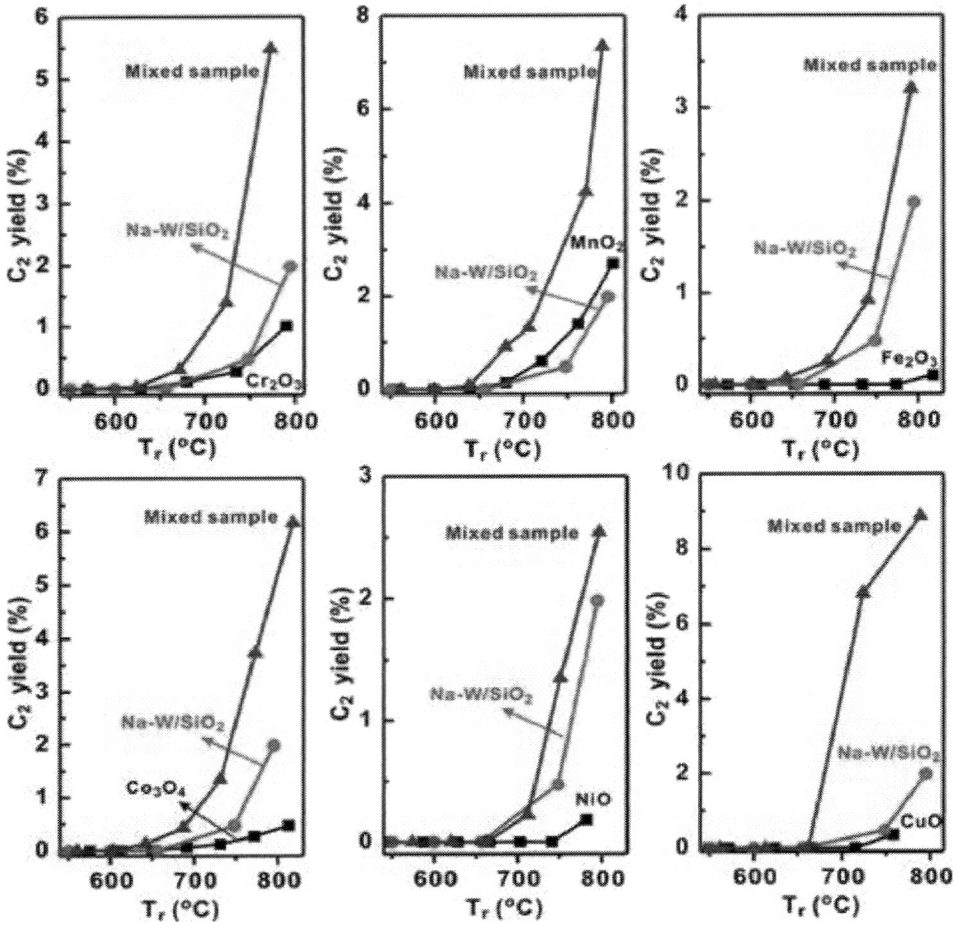

Figure 3. OCM performance of different transition metal oxides and the mixed samples. 0.1 g transition metal oxides nanoparticles (40 - 60 mesh) or mixed sample (0.1g transition metal oxides mixed with 0.1 g Na-W/SiO_2 powder), CH_4¡ÃO_2¡ÃN_2=3¡Ã1¡Ã2.6; 500 - 800 °C, total flow = 66 mL min^{-1}.

In addition, significant improvement for C2 yield has been observed for most transition metal oxides after mixing transition metal oxides particles (Cr_2O_3, MnO_2, Fe_2O_3, Co_3O_4, NiO and CuO) with Na-W/SiO_2 compared with pure transition metal oxides as shown in Figure 3. The higher selectivity and conversation compared to transition metal oxides further verified that coupling of methyl radical can partly occur on the surface of catalyst.

In this study, we found the C2 selectivity of La_2O_3 nanoparticle at low temperature can further improve after mixing with Na-W/SiO_2. These improvements resulted from the coupling site of W-O-Si interface. To summarize, we confirm that the catalyst can play an important role on coupling of methyl radical, which may work as a bifunctional catalysis system exceeding C2 yield limit in traditional heterogeneous-homogeneous mechanism. A new way to explore efficient OCM catalyst is suggested.

(1) Surendra, K. C.; Takara, D.; Hashimoto, A. G.; Khanal, S. K. *Renew Sust*

Energ Rev **2014**, *31*, 846.

(2) Weiland, P. *Appl Microbiol Biot* **2010**, *85*, 849.

(3) Agarwal, N.; Freakley, S. J.; McVicker, R. U.; Althahban, S. M.; Dimitratos, N.; He, Q.; Morgan, D. J.; Jenkins, R. L.; Willock, D. J.; Taylor, S. H.; Kiely, C. J.; Hutchings, G. J. *Science* **2017**, *358*, 223.

(4) Wang, P.; Zhao, G.; Wang, Y.; Lu, Y. *Science Advances* **2017**, *3*.

(5) Lunsford, J. H. *Angew Chem Int Edit* **1995**, *34*, 970.

(6) Amenomiya, Y.; Birss, V. I.; Goledzinowski, M.; Galuszka, J.; Sanger, A. R. *Catal Rev* **1990**, *32*, 163.

(7) Guo, X. G.; Fang, G. Z.; Li, G.; Ma, H.; Fan, H. J.; Yu, L.; Ma, C.; Wu, X.; Deng, D. H.; Wei, M. M.; Tan, D. L.; Si, R.; Zhang, S.; Li, J. Q.; Sun, L. T.; Tang, Z. C.; Pan, X. L.; Bao, X. H. *Science* **2014**, *344*, 616.

(8) Gao, J.; Zheng, Y. T.; Jehng, J. M.; Tang, Y. D.; Wachs, I. E.; Podkolzin, S. G. *Science* **2015**, *348*, 686.

(9) Keller, G. E.; Bhasin, M. M. *J. Catal.* **1982**, *73*, 9.

(10) DeBoy, J. M.; Hicks, R. F. *Industrial & Engineering Chemistry Research* **1988**, *27*, 1577.

(11) Song, J.; Sun, Y.; Ba, R.; Huang, S.; Zhao, Y.; Zhang, J.; Sun, Y.; Zhu, Y. *Nanoscale* **2015**, *7*, 2260.

(12) Huang, P.; Zhao, Y.; Zhang, J.; Zhu, Y.; Sun, Y. *Nanoscale* **2013**, *5*, 10844.

(13) Xueping, F.; Shuben, L.; Jingzhu, L.; Jingfang, G.; Dexin, Y. *Journal of Molecular Catalysis(China)* **1992**, *6*, 255.

(14) Li, S. *Journal of Natural Gas Chemistry* **2003**, *12*, 1.

(15) Zavyalova, U.; Holena, M.; Schlögl, R.; Baerns, M. *ChemCatChem* **2011**, *3*, 1935.

(16) Latimer, A. A.; Kulkarni, A. R.; Aljama, H.; Montoya, J. H.; Yoo, J. S.; Tsai, C.; Abild-Pedersen, F.; Studt, F.; Norskov, J. K. *Nat Mater* **2017**, *16*, 225.

(17) Lunsford, J. H. *Catal. Today* **2000**, *63*, 165.

(18) Schwarz, H. *Angew. Chem. Int. Ed.* **2011**, *50*, 10096.

(19) Campbell, K. D.; Morales, E.; Lunsford, J. H. *J. Am. Chem. Soc.* **1987**, *109*, 7900.

(20) Arndt, S.; Laugel, G.; Levchenko, S.; Horn, R.; Baerns, M.; Scheffler, M.; Schlogl, R.; Schomacker, R. *Catal Rev* **2011**, *53*, 424.

(21) Arutyunov, V. S.; Strekova, L. N. *J Mol Catal a-Chem* **2017**, *426*, 326.

545417 Microwave-Assisted Natural Gas conversion in presence of Low Rank Coal

Monday, June 3, 2019: 5:27 PM
Texas Ballroom EF (Grand Hyatt San Antonio)

Victor Abdelsayed[1,2], Mark W. Smith[2], Dushyant Shekhawat[3] and Sonia Hammache[4], (1)AECOM, Morgantown, WV, (2)National Energy Technology Laboratory, Morgantown, WV, (3)US Department of Energy, National Energy Technology Laboratory, Morgantown, WV, (4)National Energy Technology Laboratory, Pittsburgh, PA

Conversion of low rank coal (LRC) into chemicals is one of the challenging chemical processes due to its low heat value, high moisture, ash, and sulfur contents. In this work, we studied the direct conversion of natural gas in presence of LRC into value added chemicals. The addition of methane during microwave-assisted pyrolysis of LRC influenced both the char structure and stability and shift the product distribution of gases and tars formed during the reaction. The initial results suggested that the presence of methane helped in reducing and breaking down heavy tars into useful aromatic compounds such as benzene and toluene. The structure-reactivity of generated chars was also investigated under different methane concentrations (0, 25, 50 and 90 % balanced in Ar) using a variety of characterization tools such as XRD, BET, SEM/EDS, dielectric properties, and Raman. Finally, the char reactivity towards combustion will be discussed.

545594 Research on the Stabilizer to Non-Catalytic Combustion Reaction in Oxidative Coupling of Methane

Monday, June 3, 2019: 5:30 PM
Texas Ballroom EF (Grand Hyatt San Antonio)

Li Wei, Qingrui Zhao and Mingsen Zhang, Department of Catalysis Science, Beijing Research Institute of Chemical Industry, Beijing, China

One of the most attractive methods for the direct conversion of methane to C_2 (ethylene and ethane) by oxidative coupling of methane (OCM) was proposed first by Keller and Bhasin in 1982. OCM was usually carried out at high temperature (above 600 °C), and the catalytic reaction and non-catalytic combustion reaction must be included in the temperature range higher than 600 °C. Therefore, it is very important to improve the catalytic reaction and inhibit the combustion reaction in catalytic and non-catalytic area. In this work, inert filler and metal mesh were selected as stabilizers to inhibit the combustion reaction. The reaction was operated in a microreactor over the temperature range of 400-800 °C, and the inner diameter and length of the microreactor were 10 mm and 530 mm. A simulated gas composed of 70.2% CH_4, 21.2% C_2H_4 and 8.6% C_2H_6 based on mol of carbon was used as feed gas. The flow rates of feed gas and O_2 were 52.5 mL/min and 17.5 mL/min, respectively. The result illustrated that the feed gas started combusting at 560 °C and O_2 was consumed completely at 650 °C in the case of using the empty microreactor. When the stabilizer was filled in the microreactor, the combustion reaction was inhibited evidently. No evident combustion reaction was observed until 700 °C when the reactor was filled with inert filler, and the temperature could reach 800 °C for metal mesh.

545743 Catalytic Dehydroaromatization of Methane: Strategies for Enhancing the Yield to Aromatics and Improving Coking Resistance

Monday, June 3, 2019: 5:33 PM
Texas Ballroom EF (Grand Hyatt San Antonio)

Sheima J. Khatib, Apoorva Sridhar and Mustafizur Rahman, Chemical Engineering, Texas Tech University, Lubbock, TX

Intensifying the use of methane as a chemical feedstock for its conversion to liquid fuels and other added value chemicals is of increasing interest since methane is the main constituent of natural gas, of which vast natural reserves are being exploited. Paradoxically, a significant amount of natural gas is wasted in flaring, due to the expensive handling and transportation of natural gas. Among the different processes that exist to convert methane to added-value liquids, there has been an emerging interest in the heterogeneous catalytic conversion of methane, CH_4, to a mixture of benzene and hydrogen, by the non-oxidative methane dehydroaromatization reaction ($6\ CH_4(g) \rightarrow C_6H_6(g) + 9H_2(g)$), which happens in one step. Our group is carrying out a fundamental and systematic study of this catalytic process with the goal to answer some fundamental questions, which can mitigate the technological challenges of this process.

Zeolite-supported molybdenum catalysts have so far been the most widely studied catalysts for the CH_4 dehydroaromatization reaction. Rapid catalyst deactivation and low conversions are disadvantages that impede commercialization of the methane dehydroaromatization process. It is agreed that Mo carbide species, originated by exposure of the Mo oxide species to the reactant methane, are responsible for methane activation, however, the nature and amount of carbide phases formed in the induction period of the reaction are not controlled.

We will present our results related to the exploration of both pretreatment conditions and addition of second metal promoters to purposefully prepare zeolite-supported metal carbide phases which present improved stability in reaction and we will relate the reaction and deactivation pathways to the structure of a specific carbide phase.

545628 Low Temperature Selective Oxidation of Methane Using Gold-Palladium Based Catalysts

Tuesday, June 4, 2019: 10:15 AM
Texas Ballroom A (Grand Hyatt San Antonio)

Nishtha Agarwal[1], Graham J. Hutchings[2], Qian He[3], Christopher Kiely[4], Sultan Althahban[5], Simon Freakley[6] and **Nicholas Dummer**[3], (1)Cardiff catalysis institute, Cardiff University, cardiff, United Kingdom, (2)School of Chemistry, Cardiff Catalysis Institute, Cardiff, United Kingdom, (3)Cardiff University, Cardiff, United Kingdom, (4)Department of Materials Science and Engineering, Lehigh University, Bethlehem, PA, (5)Lehigh University, Bethlehem, PA, (6)Department of Chemistry, University of Bath, Bath, United Kingdom

Abstracts:

Selective oxidation of methane to methanol with molecular oxygen using aqueous Au-Pd colloids at low temperature

N. Agarwal[1], S. Freakley[1], Sultan M. Althahban[2], C.J. Kiely[2], G.J. Hutchings[1]
(1) *Cardiff Catalysis Institute, School of Chemistry, Cardiff University, Cardiff, CF103AT, UK*
(2) *Department of Materials Science and Engineering, Lehigh University, PA 18015, USA*

In industry, CH_4 is used as a feedstock by first indirectly converting it to methanol (CH_3OH) via the production of synthesis gas ($CO + H_2$) at high temperatures and pressures, an expensive and energy intensive process. The direct oxidation of CH_4 to CH_3OH, which is challenging as over oxidation must be avoided, has been the subject of intensive study for many decades. CH_4 oxidation using supported AuPd nanoparticles (NPs) under mild aqueous conditions using H_2O_2 as oxidant at 50°C has been reported (1). The reaction proceeded through a radical mechanism. However, the relatively high cost of H_2O_2 for even stoichiometric oxidation of CH_4 makes it difficult to be economically viable. Incorporation of O_2 into the primary oxidation products would represent substantial progress toward a feasible CH_4 to CH_3OH process. Thus, this reaction has been performed here under mild conditions using colloidal AuPd NPs in the presence of both H_2O_2 and O_2, with the selectivity to primary products reaching >90% with minimal CO_2 produced.

Bimetallic Au-Pd nanoparticles were formed by standard colloidal method by using polyvinylpyrrolidone (PVP) as the stabilizing agent and $NaBH_4$ as the reducing agent. In the methane oxidation experiments we first used H_2O_2 as an oxidant at 50°C and we found that when the unsupported AuPd colloid was used as a catalyst the level of decomposition of H_2O_2 was much lower than when the AuPd colloid was supported on TiO_2. In view of this the products observed with the unsupported catalyst were substantial (Fig. 1). However, when molecular oxygen was added to the reaction gases the amount of products observed increased markedly. Use of isotopically labelled $^{18}O_2$ showed that up to 70% of the methanol contained oxygen from the molecular oxygen. In this reaction the H_2O_2 acts as an initiator but more oxygenated products were formed than H_2O_2 consumed, suggesting that the controlled breakdown of H_2O_2 activates methane which subsequently incorporates molecular oxygen through a radical process as shown in scheme.

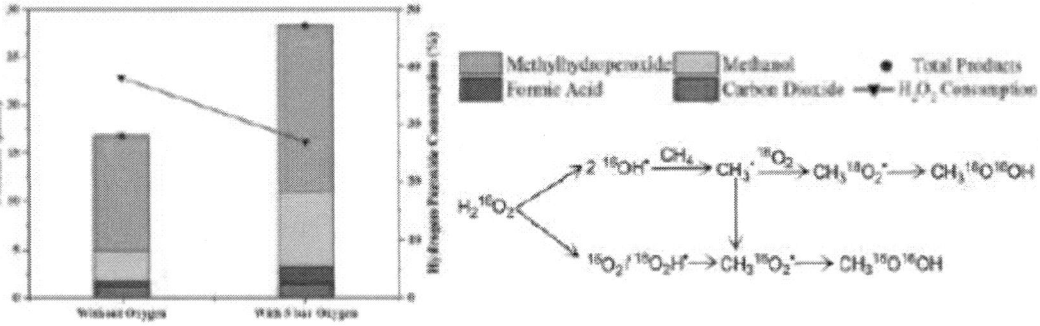

Figure 1. Conversion of methane using unsupported AuPd colloid with oxygen. And proposed reaction mechanism. Test conditions: 0.5 h, 50ºC, H_2O_2 amount: 1000 μmol, 1500 rpm, $P(CH_4)$ = 30 bar, $P(O_2)$ = 5 bar, 6.6 μmol metal per reaction

Fresh and used catalyst were characterized using electron microscopy and XPS. Particles were found to be alloyed with icosahedral structure with no size dependent composition variation and no single atoms or clusters were detected in fresh and used catalyst. The catalyst was found to be stable for longer reactions and reuse.

The reaction proceeded with same efficiency at room temperature and the activation energy for the reaction was found to be 39 kJ/mol. Using unsupported AuPd colloids with H_2O_2 as an initiator together with O_2, methane can be oxidized to methanol where the oxygen in the methanol is derived from O_2.

References

1- N. Agarwal et al. *Science* **2017**, 358, 223.
2- M. H. Ab Rahim et al. *Angew. Chem. Intl. Ed.* **2013**, 52, 1280.

Corresponding author email: agarwaln12@cardiff.ac.uk

- Abstract Text

545831 Selective Oxidation, Reforming and Combustion of Methane over NiO/Ni on Ceria-Zirconia

Tuesday, June 4, 2019: 10:39 AM
Texas Ballroom A (Grand Hyatt San Antonio)

Yimeng Lyu[1], Jennifer Jocz[1], Rui Xu[1], Eli Stavitski[2] and **Carsten Sievers**[1],
(1)School of Chemical & Biomolecular Engineering, Georgia Institute of Technology, Atlanta, GA, (2)BrookHaven National Laboratory, Upton, NY

Significant quantities of methane occur in remote locations where the economy of scale for commercial reformers is not feasible. Selective oxidation of methane offers an attractive approach to convert these stranded methane reserves to condensable alcohols that can be transported to a central processing facility in pipelines. This contribution analyzes the potential of well dispersed NiO clusters on a redox-active ceria zirconia (CZ) support as catalyst for such a process.

Ceria-zirconia (CZ) is capable of stabilizing well-dispersed metal and metal oxide clusters, activating O_2 and storing activated oxygen species. The combination of these properties makes it a promising support for catalysts for methane conversion in oxidative environments. Pyridine adsorption followed by IR spectroscopy and EDX mapping indicate that most NiO clusters only contain a small number of Ni atoms. The synergy between supported NiO clusters and CZ was demonstrated by in-situ XANES experiments. Specifically, CZ is capable of supplying NiO with oxygen until a reservoir of active oxygen species is depleted. Combined with the well-known activity of CZ to activate O_2, these results illustrate CZ can facilitate oxidation reactions of hydrocarbon moieties on the supported NiO clusters

Depending on the pretreatment and reaction conditions, NiO/CZ or Ni/CZ can convert methane by selective oxidation, reforming and combustion. Selective oxidation to alcohols can be achieved at moderate temperatures when nickel is present in NiO clusters and the feed is rich in methane and steam but contains limited amounts of oxygen. In the presence of larger amounts of O_2, complete combustion becomes the dominant reaction, whereas reduced Ni clusters favor reforming reactions. The strong metal-support interactions between the different components of this catalyst stabilize NiO particles that appear to be too small to catalyze the formation of graphitic coke or carbon nanotubes. The dispersion of nickel can be increased further by specific synthesis techniques like strong electrostatic adsorption, co-precipitation, and combustion synthesis.

545311 DRIFTS Studies on Zeotypes: Towards Fundamental Understanding of the Direct Conversion of Methane to Methanol

Tuesday, June 4, 2019: 11:03 AM
Texas Ballroom A (Grand Hyatt San Antonio)

Simone Creci, Per-Anders Carlsson and Magnus Skoglundh, KCK - Competence Centre for Catalysis, Chalmers University of Technology, Gothenburg, Sweden

The direct conversion of methane to methanol, DCMM, represents the "holy grail" for catalysis of the last two decades [1]. The oxidation of CH_4 to value added products, e.g. CH_3OH, *via* non-syngas based techniques is desirable to obtain higher energy storage capacity and more sustainable fuels. The breakage of the strong C-H bond in the CH_4 molecule and the easy over-oxidation of CH_3OH represent the main thermodynamic challenges in the DCMM [2]. In nature, some enzymes, particularly the methane monooxygenases (MMO), are known for selectively convert methane to methanol, although with low conversion rates [3]. Metals containing zeolites seem to show sites similar to the active sites of these enzymes and thus have been intensively investigated for the DCMM [4]. The DCMM using zeolites as catalysts involves commonly four steps: addiction of metals to the catalyst, formation of the active site by oxidation at high temperature, reaction with methane at low temperature and extraction of methanol by solvents, *e.g.* water. It has been shown that most of the reaction products remain adsorbed on adsorption sites, making the extraction of methanol the limiting step [2].

Zeolites are microporous aluminosilicates and their crystal lattice is based on the SiO_4^{4-} tetrahedron. The net negative charge due to substitution of Si^{4+} with Al^{3+} allows to host a number of elements in extra-framework positions. Nevertheless, Al-free zeotypes can be synthesized with elements other than Al in framework positions [5]. In 2012, Hammond *et al.* obtained remarkable CH_4 conversion and selectivity towards CH_3OH using a mixture of Al-free zeotypes as catalysts and liquid H_2O_2 as oxidant. In particular, the catalysts consisted of Fe and Cu containing silicalite, the Al-free parent of the commercial ZSM-5 zeolite with MFI type of framework structure. Fe and Cu were added *prior* to and *post* crystallization, respectively, resulting in framework and extra-framework metal species [6]. The lower acidity of the Fe-silicalite zeotype compared to Fe-exchanged ZSM-5 zeolite is believed to facilitate CH_3OH desorption due to the higher hydrophobicity of the Al-free zeotype.

This study aims to give a fundamental understanding of metal containing zeotypes with MFI type of framework structure suitable for the direct conversion of methane to methanol. The zeotypes are hydrothermally synthesized and characterized with diffraction, physisorption and spectroscopy techniques. Furthermore, the tendency of CH_3OH to desorb from the internal surfaces of the catalyst is detected by temperature programmed desorption of methanol, CH_3OH TPD, and the evolution of surface species is followed using DRIFTS. In a standard synthesis procedure, the silica precursor is mixed with the structure directing agent, SDA, and the metal precursor in aqueous medium. The so obtained gel is crystallized in autoclave and the products are washed, filtered and dried. The SDA is then removed by calcination and the H^+-form of the zeotype is obtained by ammonium exchange and further calcination. The MFI type of framework structure is confirmed by PXRD and more insights about the incorporation of metals in the framework are given by N_2 physisorption and IR spectroscopy.

1. Čejka, G. Centi, J. Perez-Pariente and J.W. Roth, *Catalysis Today*, **179** (2012), 2-15.
2. Taifan, W. & Baltrusaitis, J. CH4 conversion to value added products: Potential, limitations and extensions of a single step heterogeneous catalysis. *Appl. Catal. B Environ.* **198,** 525–547 (2016).
3. Zakaria & Kamarudin. Direct conversion technologies of methane to methanol: An overview. *Renew. Sustain. Energy Rev.* **65,** 250–261 (2016).
4. Kondratenko, E. V. *et al.* Methane conversion into different hydrocarbons or oxygenates: current status and future perspectives in catalyst development and reactor operation. *Catal. Sci. Technol.* **7,** 366–381 (2017).
5. Elaine, L. E. S. & A.Moore, E. *SOLID STATE CHEMISTRY: An Introduction. Taylor & Francis Group* (2005).
6. Hammond, C. *et al.* Direct catalytic conversion of methane to methanol in an aqueous medium by using copper-promoted Fe-ZSM-5. *Angew. Chemie - Int. Ed.* **51,** (2012).

546630 Selective Oxidation of Methane to Methanol By Fe and Cu-Exchanged Zeolites: A Spectroscopic Investigation

Tuesday, June 4, 2019: 11:27 AM
Texas Ballroom A (Grand Hyatt San Antonio)

Dieter Plessers, Center for Surface Science and Catalysis, KU Leuven, Leuven, Belgium

For decades Fe-exchanged zeolites such as FeBEA*, FeFER and FeZSM-5 are known to be active in the direct, selective oxidation of methane to methanol at room temperature, an interesting reaction given the emerging feedstock transition from oil to gas. Recently our group elucidated the structure of the α-Fe site and the reactive α-O species, formed after oxidation with N_2O, on FeBEA*, using site-selective spectroscopic techniques (Nature 536, 317–321, 2016). α-Fe was assigned to a mononuclear high-spin S=2 square planar Fe(II) species and α-O to a high-spin S=2 square pyramidal Fe(IV)=O species. These α-species are stabilized in a six-membered ring (6MR), containing an Al-Si-Si-Al sequence, i.e. paired Al atoms (see Figure).

Recently we proved that the chabazite (CHA) zeolite topology is also able to stabilize α-Fe and α-O species (JACS, 140 (38), 12021–12032, 2018). Spectroscopy (UV-Vis-NIR and Mössbauer) combined with theoretical models (DFT and CASPT2) proved that the α-species are again stabilized in a six-membered ring (6MR), containing an Al-Si-Si-Al sequence (see Figure). The simple CHA topology (only 1 type of T-atom) is shown to mitigate the heterogeneity of iron speciation found on other Fe-zeolites, with Fe_2O_3 the only identifiable phase other than α-Fe formed in Fe-CHA at more elevated Fe loadings. Finally, through a comparison between α-Fe in Fe-CHA and Fe-*BEA, the topology's 6MR geometry is found to influence the structure and the ligand field and consequently the spectroscopy of the α-Fe site in a predictable manner.

In our current research we are investigating the different steps of the reaction cycle in more detail, both mechanistically and kinetically (N_2O activation, reaction path of α-O with methane) to get an even better understanding of the requirements for an efficient reaction cycle. Then we are using this information to try increasing the methanol yield by for instance investigating the introduction method of Fe in the zeolite.

Alongside Fe-exchanged zeolites, Cu-exchanged zeolites are another promising candidate for direct methane oxidation to methanol, attracting a lot of research since 2005. Our group managed to identify and characterize a mono-(μ-oxo) dicopper(II) species (Cu-O-Cu) in ZSM-5 zeolites (PNAS, 106 (45), 18908-18913, 2009) and two slightly differing Cu-O-Cu species in mordenite (MOR) zeolites (JACS, 137 (19), 6383–6392, 2015) as active sites for methane conversion at temperatures around 423 K (see Figure). These species can be formed with either O_2 or N_2O as the oxidant. Their structure (bond angle, bond lengths) differs slightly based on the location in the zeolite channels and cavities. This leads to different reactivity with methane. In addition, we have shown that the zeolite framework itself has also a relevant influence on the reaction coordinate (confinements effect). This is comparable to the workings of an enzyme, where second-sphere effects are directing the approach of methane to the active sites and stabilize the transition state.

Currently we are hunting for different copper active sites in other zeolite frameworks (or porous structures in general). We are also trying to understand the influence of every step in the activation procedure on the copper speciation in the zeolite. To this end we are again using a combination of site-selective spectroscopy (DR-UV-Vis, resonance Raman, EPR...) backed-up by computational studies, and our in-house knowledge of zeolite properties and zeolite synthesis.

In conclusion we hope to present our most recent research on Fe- and Cu-exchanged zeolites for selective methane oxidation to methanol on the 12th Natural Gas Conversion Symposium.

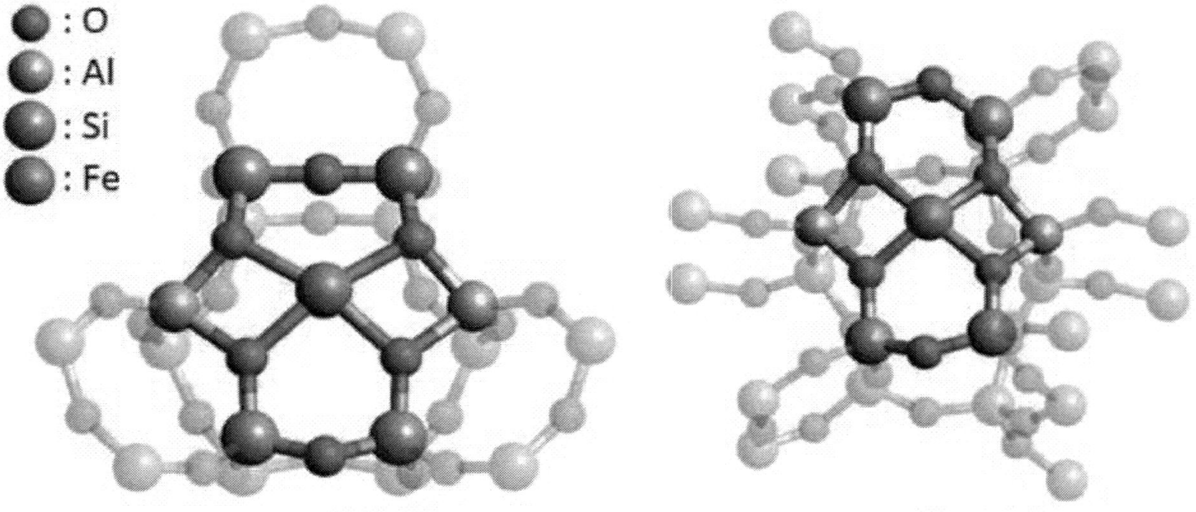

α-Fe in CHA α-Fe in BEA*

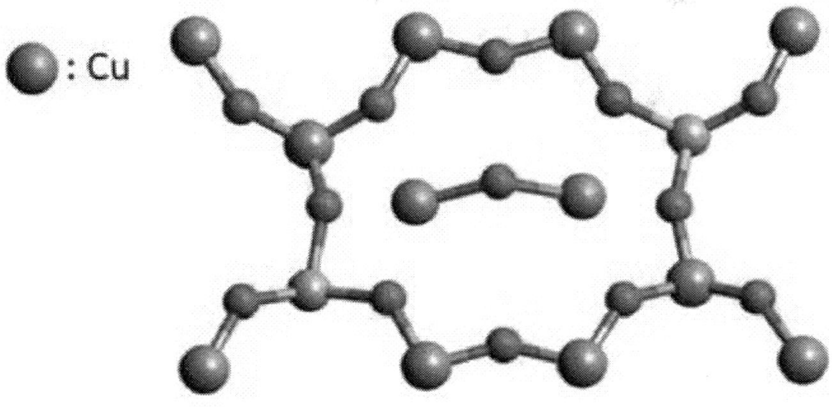

Cu-O-Cu in MOR

546235 Identification of the Reaction Pathway and Active Site for the Catalytic Conversion of Methane to Methanol over Copper-Exchanged Zeolites

Tuesday, June 4, 2019: 11:51 AM
Texas Ballroom A (Grand Hyatt San Antonio)

Kimberly Dinh[1], Mark Sullivan[1], Pedro Serna[2], Randall Meyer[3], Mircea Dincă[4] and Yuriy Roman[1], (1)Chemical Engineering, Massachusetts Institute of Technology, Cambridge, MA, (2)ExxonMobil, (3)ExxonMobil, Annandale, NJ, (4)Department of Chemistry, Massachusetts Institute of Technology, Cambridge, MA

The selective oxidation of methane to value-added chemicals remains a grand challenge in catalysis. Methane monooxygenase enzymes selectively convert methane to methanol at room temperature, but these enzymes suffer from limited temperature stability and are difficult to purify. As inorganic analogs of these enzymes, small metal clusters and single metal atoms (Fe, Cu, Rh) in zeolites are capable of converting methane to methanol directly. Recently, we reported the first instance of a continuous, gas phase process for the direct conversion of methane to methanol at high selectivity using only Cu-exchanged zeolites and gaseous CH_4, O_2, and H_2O, but the system remains largely unexplored. This work represents the first kinetic investigation and characterization of Cu-based catalysts for the steady state conversion of methane to methanol in zeolitic catalysts.

Rigorous kinetic experiments on a series of copper-exchanged zeolites show that the rate-limiting step is the C-H bond scission of methane, carbon dioxide stems from the sequential overoxidation of methanol, re-oxidation of the Cu active site is kinetically irrelevant, and water is required for methanol desorption. We then utilized *in situ* NH_3 titration experiments coupled with infrared and x-ray absorption spectroscopy to identify a $[Cu\text{-}O\text{-}Cu]^{2+}$ motif as the active site for H-abstraction from CH_4, and demonstrate the necessity of zeolitic protons for both methanol desorption and copper active site formation. We finally hypothesize that zeolitic protons facilitate the diffusion of isolated, hydrated Cu species throughout the zeolite framework, allowing for the formation of dimeric Cu active sites.

545911 Methane to Methanol Conversion over Cu-Zeolites

Tuesday, June 4, 2019: 1:30 PM
Texas Ballroom A (Grand Hyatt San Antonio)

Elisa Borfecchia[1], Dimitrios K. Pappas[1], Michael M. Dyballa[1], Kirill A. Lomachenko[2], Andrea Martini[3,4], Gloria Berlier[4], Bjornar Arstad[5], Carlo Lamberti[3,6], Silvia Bordiga[1,4], Unni Olsbye[1], Pablo Beato[7] and **Stian Svelle**[1], (1)Department of Chemistry, University of Oslo, OSLO, Norway, (2)European Synchrotron Radiation Facility, Grenoble, France, (3)The Smart Materials Research Institute, Rostov-on-Don, Russian Federation, (4)Department of Chemistry and NIS Centre, University of Turin, Torino, Italy, (5)Sintef, Oslo, Norway, (6)Department of Physics, University of Turin, Torino, Italy, (7)Atomic-Scale Analysis Department, Haldor Topsoe A/S, Kgs. Lyngby, Denmark

Methane to methanol conversion over Cu-zeolites

A process allowing energy-effective methane to methanol (MTM) conversion would represent a major breakthrough for the chemical industry. Cu-exchanged zeolites have been shown to possess active Cu sites able to cleave the C−H bond of methane at temperatures ≤ 200 °C, enabling the stoichiometric transformation of methane into methanol [1]. The conversion is performed through a stepwise process, involving high-temperature activation in O_2 to generate the active sites, methane loading at 200 °C, and steam-assisted methanol extraction.

We have combined laboratory performance testing with in situ/operando Cu K-edge XAS to establish structure-activity relationships for the MTM conversion over Cu-CHA [2] and Cu-MOR [3] zeolites. By operando XAS, we tracked the oxidation state and average coordination of Cu ions during each step of the process and explored the impact of different pretreatments and compositional characteristics by in situ XAS.

High-temperature reaction with O_2 is evidenced as a key requirement to form the active Cu(II) sites, which then undergo reversible redox chemistry during the CH_4-loading and CH_3OH extraction steps. For Cu-CHA, we identified a positive linear correlation between the methanol productivity and the composition-dependent self-reducibility under high-temperature treatment in He, allowing us to rationalize the impact of the sample composition on the productivity for the MTM conversion. The most recent research efforts on Cu-MOR have highlighted the crucial role of operating at consistent conditions for both spectroscopy and performance testing. Having fulfilled this requirement, aided by high-energy resolution XANES and multivariate analysis [4], we unambiguously assessed the nuclearity of the active Cu site in the investigated series of Cu-MOR materials. These studies highlights the potential of the combination between XAS and testing at consistent conditions and paves the way for rationalized material synthesis to improve the reaction performance.

[1] E. Borfecchia et al., Chem. Soc. Rev. (2018) in press, doi: 10.1039/c8cs00373d.

[2] D. K. Pappas et al., J. Am. Chem. Soc., 139 (2017) 14961.
[3] D. K. Pappas et al., J. Am. Chem. Soc., under review.
[4] A. Martini et al., Chem. Sci. 8 (2017) 6836.

546228 Binuclear Transition Metal Ion Centers in Zeolites for Selective Oxidation of Methane

Tuesday, June 4, 2019: 1:54 PM
Texas Ballroom A (Grand Hyatt San Antonio)

Jiri Dedecek[1], Edyta Tabor[1], Mariia Lemishka[1,2], Kinga Mlekodaj[1], Stepan Sklenak[1] and Zdenek Sobalik[1], (1)Department of Structure and Dynamics in Catalysis, J. Heyrovský Institute of Physical Chemistry of the CAS, Prague, Czech Republic, (2)Faculty of Chemical Technology, University of Pardubice, Pardubice, Czech Republic

The recent development of the technology of shale gas production results in the dramatic increase of the interest on the utilization of natural gas. One of promising ways represents selective oxidation of methane to methanol and other valuable liquid products. Methane steam reforming with breaking the C-H bonds to syngas followed by the re-formation of C-H bonds into hydrocarbons is energetically and technically highly demanding and costly. Thus, a direct route of methane selective oxidation to methanol would be a dramatic step ahead in the methane processing. Nevertheless, this process requires development of highly effective catalysts. Iron containing zeolites represent important model catalysts for this reaction.

Interaction of Fe(II) containing zeolites with N_2O results in the formation of active oxygen form (so called α-oxygen), which can result in the formation of molecular products (N_2 and O_2) of N_2O decomposition process or can be used as oxidant in the selective oxidation of methane to methanol.[1,2] However, Fe(II) species active in the formation of α-oxygen in zeolite matrices are difficult to be prepared and they are unstable both under conditions of catalyst preparation and of N_2O decomposition / methane oxidation. Recently, it has been shown that the special arrangement of Fe(II) ions in zeolite matrix in the form of binuclear Fe(II) ion species results in significant facilitation both of N_2O decomposition and α-oxygen stabilization. This begs the question if also other transition metal cations embedded in the zeolite matrix in the form of binuclear M(II) structure are able to effectively decompose N_2O and form α-oxygen formation and subsequently oxidize methane to methanol.

This study aimed in targeted preparation of the binuclear Co(II), Ni(II), Mn(II) structures in zeolite matrix and testing their activity in N_2O decomposition, formation of active α-oxygen and selective oxidation of methane.

The catalysts with high loading of binuclear structures (Me = Mn, Co, and Ni) were prepared using ion exchanged method (Mn and Co) or impregnation (Ni). The presence of divalent cations (Mn, Co, and Ni) in cationic position of the zeolite matrix was confirmed by FTIR spectroscopy in the region of anti-symmetric T-O-T vibrations (1000-860 cm^{-1}) of the zeolite framework. FTIR results showed that in all prepared Me-zeolites, the majority of metal cations has to be present as binuclear M(II) cations due to the dominant presence of bare M(II) ions and number of possible sites for divalent M(II) ions. *In-situ* FTIR studies confirmed the interaction of M(II)-zeolites with N_2O at 250°C, reflected in the shift of the skeletal vibrations, which was assigned to atomic oxygen ($[M(III)\text{-}O^-]^{2+}$) stabilized on metal cations (Co, Mn or Ni). The introduction of CH_4 at 250 °C to oxidized M-zeolites containing active oxygen species $[M(III)\text{-}O^-]^{2+}$ resulted in the significant shift of the band of T-O-T vibrations reflecting presence of M(II) ions with adsorbed guest molecule. This result clearly confirms the complete reduction of the $[M(III)\text{-}O^-]^{2+}$ species by CH_4. Moreover, the FTIR spectra of M-zeolites ofter oxidation and methane treatment showed new bands at 2920 and 2832 cm^{-1}, which are typical for vibrations of methoxyl group, and the bands at 2979 and 1371cm^{-1}, characteristic for methanol vibrations.

The obtained results clearly show that binuclear transition metal species (Mn, Ni and Co) stabilized in zeolite matrix can split N_2O. This interaction results in the stabilization of active oxygen on the metal cation and N_2 formation. This finding is in sound agreement with the results of *ab initio* calculations which suggest dramatically lower energy barrier of N_2O decomposition over binuclear M(II)structures. The active oxygen formed then easily oxidizes methane to methanol at 250 °C.

Refences:

[1] Snyder, B. E. R.; Vanelderen, P.; Bols, M. L.; Hallaert, S. D.; Böttger, L. H.; Ungur, L.; Pierloot, K.; Schoonheydt, R. A.; Sels, B. F.; Solomon, E. I., Nature **2016**, 536 (7616), 317-321,.

[2] Sklenak, S.; Andrikopoulos, P. C.; Boekfa, B.; Jansang, B.; Novakova, J.; Benco, L.; Bucko, T.; Hafner, J.; Dedecek, J.; Sobalik, Z., J. Catal. **2010**, 272 (2), 262-274.

545829 Impact of Al Distribution in the Cu-Exchanged CHA-Type Zeolite on the Catalytic Performance in CH_4 Conversion

Tuesday, June 4, 2019: 2:18 PM
Texas Ballroom A (Grand Hyatt San Antonio)

Toshiyuki Yokoi[1], Ryota Osuga[2], Yusuke Kunitake[2] and Junko N. Kondo[2],
(1)Institute of Innovative Research, Tokyo Institute of Technology, Yokohama, Japan, (2)Tokyo Institute of Technology, Yokohama, Japan

1. Introduction

Methane is a highly abundant and inexpensive source of fuel and chemicals. The development of novel technologies that can convert methane easily into chemicals has strongly been desired. However, its kinetic inertness and low reactivity limit the industrial utilization of methane. Recently, it was reported that Cu-exchanged zeolites are a promising catalyst for the catalytic conversion of methane. We are also tackling the development of novel zeolite-based catalysts for direct conversion of methane into methanol followed by lower olefins.

Meanwhile, recently, we have developed a facile method for preparing the CHA-type zeolite with controlling the Al distribution, *i.e.*, proportion of Q^4(2Al)/Q^4(1Al) ratio, where Q^4(nAl) is $Si(OSi)_{4-n}(OAl)_n$. In this work, the impact of Al distribution in the Cu-exchanged CHA-type zeolite on the catalytic performance in CH_4 conversion was investigated.

2. Experimental

The CHA-type aluminosilicate zeolites with different proportions of Q^4(2Al)/Q^4(1Al) ratios were synthesized in the presence of *N,N,N*-trimethyl-1-adamantammonium cation ($TMAda^+$) from the different starting materials including fumed silica, aluminum hydroxide, and the FAU-type zeolite (JRC-Z-Y5.5, Si/Al = 2.8), with their proportions varied. Cu-ion exchange was carried out using $Cu(NO_3)_2 \cdot 3H_2O$.

The catalytic reaction was performed in a fixed bed reactor. The flow rates of the reactants were $CH_4/O_2/Ar$ = 16/4/5 (SCCM). The catalyst amount was 100 mg. The reaction temperature was varied ranging from 400 to 600 °C, and the reaction time at each temperature was 1 min. The products including CO and CO_2 were analyzed by GC-TCD, and other hydrocarbon products were analyzed by GC-FID.

3. Results and discussion

The ^{29}Si MAS NMR spectra clearly showed that thus obtained CHA-type zeolites had different Al distributions depending on the proportion of the **FAU**-type zeolite as the starting material. The use of the FAU-type zeolite with a high proportion led to the high proportion of Q^4(2Al). Here, representative catalysts were designated as "Q^4(1Al)-rich" and "Q^4(2Al)-rich", respectively. The Si/Al ratios of Q^4(1Al)-rich and Q^4(2Al)-rich were 9 and 13, respectively, and the Cu content for both catalysts were almost similar, ca. 1.4 wt%. The Cu state was evaluated by UV.-vis technique; both catalysts gave only one peak at 220 nm, indicating Cu species were highly isolated.

Both catalysts showed similar CH_4 conversion, and it was increased along with the temperature. However, there was a significant difference in the products selectivities between Q^4(1Al)-rich and Q^4(2Al)-rich catalysts. The yield of CO_2 on Q^4(2Al)-rich catalyst was higher than that on Q^4(1Al)-rich. The yields of target products, CH_3OH, DME, C_2H_6 and C_2H_4 on Q^4(1Al)-rich were higher than those on Q^4(2Al)-rich. These results clearly suggest that the Al distribution affected the catalytic performance in CH_4 conversion; Q^4(1Al)-rich catalysts are advantageous in terms of the production of target products from CH_4.

The Cu state was also investigated by NO-adsorbed FT-IR technique as follow [4]. First, the catalysts were evacuation at 723 K, and then NO molecules was adsorbed onto the catalysts at 293 K. The peaks around 1900-1950 cm^{-1} are derived from to NO adsorbed on Cu^{2+} species, and ones around 1810 cm^{-1} are derived from NO adsorbed on Cu^{+} species. There was a marked change in the peaks derived from NO adsorbed on Cu^{+} species; Q^4(1Al)-rich first gave a peak at 1809 cm^{-1}, but another peak at 1816 cm^{-1} was observed when NO loading was increased. However, Q^4(2Al)-rich did not exhibit such change. Thus, we have considered that the Al distribution would affect the state of Cu species, resulting in different catalytic performance.

References

1. J. Wulfers, S. Teketel, B. Ipek, R. F. Lobo, Chem. Commun. 51 (2015) 4447.
2.
3. Nishitoba, N. Yoshida, J. N. Kondo, T. Yokoi, in preparation.
4. Zhang, J.-S. McEwen, M. Kollár, F. Gao, Y. Wang, J. Szanyi, C. H.F. Peden ACS Catal. 4 (2014) 4093.

545715 Selective Oxidation of Methane to Methanol on Cu/SBA-15 Via a Stepwise Manner

Tuesday, June 4, 2019: 2:42 PM
Texas Ballroom A (Grand Hyatt San Antonio)
Ha V. Le, Functional Materials, Institute of Chemistry-Functional Materials, Technische Universitat Berlin, Berlin, Germany

Selective oxidation of methane to methanol on Cu/SBA-15 via a stepwise manner

Ha V. Le[1,2], Samira Parishan[3], *Anton* Sagaltchik[4], Hamideh Ahi[4,5], Annette Trunschke[5], Reinhard Schomäcker[3], Arne Thomas[1]

[1] Institute of Chemistry-Functional Materials, Technische Universitat Berlin, BA2, Hardenbergstraße 40, 10623 Berlin, Germany

[2] Faculty of Chemical Engineering, Ho Chi Minh City University of Technology (VNU–HCM), Building B2, Ly Thuong Kiet Street 268, District 10, 700910 Ho Chi Minh City, Vietnam

[3] Institute of Chemistry-Technical Chemistry, Technische Universitat Berlin, TC8, Straße des 17. Juni 124, 10623 Berlin, Germany

[4] BasCat-UniCat BASF Joint Lab, Technische Universitat Berlin, EW K 01, Hardenbergstraße 36, 10623 Berlin, Germany

[5] Department of Inorganic Chemistry, Fritz Haber Institute of the Max Planck Society, Faradayweg 4-6, 14195 Berlin, Germany

Abundant and cheap resources containing methane such as natural gas, methane hydrates, and biogas have been considered as promising alternatives to decline the dependence of chemical and energy industries on crude oil. However, unsafe and costly transportation and storage, and high chemical inertness of methane cause an underutilization of these resources at present.[1, 2, 3] The direct conversion of methane to more energy-dense liquid derivatives such as methanol is an economically efficient protocol to promote the application of methane. Although great interest and numerous efforts have been devoted to the hydroxylation of methane, unfortunately, the development of a commercially viable catalytic process with a high yield of methanol remains unsolved.[4]

Inspired by prototype natural machines, namely, Fe- and Cu-dependent methane monooxygenase enzymes (MMOs) in methanotrophic bacteria, in the last decades, several Cu- and Fe-exchanged zeolites have been applied as heterogeneous catalysts for the methane-to-methanol conversion due to their ability to activate methane at low temperatures and reveal high selectivity to methanol. This is typically achieved in a stepwise process, in which the catalyst is first oxidized in O_2 or N_2O at high temperatures (> 200 °C), and then reacts with methane at lower temperatures (≤ 200 °C). The final transformation of the formed intermediate bound on the catalyst surface to methanol should be performed via a treatment with water.[5-9] As the so far most active catalysts, Cu/zeolites have been intensively studied to identify possible structures of active Cu sites and the catalytic mechanisms, finally obtaining improved performances.[10, 11]

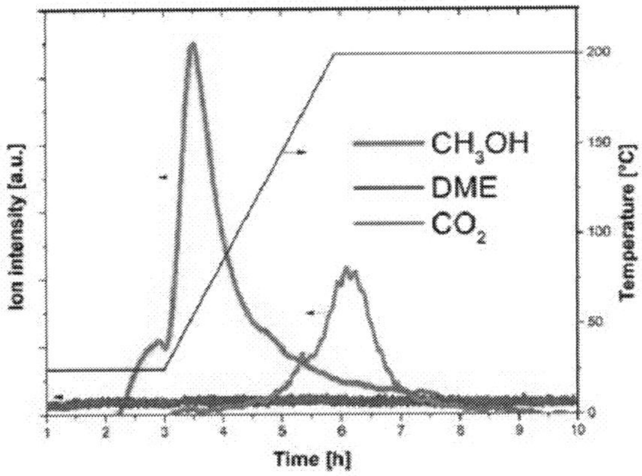

Figure 1. Mass spectral (MS) signals of products after interaction of Cu/SBA-15 with methane at 200 °C, followed by online extraction in a water-saturated He flow.

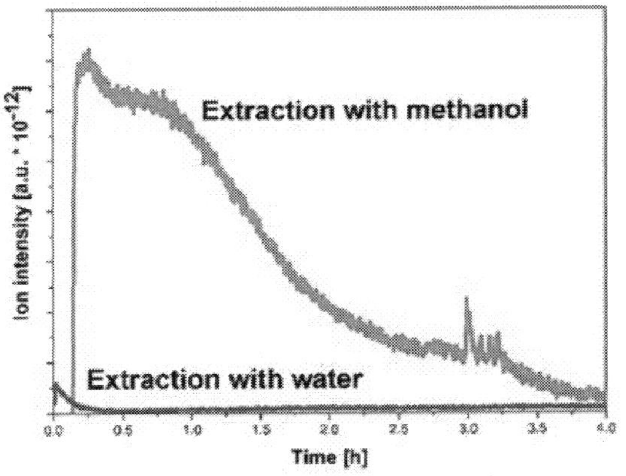

Figure 2. MS signals of DME extracted from Cu/SBA-15 in the online stage at room temperature using different solvents.

To expand the Cu-containing catalytic material class for the O_2-mediated hydroxylation of methane, instead of zeolite, a highly ordered mesoporous silica type, i.e. SBA-15, was used as a support for Cu-based species. Our study proves that highly dispersed CuO species on SBA-15 are able to react with methane and subsequently produce methanol with high selectivity (> 84%) via water-assisted extraction (Figure 1). Furthermore, it is confirmed that the main intermediate formed after interaction of the catalyst with methane is indeed a methoxy species, which can be further converted to methanol or dimethyl ether (DME) upon extraction with water or methanol, respectively (Figure 2).[12]

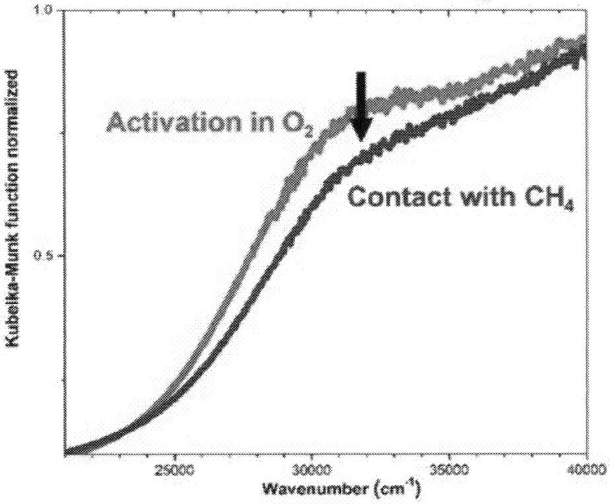

Figure 3. *In situ* UV-vis spectra of CuO/SBA-15 after activation in O_2 at 550 °C and subsequent reaction with methane at 200 °C.

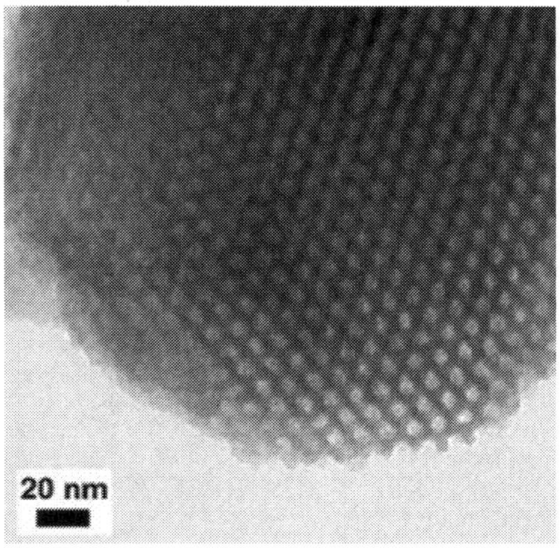

Figure 4. TEM image of Cu/SBA-15.

The reactivity of the Cu-based species on SBA-15 with methane was further demonstrated by *in situ* UV-vis spectroscopic measurements, in which intensity of the band centered at 32 000 cm^{-1} increased by activation in O_2 at 550 °C and significantly decreased upon contact with methane (Figure 3) at 200 °C. Unlike the Cu-oxo complexes in Cu/zeolites previously proposed as the active sites for activation of methane, the catalytic activity of Cu/SBA-15 is attributed to small CuO nanoclusters that are highly dispersed throughout the SBA-15 framework (Figure 4). Based on this demonstration, it might be worthwhile to further test various porous silica forms or other materials as supports for Cu-base species for the challenging direct conversion of methane to methanol.[12]

References

[1] C. Hammond, S. Conrad, and I. Hermans, *ChemSusChem* **2012**, *5*, 1668.

[2] P. Schwach, X. Pan, and X. Bao, *Chem. Rev.* **2017**, *117*, 8497.

[3] P. Tomkins, M. Ranocchiari, and J. A. van Bokhoven, *Acc. Chem. Res.* **2017**, *50*, 418.

[4] V. L. Sushkevich, D. Palagin, M. Ranocchiari, and J. A. van Bokhoven, *Science* **2017**, *356*, 523.

[5] J. M. Bollinger Jr, *Nature* **2010**, *465*, 40.

[7] R. Banerjee, Y. Proshlyakov, J. D. Lipscomb, and D. A. Proshlyakov, *Nature* **2015**, *518*, 431.

[8] M. H. Groothaert, P. J. Smeets, B. F. Sels, P. A. Jacobs, and R. A. Schoonheydt, *J. Am. Chem. Soc.* **2005**, *127*, 1394.

[8] E. V. Starokon, M. V. Parfenov, L. V. Pirutko, S. I. Abornev, and G. I. Panov, *J. Phys. Chem. C* **2011**, *115*, 2155.

[9] N. V. Beznis, A. N. C. van Laak, B. M. Weckhuysen, and J. H. Bitter, *Microporous Mesoporous Mater.* **2011**, *138*, 176.

[10] M. Ravi, M. Ranocchiari, and J.A. van Bokhoven, *Angew. Chem., Int. Ed.* **2017**, *56*, 16464.

[11] H. V. Le, S. Parishan, A. Sagaltchik, C. Göbel, C. Schlesiger, W. Malzer, A. Trunschke, R. Schomäcker, and A. Thomas, *ACS Catal.* **2017**, *7*, 1403.

[12] H. V. Le, S. Parishan, A. Sagaltchik, H. Ahi, A. Trunschke, R. Schomäcker, and A. Thomas, *Chem. Eur. J.* **2018**, *24*, 12592.

Mechanistic Insights into the Conversion of Methane to Methanol over Cu–ZSM-5 Zeolite

Jian-Feng Wu, Xu-Dong Gao, Long-Min Wu, Si-Min Yu, Shi Bai, Wei Wang*
State Key Laboratory of Applied Organic Chemistry, College of Chemistry and Chemical Engineering, Lanzhou University, Lanzhou, Gansu 730000, P. R. China
*wang_wei@lzu.edu.cn

Introduction

Methane is the main component of nature gas, which is considered as a potential alternative energy to oil. In industry, however, the direct conversion of methane to syn gas ($CO + H_2$) needs harsh reaction conditions. Followed by, the syn gas needs a further step to synthesize methanol as a versatile molecule. Therefore, the direct conversion of methane to methanol is highly desired.[1,2]

Inspired by the particulate methane monooxygenase (pMMO) enzymes in nature, researchers found that copper modified zeolites could selectively convert methane to methanol under mild conditions.[3] Understanding the mechanism of methane to methanol process involved the structure of active sites and the reaction mechanism is of paramount importance. The active sites have been proposed as mononuclear copper, binuclear copper and trinuclear copper species based on zeolite topologies and synthesis conditions.[4] However, less attention has been devoted to the study of the reaction mechanism.

Materials and Methods

The Cu–ZSM-5 catalyst was synthesized by ion-exchange of zeolite Na-ZSM-5 with $Cu(NO_3)_2$. The Cu–ZSM-5 catalyst is treated in a vacuum (< 10^{-2} mbar) at 763 K for 43 h before loading ^{13}C-labeled $^{13}CH_4$ or CH_4 (~50 mbar). After the reaction at 423 K for 1.5 h, the Cu–ZSM-5 catalyst was subjected to the ^{13}C MAS Solid-State NMR measurements. *In situ* EPR experiments were also conducted to elucidate the reaction mechanism at the same reaction conditions.

Results and Discussion

Upon the *in situ* heating of CH_4 on Cu–ZSM-5 zeolite at 423 K, the signal intensity of the Cu^{2+} ion decreased. Meanwhile, we do not observe the Cu(0) signal, indicating the reduction of the Cu^{2+} ion to the Cu^{+} ion upon the reaction with CH_4.

Upon the reaction of CH_4 on Cu–ZSM-5 zeolite at 423 K, two signals at 59 and 53 ppm were observed besides the CH_4 signal at -10 ppm, indicating the activation of methane. The signal intensity of 53 ppm decreased upon the vacuum treatment, while the signal intensity of 59 ppm maintained the same. Therefore, the 53 ppm signal can be assigned to adsorbed methanol, and the 59 ppm signal can be assigned to surface methoxy species. After the hydration of sample b, copper formate species at 171 ppm was formed, indicating the surface methoxy species should bond to the Cu species. Further heating CH_4 on Cu–ZSM-5 zeolite at 573 K will yield fully oxidized CO_2 as a product.

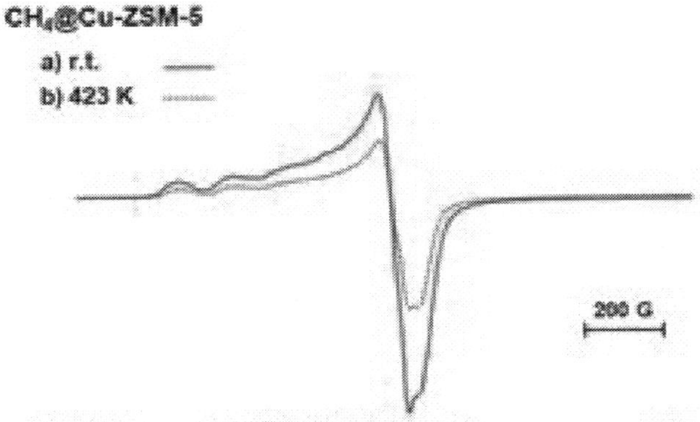

Figure 1. The reaction of CH_4 on Cu–ZSM-5 zeolite measured by *in situ* EPR. The EPR spectra were collected at a) room temperature (r.t.) and b) 423 K.

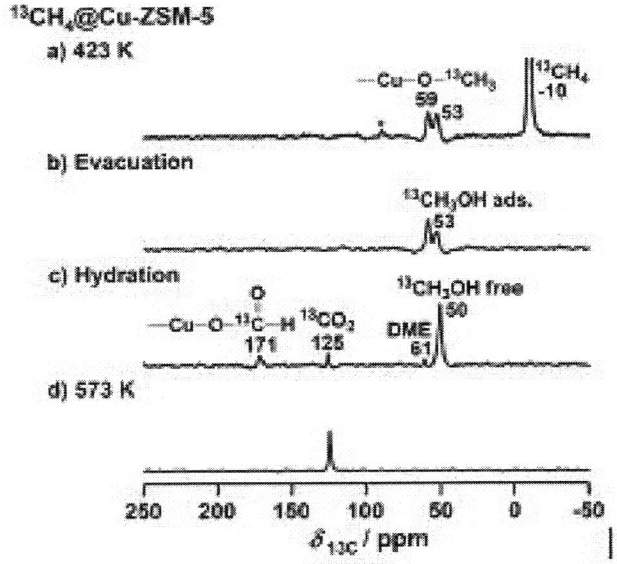

Figure 2. The ^{13}C HPDEC MAS NMR results provide the evidence of the formation of copper methoxy species upon the reaction of CH_4 on Cu–ZSM-5 zeolite.

Significance

The mechanism of the transformation of methane into methanol on a Cu–ZSM-5 zeolite at the temperature of 423 K was investigated by *in situ* EPR and Solid-State NMR. EPR results indicate the reduction of Cu^{2+} to Cu^{+} during the reaction of CH_4 on Cu–ZSM-5 zeolite. Solid-state NMR experiments demonstrate the formation of copper methoxy species and adsorbed methanol on Cu–ZSM-5 zeolite at 423 K. This work provides valuable mechanistic insights into the methane to methanol process and shed light on new catalyst synthesis.

References

1. Crabtree, R. H. *Chem. Rev.* **1995**, *95*, 987-1007.
2. Lunsford, J. H. *Catal. Today* **2000**, *63*, 165-174.
3. For example, Groothaert, M. H.; Smeets, P. J.; Sels, B. F.; Jacobs, P. A.; Schoonheydt, R. A. *J. Am. Chem. Soc.* **2005**, *127*, 1394-1395.
4. For example, Snyder, B. E. R.; Bols, M. L.; Schoonheydt, R. A.; Sels, B. F.; Solomon, E. I, *Chem. Rev.* **2018**, *118*, 2718-2768.

545733 Activation of Oxygen - Key to the Oxidative Coupling of Methane (OCM) Reaction

Tuesday, June 4, 2019: 4:00 PM
Texas Ballroom A (Grand Hyatt San Antonio)

Lukas Thum, Dept. of Chemistry, TU Berlin, Berlin, Germany, **Prof. Dr. Reinhard Schomäcker**, Institute of Chemistry, TU Berlin, Berlin, Germany, Annette Trunschke, Inorganic Chemistry, Fritz Haber Institute, Berlin, Germany and Robert Schlögl, Department of Inorganic Chemistr, Fritz Haber Insitute of the Max Planck Society, Berlin

Activation of Oxygen - Key to the Oxidative Coupling of Methane (OCM) Reaction

Since the 1980's, many materials have been investigated as suitable catalysts for the OCM reaction with mediocre success so far. The major challenge is the direction of the reaction towards the desired higher hydrocarbons while suppressing the thermodynamically favored deep oxidation towards carbon monoxide and carbon dioxide. An apparent yield barrier of 30% has yet to be surpassed.[1] Non reducible basic oxides such as alkaline earth oxides as well as rare earth oxides have been found as promising catalyst materials, finding first application in an OCM process by siluria technology. Even though the bond dissociation energy of methane is large with 423 kJ/mol, controlling the activation of oxygen is much more important to direct the selectivity towards the desired products. In case of basic non reducible oxides e.g. CaO the capability of the material to dissociatively activate oxygen appears to be directly linked to the OCM reactivity, making it very likely the bottleneck of the reaction. The strong basicity which is needed for the reaction is also a major weakness of those materials, forming stable hydroxides and carbonates incapable of activating oxygen even at high temperatures. This is not observed with another promising catalyst for the OCM reaction, the $Na_2WO_4/Mn_xO_y/SiO_2$ system, showing a good overall catalytic performance. With Mn_xO_y as an oxygen reservoir this system also offers the possibility of a chemical looping process concept, showing two major prospects: the absence of gas phase oxygen suppressing gas phase reaction and the control of the oxygen species on the catalysts surface by the removal of weakly adsorbed species.[2]

References

(1) Zavyalova, U.; Holena, M.; Schlögl, R.; Baerns, M. Statistical Analysis of Past Catalytic Data on Oxidative Methane Coupling for New Insights into the Composition of High-Performance Catalysts. *ChemCatChem* **2011**, *3* (12), 1935–1947.

(2) Fleischer, V.; Simon, U.; Parishan, S.; Colmenares, M. G.; Görke, O.; Gurlo, A.; Riedel, W.; Thum, L.; Schmidt, J.; Risse, T.; et al. Investigation of the role of the Na2WO4/Mn/SiO 2 catalyst composition in the oxidative coupling of methane by chemical looping experiments. *J. Catal.* **2018**, *360*, 102–117.

551210 Experimental and Theoretical Mechanistic Insights into the Oxidative Coupling of Methane with Soft Oxidants

Tuesday, June 4, 2019: 4:48 PM
Texas Ballroom A (Grand Hyatt San Antonio)

Tobin J. Marks, Department of Chemistry, Northwestern University, Evanston, IL and **Shanfu Liu**, Chemistry, Northwestern University, Evanston, IL

ABSTRACT BODY:
Abstract: The catalytic conversion of methane to high-value molecular building blocks should in principle be possible using abundant soft oxidants that exhibit low thermodynamic driving forces for over-oxidation, thereby increasing the selectivity to valuable C_2 products such as ethylene. Previous studies of methane oxidative coupling using gaseous S_2 (SOCM) demonstrated ethylene selectivities up to 33% and C_2H_4/C_2H_6 ratios of 9-12 over an Fe_3O_4–derived catalyst at 950 °C. Here we report detailed experimental and theoretical mechanistic studies that examine the SOCM process for CH_4 to C_2H_4 over the Fe_3O_4-derived catalyst. Kinetic analyses show that ethylene is produced as a primary product in methane activation and likely proceeds via the coupling of CH_2 intermediates over Fe-S sites on the sulfided Fe_3O_4 surface. Hydrocarbon reactivity with S_2 increases in the order $CH_4 < C_2H_4 < C_2H_2 < C_2H_6$, while CS_2 is found to be unreactive. The reaction rate is found to be first-order in both CH_4 and S_2, with a CH_4/CD_4 kinetic isotope effect of 1.86 ± 0.17. The measured (66±8 kJ/mol) and calculated (44 kJ/mol via dimeric S_2 sites, 94 kJ/mol for monomeric S sites) apparent activation barriers for methane conversion are significantly lower than the previously reported activation energies for nonoxidative CH_4 coupling or for the oxidative coupling of CH_4 with O_2. The present lower barriers are attributed to the strong binding between both methyl and hydrogen in the transition state and the active sulfur on the catalytic surface. The undesirable CS_2 product appears to form predominantly from CH_4 over-oxidation via a series of C-H activation and S-addition steps over adsorbed sulfur sites, whereas over-oxidation of the C_2 products to CS_2 is of minor significance.

543688 Catalyst Development for Autothermal Adiabatic Operation of Oxidative Coupling of Methane

Tuesday, June 4, 2019: 5:12 PM
Texas Ballroom A (Grand Hyatt San Antonio)

Wugeng Liang[1], Luanyi Li[1], Hector Perez[1], Sarsani Sagar[1], Tian Gu[1], David West[1] and Balakotaiah Vemuri[2], (1)Corporate Research & Development, SABIC, Sugar Land, TX, (2)Department of Chemical and Biomolecular Engineering, University of Houston, Houston, TX

Due to the exothermic nature of oxidative coupling of methane (OCM) reaction, conventional fixed bed reactors are not economically feasible for the process. Based on our analysis, the feasible reactor design will be the autothermal adiabatic reactor with feed temperature near ambient temperature. For reactor to operate in this mode, there are special requirements from the catalyst.

High activity is required so that the near ambient temperature operation can be carried out. The catalyst selectivity is always very important for the reaction. In addition, under the adiabatic operation, there will be a temperature rise in the reactor, therefore, it requires the catalyst to have high selectivity across this temperature range, not only at a narrow temperature range; that is, it requires of high selectivity in a broad temperature range. Last, the catalyst should also have good stability.

Catalyst development were based on OCM reaction mechanism. Oxides which have high methyl radical formation rates are used to enhance catalyst activity. On the other hand, oxides which have low methyl radical stick coefficients are used to enhance catalyst selectivity. Special crystal structure is designed to enhance product diffusion, which will benefit catalyst selectivity. The stability of the catalyst comes from the thermal stability of oxides used under the reaction conditions.

New catalysts system developed demonstrated stable performances which meet or beat the requirement for autothermal, near ambient and adiabatic operation.

551530 Crystallization of metal oxides with well-defined morphologies for converting methane to high-value chemicals

Tuesday, June 4, 2019: 5:36 PM
Texas Ballroom A (Grand Hyatt San Antonio)

Mariano Susman, University of Houston, Houston, TX, Hien N. Pham, Department of Chemical and Biological Engineering, University of New Mexico, Albuquerque, NM, University of New Mexico, Albuquerque, NM, Abhaya Datye, Chemical and Biological Engineering, Center for Microengineered Materials, University of New Mexico, Albuquerque, NM, S. Chinta, SABIC Technology Center, Houston, TX and Jeffrey D. Rimer, Chemical and Biomolecular Engineering, University of Houston, Houston, TX

In metal oxide catalysis, the performance of the catalyst often depends on the exposed crystal facets. For challenging industrial processes such as oxidative coupling of methane (OCM), it has been reported that MgO particles exposing a high ratio of {111} surface areas yield higher C2 selectivity than reactions over MgO(100) cubic facets.1 Thus, suitable strategies for maximizing the exposure of specific facets and attaining controlled morphologies of metal oxide crystals for OCM studies are needed. Moreover, high-index facets are often desirable for a broad range of chemical reactions given the large number of undercoordinated metal sites that are available as active sites for catalysis. Here, we will discuss the crystallization of structured metal oxide materials via molten salt syntheses (MSS), where the main driving force(s) and factors that determine particle morphology have largely remained elusive. We show that the formation of polar facets is facilitated in salt media when the metal oxide is generated via a liquid-to-solid reaction (with intermediates in the molten state).2 The presence of residual water and ions impact oxide crystallization in ways that still remain unknown, but are not necessarily governed by adsorption- stabilization processes. Correlations between metal oxide catalyst faceting and corresponding OCM performance will be presented.

(1) Hargreaves, J. S. J.; Hutchings, G.J.; Joyner, R.W.; *Catal. Today* **1990**, 6, 481–488.

(2) Susman, M. D.; Pham, H. N.; Datye, A. K.; Chinta, S.; Rimer, J. D.; *Chem. Mater.* **2018**, 30, 2641–2650.

549437 Low-Temperature One-Step Conversion of Methane to Methanol and Acetic Acid with Supported Rhodium Single Cations Using Molecular Oxygen

Wednesday, June 5, 2019
Texas Ballroom Prefunction Area (Grand Hyatt San Antonio)

Mengwei Li[1], Junjun Shan[2] and Maria Flytzani-Stephanopoulos[1], (1)Department of Chemical and Biological Engineering, Tufts University, Medford, MA, (2)NICE AMERICA RESEARCH, Inc, Mountain View, CA

Low-temperature one-step conversion of methane to methanol and acetic acid with supported rhodium single cations using molecular oxygen

Mengwei Li, Junjun Shan†, Maria Flytzani-Stephanopoulos*

Department of Chemical and Biological Engineering, Tufts University, MA 02155, USA

†Current address: NICE America Research, Inc., Mountain View, California 94043, USA

**Corresponding author: maria.flytzani-stephanopoulos@tufts.edu*

Methane, the main component of natural gas, is the most stable hydrocarbon. The first C-H bond in methane possesses bond energy of ~104 kcal/mol. Due to the recent breakthrough in hydraulic fracturing, there has been renewed interest in methane conversion to valuable liquid oxygenates. The industrial process to synthesize oxygenates from methane involves first converting methane to syngas and then perform methanol synthesis and methanol carbonylation as separate steps. This process is energy intensive and not economical for small scale production. On the other hand, one-step conversion which utilizes oxidation reaction at mild conditions, offers an alternative small-scale process. The potential application of this process to stranded natural gas is highly desirable.

The direct catalytic conversion of methane to methanol and other liquid oxygenates with high yield has long been considered the holy grail of catalysis. The first C-H bond requires either harsh conditions or highly energized species to activate, while the desired product is prone to over oxidation under the same conditions. Recently, extensive studies have been directed to methane-to-oxygenates using heterogeneous catalyst at temperatures below 250 C, including low temperature gas phase reaction [1], electrophilic C-H bond activation in acidic media [2] and aqueous phase conversion using peroxide [3]. However, these systems either require expensive oxidant, e.g. peroxide or sulfuric acid that poses economic and environmental concerns, or the rate is too low to be considered applicable. Here we report that supported rhodium single cations catalyze the selective one-step oxidation of methane to methanol and acetic acid using molecular oxygen as oxidant, carbon monoxide as additive gas, in aqueous solutions under mild conditions (110-150 C, 20-30 bar) [4]. Liquid oxygenates are formed with rate of 3000 μmol/g/h and conversion up to 4% when anchoring rhodium single cations on zeolite. The Brønsted acidity of zeolite is required for the carbonylation of methane to form acetic acid while the direct oxygen insertion route to form methanol can proceed on non-acidic oxide support.

[1] Narsimhan, K., Iyoki, K., Dinh, K., and Román-Leshkov, Y. *ACS Cent. Sic.* 2, 424 (2016).

[2] Soorholtz, M., White, R. J., Zimmermann, T., Titirici, M., Antonietti, M., Palkovits, R. and Schüth, F. *Chem. Commun.* 49, 240 (2013).

[3] Hammond, C., Forde, M. M., and Ab Rahim, M. H., et al. *Angew. Chem. Int. Ed.* 51, 5129 (2012).

[4] Shan, J., Li, M., Allard, L. F., Lee, S. and Flytzani-Stephanopoulos, M. *Nature* 551, 605 (2017).

546097 Direct Gas-Phase Conversion of Methane into Methanol with Water

Wednesday, June 5, 2019
Texas Ballroom Prefunction Area (Grand Hyatt San Antonio)

Jong Kyu Kang[1], Sae Ha Lee[1], Yu-ri Jeong[1] and **Eun Duck Park**[1,2], (1)Department of Energy Systems Research, Ajou University, Suwon, Korea, Republic of (South), (2)Department of Chemical Engineering, Ajou University, Suwon, Korea, Republic of (South)

The direct formation of methanol from methane and steam was confirmed over Cu ion-exchanged zeolites. Furthermore, the continuous production of methanol was accomplished by co-feeding methane and steam over Cu-zeolites. The methanol production rate in this continuous process was comparable to those reported previously in the stepwise process in which activation, methane reaction, and extraction of methanol were carried out separately.

544114 Selective Oxidation of Methane to Oxygenates in Aqueous Hydrogen Peroxide Using Bimetallic Au-Pd Supported By Different Nanostructured Materials

Wednesday, June 5, 2019
Texas Ballroom Prefunction Area (Grand Hyatt San Antonio)
Saeed Alshihri, King Abdulaziz City for Science and Technology, Riyadh, Saudi Arabia

Selective Oxidation of Methane to Oxygenates in Aqueous Hydrogen Peroxide using bimetallic Au-Pd supported by different nanostructured materials

Saeed Al-Shihri[a]*, Christian Richard[b], Hamid Al-Megren[a], David Chadwick[b]

[a] National Petrochemical Technology Center, Materials Science Research Institute, King Abdulaziz City for Science and Technology (KACST), P. O.Box6086,Riyadh11442,SaudiArabia

[b]Imperial College London, Chemical Engineering, London, UNITED KINGDOM

*Corresponding author: alshihri@kacst.edu.sa,

Key words: methane oxidation, Gold-palladium, Nanostructured catalysts, methanol, catalysis.

The direct selective oxidation of methane in aqueous hydrogen peroxide solution has been investigated using four different supported Au/Pd catalysts. The following supports: titanium oxide (anatase), cerium oxide (nanoparticles), titanate nano tubes (TiNT) and ceria nanorods (CNR), were all used and their influence on the product distributions of MeOOH and MeOH evaluated. The catalysts have been characterized by a range of techniques including: XRD, XRF, BET, TEM. The reaction was carried out in a purpose-made glass liner reactor and the liquid was stirred using a glass-coated magnetic bar to minimize the possibility of metal leaching into solution from reactor components. Au-Pd over ceria nanorods (CNR) was found to be the most effective catalyst, and the main product to be formed was formic acid, which is not in agreement with a previous work. In comparison to ZSM-5(50) catalysts (total productivity was 24.4 mol kgcat-1h-1) reported in our previous study [1], the productivity is low which could be due to the high decomposition of H_2O_2 over Au-Pd.

[1] S. Al-Shihri, C.J. Richard, D. Chadwick, ChemCatChem 9 (2017) 1276–1283.

544692 High Efficiency Gas-to-Liquids through Condensation of Small Hydrocarbons Using Corona Discharge Processes

Wednesday, June 5, 2019
Texas Ballroom Prefunction Area (Grand Hyatt San Antonio)

Yu Miao[1], Alexandre Yokochi[1], Goran Jovanovic[2], Nick AuYeung[2], Annette von Jouanne[1], Ryan Collin[1], Ian Reddick[2], Adam Shareghi[2] and Andrew Traverso[2], (1)School of Engineering and Computer Science, Baylor University, Waco, TX, (2)School of Chemical, Biological and Environmental Engineering, Oregon State University, Corvallis, OR

The conversion of small hydrocarbons to longer ones that are liquids at room temperature is the general focus of gas to liquid (GTL) technologies. Recently, the co-authors have been investigating the direct GTL conversion through direct condensation of short hydrocarbons to longer ones using low energy corona discharges.

In this process, low energy electron impact ionization through the use of a non-thermal plasma generated by an atmospheric pressure electric corona discharge is used to generate reactive radicals in the gas phase that subsequently recombine to produce the longer hydrocarbons. By implementing this approach in a micro-chemical reactor we can take advantage of strong electric field gradients and efficient heat and energy management resulting in the stabilization of the corona discharge in a very non-thermal plasma regime.

Both single-discharge and multi-discharge reactors have been built, and their performance demonstrates the chemical conversion concept presented. A brief investigation of important process parameters shows that the power level per discharge, the discharge gap, and the reactive gas flow rate is particularly important. A high energy efficiency of up to 85% (electrical to chemical energy) can be achieved, and methane conversion can reach up to 25% with high selectivity (80~90%) towards C2+ hydrocarbons.

545465 An Automatically Generated Pressure-Dependent PAH Mechanism for Rich-Methane Combustion

Wednesday, June 5, 2019
Texas Ballroom Prefunction Area (Grand Hyatt San Antonio)

Te-Chun Chu[1], Zachary J. Buras[1], Patrick Oßwald[2], Mengjie Liu[1], Mark J. Goldman[1] and William H. Green[1], (1)Department of Chemical Engineering, Massachusetts Institute of Technology, Cambridge, MA, (2)Institute of Combustion Technology, German Aerospace Center (DLR), Stuttgart, Germany

The need for an accurate model for rich-methane combustion relevant to High Temperature Pyrolysis and partial oxidation processes has grown as the supply of natural gas has increased in recent years. In addition to the chemistry of forming dominant products (CO and H_2) or valuable chemicals (acetylene and ethylene), the model must accurately predict formation of undesired incomplete combustion byproducts, e.g. polycyclic aromatic hydrocarbons (PAH) and soot. To meet this need, a detailed chemical mechanism has been generated in this work focusing on high-temperature combustion of rich-methane up to three-ring PAH formation.

The Reaction Mechanism Generator (RMG) software generated a pressure-dependent chemical mechanism for specified conditions to investigate chemistry in rich-methane combustion. Pressure-dependent rate coefficients are estimated by RMG using master equation methods from potential energy surfaces computed from first-principles. The accuracy of the new chemical kinetic model is tested against the experiments done by Köhler et al.,[1] where major species along with intermediates and aromatics were measured by molecular beam mass spectrometry in a flow reactor between 1100-1800 K. The model predictions are in excellent agreement with the experimental results. Major species are predicted within the experimental uncertainties at various conditions. Acetylene, which is over-predicted by existing mechanisms like GRI 3.0[2] and USC-II[3], is predicted more accurately by the new mechanism. The predictions of benzene, indene, naphthalene, and acenaphthylene are highly accurate in terms of both magnitude and temperature dependence, which is ideal for further modeling of soot formation. This new model, which can predict detailed natural gas chemistry of industrial relevance from first-principles calculations, and the process to develop it are valuable resources for the growing natural gas industry.

1. Köhler, M. et al., Chemical Engineering Science 2016, 139, 249-260.

2.Gregory P. Smith et al., GRI-Mech 3.0. http://www.me.berkeley.edu/gri_mech/.

3. Hai Wang, X. Y. et al., USC Mech Version II. High-Temperature Combustion Reaction Model of H2/CO/C1-C4 Compounds. 2007.

545502 Catalytic Conversion of Methane into Oxygenated Compounds over $H_4PMo_{10}WVO_{40}$ Catalyst

Wednesday, June 5, 2019
Texas Ballroom Prefunction Area (Grand Hyatt San Antonio)

Djaouida Allam, Chemistry, University, Laboratoire de Génie Chimique et Chimie Appliquée Hasnaoua I, Un, Algeria

The partial oxidation of methane into oxygenated compounds over $H_4PMo_{10}WVO_{40}$ catalyst was investigated at different reaction temperatures and the catalytic activity was examined. The products obtained were methanol, formaldehyde and carbon oxides. The results obtained have shown that IR, UV, NMR spectroscopy and X-ray diffraction indicate for $H_4PMo_{10}WVO_{40}$, the presence of a single crystalline phase and IR spectra typical of the Keggin anion. After reaction $H_4PMo_{10}WVO_{40}$ were decomposed into P_2O_5, V_2O_5, MoO_3 and WO_3.

Mild oxidation of Methane to C_1-C_2 oxygenates over mononuclear complexes reaction-induced on activated carbon

Siquan Feng[a,c], Bing Li[a,c], Xiangen Song[a], Yunjie Ding[a,b*]

[a]Dalian National Laboratory for Clean Energy and [b]State Key Laboratory of catalysis, Dalian Institute of Chemical Physics, Chinese Academy of Sciences, Dalian, 116023 China
[c]University of Chinese Academy of Sciences, Beijing, 100049, China

Abstract: The precious metal mononuclear complex with its unique traits of 100% atom utilization, simple structure model for mechanism study, functioning as a bridge between the heterogeneous and homogeneous catalysis, have been draw much interest in academic or industry field. However, the problem of deactivation induced by metal agglomeration is always in company with the mononuclear complexes, especially when the mononuclear complexes with high loading were operated under harsh reaction conditions. Therefore, atomically re-dispersion of supported noble metal nanoparticle is of much significance for the preparation and regeneration of the mononuclear complexes. Herein, we show a preparation method for the supported mononuclear complexes via the atomically re-dispersion of their nanoparticles on activated carbon (AC) by the treatment of CO/CH_3I and applications in a mild oxidation of methane to C_1-C_2 oxygenates and the results comprised with the results in literatures as follows:

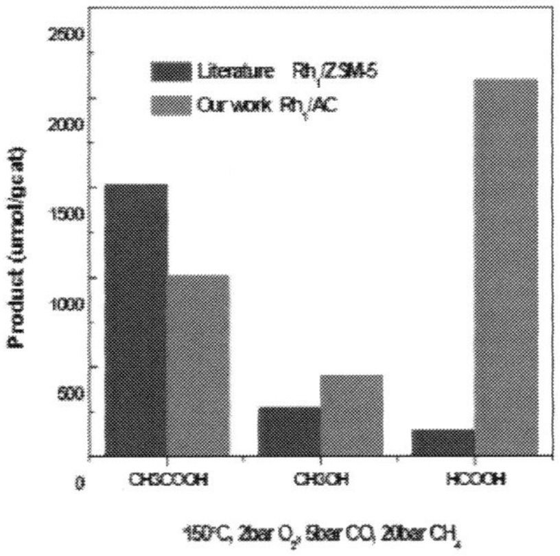

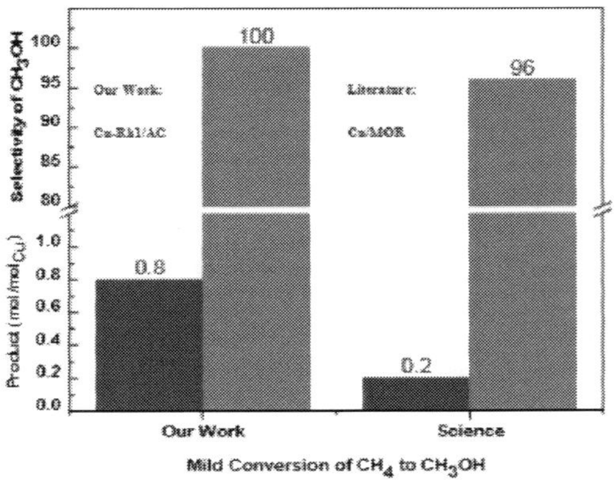

Shan, J., et al. "Mild oxidation of methane to methanol or acetic acid on supported isolated rhodium catalysts." *Nature* 551 (2017):605-608.

Sushkevich, V. L. "Selective Anaerobic Oxidation of Methane Enables Direct Synthesis of Methanol". *Science* 2017, 356, 523-527

545802 Single Rhodium Atoms Anchored in Micropores for Efficient Transformation of Methane to Methanol and Acetic Acid Under Mild Conditions

Wednesday, June 5, 2019
Texas Ballroom Prefunction Area (Grand Hyatt San Antonio)

Yu Tang[1], **Franklin (Feng) Tao**[2] and Yuting Li[1], (1)University of Kansas, Lawrence, KS, (2)Departments of Chemical Engineering and Chemistry, The University of Kansas, Lawrence, KS

Catalytic transformation of CH_4 under a mild condition is significant for efficient utilization of shale gas under the circumstance of switching raw materials of chemical industries to shale gas. Here, we report the transformation of CH_4 to acetic acid and methanol through coupling of CH_4, CO and O_2 on single-site Rh_1O_5 anchored in microporous aluminosilicates in solution at ≤150 °C. The activity of these singly dispersed precious metal sites for production of organic oxygenates can reach about 0.10 acetic acid molecules on a Rh_1O_5 site per second at 150 °C with a selectivity of ~70% for production of acetic acid. It is higher than the activity of free Rh cations by >1000 times. Computational studies suggest that the first C–H bond of CH_4 is activated by Rh_1O_5 anchored on the wall of micropores of ZSM-5; the formed CH_3 then couples with CO and OH, to produce acetic acid over a low activation barrier. This heterogeneous catalytic process opens a new route to synthesize acetic acid through direct utilization of methane under a mild condition at 150 °C or lower by using a low-cost oxidant, O_2 or air instead of current industrial process of synthesizing acetic acid through carbonylation of methanol.

543879 A Study of Selective Methane Chlorination Reaction Using Solid Superacid Metal Oxide Catalysts

Wednesday, June 5, 2019
Texas Ballroom Prefunction Area (Grand Hyatt San Antonio)

Jip Kim[1,2], Youngmin Kim[3], Kyoung-Su Ha[2] and Ho-Jeong Chae[1], (1)Carbon Resources Institute, Korea Research Institute of Chemical Engineering, Daejeon, Korea, Republic of (South), (2)Department of Chemical and Biomolecular Engineering, Sogang University, Seoul, Korea, Republic of (South), (3)Korea Institute of Machinery and Materials, Daejeon, Korea, Republic of (South)

Currently, the petrochemical industry uses naphtha as main raw material. Naphtha has the advantage of being able to produce not only olefin products but also non-olefin products, but it is more expensive than natural gas and coal raw materials. Especially, the domestic petrochemical industry is highly based on basic oil, so it is necessary to reduce the risk through diversification of raw materials. As an alternative plan, C1 gas like methane can be utilized. Methane is a valuable resource which can be used as a basic chemical feedstock or a transportation fuel through chemical conversion processes. And currently, typical and commercialized processes are mostly indirect conversion technologies through synthesis gas, which requires high energy and multi-step process. On the other hand, the direct conversion technology without passing through the syngas has attracted a great deal of attention because it uses low energy and does not require a multi-step process, which is economical. In this study, selective production of methyl chloride was performed by methane chlorination reaction which does not require additional oxygen process. In this study, we prepared solid superacid metal oxide catalysts such as Sulfated Zirconia Oxide, SZ/SBA-15, Combustion SZ, Sulfated Tin Oxide. The prepared catalysts were characterized by XRD, BET, FT-IR, NH_3-TPD and Pyridine-IR analysis. Also, the reaction characteristics of the catalyst were investigated through methane chlorination reaction.

551189 Historical Review and Recommendations on the Oxidative Coupling of Methane

Thursday, June 6, 2019: 10:30 AM
Texas Ballroom D (Grand Hyatt San Antonio)
Anne Gaffney, Idaho National Laboratory, Idaho Falls, ID

The oxidative coupling of methane (OCM) to higher hydrocarbons has been under investigation for over fifty years. Some of the earliest reports were published by major chemical companies, including Exxon Research and Engineering Company [1] and Union Carbide [2]. Compared to non-oxidative routes, the major advantage of using oxygen as a reactant with methane is that the reaction is exothermic and not limited by thermodynamic constraints. Yet, the greatest challenge is to selectively control the carbon-hydrogen bond breaking process leading to the carbon-carbon coupling and to mitigate the free-radical chemistry leading to carbon oxide formation both at the catalyst surface and in the gas phase. This presentation will summarize the advances in OCM technology over the past fifty+ years and high light recent advances that might offer economical and technological advantages.

1. Mitchell, H. L., III, and Waghorne, R. H., U.S. Patents 4,172,810 (1979); 4,205,194 (1980); and 4,239,658 (1980) to Exxon Res. And Eng. Co.
2. Keller, G. E., and Bhasin, M. M., J. Catal. **73**, 9 (1982).

544657 New Perspectives to "Old" Catalysts: Impact of Reactor Operating Mode on the Performance Ranking for Oxidative Coupling of Methane

Thursday, June 6, 2019: 11:18 AM
Texas Ballroom D (Grand Hyatt San Antonio)

Laura Pirro[1], Pedro S.F. Mendes[1], Ana Obradovic[1], Bart D. Vandegehuchte[2], Guy B. Marin[1] and Joris W. Thybaut[1], (1)Laboratory for Chemical Technology, Ghent University, Ghent, Belgium, (2)Total Research and Technology Feluy, Seneffe, Belgium

Industrial-scale implementation of isothermal reactors for Oxidative Coupling of Methane (OCM) is hampered by the high exothermicity of the reaction; on the other hand, decades of lab-scale catalyst screening relied on (pseudo) isothermal data. However, the extrapolation of isothermal performances towards the industrially-relevant adiabatic operation is catalyst specific, to such an extent that the catalyst performance ranking might no longer hold.

In order to gain insight into the changes in OCM performance when proceeding from isothermal to adiabatic operation, over 150 catalysts were screened via microkinetic simulations. The adopted kinetic model [1] is based on 16 catalyst descriptors, it comprises 39 gas-phase and 26 surface reversible reactions and was embedded in a 1D heterogeneous fixed-bed reactor model.

Catalysts were ranked according to C_{2+} yields simulated at isothermal and adiabatic mode and at varying operating conditions. Some active but not very selective catalysts, such as promoted La_2O_3, were ranked lower compared to more selective Na-Mn-W/SiO_2 in isothermal operation. In adiabatic reactors, however, the difference between these catalyst groups was much less pronounced and, in specific conditions, the ranking was even inverted.

The comparison of C_{2+} selectivity as function of CH_4 conversion and Rate of Production Analysis elucidated the origin for the performance diversity among catalysts in the two operating modes. The interplay between two phenomena, namely the increase in maximum achievable selectivity with temperature [2] and the less pronounced adiabatic temperature rise with increasing selectivity [3], is crucial in establishing the ultimate catalyst ranking.

Via incorporation of this descriptor-based microkinetic model into a process simulator, the way is being paved for a catalyst-tailored OCM process design.

[1] Alexiadis V.I. et al., *Appl. Catal. B* 150 (2014) 496.

[2] Sinev M.Y. et al., *J. Nat. Gas Chem.* 18 (2009) 273.

[3] Balakotaiah V., West D.H., *Curr. Opin. Chem. Eng.* 5 (2014) 68.

546222 Process Intensification for Oxidative Coupling of Methane

Thursday, June 6, 2019: 11:42 AM
Texas Ballroom D (Grand Hyatt San Antonio)

Laurien A. Vandewalle[1], Manuel Nunez Manzano[2], Kevin M. Van Geem[1] and **Guy B. Marin**[1], (1)Laboratory for Chemical Technology, Ghent University, Ghent, Belgium, (2)Laboratory for Chemical Technology, Ghent University (Belgium)

Oxidative coupling of methane (OCM) is considered one of the most promising routes to directly convert methane into more valuable hydrocarbons. As a result of an increased methane supply since 2008 (because of the exploitation of shale and stranded gas reserves) and a significantly dropped methane price, the petrochemical industry is actively investigating the OCM process for commercial/industrial application. The uncertain economics related to the tradeoff between conversion and C2 selectivities is an important reason why OCM is currently not industrially applied. In the last decades, numerous studies have focused on developing a viable catalyst that has the potential to improve the low C2 yields. However, so far, the research on catalyst development has not led to any major breakthrough. One of the reasons is that next to the catalyst aspects, reactor design is of crucial importance for OCM. The lack of an appropriate reactor is one of the primary reasons why OCM has not been commercialized at industrial scale.

Previous studies based on bifurcation theory have shown that the key features of an ideal OCM reactor are high thermal backmixing (i.e. high effective thermal conductivity) and low species backmixing (i.e. narrow residence time distributions). Narrow residence time distributions, i.e. plug flow behavior, is necessary to control and maximize the selectivity towards the intermediate products ethane and ethylene. High effective thermal conductivity creates the opportunity to exploit the bifurcation behavior and operate an OCM reactor autothermally, in this way utilizing the reaction heat in the best possible way. Both these characteristics can be obtained in the gas-solid vortex reactor (GSVR) that is studied in this work.

Gas-solid fluidized beds are generally known for their enhanced heat and mass transfer characteristics and are therefore widely used for both reactive and non-reactive processes in the chemical industry. The efficiency of heat and mass transfer in fluidized beds is determined by the relative velocity, the so-called slip velocity, between the gas and the solid phase. As a consequence of the balance between the drag force and the gravitational force, the slip velocity in a conventional gravitational fluidized bed is limited by the terminal free-fall velocity of the particles. Higher gas velocities give rise to the formation of bubbles and slugs, which negatively affect the heat and mass transfer efficiency as the extensive gas bypass decreases the gas-solid contact. If the gas flow rate is increased even more, particle entrainment is unavoidable. Higher gas throughput, more uniform fluidization, higher slip velocities, and hence better heat and mass transfer, can be achieved by using a centrifugal force instead of the gravitational force. A centrifugally fluidized bed thus emerges as an excellent candidate for process intensification. There are two possibilities to achieve a centrifugal fluidized bed: using a rotating fluidized bed (RFB), where the particles are set in motion by rotating the operating vessel itself, or using a gas-solid vortex reactor (GSVR), where the particles are introduced in a swirling flow field of azimuthally injected gas in a static vessel. In a RFB, the tangential and radial velocity components can be varied independently, as the rotational velocity of the vessel and the injected gas flow rate can be controlled in a separate way. The downside of using RFBs however is the use of mechanically moving parts which imply the risk for mechanical abrasion. In a GSVR the fluidizing gas is injected tangentially through multiple gas inlet slots in the cylindrical wall of the fluidization chamber. The swirling gas transfers tangential momentum to the particles, which in turn start rotating and experience an outward centrifugal force. A fluidized state is reached when the radially inward drag force exerted by the gas overcomes the apparent weight of the solids in the centrifugal field. In contrast to the RFB, it is impossible to control the particle velocity components in a GSVR independently. Nevertheless, the GSVR is preferred to the RFB as the absence of moving parts makes the unit more appropriate for scale-up. At this moment, the first reactive GSVR unit is being tested at the Laboratory for Chemical Technology (LCT). Because of its excellent heat transfer characteristics and narrow residence time distributions, this reactor emerges as an ideal technology for OCM.

The reactive GSVR setup at the LCT consists of a cylindrical unit positioned along a vertical axis with eight gas injection slots of 1 mm width, equally distributed over the circumferential wall and tangentially inclined at a 10° angle. A reactor diameter of 80 mm and a reactor length of 15 mm, are defined. The design of the bottom end wall and exhaust profiles was based on preliminary calculations and CFD simulations. The design of the reactor is such that the inlet slots are easy to replace, so that the number and the angle of the inlet slots can be altered for parametric studies etc. For a given gas flow rate, the thickness of the injection slots determines the magnitude of the gas injection velocity, while the injection angle determines its radial and tangential components. The GSVR is currently operated in semi-batch mode. Solid (catalyst) particles are already present in the inner chamber at the beginning of an experiment. Then the hot fluidizing gas is introduced until a pseudo-steady-state swirling gas-particle flow is established in the reactor. Both PIV and infrared measurements are available for validation of CFD results. Preliminary reactive OCM test experiments are being performed as well. The product composition is analyzed using gas chromatography (GC).

In this work, the open-source CFD package OpenFOAM is used to simulate an adiabatic gas-solid vortex reactor using an Euler-Euler approach. A detailed microkinetic model is used, which was developed at the Laboratory for Chemical Technology for different types of catalyst (Sn-Li/MgO, Mn/Na2WO4/SiO2, Sr/La2O3). The model contains 26 elementary reactions on the catalyst surface and 39 gas phase reactions. Cantera is used as mechanism interpreter and linked to a newly developed OpenFOAM solver. The simulation method is validated by comparison with experimental data. The effect of operating conditions and reactor geometry is evaluated. The CFD simulations show that the narrow residence time distributions and efficient heat management in the GSVR can give rise to C2 yields as high as 25 %.

552133 Packed foams: a novel reactor configuration for intensification of the methane steam reforming process

Thursday, June 6, 2019: 12:06 PM
Texas Ballroom D (Grand Hyatt San Antonio)

Riccardo Balzarotti[1], Matteo Ambrosetti[1], Tommaso Selleri[1], **Alessanda Beretta**[1], Gianpiero Groppi[1] and Enrico Tronconi[2], (1)Politecnico di Milano, Dipartimento di Energia, Via La Masa, Milan, Italy, (2)energy, politecnico di milano, milano, Italy

Packed foams: a novel reactor configuration for intensification of the methane steam reforming process

Riccardo Balzarotti[1], Matteo Ambrosetti[1], Tommaso Selleri[1], Alessanda Beretta[1], Gianpiero Groppi[1] and Enrico Tronconi[1,*]

[1] Politecnico di Milano, Dipartimento di Energia, Via La Masa, 34 - 20156 Milano (Italy)

**enrico.tronconi@polimi.it*

Structured catalysts are widely considered a valuable solution for the intensification of non-adiabatic chemical processes, where reactor performance and process scalability are limited by radial heat transfer [1]. The adoption of highly conductive metallic internals has been proposed to enhance heat transfer in fixed bed reactors with respect to conventional packed beds [2]. These metallic structures are available in various shapes (i.e. honeycomb monoliths, open cell foams) and they are usually made catalytically active by washcoating a thin layer of catalytic material onto their surface [3]. Despite the high potential of coated structured supports, some drawbacks such as smaller catalyst inventory with respect to packed beds and issues associated with washcoat adhesion and catalyst loading/unloading have discouraged the application of this technology at the industrial scale [4].

In this work, a novel fixed bed reactor configuration is proposed and tested for the steam reforming of CH_4. The concept consists in filling the void volume of conductive open-cell foams with small catalytic pellets. As demonstrated for exothermic processes [5], the aim of using this new reactor layout is to enhance the radial heat transfer, thanks to the high thermal conductivity of the solid interconnected matrix. In principle, this could be obtained with negligible losses in terms of catalyst inventory with respect to the traditional packed bed configuration.

Experimental tests were performed using a catalyst in the form of alumina egg-shell particles. The catalyst was prepared by incipient wetness impregnation using alumina particles with a diameter equal to 0.6 mm. Rhodium was used as active metal, aiming to a metal content of 0.3 % wt. with respect to the carrier. In addition, a dedicated 2D two-phase model of the packed foam reactor has been developed, assuming: i) a pseudo homogeneous phase comprising both solid catalytic pellets and flowing gas, as traditionally done for packed bed reactor models; ii) a solid phase to account for the additional contribution of the conductive open-cell foam to heat and mass transfer. For the gas-pellet phase we solve material and energy balances including the effects of axial and radial diffusion as well as a reaction term while, on the solid phase, the heat exchange between gas and solid and the conduction through the foam matrix are accounted for. The model incorporates an already validated kinetic scheme for SR and WGS reactions [6].

In the case of the packed foam layout, three different cellular matrices were tested, namely a 12 PPI FeCrAlY foam (Porvair) and two copper foams, with 10 and 40 PPI, respectively (ERG Aerospace Corporation). In the case of FeCrAlY material (labelled as FeCrAl10 in the following), cell size and void fraction were equal to 5.2 mm and 0.92, respectively. In the case of copper foams, the 10 PPI foam (labelled as Cu10 in the following) had similar properties to the FeCrAlY foam (i.e. porosity of 0.91 and cell size of 4.6), while the 40 PPI copper foam (Cu40 in the following) had values of 0.88 and 2 mm for void fraction and cell size, respectively. For all samples, diameter and height of the foam were set at 29 mm and 25 mm. In order to collect temperature profiles in the axial direction, holes were drilled for the thermocouple-wells. Temperature profiles were in fact recorded longitudinally across the catalytic bed in three different radial positions, namely at the centerline, at 8 mm from the center and at the external wall. In the packed foam configuration, the foam void volumes were filled with catalytic particles, as shown in Fig. 1.

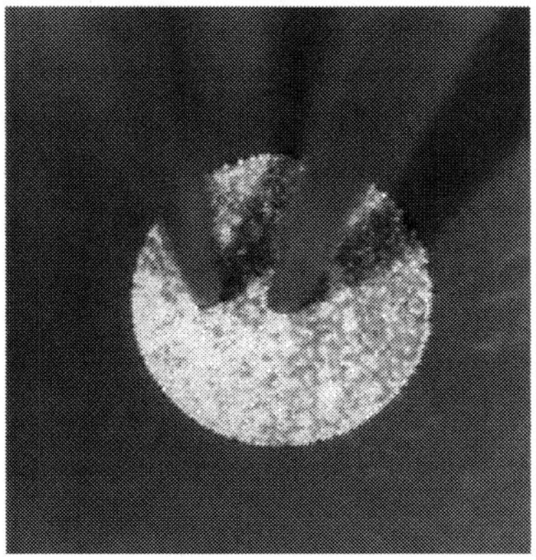

Fig. 1. Close-up view of a packed copper foam into the reactor

For comparison, tests in a conventional packed bed system were also performed. In this case, the same mass of catalytic particles (5.75 g) was loaded in the reactor, mixed with SiC particles (SiC to catalyst weight ratio of 1.54) in order to have the same reactor volume as for the packed foam layout.

Catalytic tests were carried out in a tubular reactor (I.D. = 29.5 mm) externally heated by a furnace; tests were performed in the 600-800 °C range. The feed consisted of a steam/CH_4 mixture with S/C ratio of 3.5. Water was condensed and separated downstream from the reactor. A small N_2 flow was mixed to the dry product stream as an internal standard. The quantification of reaction products was performed using an on line micro-GC (model GCX by Pollution) equipped with MolSieve and Porapack columns connected to TCD detectors.

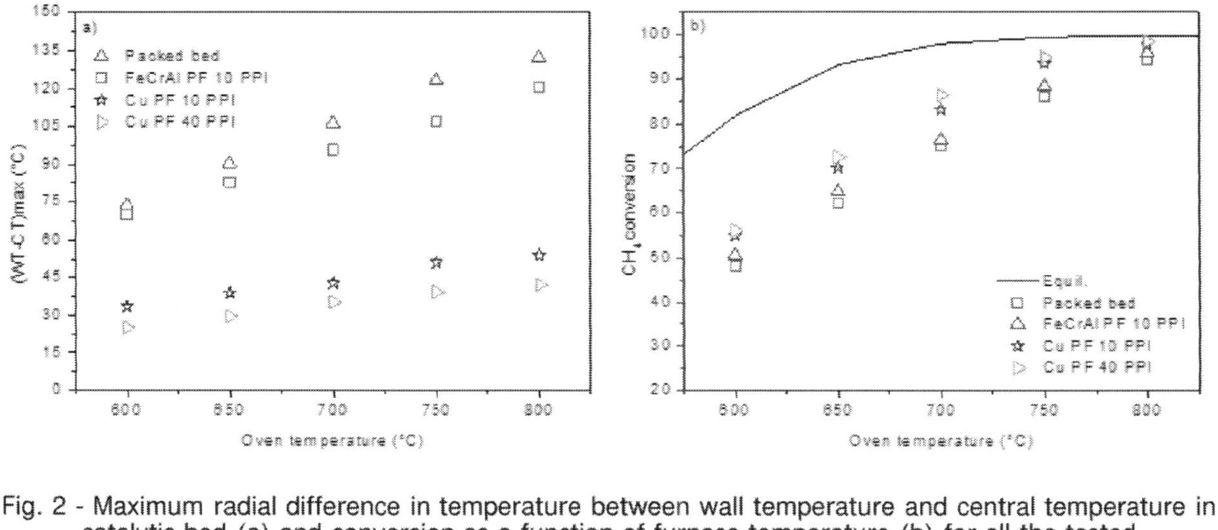

Fig. 2 - Maximum radial difference in temperature between wall temperature and central temperature in catalytic bed (a) and conversion as a function of furnace temperature (b) for all the tested configurations

At the GHSV of 10,000 h^{-1}, the introduction of a conductive matrix resulted in a drop of the maximum temperature difference between the wall and the center of the reactor (Fig. 2-a). Such a decrease was more evident when copper foams were used, thanks to the higher thermal conductivity of copper.

Additionally, an improvement of CH_4 conversion was observed in the packed foam configurations when compared with the packed bed at any furnace temperature (Fig. 2-b). Although small, this improvement is significant, given the irreducible role of thermodynamics at these high temperatures: it can be ascribed to the reduced heat transfer resistances from the reactor wall to the catalyst, which enabled smaller temperature gradients along the radial direction thanks to the conductive heat transfer mechanism favored by the presence of the metallic open cellular structure (Fig. 1).

References

[1] J. Gascon, J. R. van Ommen, J. A. Moulijn, F. Kapteijn, Catal. Sci. Technol. 5 (2015) 807-817

[2] E. Tronconi, G. Groppi, C.G. Visconti, Curr. Opin. Chem. Eng. 5 (2014) 55-67

[3] A. Montebelli, C.G. Visconti, G. Groppi, E. Tronconi, C. Cristiani, C. Ferreira, S. Kohler, Catal. Sci. Technol., 4 (2014) 2846-2870

[4] C.G. Visconti, G. Groppi, E. Tronconi, Cat. Today 273 (2016) 178-186

[5] L. Fratalocchi, C.G. Visconti, G. Groppi, L. Lietti, E. Tronconi, Chem. Eng. J. 349 (2018) 829-837

[6] A. Donazzi, A. Beretta, G. Groppi, P. Forzatti, J. Catal. 255 (2008) 241-258

Acknowledgment: This project has received funding from the European Research Council (694910/INTENT). Co-funding from Microgen30 project (Italian Ministry of Industrial Development) is also acknowledged.

552132 Density Functional Theory Analysis of Propane Dehydrogenation on Palladium Alloys

Monday, June 3, 2019: 10:15 AM
Republic ABC (Grand Hyatt San Antonio)

Ranga Rohit Seemakurthi, Yinan Xu, Brandon C. Bukowski, Fabio H. Ribeiro and Jeffrey Greeley, Charles D. Davidson School of Chemical Engineering, Purdue University, West Lafayette, IN

Alkane dehydrogenation has the potential to convert large amounts of ethane and propane available in shale gas to more valuable alkenes and hydrogen. Bimetallic alloy catalysts have been used for this purpose, as they have been found to achieve higher selectivity than their pure metal counterparts.[1] In addition to the industrially used PtSn catalyst, work by our collaborators have shown that various other Pt, Pd alloys (In, Zn) are also highly selective and active for propane dehydrogenation (PDH).[2,3] This improvement in performance has been attributed to the electronic and geometric effects due to alloying, but the exact mechanistic details remain poorly understood.

In the present study, first-principles periodic density functional theory (DFT) calculations, in conjunction with microkinetic modelling, are used to elucidate the characteristics of nanoparticle alloys that underlie their selectivity and improvement in turnover rates. We focus, in particular, on PdIn alloys, for which the experimental results from our collaborators have shown to have high selectivity to propylene. We additionally investigate reaction pathways and catalyst structures that lead to deactivation and coke formation, and we leverage these insights to point towards electronic and geometric descriptors for catalyst activity, selectivity and stability trends across a wider space of alloy catalysts.

We describe a comprehensive thermodynamic analysis on pure Pd (111) and PdIn (110) alloys, which demonstrates that the binding of all of the reaction intermediates is weaker on the alloy compared to the pure metal. Comparison of desorption energy with the dehydrogenation barrier, the proposed selectivity descriptor on Pt alloys,[4] suggests that the desorption of propylene is far more favorable than dehydrogenation on the PdIn alloy compared to Pd. These results point to an explanation for the measured selectivity of the alloy. To complement the thermodynamic analysis, we also calculate activation barriers for the elementary reactions (C-H and C-C bond breaking), including deep dehydrogenation pathways of propylene. We include these results in a comprehensive microkinetic model, and we use this model to predict likely coverages of spectator coking species on the pure metal and alloy surfaces on both terraces and step defects. We then evaluate the effect of these spectator species on the thermodynamic and kinetic barriers of the reaction mechanism, and we discuss how the results can be used to formulate descriptors for subsequent catalyst screening studies.

References:

1. Sattler, J. J. H. B., Ruiz-Martinez, J., Santillan-Jimenez, E. & Weckhuysen, B.

M. Catalytic Dehydrogenation of Light Alkanes on Metals and Metal Oxides. *Chem. Rev.* 114, 10613–10653 (2014).

2. Wegener, E. C. *et al.* Structure and reactivity of Pt–In intermetallic alloy nanoparticles: Highly selective catalysts for ethane dehydrogenation. *Catalysis Today* 299, 146–153 (2018).
3. Wu, Z. *et al.* Pd–In intermetallic alloy nanoparticles: highly selective ethane dehydrogenation catalysts. *Catalysis Science & Technology* 6, 6965–6976 (2017).
4. Nykänen, L. & Honkala, K. Selectivity in Propene Dehydrogenation on Pt and Pt_3Sn Surfaces from First Principles. *ACS Catal.* 3, 3026–3030 (2013).

550798 Subsurface Effects on Supported 2 Nm Pt_3mn Nanoparticles for Catalytic Propane Dehydrogenation

Monday, June 3, 2019: 10:39 AM
Republic ABC (Grand Hyatt San Antonio)

Zhenwei Wu[1], Brandon C. Bukowski[2], Zhe Li[3], Cory Milligan[1], Lin Zhou[4], Tao Ma[4], Yue Wu[3], Yang Ren[5], Fabio H. Ribeiro[2], W. Nicholas Delgass[1], Jeffrey Greeley[2], Guanghui Zhang[1] and Jeffrey T. Miller[1], (1)Davidson School of Chemical Engineering, Purdue University, West Lafayette, IN, (2)Charles D. Davidson School of Chemical Engineering, Purdue University, West Lafayette, IN, (3)Chemical and Biological Engineering, Iowa State University, Ames, IA, (4)Department of Materials Science, Ames Laboratory, Ames, IA, (5)X-Ray Science Division, Argonne National Laboratory, Argonne, IL

Subsurface Effects on Supported 2 nm Pt_3Mn Nanoparticles for Catalytic Propane Dehydrogenation

Zhenwei Wu[1] Brandon C. Bukowski[1], Zhe Li[2], Cory Milligan[1], Lin Zhou[3], Tao Ma[3], Yue Wu[2], Yang Ren[4], Fabio H. Ribeiro[1], W. Nicholas Delgass[1], Jeffrey Greeley[1], Guanghui Zhang[1, *], Jeffrey T. Miller[1, *]

[1]Davidson School of Chemical Engineering, Purdue University, West Lafayette, IN 47907 (USA)

[2]Department of Chemical and Biological Engineering, Iowa State University, Ames, IA 50011 (USA)

[3]Department of Materials Science, Ames Laboratory, Ames, IA 50012 (USA)

[4] X-Ray Science Division, Argonne National Laboratory, Argonne, IL 60439 (USA)

*gzhang@dlut.edu.cn, mill1194@purdue.edu

Introduction

Supported multimetallic nanoparticles (NPs) are highly tunable by control of their composition and structure. The geometric and electronic effects related to the NP surface have been well studied since the early work of Sinfelt. [1, 2] It has been, however, very difficult to identify any subsurface effects on NPs, even while changing the subsurface metal atoms has been observed to have a large impact on single crystal model catalysts [3]

The challenges to study the effect of subsurface layers on NPs lie in the difficulties in preparing and characterizing atomically precise and high temperature stable NPs with the same surface structure but different subsurface phase. Recently it was shown that stable core-shell NPs can be achieved by intermetallic compound phases which feature precisely ordered arrays of different metals with strong heteroatomic bonds. An intermetallic surface or subsurface maintain ordering even at high temperatures. [4] Herein, we investigate the effect of subsurface composition on supported 2 nm Pt_3Mn NPs for catalytic selectivity of C-H vs C-C bond activation during propane dehydrogenation at 550 ¡C.

Materials and Methods

Catalyst synthesis was conducted through diffusion-controlled Pt_3Mn intermetallic compound NPs formation. The positively charged metal precursors was electrostatically adsorbed onto the deprotonated SiO_2 support surface under controlled pH, leading to uniform and small NP size. The catalyst structure was characterized by *in situ* synchrotron X-ray diffraction (*in situ sXRD*), *in situ* X-ray absorption spectroscopy (*in situ* XAS), STEM imaging and EDS analysis. The catalytic performance was evaluated for light alkane dehydrogenation at 550 ¡C in a plug-flow tube reactor. The heats of CO adsorption were measured experimental on a Micromeritics Autochem 2920 automated catalyst characterization system. All DFT calculations are performed using VASP with PBE exchange-correlation function.

Results and Discussion

Two different Pt-Mn NP catalysts were prepared, namely Pt_3Mn-s (core-shell NPs with Pt_3Mn shell on Pt core), and Pt_3Mn (full-body intermetallic NPs with Pt_3Mn shell on Pt_3Mn core). Formation of partial and full Pt_3Mn phase were confirmed from *in situ* synchrotron XRD by different peak shifts compared to Pt and *in situ* EXAFS fits. Ordering in the Pt_3Mn intermetallic phase is identified through the superlattice diffraction. For Pt_3Mn-s, its Pt core Ð Pt_3Mn shell structure is discerned from a physical mixture of the two phases according to EDS and a difference EXAFS analysis showing that the surface Mn concentration is consistent with Pt_3Mn phase while the core is Pt.

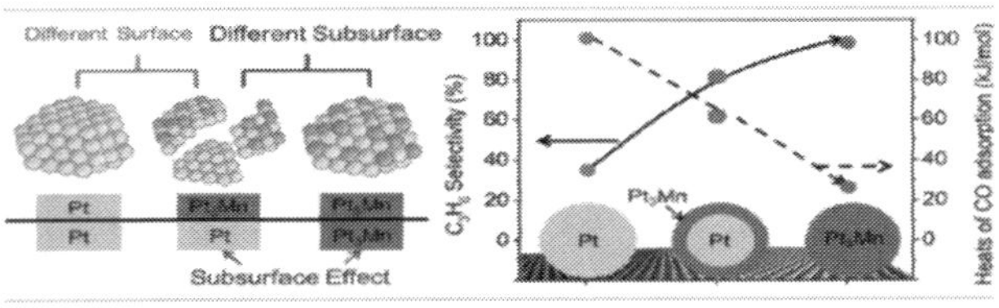

Figure 1. A schematic of the three catalysts investigated: Pt, Pt_3Mn-s and Pt_3Mn, and their propylene selectivity during propane dehydrogenation vs experimental heats of CO adsorption

The effect of NP subsurface composition on selectivity of C-H over C-C bond activation was evaluated for propane dehydrogenation at 550 ¡C. At 20 % conversion with co-feeding hydrogen, the propylene selectivity is improved from 35 % for Pt, to 82 % for Pt Mn-s and further to 98 % for Pt Mn, showing separately the impact of surface phase vs subsurface phase on these supported 2 nm NPs. The improvement is correlated with significant reduction in reactants adsorption strength as shown with the experimental heats of CO adsorption. DFT derived binding energies of CO, H and C_xH_y show consistent trends for the three samples. Calculations on different model structures suggest that the decrease in binding energy is mainly due to the Pt_3Mn phase in the subsurface. Weaker binding of olefins promotes their desorption compared to further dehydrogenation or C-C bond breaking, improving their selectivity.

Significance

This work highlights the importance of the subsurface composition to the surface chemistry for industry relevant supported NP catalysts. The subsurface electronic effect can be precisely controlled via a diffusion limited synthesis of (surface) intermetallic phases with a wide range of possible compositions, promising versatile catalyst design for various catalytic processes. The Pt_3Mn NP catalyzes stable dehydrogenation with high selectivity as well as conversion even at 750 ¡C.

References

1. Bond, G. C.: Supported metal catalysts: some unsolved problems. *Chem. Soc. Rev.* 1991, 20, 441-475

2. Sinfelt, J. H.: Catalysis by alloys and bimetallic clusters. *Acc.Chem.Res* 1977, *10*, 15-20.

3. Greeley, J.; Mavrikakis, M.: Alloy catalysts designed from first principles. *Nat. Mater.* 2004, *3*, 810

4. Gallagher, J. R.; Childers, D. J.; Zhao, H.; Winans, R. E.; Meyer, R. J.; Miller, J. T.: Structural evolution of an intermetallic Pd-Zn catalyst selective for propane dehydrogenation. *Phys. Chem. Chem. Phys.* **2015**, *17*, 28144-28153.

551393 Metal Phosphides for Selective Alkane Dehydrogenation

Monday, June 3, 2019: 11:03 AM
Republic ABC (Grand Hyatt San Antonio)

Jessica A. Muhlenkamp, Jeonghyun Ko, William F. Schneider and Jason C. Hicks, Chemical and Biomolecular Engineering, University of Notre Dame, Notre Dame, IN

Increased shale gas production in the United States has resulted in an abundance of light alkanes, which can be converted to light alkenes via catalytic dehydrogenation and used as building blocks for transportation fuels and chemicals. The challenges of alkane dehydrogenation are (1) to suppress coke formation and (2) to replace current expensive Pt-based catalysts. One particular class of understudied, yet highly tunable catalysts are transition metal phosphides due to their inherent thermal stability and the ability to structurally and electronically tune their catalytic performance for hydrogenation and dehydrogenation reactions.[1,2]

In this contribution, we will discuss recent experimental and computational methods to evaluate a range of metal phosphides for ethane and propane dehydrogenation. To date, we have evaluated a number of metal phosphides supported on SBA-15 for both propane and ethane dehydrogenation using time on stream experiments in a flow reactor. Catalysts have been characterized using XANES, EXAFS, XPS, TEM, N_2 physisorption, as well as other techniques in order to correlate structure with catalytic performance. We have determined that P addition vastly improved the alkene selectivity from Ni to Ni-P, going from 0% to > 99% selectivity for propane dehydrogenation. The P is most likely acting in a manner similar to Sn in current industrial Pt-Sn dehydrogenation catalysts, helping to prevent the formation of large Ni ensembles which favor side reactions such as hydrogenolysis. In order to better understand the effects of P addition we employed density functional theory (DFT) calculations to compare the reactivity for dehydrogenation of Ni_2P with that of Ni. Firstly, we compared adsorption energies and site preferences of C1 and C2 hydrocarbon species on Ni(111) and Ni_2P(001). They generally exhibit weaker adsorption strength with Ni_2P compared to Ni. Certain species, however, tend to favor binding at P sites over Ni sites, providing further stabilization. We then predicted ethylene selectivity by comparison of the activation barrier for ethylene dehydrogenation and ethylene desorption energy on both surfaces. While both surfaces bind ethylene nearly isoenergetically, ethylene dehydrogenation to vinyl is considerably more activated on Ni than on Ni_2P. These results imply a higher selectivity towards ethylene on Ni_2P over Ni.

As a means to enhance the stability of Ni-P catalysts, we evaluated the effects of Cs doping on the catalyst performance. For ethane dehydrogenation, Cs-Ni-P improved the ethylene selectivity over Ni-P, going from 67% for Ni-P to >99% for Cs-Ni-P due to the reduced acidity of the catalyst which also reduces side reactions.[3] We have also evaluated the performance of other mono- and bimetallic transition metal phosphides for dehydrogenation reactions. When comparing Mo-P and Ni-Mo-P for propane dehydrogenation, we found that Ni-Mo-P displayed a higher propylene selectivity than Mo-P, >99% and 97%, respectively. Both Fe-P and Ni-Fe-P showed selectivity to propylene of >99% as well. Using a first order deactivation model, the stability of the bimetallic materials was significantly higher than the monometallic catalysts. The observed enhancement is most likely due to both structural and electronic differences that result from the incorporation of Ni.

Overall, our results provide evidence that a variety of metal phosphides are selective dehydrogenation catalysts, and that bimetallic compositions can further enhance the alkene selectivity and stability of metal phosphides. This presentation will highlight our current efforts in identifying new catalyst compositions for light alkane dehydrogenation reactions.

References

(1) S. T. Oyama and T. Gott et al., Catal. Today. 2009, 143, 94–107. (2) Y. Xu and H. Sang et al., Applied Surface Science. 2014, 316, 163-170. (3) Sattler, J.J.H.B et al., Chem. Rev. 2014, 114, 10613-10653.

546501 Regenerable Single Atom Catalysts for Propane Dehydrogenation

Monday, June 3, 2019: 11:27 AM
Republic ABC (Grand Hyatt San Antonio)

Griffin A. Canning[1], Ryan Alcala[1], Haifeng Xiong[1], Megan Senn[1], Nicole LiBretto[2], Jeffrey T. Miller[2], Fabio H. Ribero[2] and Abhaya Datye[1], (1)Chemical and Biological Engineering, Center for Microengineered Materials, University of New Mexico, Albuquerque, NM, (2)Charles D. Davidson School of Chemical Engineering, Purdue University, West Lafayette, IN

Single atom catalysts are currently of great interest due to high selectivity, reactivity and excellent atom efficiency. Here we have explored the use of single atom catalysts for high temperature reactions involving alkane conversions. The work is motivated by the availability of natural gas due to advancements in fracking and drilling technologies. The increased production has led to lower costs, which are beneficial in lowering the carbon footprint of fuels when methane is used to replace coal for electricity production. Ethane and other natural gas liquids (NGLs) recovered from shale with methane, however, have limited markets resulting in export of US produced ethane to Europe starting in 2016. These NGLs are attractive as feedstocks for chemical production. The conversion of these NGLs to value added products involves dehydrogenation as the first step. This has traditionally been done through steam cracking or catalytic dehydrogenation, both of which require large plants. Platinum based catalysts require regeneration which requires significant catalyst handling and the use of corrosive, halogen based treatments. The use of single atom catalysts provides an alternative approach which also alleviates the problem of coke formation which is less likely on single sites.

In previous work we have demonstrated the ease of synthesis of single atom Pt catalysts via vapor phase synthesis on industrial catalyst supports[1]. These single atom catalysts are regenerable using oxidizing conditions at modest temperatures[2]. The focus of the present work is the use of single atom dopants as modifiers to improve the selectivity and stability of the single atom catalysts for alkane conversions. Characterization via Aberration Corrected STEM and X-ray absorption spectroscopy helps us advance the applications of SACs to industrially relevant reactions.

(1) Jones, J.; Xiong, H.; DeLaRiva, A. T.; Peterson, E. J.; Pham, H.; Challa, S. R.; Qi, G.; Oh, S.; Wiebenga, M. H.; Hernández, X. I. P.; et al. Thermally Stable Single-Atom Platinum-on-Ceria Catalysts via Atom Trapping. *Science (80-.).* **2016**, *353* (6295), 150–154.

(2) Xiong, H.; Lin, S.; Goetze, J.; Pletcher, P.; Guo, H.; Kovarik, L.; Artyushkova, K.; Weckhuysen, B. M.; Datye, A. K. Thermally Stable and Regenerable Platinum–Tin Clusters for Propane Dehydrogenation Prepared by Atom Trapping on Ceria. *Angew. Chemie - Int. Ed.* **2017**, *56* (31), 8986–8991.

546681 Improved Stability of Ptsn for Propane Dehydrogenation By Boron-Modified ZrO_2 Support

Monday, June 3, 2019: 11:51 AM
Republic ABC (Grand Hyatt San Antonio)

Zhonghai Ji, **Xiulian Pan** and **Xinhe Bao**, Dalian Institute of Chemical Physics, Chinese Academy of Sciences, Dalian, China

Improved stability of PtSn for propane dehydrogenation by boron-modified ZrO_2 support

Zhonghai Ji, Xiulian Pan, Xinhe Bao,

State Key Laboratory of Catalysis, Dalian Institute of Chemical Physics, Chinese Academy of Sciences, Dalian 116023, People's Republic of China. Email: panxl@dicp.ac.cn; xhbao@dicp.ac.cn

PtSn based catalysts are widely used in propane dehydrogenation (PDH)[1-5]. Since it involves noble metal, it is highly desirable to reduce the usage of Pt without degrading the catalytic performance. However, the catalyst stability is apt to deterioration with decreasing Pt content [1]. In this contribution, we demonstrate that the B modified ZrO_2 as a support can enhance the stability of PtSn catalyst.

Boron modified ZrO_2 (B-ZrO_2) was prepared by co-precipitation. PtSn/B-ZrO_2 and PtSn/Al_2O_3 was prepared by the incipient-wetness impregnation method. Reaction was conducted under nonoxidative conditions at 550 ^oC, weight hourly space velocity (WHSV) of propane 3 h^{-1}. Results in Fig. 1 show that the initial propane conversion is 32.4% for PtSn/ZrO_2, 36.0% for PtSn/B-ZrO_2 and 42.2% for PtSn/Al_2O_3. However, the propylene selectivity is lowest on PtSn/Al_2O_3. The deactivation rate constant (k_d) is 0.08 h^{-1} over PtSn/ZrO_2 and 0.04 h^{-1} over PtSn/Al_2O_3 whereas it is only 0.01 h^{-1} over PtSn/B-ZrO_2. The stability of PtSn/B-ZrO_2 is obviously better than that of 0.35wt% PtSn/Al_2O_3 and PtSn/ZrO_2. Characterization by NH_3-TPD and pyridine IR shows that Lewis acid is present on B-ZrO_2. XPS indicates that the presence of even a small amount of boron stabilizes the oxidation state of Sn, which may play an important role. More detailed discussions will be given in the presentation.

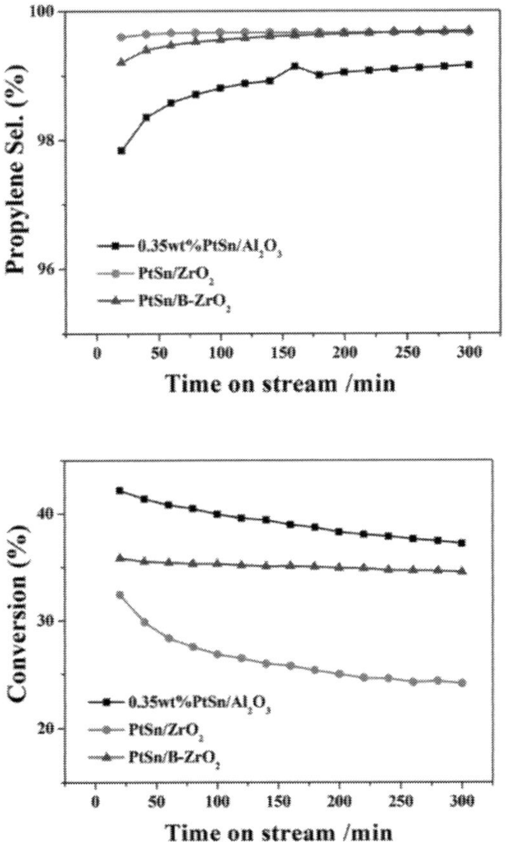

Fig. 1 Catalytic performance in propane dehydrogenation on various catalysts.

References

[1] X. Liu, W.-Z. Lang, L.-L. Long, C.-L. Hu, L.-F. Chu, Y.-J. Guo, Chem. Eng. J. 247 (2014). 183-192.

[2] J. Li, J. Li, Z. Zhao, X. Fan, J. Liu, Y. Wei, A. Duan, Z. Xie, Q. Liu, J. Catal., 352 (2017). 361-370.

[3] A. Iglesias-Juez, A.M. Beale, K. Maaijen, T.C. Weng, P. Glatzel, B.M. Weckhuysen, J. Catal., 276 (2010). 268-279.

[4] G.-Q. Ren, G.-X. Pei, Y.-J. Ren, K.-P. Liu, Z.-Q. Chen, J.-Y. Yang, Y. Su, X.-Y. Liu, W.-Z. Li, T. Zhang, J. Catal., 366 (2018).115-126.

[5 L. Shi, G.-M. Deng, W.-C. Li, S. Miao, Q.-N. Wang, W.-P. Zhang, A.-H. Lu, Angew. Chem.

Int. Ed.ion, 54 (2015). 13994-13998.

545962 Boron-Based Materials As Highly Selective Catalysts for the Oxidative Dehydrogenation of Alkanes

Monday, June 3, 2019: 1:30 PM
Texas Ballroom A (Grand Hyatt San Antonio)

Ive Hermans, Department of Chemical and Biological Engineering, University of Wisconsin-Madison, Madison, WI

The oxidative dehydrogenation of propane (ODHP) as a method of "on-purpose" propylene production has been broadly studied as an alternative to cracking and dehydrogenation.[1] However, even after decades of research establishing supported vanadium oxide as the state-of-the-art catalyst for ODHP,[2,3] selectivity to the olefin product remains too low to be commercially attractive because of facile over-oxidation of propylene into CO and CO_2 (CO_x). During this presentation I will bring the story of how we discovered that boron and boron containing materials (like BN, B_4C, NiB, amongst others) are highly selective catalysts for the ODHP reaction.[6,7] Kinetic and spectroscopic insights will be combined to shine light on potential reaction mechanisms.[8,9] FTIR, Raman, XPS, XAS and MAS NMR spectroscopy all indicate that the surface of all tested boron-containing materials becomes partially oxidized, forming an active layer of BO_xH_y species. The reactivity of this surface layer is likely influenced by the properties of the bulk material, opening new possibilities for future catalyst development.

References

1. T. Degnan, *Focus on Catalysts***2016**, *2*, 1.
2. C. A. Carrero, R. Schlogl, I. Wachs, R. Schomaecker, *ACS Catal.***2014**, 4, 3357.
3. J. T. Grant, C. A. Carrero, A. M. Love, R. Verel, I. Hermans, *ACS Catal.***2015**,*5*, 5787.
4. A. M. Love, C. A. Carrero, A, Chieregato, J. T. Grant, S. Conrad, R. Verel, I. Hermans,*Chem. Mater.* **2016**, *28*(8), 5495.
5. J. T Grant, C. A. Carrero, A. M. Love, R. Verel, I. Hermans,*ACS Catalysis***2015**, *5*(10), 5787.
6. J. Grant, C. Carrero, F. Goeltl, J. Venegas, P. Mueller, S. Burt, S. Specht, W. McDermott, A. Chieregato, I. Hermans, *Science***1016**, *354*, 1570.
7. J. Grant, C. Carrero, J. Venegas, A. Chieregato, I. Hermans, US Patent Application 15/260,649.
8. J. M. Venegas, J. T. Grant, W. P. McDermott, S. P. Burt, J. Micka, C. A. Carrero, I. Hermans, *ChemCatChem***2017**, (9), 2118-2127.
9. J. T. Grant, W. P. McDermott, J. M. Venegas, S. P. Burt, J. Micka, S. P. Phivilay, C. A. Carrero, I. Hermans, *ChemCatChem***2017**, (9), 3623-3626.

545556 Mixed Oxide Redox Catalysts for Chemical Looping Oxidative Dehydrogenation of Light Alkanes

Monday, June 3, 2019: 1:54 PM
Texas Ballroom A (Grand Hyatt San Antonio)

Fanxing Li, Department of Chemical and Biomolecular Engineering, North Carolina State University, Raleigh, NC

The supplies of light alkanes, e.g. ethane and propane, have increased considerably over the recent years resulting from the shale gas revolution. Although there are well-established commercial processes for alkane conversions, they tend to be capital and energy intensive due to the endothermicity of cracking or dehydrogenation reactions, complexity in product separations, and/or energy consumptions for cryogenic air separation. The complexity of these conventional processes also limit their applications to convert distributed shale gas resources. Therefore, novel approaches that can intensify light alkane conversion processes while reducing the energy consumptions and carbon emissions are highly desirable.

This presentation discusses a chemical looping – oxidative dehydrogenation (CL-ODH) scheme which can simplify the conversion of light alkanes to olefins. The chemical looping process utilizes an oxygen carrier, also known as redox catalysts, to convert shale gas components via an indirect reduction-oxidation (redox) scheme: the redox catalyst first releases its lattice oxygen to convert, e.g. ethane into ethylene; the oxygen depleted redox catalyst is then regenerated with air while releasing heat. The CL-ODH allows intensified, auto-thermal operation without the needs for air separation. As such, over 80% reduction in CO_2 emission can be realized. In this presentation, the design principles for the redox catalysts and the underlying mechanisms for ethane CL-ODH are discussed. Results from second law analysis will also be presented to illustrate the benefits of CL-ODH relative to conventional approaches. Examples of CL-ODH for the conversions of C3+ alkanes will also be provided.

549459 Stability of Supported Metal Oxide Catalysts for the Non-Oxidative Dehydrogenation of Propane: An *in Situ/Operando* Raman Study

Monday, June 3, 2019: 2:18 PM
Texas Ballroom A (Grand Hyatt San Antonio)
Carlos A. Carrero, Department of Chemical Engineering, Auburn University, Auburn, AL

Stability of supported metal oxide catalysts for the non-oxidative dehydrogenation of propane: An *in situ/operando* Raman Study

*Jorge Moncada, Lindsey Parsons, Carlos A. Carrero**

Department of Chemical Engineering, Auburn University, Auburn, Alabama 36849, United States

* cac0134@auburn.edu

In this study, we investigate the activity, selectivity, and stability of VO_x/Al_2O_3 and $(VO_x)_m$-$(M_2O_y)_n/Al_2O_3$ (M_2 = Mn, Mo, Zr, Ta) catalysts in the non-oxidative dehydrogenation of propane. We will highlight the promoting effect of M_2O_y for tuning the type and amount of carbon produced during the catalyst's deactivation process. Among the current on-purpose processes to produce light olefins, either platinum or chromium supported catalyst are the most used primarily because of their significant activity and selectivity [1]. However, both the high price and environmental restrictions demand the discovery of alternative catalytic materials to produce light olefins [2]. Both vanadium and molybdenum oxide supported catalysts have already shown promising results in the dehydrogenation of light olefins, [1, 3, 4] although improvements are still needed. The main challenge to overcome in the non-oxidative dehydrogenation of light hydrocarbons is the catalyst's stability during both the reaction and the regeneration process. The chemical nature of the carbon formed during the catalyst's deactivation is directly related to the thermal stability of such carbonaceous species and, therefore, to the regeneration process [5]. *In situ/operando* Raman spectroscopy is a suitable technique to identify and monitor changes of the type of carbon produced during catalytic processes [5]. Our preliminary results indicate that basic metals oxides, such as manganese, promotes the formation of principally amorphous carbon while acidic promoters lead to more graphitic carbon. By monitoring the combustion products (CO and CO_2) using *operando* Raman spectroscopy and Raman-spectrokinetics [6], we aim to gain further insights both in the thermal stability and the chemical nature of the carbonaceous species either deposited on the catalyst surface or reacting with the catalytic component(s) forming, for instance, metal carbides.

References

[1] Liu G, Zhao Z-J, Wu T, Zeng L, Gong J. Nature of the Active Sites of VOx/Al2O3

Catalysts for Propane Dehydrogenation. ACS Catalysis 2016; 6 (8):5207-14.

[2] Sattler JJ, Ruiz-Martinez J, Santillan-Jimenez E, Weckhuysen BM. Catalytic dehydrogenation of light alkanes on metals and metal oxides. Chem Rev 2014; 114 (20):10613-53.

[3] Gong J, Zhao Z, Wu T, Xiong C, Sun G, Mu R, et al. Hydroxyl-mediated Non-oxidative Propane Dehydrogenation over VO_x/Al_2O_3 Catalysts with Promoted Stability. Angew Chem Int Ed Engl 2018;Advance online publication.

[4] Zhao H, Song H, Chou L, Zhao J, Yang J, Yan L. Insight into the structure and molybdenum species in mesoporous molybdena–alumina catalysts for isobutane dehydrogenation. Catalysis Science & Technology 2017; 7 (15):3258-67.

[5] An H, Zhang F, Guan Z, Liu X, Fan F, Li C. Investigating the Coke Formation Mechanism of H-ZSM-5 during Methanol Dehydration Using Operando UV–Raman Spectroscopy. ACS Catalysis 2018; 8 (10):9207-15.

[6] Moncada J, Adams W, Thakur R, Julin M, Carrero C. A, Developing a Raman-spectrokinetic approach to gain insights on the structure-reactivity relationship of supported metal oxide catalysts, ACS Catalysis; 2018; 8, 8976

545752 Supported Liquid Metals - a New Class of Robust Dehydrogenation Catalysts

Monday, June 3, 2019: 2:42 PM
Texas Ballroom A (Grand Hyatt San Antonio)

Nicola Taccardi[1], Mathias Grabau[2], Jonas Debuschewitz[1], Monica Distaso[3], Marco Brandl[4], Rainer Hock[4], Florian Maier[2], Christian Papp[2], Jannis Erhard[5], Christian Neiss[5], Wolfgang Peukert[3], Andreas Görling[5], Moritz Wolf[1], Marco Haumann[1], Hans-Peter Steinrück[2] and Peter Wasserscheid[1,6], (1)Institute of Chemical Reaction Engineering, University of Erlangen-Nuremberg, Erlangen, Germany, (2)Institute of Physical Chemistry II, University of Erlangen-Nuremberg, Erlangen, Germany, (3)Institute of Particle Technology, University of Erlangen-Nuremberg, Erlangen, Germany, (4)Institute of Crystallography and Structural Physics, University of Erlangen-Nuremberg, Erlangen, Germany, (5)Institute of Theoretical Chemistry, University of Erlangen-Nuremberg, Erlangen, Germany, (6)Helmholtz Institute Erlangen-Nürnberg for Renewable Energy, Erlangen, Germany

Our contribution demonstrates that Supported Catalytically Active Liquid Metal Solutions (SCALMS) offer the potential for new and unusual catalytic reactivity in propane dehydrogenation. SCALMS are composed of catalytically active liquid alloy droplets on a porous support.[1] In contrast to conventional supported liquid phase catalysis, the catalytic reaction in SCALMS occurs only at the highly dynamic liquid metal/gas interface, as the liquid metal does not provide any relevant reactant solubility.

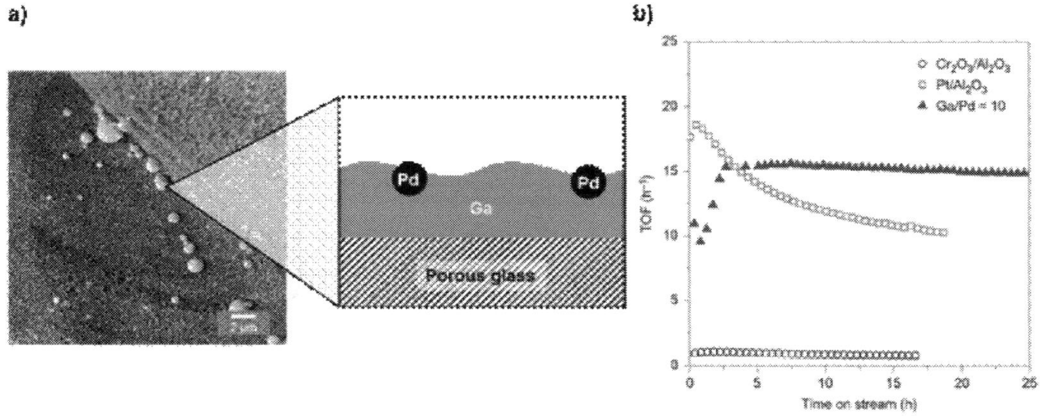

Figure 1. a) SEM image of Pd-Ga decorated porous glass and schematic representation of the microscopic nature of the material. b) Catalytic performance of standard heterogeneous and PdGa SCALMS catalysts in butane dehydrogenation.

Recently, we reported the use of Ga-rich Ga/Pd mixtures (Ga/Pd ratio> 10) on porous glass in the highly endothermal n-butane dehydrogenation [1]. The liquid nature of the supported alloy droplet under the reaction conditions was confirmed through a combination of XRD, SEM, XPS, and ab initio dynamics calculations. Most remarkably, these SCALMS materials outperformed commercial dehydrogenation catalysts (Pt/Al_2O_3, Cr_2O_3/Al_2O_3) without any material or process optimisation. The most remarkable finding, however, was that coking, the typical deactivation mechanism for this type of high temperature hydrocarbon chemistry under reductive conditions [2-4] is largely suppressed with SCALMS systems. As explanation for this unexpected behaviour our molecular dynamics (MD) calculations suggested that the topmost layer of the Ga-rich Ga/Pd alloy is depleted in Pd, but the layer directly underneath is enriched.

[1] N. Taccardi, M. Grabau, et int., C. Papp, W. Peukert, A. Görling, H.-P. Steinrück, P. Wasserscheid, *Nat. Chem.* 9 (2017) 862-867.

[2] J.R. Rostrup-Nielsen, *J. Catal.* 1984, 85, 31.

[3] V.L. Kuznetsov, A.N. Usol'tseva, Y.V. Butenko, *Kin. Catal.* 2003, 44, 726.

[4] J.J.H.B. Sattler, J. Ruiz-Martinez, E. Santillan-Jimenez, B.M. Weckhuysen, *Chem. Rev.* 2014, 114(20), 10613.

546000 New Insight into Activation of Light Paraffin: A Case Study of Electrochemical Deprotonation of Ethane at Reduced Temperatures

Monday, June 3, 2019: 3:06 PM
Texas Ballroom A (Grand Hyatt San Antonio)

Dong Ding, Wei Wu and Lucun Wang, Biological and Chemical Processing, Idaho National Laboratory, Idaho Falls, ID

The shale gas revolution in the United States in recent years has resulted in an oversupply of methane and ethane, the major components of natural gas and natural gas liquids, respectively, and this trend will continue for the foreseeable future. Due largely to these cheap resources, the chemical industry has moved manufacturing plants back to the United States, creating thousands of jobs, while enjoying healthy margins compared to international rivals that typically use more expensive oil-based naphtha as feedstock. However, the ethylene production from ethane or naphtha is energy intensive and represents the single most energy consuming process in the chemical industry. To fully exploit the potential of ethane as a feedstock, low-thermal-budget and low-carbon footprint processes are required. We demonstrated that co-production of ethylene and hydrogen through an electrochemical deprotonation process at intermediate temperatures with significant reduction in process energy and CO_2 emission, compared to the commercial ethane steam cracking process. If the heating value of produced hydrogen is taken into account, the electrochemical deprotonation process actually has a net gain in processing energy.

545659 Dow Fluid Catalytic Dehydrogenation for on-Purpose Propylene Production

Monday, June 3, 2019: 4:00 PM
Texas Ballroom A (Grand Hyatt San Antonio)
Lin Luo and Matt Pretz, Hydrocarbons R&D, The Dow Chemical Company, Freeport, TX

Shale gas produced an imbalance in propylene supply and demand. This has driven the implementation of several projects using available propane dehydrogenation commercial technology. These technologies were adapted from existing processes. These adapted technologies are not tailored to efficiently manage the inherent challenges of catalyst activity, heat input, reaction equilibrium and unselective thermal reactions. Therefore, the Dow Chemical Company has developed a Fluid Catalytic Dehydrogenation Process that is designed from the standpoint of an operating company to efficiently solve the problems inherent in catalytic dehydrogenation processes.

The new technology is a circulating fluid bed technology based on existing Fluid Catalytic Cracking (FCC) technology. The Dow process utilizes a proprietary catalyst with a proprietary reactor and regenerator design to achieve 45% propane conversion at approximately 93 mol% selectivity to propylene. The simplicity of the system enables a capital savings greater than 20% versus other commercial offerings. In addition, the higher conversion at moderate pressure reduces the energy requirement per pound of propylene.

This new technology can be fully integrated with existing ethylene crackers in parallel with new or existing furnaces. In this way, a producer can increase production in an existing light hydrocarbon facility or build a new facility with exactly the amount of ethylene and propylene product that is desired based on ethane and propane feeds. Dow is currently planning a project to implement this technology in an existing cracker at commercial scale and would license the technology upon request.

545405 Deciphering the Reaction Network of the Selective Oxidation of Butane to Maleic Anhydride over $(V_{1-x}W_x)OPO_4$ Catalysts

Monday, June 3, 2019: 4:24 PM
Texas Ballroom A (Grand Hyatt San Antonio)

Jingxiu Xie[1], Felix Pohl[1], Christian Schulz[1], Knut Wittich[2], Ralph Kraehnert[1], Benjamin Frank[1], Robert Glaum[2] and Frank Rosowski[1,3], (1)BasCat – Unicat BASF JointLab, Berlin, Germany, (2)Institut für Anorganische Chemie, Rheinische Friedrich-Wilhelms-Universität, Bonn, Germany, (3)Process Research and Chemical Engineering, BASF SE, Ludwigshafen, Germany

Selective oxidation of lower alkanes (ethane, propane, and *n*-butane) converts these less reactive molecules to valuable chemicals in a resource-efficient and environmentally friendly approach. The selective oxidation of *n*-butane to maleic anhydride (MAN) is an important industrial process with a worldwide capacity of more than 1 million metric tons per year. The industrial vanadyl pyrophosphate (VPP) catalyst reaches a MAN yield of ~60% and research thus far focused on adding electronic promoters to improve its catalytic performance. To investigate potential catalysts beyond the extensively studied and optimized VPP, novel classes of catalysts with different crystal structures to VPP need to be designed.

We recently identified a new class of catalysts based on the tungsten-doped α_{II} $VOPO_4$-structure. Single phase microcrystalline powders of $(V_{1-x}W_x)OPO_4$ are accessible by solution combustion synthesis . X-ray powder diffraction before and after catalytic testing confirmed that these catalysts were phase-stable during catalytic testing ($C_4H_{10}/O_2/H_2O$ = 2/20/3 vol-%, GHSV = 2000 h ^{1}, TOS = 1000 h). By variation of the V/W ratio the $V_{0.8}W_{0.2}OPO_4$ was found to be most active and selective towards MAN. A parameters field study with $V_{0.8}W_{0.2}OPO_4$ was subsequently carried out to obtain mechanistic insights and its reaction network was compared with those of highly selective reference VPP and non-selective V_2O_5, respectively. Parameters variation included feed concentrations (*n* butane and oxygen), gas hourly space velocity, temperature, and co-feeds (ethene, CO, and H_2O).

For all catalysts, MAN, CO, and CO_2 selectivities totaled more than 95% but the product distributions were different, as shown in Figure 1a. At 420 °C, 1 bar, $C_4H_{10}/O_2/H_2O$ = 2/20/3 vol-%, GHSV = 2000 h^{-1}, the selective VPP showed 62% MAN selectivity at 95% *n*-butane conversion and CO/CO_2 = 1.2. Instead, non-selective V_2O_5 showed negligible MAN selectivity at 22% *n*-butane conversion and CO/CO_2 = 2.1. $V_{0.8}W_{0.2}OPO_4$ displayed similarities to both VPP and V_2O_5, showing 18% MAN selectivity at 32% *n*-butane conversion and CO/CO_2 = 3. The high CO/CO_2 ratio of $V_{0.8}W_{0.2}OPO_4$ suggests that CO_x is formed by primary oxidation of *n*-butane instead of secondary oxidation of MAN (Figure 1b). Minor products also showed different trends for the catalysts, suggesting that $(V_{1-x}W_x)OPO_4$ and VPP have different reaction networks. For instance, acrylic acid is a significant byproduct for VPP but not for $V_{0.8}W_{0.2}OPO_4$. To confirm reaction pathways identified by steady state parameter field studies for $V_{0.8}W_{0.2}OPO_4$ and VPP catalysts, transient pulse experiments were carried out at 420 °C, 1 bar, $C_4H_{10}/O_2/H_2O$ = 2/20/3 vol-%, GHSV = 2000 h^{-1}. Important minor products including furan, 1-butene, acetaldehyde, acrylic acid, acetic acid were individually pulsed through the reactor. The identified reaction pathways give a deeper insight into the reaction network of each catalyst. The comparison of reaction networks of $V_{0.8}W_{0.2}OPO_4$ with selective VPP and non-selective V_2O_5 catalysts is expected to lead to further understanding towards catalysts for the selective oxidation of *n*-butane to MAN.

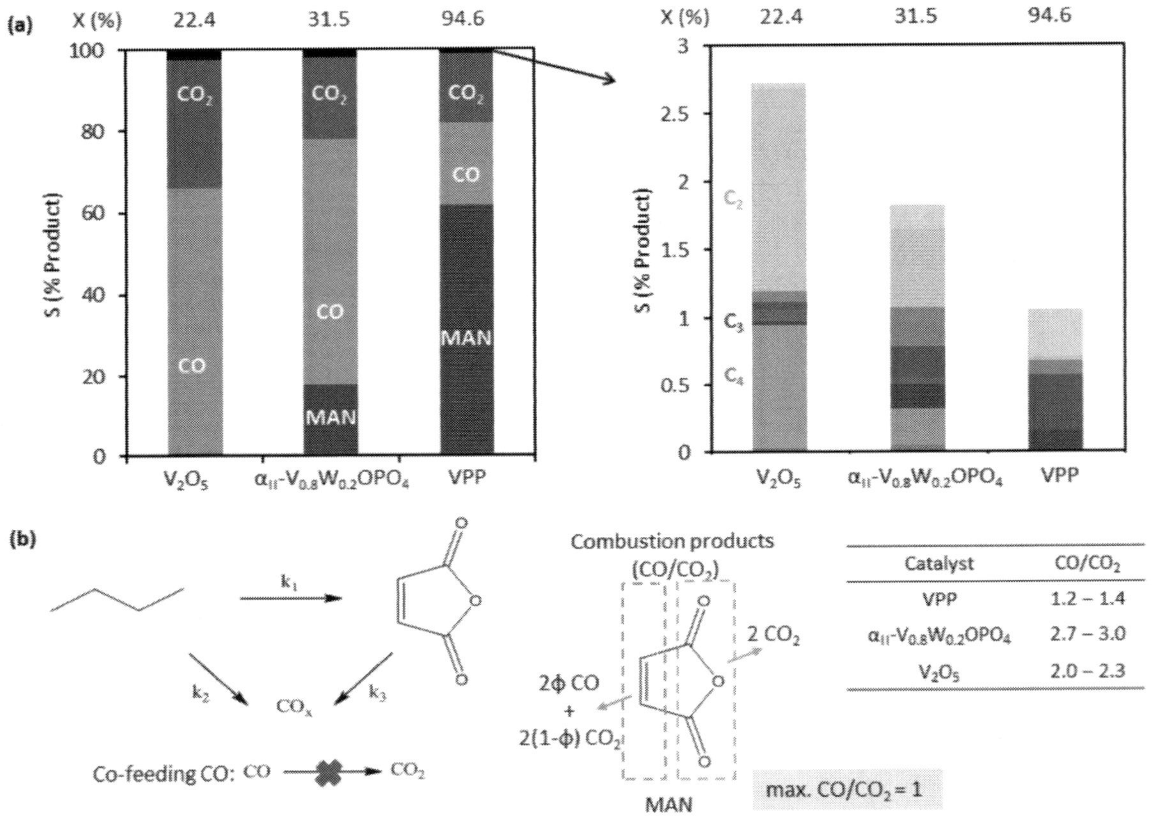

Figure 1. (a) Product distribution of V_2O_5, α_{II}-$V_{0.8}W_{0.2}OPO_4$, and VPP at 1 bar, 420 °C, 2 %v C_4H_{10}, 20 %v O_2, 3 %v H_2O, 2000 h^{-1} and (b) classic triangle reaction network of n-butane to maleic anhydride.

551433 Light Alkene Oligomerization for Shale Gas Upgrading: A Microkinetic Model

Monday, June 3, 2019: 4:48 PM
Texas Ballroom A (Grand Hyatt San Antonio)

Sergio Vernuccio[1], Elsa Koninckx[2], Elizabeth E. Bickel[3], Han-Ting Tseng[4], Rajamani Gounder[3], Fabio Ribeiro[5] and Linda J. Broadbelt[1], (1)Chemical and Biological Engineering, Northwestern University, Evanston, IL, (2)Department of Chemical Engineering, Northwestern University, Evanston, IL, (3)Charles D. Davidson School of Chemical Engineering, Purdue University, West Lafayette, IN, (4)Davidson School of Chemical Engineering, Purdue University, West Lafayette, IN, (5)Purdue University, W LAFAYETTE, IN

Introduction

In the last decade, large scale upgrading of shale gas has been considered a major issue in the US energy industry and has, for this reason, become a very popular research area. Unlike natural gas, shale gas contains higher concentrations of hydrocarbons other than methane, such as ethane and propane. Conversion of these light gases to liquid products would open up new opportunities for their use as valuable chemicals or fuels (Ka et al., 2009). Dehydrogenation of these hydrocarbons is a common process leading to light olefins that can be, in a second process, oligomerized over proton-exchanged zeolites.

The present work aims at developing and validating a microkinetic model to simulate the oligomerization process of light olefins (ethene and propene) over a proton-exchanged zeolite H-ZSM-5. Despite the industrial relevance of these processes, a detailed characterization of the kinetic parameters governing the process has not been proposed in the literature. All of the preceding results are related to "pathways-level models", consisting of the lumping of several reactions in a single one describing the conversion of a reagent into a product and disregarding any reaction intermediates. Thus, due to the rough lumping by carbon number, these models are not detailed enough to predict the product distribution and have not been demonstrated to predict the evolution of the system under a wide range of operating conditions.

Methods

The oligomerization of light alkenes over acidic zeolites is characterized by an extremely complex product distribution due to the large diversity of ionic species formed as reaction intermediates during the process. However, this large reaction network can be condensed into a limited number of reactions occurring with similar chemistry (reaction families). The acid-catalysed oligomerization of an olefin can be fully described using five main reaction families (protonation, deprotonation, oligomerization, β-scission and hydride transfer) and several isomerization steps. The species present in the reacting system were represented using bond and electron (BE) matrices based on graph theory. A unique mathematical operator was specified for each reaction family (Broadbelt et al., 1994) and the reaction mechanism was automatically created by applying these operators to the reactants and their progeny.

The kinetic constants governing the process were assumed to obey an Arrhenius temperature dependence. The frequency factors were estimated directly from transition state theory knowing the entropy changes between the reactants and the corresponding transition states. These parameters were calculated for the ionic species characterized in the literature as alkoxides or carbenium ions. The activation energies were related to the heats of reaction of each elementary step through the Evans-Polanyí relationship. Group additivity approaches were used to express the heats of reactions based on the heats of formation of the reacting species (Bjorkman et al., 2014).

Results

The reaction network of olefin oligomerization is initiated via formation of an ionic species by protonation of the double bond of the reactant. The reverse reaction, deprotonation, is a termination step that desorbs the alkenes from the surface of the catalyst returning a proton to the zeolite surface. Oligomerization proceeds through addition of an alkene to an ionic species and consequent formation of a new sigma bond. β-scission breaks the bond between the carbon atom in the β-position with respect to the carbocationic center, representing the reverse step of oligomerization. The application of these reaction families was restricted to a limited number of reactants to generate molecular species with relatively low carbon number ($\leq$ 12) to control the growth of the oligomerization mechanism (Susnow et al., 1997).

The resulting kinetic networks were coupled with the design equations of a PFR to build a continuum kinetic model. The proposed model was solved according to proper initial conditions to simulate reaction kinetics, product yields and selectivity. The proposed mathematical model, reliably validated, is able to predict the kinetic behaviour of the oligomerization system given its temperature, reactant pressure and reactant space velocity.

Conclusions

This paper presents a microkinetic analysis to describe the kinetics of light olefin oligomerization over acidic zeolites. The developed mathematical model represents a powerful tool to reproduce and predict conversion, selectivity and product distributions obtained from experimental operation under conditions relevant to industrial practice.

References

Ka, B., Pe K. (2009) *Oil Gas J. 107*, 50–55.

Labinger, J. A., Leitch, D. C., Bercawl, J. E., Deimund, M. A., Davis, M. E. (2015) *Top. Catal. 58*, 494–501.

Broadbelt, L. J., Stark, S. M., Klein, M. T. (1994*) Ind. Eng. Chem. Res. 33*, 790–799.

Bjorkman, K. R., Sung, C.-Y., Mondor, E., Cheng, J., C., Jan, D.-Y., Broadbelt, L. J.,

(2014) *Ind. Eng. Chem. Res. 53*, 19446-19452.

Susnow, R. G., Dean, A. M., Green, W. H., Peczak, P., Broadbelt, L. J., J. (1997) *J. Phys. Chem. A 101*, 3731–3740.

544993 Microkinetic Modeling of Oxidative Dehydrogenation of Ethane to Ethylene at High Temperatures

Monday, June 3, 2019: 5:30 PM
Texas Ballroom A (Grand Hyatt San Antonio)

Hilal Ezgi Toraman, Delaware Energy Institute, Department of Chemical and Biomolecular Engineering, University of Delaware, Newark, DE and Dionisios G. Vlachos, Chemical and Biomolecular Engineering, University of Delaware, Newark, DE

The large availability of low cost shale gas reserves has generated interest for the production of high value-added chemicals. Although methane is the main component in natural gas (often more than 85%), ethane is also available due to the large amount of natural gas processed. Olefin production occurs via steam cracking, a high energy and capital intensity process. The on-purpose conversion technologies are widely exploited for more efficient utilization of energy and resources. In this regard, the short-contact-time reactor is a promising concept for autothermal conversion of light paraffins into their corresponding olefins in compact reactors at high temperatures. In this work, we use detailed modeling to understand the process. Our simulations show that gas phase reactions occur within less than 1 ms at temperatures above 1173 K. This suggests that gas phase reactions are unavoidable for the process of interest. It has been experimentally shown that ethylene yields, approximately 60%, comparable to steam cracking can be obtained by partial oxidation of ethane at 1223 K with a contact time of ca. 1 ms using platinum-tin catalyst.[1] Different mechanisms from purely heterogeneous to combination of both heterogeneous and homogeneous and purely homogeneous reaction mechanisms have been proposed in the literature. In this study, we model this process by a detailed gas-phase reaction mechanism including more than 100 species and over 2500 reactions. Microkinetic simulations with and without oxygen bracket the experimental ethylene selectivity-ethane conversion profiles. This provides an indication for the role of platinum catalyst as a heat generator driving the homogeneous reactions in the gas phase and relatively lower reforming capability of platinum to quench the gas-phase reactions. Overall, the present study exemplifies the importance of gas phase chemistry at high temperature/short contact time of alkane oxidative dehydrogenation.

1. Bodke, A. S.; Olschki, D. A.; Schmidt, L. D.; Ranzi, E. High selectivities to ethylene by partial oxidation of ethane. Science 1999, 285 (5428), 712.

545472 High-Performance Ni-Foam-Structured Nb_2O_5-NiO Nanocomposite Catalysts for Oxidative Dehydrogenation of Ethane to Ethylene

Monday, June 3, 2019: 5:33 PM
Texas Ballroom A (Grand Hyatt San Antonio)

Zhiqiang Zhang, **Guofeng Zhao**, Jian Zhu, Ye Liu and Yong Lu, Department of Chemistry, East China Normal University, Shanghai, China

We report a promising Nb_2O_5-NiO/Ni-foam catalyst for the ODE reaction, which achieves an ethylene productivity of 0.43 $g_{Ethylene}\ g_{cat}^{-1}\ h^{-1}$ at 60% ethane conversion with 68% ethylene selectivity for a feed of $C_2H_6/O_2/N_2$=1/1/8 at 410 °C and a high GHSV of 9000 $cm^3\ g^{-1}\ h^{-1}$, and showed good stability due to high resistance to sintering. Such catalyst engineered from nano- to macro-scale is only obtainable by endogenous growth of $NiC_2O_4 \cdot 2H_2O$ onto Ni-foam followed by (niobium ammonium oxalate)-modification and subsequent pyrolysis treatment. In this case, the Nb_2O_5 lumps were formed to have the NiC_2O_4-derived NiO nanorods partially covered up, and therefore induced helpful weak interaction between Nb_2O_5 and NiO while leading to a remarkable reduction of the nonselective O_2 species but almost not touching the selective O species. External and partial coverage of the NiC_2O_4-derived NiO nanorods with Nb_2O_5 lumps, coupled with foam-structure enhanced heat transfer, is paramount for high-efficiency ethylene production.

546529 Oxidative Dehydrogenation of Propane over Pt-Sn/Si-Beta Catalysts: Key Role of Pt-Sn Interaction

Monday, June 3, 2019: 5:39 PM
Texas Ballroom A (Grand Hyatt San Antonio)

Lanlan Sun, School of Materials Science and Engineering & National Institute for Advanced Materials, Nankai university, Tianjin, China

The selective oxidative dehydrogenation of propane (ODHP) represents an attracting route for propylene production. Despite of significant achievements made so far, eligible ODHP catalysts with high propylene selectivity at acceptable propane conversions are still being explored.

Herein, we report the construction of bimetallic Pt-Sn catalysts supported on dealuminated Si-beta zeolite, which exhibit remarkable performance in ODHP reaction (75% propylene selectivity at 37% propane conversion, and 1.22 g_{C3H6}/(g*h)propylene productivity at 773 K). With silanols created from zeolite dealumination as anchoring sites, the homogeneous distribution of Pt and Sn species is guaranteed and their close contact is clearly demonstrated by electron microscopy analyses. The key role of Pt-Sn interaction in ODHP is therefore established.

Characterization results from temperature-programmed reduction by hydrogen (H_2-TPR) and X-ray photoelectron spectroscopy (XPS) indicated the interaction between Pt and Sn species, and the electron transfer from SnO_2 to Pt species was clearly demonstrated by XPSand FTIR spectra of CO adsorption. Temperature-programmed desorption experiments revealed that the propane adsorption was significantly strengthened while the propylene adsorption weakened on bimetallic Pt-Sn/Si-beta catalyst, which accordingly resulted in its remarkable performance in ODHP reaction. The TEM analyses on the catalyst before and after reaction are very impressive. The spatial separation of Pt and Sn species was observed on used Pt-Sn/Si-beta catalyst and it was responsible for the catalytic deactivation.

546817 A Fresh Look at Ni-Based Catalysts for the Oxidative Dehydrogenation of Ethane

Monday, June 3, 2019: 5:45 PM
Texas Ballroom A (Grand Hyatt San Antonio)

Kara Stowers[1], Justin Park[2], Kelsey Canizales[2] and Morris D. Argyle[3], (1)Chemistry and Biochemistry, Brigham Young University, Provo, UT, (2)Brigham Young University, Provo, UT, (3)Chemical Engineering, Brigham Young University, Provo, UT

As shale oil deposits continue to yield large amounts of light paraffins, technologies must account for how to obtain routes for the production of light olefin feedstocks previously obtained through steam pyrolysis or cracking. A very important feedstock is ethylene, which has a market need of 150 million tons/year. Standard operating reaction conditions for obtaining ethylene often require enormous amounts of energy with operating temperatures higher than 900°C. Low temperature oxidative dehydrogenation not only operates at temperatures of 500°C or lower, but also introduces oxygen as a co-reagent for the conversion of ethane to ethylene.

Although many catalysts are viable for the oxidative dehydrogenation process, the most promising are metal-based oxides. Nickel based catalysts have shown great promise at low temperatures for conversion. These catalysts are sensitive to the structure of the nickel at the surface and metal-support interactions as well as the physical properties of the supports. We have also found that the amount of metal in the catalyst and the preparation method make a difference in yield of ethylene. Accepted conversion rates for ethane to ethylene are 1 kg ethylene/kg catalyst-hour. We aim to improve this conversion rate by optimizing the nickel based catalyst and finding new pathways for decreasing energy costs for the reaction.

In this study, we report the optimization of nickel based catalysts through *i*) support properties *ii*) dopants and additives and *iii*) operating parameters for the reaction. Our findings deconvolute the effect of support surface area and pore volume from acid/base properties. We also report on the effect of additives in light of the oxygen storage capacities of the catalysts. All reaction conditions for oxidative dehydrogenation were performed in the temperature range of 300-500°C. We also discuss the role and effect of co-feeds such as CO_2 as an additive for these reactions using a Ni-based catalyst. At higher temperatures, over-oxidation to CO_2 outcompetes the rate of ethylene formation, however, we can offset this through further catalyst optimization.

544958 Status and Prospects of the Halogen-Mediated Catalytic Processes for Natural Gas Upgrading

Tuesday, June 4, 2019: 1:30 PM
Republic ABC (Grand Hyatt San Antonio)

Guido Zichittella[1], Vladimir Paunovic[2] and **Javier Perez-Ramirez**[1], (1)Department of Chemistry and Applied Biosciences, ETH Zurich, Zurich, Switzerland, (2)ETH Zurich, Zurich, Switzerland

Status and Prospects of the Halogen-Mediated Catalytic Processes for Natural Gas Upgrading

±

Institute for Chemical and Bioengineering, ETH Zurich, Switzerland

Development of catalytic technologies enabling the direct functionalization of light alkanes, main components of abundant natural gas, into valuable and transportable liquid chemicals and fuels is quite possibly the winning strategy to transit from the oil to the renewables era.[1] Halogen mediated activation of methane, ethane, and propane might offer the potential to confront this quest head on using compact chemical plants amenable to decentralization.[1,2] Nevertheless, its industrialization has been constrained by several hurdles related to the necessity of achieving a complete halogen recycle, maximizing the yield of the intermediate haloalkanes, and tailoring the performance in their subsequent coupling into commodities. In this lecture, we will discuss the recent advances in catalytic processes aiming to overcome these challenges, which is accomplished through the synergy between catalyst synthesis, advanced *operando*characterization tools, and *ab-initio* calculations that enable to derive structure performance relationships, guiding optimal catalyst design.[1-3] Building on this progress, environmental and economic footprints of halogen based technologies will be analyzed with respect to alternative routes for natural gas valorization, together with future research directions that can kindle scientific advancements with societal impact.

References

[1] R. Lin, A. P. Amrute, J. Pérez-Ramírez, *Chem. Rev.***2017**,*117*, 4182 4247.

[2] a) M. Moser, I. Czekaj, N. Lõpez, J. Pérez-Ramírez, *Angew. Chem. Int. Ed.***2014**, *53*, 8628 8633; b) V. Paunović, G. Zichittella, M. Moser, A. P. Amrute, J. Pérez-Ramírez, *Nat. Chem.***2016**, *8*, 803 809; c) V. Paunović, G. Zichittella, R. Verel, A. P. Amrute, J. Pérez-Ramírez, *Angew. Chem. Int. Ed.***2016**,*55*, 15619 15623; d) V. Paunović, R. Lin, M. Scharfe, A. P. Amrute, S. Mitchell, R. Hauert, J. Pérez-Ramírez, *Angew. Chem. Int. Ed.***2017**,*56*, 9791 9795; e) G. Zichittella, N. Aellen, V. Paunović, A. P. Amrute, J. Pérez Ramírez, *Angew. Chem. Int. Ed.***2017**, *56*, 13670 13674.

[3] a) G. Zichittella, V. Paunović, A. P. Amrute, J. Pérez-Ramírez, *ACS Catal.* **2017**, *7*, 1805 1817; b) V. Paunović, G. Zichittella, S. Mitchell, R. Hauert, J. Pérez-Ramírez, *ACS Catal.* **2018**, 291 303; c) V. Paunović, P. Hemberger, A. Bodi, N. López, J. Pérez-Ramírez, *Nat. Catal.* **2018**, *1*, 363–370.

Corresponding author email: jpr@chem.ethz.ch

545793 Conversion of Methane, Ethane and Propane to Form Alkyl Esters: Oxy-Esterification As a Strategy for Selective Partial Oxidation

Tuesday, June 4, 2019: 2:09 PM
Republic ABC (Grand Hyatt San Antonio)

T. Brent Gunnoe[1], John Groves[2], William A. Goddard III[3], Nichole Schwartz[4], Nicholas Boaz[5], Steven Kalman[6], Jonathan Goldberg[7], George Fortman[6], Ross Fu[8] and Robert Nielsen[8], (1)Department of Chemistry, University of Virginia, Charlottesville, VA, (2)Princeton U., Princeton, NJ, (3)Chemistry and Chemical Engineering Division, California Institute of Technology, Pasadena, CA, (4)University of Virginia, (5)Department of Chemistry and Princeton Institute for the Science and Technology of Materials, Princeton, Princeton, NJ, (6)Department of Chemistry, University of Virginia, (7)Princeton University, (8)California Institute of Technology

The selective and low temperature conversion of light alkanes (methane, ethane and propane) to functionalized products has been a long-standing challenge. In protic media, the reaction of light alkanes with iodine oxides results in the conversion to mono-functionalized products with alkyl trifluoroacetates as the primary products. The transformations are effective over a broad range of temperatures (100 - 235 ºC) and pressures (240-6900 kPa) in non-superacidic media. Alkane conversions between 15% and 40% with selectivity for mono-functionalized products between 85% and 95% have been achieved. The presentation will focus on mechanistic studies that incorporate selectivity patterns and kinetic isotope effects for conversion of light alkanes and model hydrocarbons (e.g., adamantane), spectroscopic identification of reaction intermediates, and computational modeling. In addition, preliminary efforts to extend the reaction to photo-mediated conversions as well as thermal processes with non-iodine based oxidants will be presented.

545664 Co-Promoting Effects of Mg and K in Ethylene Oxychlorination

Tuesday, June 4, 2019: 2:33 PM
Republic ABC (Grand Hyatt San Antonio)

Hongfei Ma[1], Endre Fenes[1], Kumar R. Rout[2], Terje Fuglerud[3] and De Chen[4], (1)Norwegian University of Science and Technology, Trondheim, Norway, (2)SINTEF, Trondheim, Norway, (3)INOVYN, Porsgrunn, Norway, (4)Chemical Engineering, Norwegian University of Science and Technology, Trondheim, Norway

Co-promoting effects of Mg and K in ethylene oxychlorination

Hongfei Ma[1], Endre Fenes[1], Kumar Ranjan Rout[2], Terje Fuglerud[3], De Chen[1]*

1 – Department of Chemical Engineering, NTNU, Trondheim, 7491, Norway

2 – SINTEF, Trondheim, 7491, Norway

3 – INOVYN Norway, Porsgrunn, 3936, Norway

*Corresponding author: de.chen@ntnu.no

1. Introduction

Ethylene oxychlorination is an industrially important process to produce vinyl chloride monomer (VCM), where $CuCl_2/\gamma\text{-}Al_2O_3$ based catalysts undergoes a reduction and oxidation process. [1] Alkali or alkali metals, such as K and Mg, are used as the promoters to improve the performance and stability of the catalysts in the industry, but the fundamental, exact mechanism details on how the promoters work in the reaction are still not clearly. In our previous works, [2, 3, 4] we have systematically researched the roles of dopants (K, Ce) on the ethylene oxychlorination $CuCl_2/\gamma\text{-}Al_2O_3$ catalyst.

$$2CuCl_2 \rightarrow 2CuCl \rightarrow Cu_2OCl_2 \rightarrow 2CuCl_2$$

In this work, we prepared the co promoted (K and Mg) catalysts to research the effect of the two promoters based on our previous reported methods. Reaction rate for reduction and oxidation were evaluated for the step transient reactions, and also for steady state. $CuCl_2$ band gap energy with different promoters were also calculated to investigate the different influences of the promoters.

2. Experimental

The catalysts prepared from the chlorides salts with the incipient wetness method, containing 5 wt% Cu and e. g. 0.4 K have a K: Cu molar ratio of 0.4 and 5 wt% Cu. The reaction for both step transient and steady state experiments were conducted at 230 °C under atmospheric pressure, with the total gas flow of 3 ml/s. While the partial pressure of C_2H_4 for step transient was 10%, and 8% for steady state experiments. All the reactions were performed in our operando set up combining UV-Vis-NIR spectrometry and mass spectroscopy.

3. Results and discussion

From XRD patterns we can know that for all the catalysts copper chlorides remain undetectable by XRD due to a high dispersion of the copper metals. In order to fully active the catalysts, three cycles of redox, also called step transient reduction, oxidation and chlorination were proceeded. And the reaction rate of reduction and oxidation were obtained from MS. Since the third step chlorination is very fast, so it is difficult to get the reaction rate. We can know that different promoters had different influences to the reduction and oxidation steps. The rate determined step could be tuned. Figure 1 shows the C_2H_4 conversion of steady state reactions. We can know that catalysts promoted by Mg had higher C_2H_4 conversion. Looking deeply on the results, we can find that the initiation point of Mg promoted catalysts was very high, while the decreasing from starting point to steady state was also the most obvious. But it was the least for co promoted catalyst. Based on our previous researches [2, 3], the performance was closed related with the concentration of Cu^{2+}, so we can see that the cooperation of K and Mg can keep the Cu^{2+}, through tuning the reduction and oxidation rate.

In order to have a deeper understanding on the influence of the promoter to the active site of $CuCl_2$, band gap energy was calculated by KMF function. The results are shown in Figure 2. We can see that the energy is much higher for Mg promoted than K promoted. It slightly increased from 0.4K to 0.1K0.3Mg and 0.4Mg. What the interesting thing is the tendency is similar with C_2H_4 conversion in steady state reactions. This means that band gap energy is a very important term in redox reactions. What's more, compared with peak positon in TPR results the tendency was reverse.

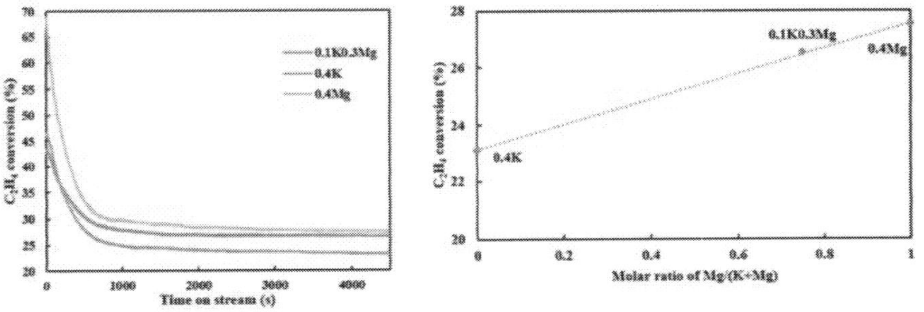

Figure 1. C2H4 conversion in steady state reaction on different catalysts, reaction conditions: T=230 °C, P_{total}= 1 atm, F_{total}= 180 ml/min, partial pressure: C_2H_4 8%, O_2 4%, HCl 16%,

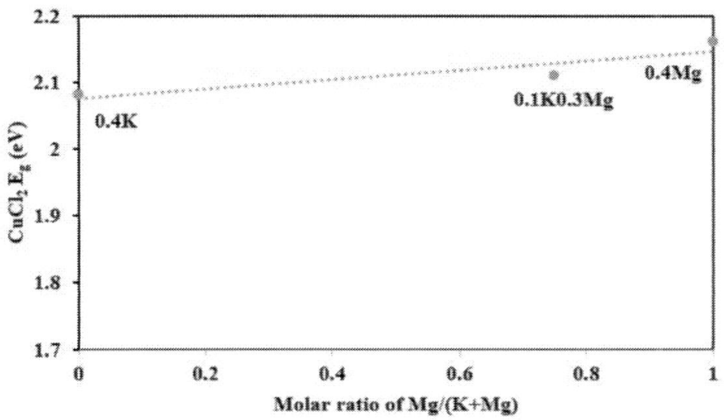

Figure 2. Band gap of $CuCl_2$ calculated by UV-Vis.

4. Conclusion

As far as we have known, for the first time a $CuCl_2$ based ethylene oxychlorination catalyst with two promoters were employed to understand the promoter effect. We utilized an operando method combined UV-Vis and MS to detect the oxychlorination cycles, including step transient and steady state conditions. Band gap energy was obtained to understand the difference of the catalysts. It also proved that UV-Vis spectroscopy is a useful method to evaluate the band gap energy in redox reactions. We can find the relation of band gap energy and the performance. More experiments and techniques will be employed to investigate deeper on the influence of co promoters to the active copper phase. Then the detailed difference will be thoroughly evaluated.

5. Reference

[1] N. B. Muddada, U. Olsbye, L. Caccialupi, F. Cavani, G. Leofanti, D. Gianolio, S. Bordiga, C. Lamberti, Phys. Chem. Chem. Phys. 12 (2010) 5605.

[2] K. R. Rout, E. Fenes, M. F. Baidoo, R. Abdollahi, T. Fuglerud, D. Chen, ACS Catal. 6 (2016) 7030.

[3] K. R. Rout, M. F. Baidoo, E. Fenes, J. Zhu, T. Fuglerud, D. Chen, Journal of Catalysis, 352 (2017) 218.

[4] M. F. Baidoo, E. Fenes, K. R. Rout, T. Fuglerud, D. Chen, Catalysis Today, 299 (2018) 164.

545488 Selective Oxidation of Propane to Propylene in the Presence of HCl over CeO2 and NiO-Modified CeO2 Nanocrystals

Tuesday, June 4, 2019: 2:57 PM
Republic ABC (Grand Hyatt San Antonio)

Jincan Kang[1], Quanhua Xie[1], Huamin Zhang[1], Jun Cheng[2], Qinghong Zhang[1] and Ye Wang[1], (1)College of Chemistry and Chemical Engineering, Xiamen University, Xiamen, China, (2)Xiamen University, Xiamen, China

Selective oxidation of propane to propylene in the presence of HCl over CeO_2 and NiO-modified CeO_2 nanocrystals

Jincan Kang, Quanhua Xie, Huamin Zhang, Jun Cheng, Qinghong Zhang, Ye Wang*

State Key Laboratory of Physical Chemistry of Solid Surfaces, Collaborative Innovation Center of Chemistry for Energy Materials, National Engineering Laboratory for Green Chemical Productions of Alcohols, Ethers and Esters, College of Chemistry and Chemical Engineering, Xiamen University, Xiamen, 361005, P. R. China

**Corresponding author:* wangye@xmu.edu.cn

Abstract: Here, we report a new strategy for the oxidative dehydrogenation of propane by O_2 with outstanding propylene yield in the presence of HCl. CeO_2 is an efficient catalyst for the conversion of C_3H_8 to C_3H_6 by (O_2 + HCl). The reaction was structure-sensitive, and the catalytic behavior depended on the exposed facet of CeO_2 nanocrystals. A propylene selectivity of 80% was achieved over an 8 wt% NiO $-CeO_2$ catalyst, offering a propylene yield of 55%. The structure−property correlation indicates that the surface oxygen vacancy and the surface chloride coverage are two crucial factors determining the activity and selectivity. The peroxide species formed by adsorption of O_2 on surface oxygen vacancies may activate chloride, generating a radical-like active chlorine species for the selective conversion of propane.

Keywords: Oxidative dehydrogenation, Propane, Propylene, CeO_2 nanocrystals, Hydrogen chloride

1. Scope

Oxidative dehydrogenation of C_3H_8 is an attractive reaction for C_3H_6 production, but the over-oxidation results in low C_3H_6 selectivity at considerable C_3H_8 conversions and the formation of undesirable CO_2. To increase the selectivity of the target product, which is more reactive than the substrate, is a challenging goal in selective oxidation catalysis. Despite some recent encouraging progress, high propylene selectivity (>70%) is still difficult to achieve at a high propane conversion (>30%).

Some studies have been devoted to the conversion of CH_4 and other lower alkanes by (O_2 + HX), but only a few studies have reported the conversion of propane by this strategy.[1,2] Besides lower olefins, RX (R = alkyl group) was also formed in the conversion of lower alkanes.[2] Here, we report the direct conversion of C_3H_8 to C_3H_6 by (O_2 + HCl) with a high single-pass yield using CeO_2-based catalyst.[3] We demonstrate that the reaction is structure-sensitive, and the modification of CeO_2 by NiO can further enhance C_3H_6 yield. The roles of HCl and the reaction mechanism will be discussed.

2. Results and discussion

We first investigated the catalytic behaviors of various metal-oxide catalysts for the conversion of C3H8 by (O2 + HCl). The result shows that CeO2 is a promising catalyst, not only because CeO2 demonstrates the highest single-pass C3H6 yield among all the catalysts examined but also because it shows high stability. The catalytic behavior of CeO2 depended on its morphology or the exposed facets The rates of C3H8 conversion and C3H6 formation decreased in the order of nanorods (exposing {110} + {100}) > nanocubes (exposing {100}) > nano-octahedra (exposing {111}) ≈ nanoparticle (exposing {111}). Thus, the {110} facet shows higher activity than the {100} facet, which was significantly higher than the {111} facet. We further compared the C3H6 selectivity at similar C3H8 conversion levels and found that the C3H6 selectivity decreased in the following sequence: nanocube > nanorod > nanooctahedron ≈ nanoparticle. Thus, the {100} facet is the most selective for C3H6 formation, followed by the {110} and {111} facets. In short, the CeO2-catalyzed conversion of C3H8 is a structure-sensitive reaction. The {110} facet is the most active for C3H8 conversion, whereas the {100} facet is the most selective for C3H6 formation.

We investigated the effect of various modifiers on the catalytic behavior of CeO2 nanorods. Among all the modifiers examined, NiO was the most efficient for promoting C3H6 formation. Both O2 and C3H8 conversions increased after the doping of NiO with a low content. The C3H6 selectivity increased gradually from 55% to 72% with an increase in NiO content to 8 wt%. At the same time, the selectivities of CO and CO_2 decreased, and the selectivity of organic chlorides kept low (< 4%).

Our studies revealed that HCl played a crucial role in the selective formation of C3H6 over CeO2 based catalysts. CO2 was the major product in the absence of HCl (Figure 1A), indicating that CeO2 and NiO−CeO2 were complete oxidation catalysts for the oxidation of C3H8 by O2. Both C3H8 conversion and C3H6 selectivity increased with the partial pressure of HCl. C3H8 conversion of ~70% and C3H6 selectivity of ~80% were attained over the 8 wt% NiO−CeO2 catalyst at a *P*(HCl) of 25 kPa (Figure 1B). The single-pass C3H6 yield reached ~55%. We further confirmed that ~98% HCl could be recovered.

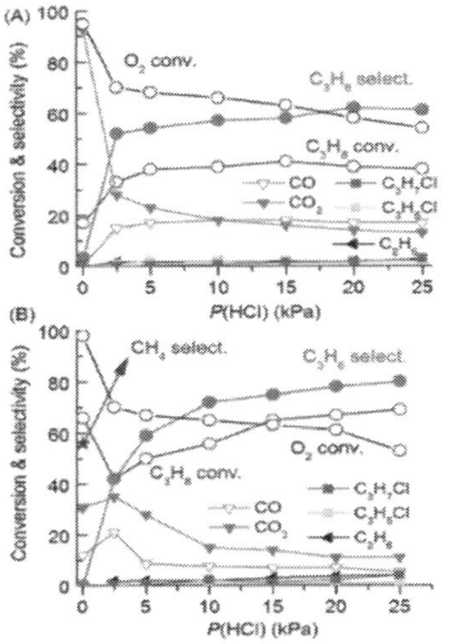

Figure 1. Effect of HCl partial pressure on the conversion of C_3H_8 by (O_2 + HCl). (A) CeO_2. (B) 8 wt% NiO−CeO_2.

We performed mechanistic studies for CeO2 and 8 wt% NiO−CeO2 catalysts (Figure 2). We uncovered that the oxidation of HCl by O2 to Cl2 (the Deacon reaction) occurred on our catalysts, but the formation of Cl2 was inhibited by the presence of C3H8. We characterized the CeO2 nanocrystals with different morphologies and the NiO−CeO2 catalysts with different NiO contents by UV-Raman and HCl chemisorption studies. The correlation of the characterization results with the catalytic behaviors suggests that C3H8 conversion activity depends on the concentration of oxygen vacancies, while the chemisorption amount of HCl determines C3H6 selectivity. In combination with DFT calculations, we propose that oxygen is adsorbed on the O-vacancy site on CeO_2 surfaces, forming O_2^{2-} species. And the HCl is activated by the O_2^{2-} species into Cl•-like species, which may be responsible for the activation of C3H8 and the selective formation of C3H6.

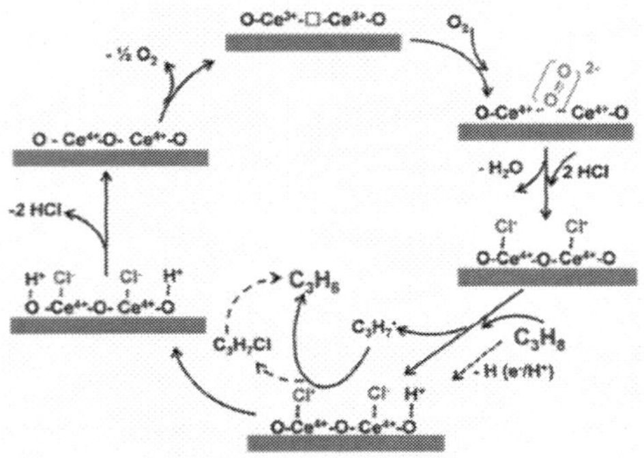

Figure 2. Proposed reaction mechanism for the oxidative dehydrogenation of C_3H_8 to C_3H_6 by (O_2 + HCl) over CeO_2-based catalysts.

3. Conclusions

CeO2 is an efficient and stable catalyst for the conversion of C3H8 to C3H6 by O2 in the presence of HCl. The reaction is structure sensitive and the catalytic behavior depends on the exposed facet of CeO2. CeO2 nanorods and nanocubes show the highest activity and the highest C3H6 selectivity, respectively. The modification of CeO2 nanorods with NiO increases catalytic performances, offering a C3H6 yield of ~55%. The structure–property correlation suggests that the concentration of surface O vacancies and the surface coverage of Cl^- are two crucial factors that determine the activity and selectivity. We propose that oxygen is adsorbed on the O-vacancy site on CeO_2 surfaces, forming O_2^{2-} species and the O_2^{2-} species oxidizes Cl^- into Cl•-like species, which accounts for the activation and selective conversion of C_3H_8 to C_3H_6.

References

1. J. He, T. Xu, Z. Wang, Q. Zhang, W. Deng, Y. Wang, *Angew. Chem. Int. Ed.* **2012**, *51*, 2438.

2. V. Paunović, G. Zichittella, M. Moser, A. P. Amrute, J. Pérez-Ramírez, *Nat. Chem.* **2016**, *8*, 803.

3. Q. Xie, H. Zhang, J. Kang, J. Cheng, Q. Zhang, Y. Wang, *ACS Catal.* **2018**, *8*, 4902.

551453 Halogen-Dependent Surface Confinement Governs Selective Alkane Functionalization to Olefins

Tuesday, June 4, 2019: 3:21 PM
Republic ABC (Grand Hyatt San Antonio)

Guido Zichittella[1], Matthias Scharfe[1], Begoña Puértolas[1], Vladimir Paunović[1], Patrick Hemberger[2], Andras Bodi[2], László Szentmiklósi[3], Núria López[4] and Javier Pérez-Ramírez[1], (1)Department of Chemistry and Applied Biosciences, ETH Zurich, Zurich, Switzerland, (2)Paul Scherrer Institute, Villigen, Switzerland, (3)Hungarian Academy of Sciences, Budapest, Hungary, (4)Institute of Chemical Research of Catalonia, Tarragona, Spain

Halogen-Dependent Surface Confinement Governs Selective Alkane Functionalization to Olefins

Guido Zichittella,[a] Matthias Scharfe,[a] Begoña Puértolas,[a] Vladimir Paunović,[a] Patrick Hemberger,[b] Andras Bodi,[b] László Szentmiklósi,[c] Núria López,[d] and Javier Pérez-Ramírez[a,*]

[a] *Institute for Chemical and Bioengineering, ETH Zurich, Switzerland*

[b] *Laboratory of Femtochemistry and Synchrotron Radiation, Paul Scherrer Institute, Switzerland*

[c] *Nuclear Analysis and Radiography Department, Hungarian Academy of Sciences, Hungary*

[d] *Institute of Chemical Research of Catalonia, The Barcelona Institute of Science and Technology, Spain*

The large availability of natural gas, enhanced by recent discoveries of new unconventional gas reservoirs, provides a strong incentive to consider this resource as a key feedstock for

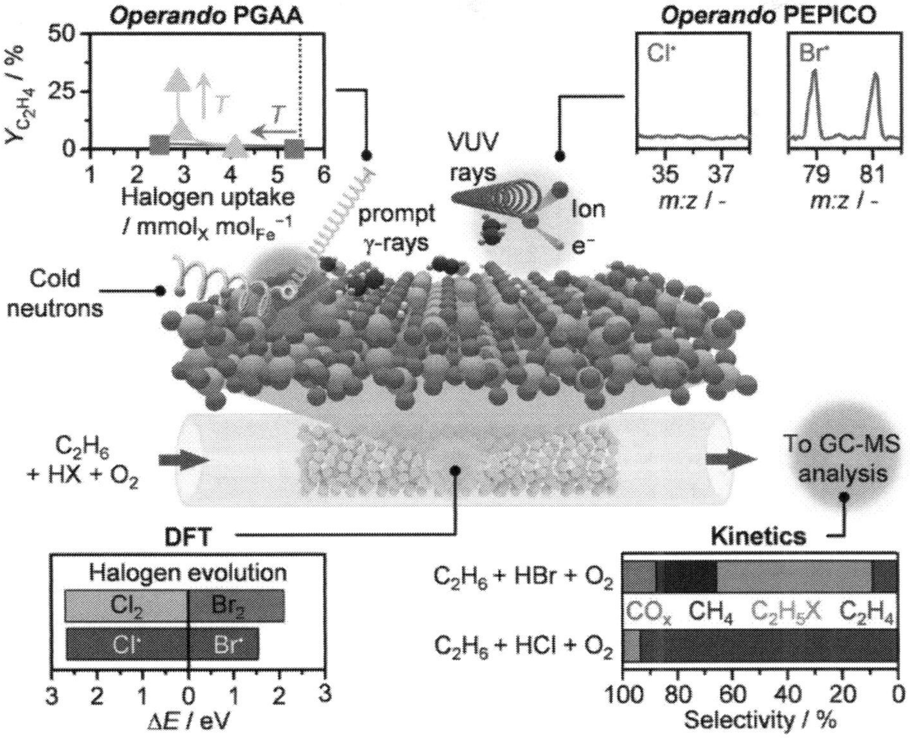

Figure 1. Approach used in this study to unravel the surface-driven and gas-phase radical-chain pathways through which alkane activation can occur in ethane oxyhalogenation over a $FePO_4$ catalyst.

energy and chemicals in the transition from the oil to the renewables era.[1] A necessary cornerstone to meet this great challenge is the in depth understanding of complex reaction mechanisms over dynamic surfaces, allowing to elucidate catalyst design criteria for selective alkane functionalization processes. A prominent example is the oxyhalogenation of ethane over iron phosphate ($FePO_4$) in which a switch from HCl to HBr results in distinct product distribution, ranging from virtually only olefins to a mixture of alkyl bromide, cracking and combustion products (**Figure 1**, bottom left). The quest for the mechanistic origin of these different selectivity patterns remains unanswered due to the inability of most characterization techniques to accurately describe the complex dynamics of catalysts under reaction conditions and to assess potential pathways that can occur in a multi phase reaction environment, where surface catalyzed reactions may coexist with gas-phase contributions.

We have resolved this "selectivity puzzle" by a multi-technique approach, combining evidence from steady-state kinetics, *operando* prompt-gamma activation analysis (PGAA) and photoelectron photoion coincidence (PEPICO) spectroscopies, ultimately rationalized by theoretical calculations (**Figure 1**). The sharp selectivity control in oxychlorination is achieved *via* a purely surface-driven functionalization of ethane into ethyl chloride, which is further dehydrochlorinated to ethylene over a Cl-Fe center, which account for ≥ 55% of the iron sites under reaction conditions (**Figure 1**, top right). In contrast, alkane activation to ethyl bromide in oxybromination mainly occurs in the gas-phase with bromine and bromine radical species that can evolve form the catalyst, as determined by PEPICO and rationalized by DFT (**Figure 1**, top left and bottom right, respectively), thus leaving a halogen-free surface that is more prone to additional cracking and combustion pathways. Our results provide a strategy to rationalize the mechanism of selective alkane functionalization, which can be extrapolated to other reactions where gas-phase contributions are of concern.

References

[1] R. Lin, A. P. Amrute, J. Pérez-Ramírez, *Chem. Rev.* **2017**, *117*, 4182 4247.

Corresponding author email: jpr@chem.ethz.ch

555887 Alkane dehydrogenation with Lewis-Bronsted acid pairs in Ga/H-ZSM-5

Tuesday, June 4, 2019: 4:00 PM
Republic ABC (Grand Hyatt San Antonio)

Lingli Ni, Moritz Schreiber, **Ricardo Bermejo-Deval** and Johannes A. Lercher, Catalysis Research Center, Techniche Universität München, Garching, Germany

Alkane dehydrogenation with Lewis-Bronsted acid pairs in Ga/H-ZSM-5

Lingli Ni, Moritz Schreiber, Ricardo Bermejo-Deval, Johannes A. Lercher

Techniche Universität München, Catalysis Research Center, 85747 Garching, Germany.

Dehydrocyclodimerization is industrially applied for converting liquefied petroleum gas (mainly C_3 and C_4 hydrocarbons) into aromatic compounds (benzene, toluene and xylenes (BTX)) and H_2. On bifunctional catalysts (e.g., Ga/H-ZSM-5) propane dehydrogenation is the rate determining step of the reaction. However, the state and the role of the Ga cation on alkane dehydrogenation has not been discerned. Here, we will identify the proximity of Ga^+ to the Brønsted acid site (BAS) and the Al distribution to be key for an efficient reaction pathway in the dehydrogenation of propane.

The addition of Ga ($Ga(NO_3)_3$ solution, incipient wetness impregnation) into H-ZSM-5 (Si/Al = 50, 31 % of Al pairs) indicated the stoichiometric replacement of one BAS by one Ga^+ cation, as in-situ IR spectroscopy showed the decrease of the Si(OH)Al stretching band (of BAS) at 3616 cm^{-1}. Therefore, the rate (1 bar, 823 K) of propane conversion and dehydrogenation increased as a function of Ga concentration, reaching a maximum at a ratio of Ga/Al = 0.5 (Figure 1). This dehydrogenation/cracking rate ratio increased from 0.63 (Ga/Al = 0) to $19.20{\cdot}10^7$ mol g^{-1} s^{-1} (Ga/Al = 0.5). However, H-ZSM-5 with Si/Al 88 (4 % Al pairs) reached the maximum rates of dehydrogenation at Ga/Al =1, showing Al pairs are necessary for the Lewis–Brønsted acid pair synergy in propane dehydrogenation. The nature and concentration of the active site were determined in situ via X-ray absorption spectroscopy (XANES), temperature-programmed reduction (TPR) and infrared (IR) spectroscopy of adsorbed pyridine.

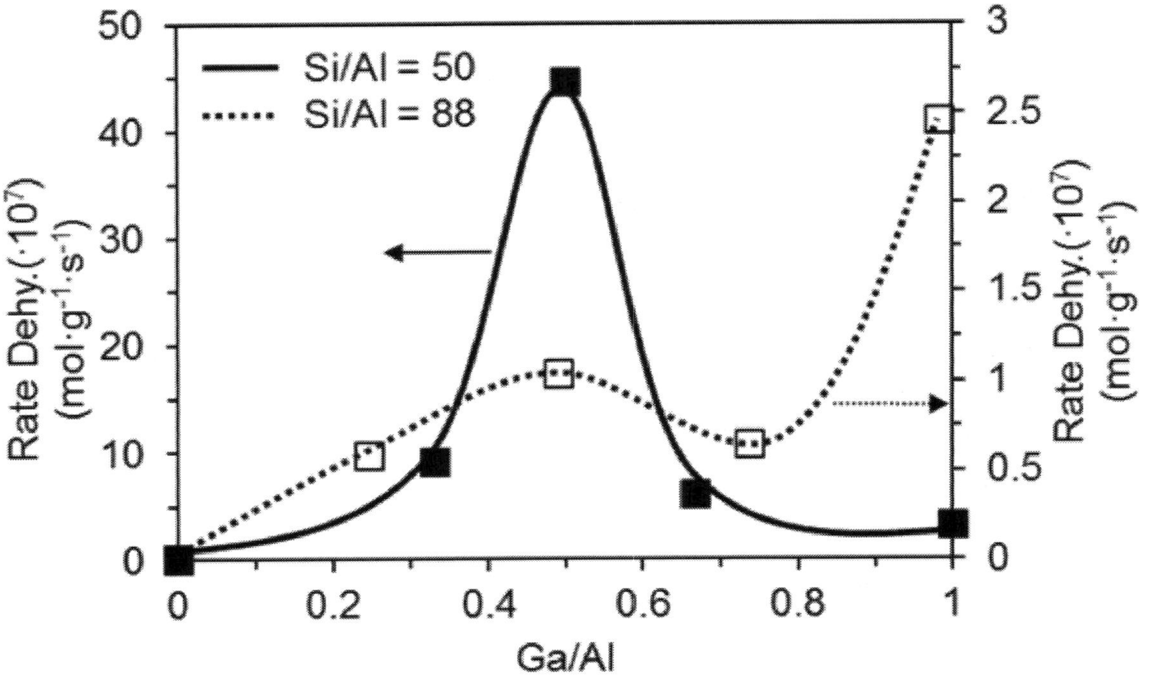

Figure 1. Dehydrogenation rates as a function of Ga/Alparent for H-MFI zeolites with Si/Al ratio of 50 and 88.

545735 Ethane Oxidative Dehydrogenation on M1 Phase Mixed Metal Oxides

Tuesday, June 4, 2019: 4:24 PM
Republic ABC (Grand Hyatt San Antonio)

Prashant Deshlahra[1], Annamalai Leelavathi[1] and Yilang Liu[2], (1)Chemical and Biological Engineering, Tufts University, Medford, MA, (2)Department of Chemical and Biological Engineering, Tufts University, Medford, MA

M1 Mixed bulk oxides containing Mo, V, Nb and Te exhibit unique catalytic properties that lead to high selectivity to ethylene at lower reaction temperatures than other oxidative processes. On most other oxides the selectivity tends to be higher at higher temperature because undesired reactions involving combustion of ethane and ethylene have lower activation enthalpies than the desired ethane dehydrogenation steps. These unique catalytic properties of M1 phase oxides are typically attributed to specific catalytic sites on the external surfaces[1] or to the one dimensional pores[2] formed from metal-oxide rings with seven metal atoms that exist in the crystal structure of the M1 phase. We will discuss our assessments of the role of the heptagonal micropores of M1 phase in regulating reactivity and selectivity by means of reactant size dependent kinetic probes and density functional theory (DFT) treatments for C_2H_6 and cyclohexane (C_6H_{12}) activations inside and outside the micropores.[3] The oxides were synthesized using hydrothermal methods and characterized using elemental analysis, X-ray diffraction measurements, electron microscopy and N_2 physisorption measurements to quantify intrapore and external surface areas.

The sizes of C_2H_6 (0.4nm) and the micropores (0.4nm) suggest a tight guest-host fit but C_6H_{12} cannot access intrapore sites because of their much larger size (0.6 nm). Thus, C_2H_6 molecules have the opportunity to react in micropores as well as external surfaces of the M1 phase crystallites, but C_6H_{12} molecules are restricted to external surfaces. Measured C_2H_6 to C_6H_{12} activation rate ratios on MoVTeNb oxides are more than two orders of magnitude larger than those measured on non-microporous vanadium oxides (VO_x/SiO_2) and those estimated by DFT on external surfaces of both oxides, suggesting that most C_2H_6 activations on MoVTeNb oxides occur inside the micropores under typical conditions. This conclusion that ethane dehydrogenation occurs predominantly inside the pores is further confirmed by measuring rate ratios on a variety of supported oxides at different surface density and structural and electronic properties, which all lead to rate ratios much less than unity. A variety of M1 phase MoV and MoVTeNb oxides are also tested, which all lead to rate ratios near unity or higher than unity. The relations among accessible micropore volumes and external surface areas are used to further confirm the role of micropores.

C_2H_6 exhibits higher activation energy than C_6H_{12} on VO_x/SiO_2, consistent with the C-H bond strengths in the two reactants. In spite of the significant stronger C-H bonds, the activation energies for C H are similar to C H on MoVTeNb oxides because micropores stabilize C-H activation transition states through van der Waals (vdW) interactions, while C_6H_{12} molecules react outside the pores and experience much weaker vdW interactions. The tightness of confinement and vdW stabilization are confirmed by slightly endothermic (representing slight steric repulsion from pore walls) and significantly exothermic C_2H_6 adsorption form DFT methods without vdW and with rigorous vdW functionals, respectively.

The selectivity to C_2H_4 in C_2H_6 oxidation is much higher on MoVTeNb than on VO_x/SiO

hydrocarbons from C_6H_{12}. These trends suggest that the ability of VO_x/SiO_2 to activate C-H bonds and resist C-O bond formation in products is similar to the external surfaces of MoVTeNb oxides but the micropores in MoVTeNb oxides are more selective for C-H activation. DFT calculations show that the tight confinement in concave pores imposed steric hindrance C-O contact necessary for C-O bond formation reactions that lead to undesired combustion products while vdW interactions lower C-H activation energies. The net result of vdW and steric forces leads to the higher measured alkene selectivity.

References

1. Zhu, Y.; Sushko, P. V.; Melzer, D.; Jensen, E.; Kovarik, L.; Ophus, C.; Sanchez-Sanchez, M.; Lercher, J. A.; Browning, N. D. Formation of Oxygen Radical Sites on $MoVNbTeO_x$ by Cooperative Electron Redistribution. *J. Am. Chem. Soc.* **2017**, *139*, 12342-12345.
2. Ishikawa, S.; Ueda, W. Microporous Crystalline Mo-V Mixed Oxides for Selective Oxidations. *Catal. Sci. Technol.* **2016**, *6*, 617-629.
3. Leelavathi, A.; Liu, Y.; Ezenwa, S.; Dang, Y.; Suib, S. L.; Deshlahra P. Influence of Tight Confinement on Selective Oxidative Dehydrogenation of Ethane on MoVTeNb Mixed Oxides. *ACS Catal.*, **2018**, *8*, 7051-7067.

545398 Introduction of Sb into Trigonal Mo3VOx Oxide and Its Crystal Transformation to Orthorhombic Mo3VOx By Heat Treatment

Tuesday, June 4, 2019: 4:48 PM
Republic ABC (Grand Hyatt San Antonio)

Satoshi Ishikawa[1], Shoma Inukai[1], Toyokazu Tanabe[2] and Wataru Ueda[1], (1)Department of material and life chemistry, Kanagawa University, Yokohama, Japan, (2)Department of materials science and engineering, National Defense Academy, Yokosuka, Japan

Abstract

Sb was introduced into trigonal Mo_3VO_x (Tri-MoVO) by an ion exchange under acidic condition and the obtained solid was heat-treated. Sb was located mainly at heptagonal channels and partly at hexagonal channels in the Tri-MoVO structure after the ion exchange. Without the introduction of Sb, Tri-MoVO was collapsed by the heat-treatment at 550 °C under N_2 flow, resulting in the formation of MoO_3, $(V_{0.07}Mo_{0.93})_5O_{14}$ and $(V_{0.97}Mo_{0.95})O_5$. On the other hand, Tri-MoVO containing Sb was transformed to orthorhombic Mo_3VO_x (Orth-MoVO) structure by the above heat-treatment. Since Sb was located mainly at the hexagonal channel of the Orth-MoVO structure after the phase transformation, we concluded that the phase transformation proceeds as to transform the heptagonal channel of Tri-MoVO structure to the hexagonal channel of Orth-MoVO structure. Thus obtained orthorhombic MoVSbO showed substantial catalytic activity for ammoxidation of propane toward acrylonitrile.

Introduction

Mo-V-Te(Sb)-Nb oxide (MoVTe(Sb)NbO) is well known as an extremely active catalyst for selective (amm)oxidation of light alkanes[1]. MoVTe(Sb)NbO is formed by an appropriate heat-treatment of an amorphous solid comprised of the constituent elements. Much effort has been devoted for a long time in order to clarify this crystal formation process since the clear understanding of the transformation process may allow the rational crystal design of MoVTe(Sb)NbO for further improvement. However, no clear conclusion has been obtained up to the date, because the amorphous nature of the MoVTe(Sb)NbO precursor is not allowed to provide clear characterization results during this crystal transformation process. Crystalline Mo_3VO_x (MoVO) has several crystal phases, including orthorhombic phase (Orth-MoVO) and trigonal phase (Tri-MoVO). Among the crystal phases, Orth-MoVO has identical crystal framework with MoVTe(Sb)NbO. Recently, we found that the crystal structure of Tri-MoVO can transform into the Orth-MoVO structure by the heat-treatment under the heat-treatment condition normally employed to form MoVTe(Sb)NbO when Sb was introduced into Tri-MoVO structure. Since the phase transformation from the Tri-MoVO structure to the Orth-MoVO structure proceeded in the crystal to crystal manner, detail investigations would be possible to clarify this phase transformation mechanism. Due to the similarity of the phase transformation between MoVTe(Sb)NbO precursor to MoVTe(Sb)NbO and Tri-MoVO to Orth-MoVO in terms of the crystal structure and the heat treatment condition, some insights would be obtainable for the crystal formation of MoVTe(Sb)NbO through the investigation of the phase transformation mechanism from Tri-MoVO to Orth-MoVO. Here, we deeply investigated the above phase transformation mechanism in order to gain insights about the crystal formation process of MoVTe(Sb)NbO for further improvement of the catalytic activity.

Experimental

Tri-MoVO was prepared by the reported procedure[2]. Sb was introduced into Tri-MoVO by ion exchange method under acidic solution. First, the desired amount of Sb_2O_3 (Sb:0.144 mmol) was completely dissolved in 80 mL of diluted HCl solution (1.2 M) at 60 °C. Then, 0.5 g of Tri-MoVO was dispersed in the Sb containing solution, followed by the stirring for 3 h at the same temperature. After the filtration, obtained solids were abbreviated as *x*Sb, where *x* indicates 100×Sb/Mo ratio measured by ICP. Heat-treatment of *x*Sb was conducted at 550 °C for 2 h under 50 mL min^{-1} of N_2 flow

Results and discussion

First, we investigated the relationship between the preparative Sb/Mo ratio and the Sb/Mo ratio of the obtained solid measured by ICP. The amount of Sb in the solid was increased with the increase of the preparative Sb/Mo ratio. However, no further increase in the solid Sb/Mo ratio was observed when preparative Sb/Mo ratio reached to 0.065. This value corresponds one Sb atom upon a unit cell of the Tri-MoVO structure. Then, in situ XRD experiments were carried out for *x*Sb. Before the heat-treatment, all of the *x*Sb showed the XRD peaks at 2θ = 4.7°, 9.0° attributable to the diffraction of (100) and (110) planes of the Tri-MoVO structure. With increasing the temperature under N_2 flow, the XRD peak intensity attributable to the (100) diffraction decreased which implied the increased electron density around the center of the three heptagonal channel sites of the Tri-MoVO structure[3]. *0*Sb (original Tri-MoVO) transformed into MoO_3 or ($Mo_{0.93}$V

heating. On the other hand, additional diffraction peaks were observed at 2θ = 6.7°, 7.9° and 9.0° when Sb was introduced into the Tri-MoVO structure. These peaks were attributable to the diffraction of (010), (120) and (210) planes of the Orth-MoVO structure. The solid Sb/Mo ratio significantly affected this phase transformation since (1) the purity of the Orth-MoVO structure was improved and (2) phase transformation temperature was decreased with the Sb/Mo ratio. The introduction of Sb was found to be crucial for the phase transformation to the Orth-MoVO structure. Through the adsorption experiments, Rietveld analysis and DFT calculation, it was found that Sb was located mainly at the heptagonal channel and partly at the hexagonal channel of the Tri-MoVO structure, while Sb in the Orth-MoVO structure was placed mainly at the hexagonal channel and partly at the heptagonal channel. In order to further make sure this phase transformation mechanism, HAADF-STEM analysis was carried out for *6.2*Sb with or without the heat treatment. During the phase transformation, the increased electron density at the center of the three heptagonal channels of the Tri-MoVO structure was observed possibly due to the location of Sb, while the location of Sb at the hexagonal channel was observed after the phase transformation to the Orth-MoVO structure. Based on the above facts, we concluded that the phase transformation proceeds as to transform the heptagonal channel of the Tri-MoVO structure into the hexagonal channel of the Orth-MoVO structure. Then, ammoxidation of propane was carried out over Tri-MoVO and *6.2*Sb heat treated either at 400 °C or 550 °C (*6.2*Sb-NT400 or *6.2*Sb-NT550, respectively). After the catalytic reaction, Tri-MoVO and *6.2*Sb-NT400 had the Tri-MoVO structure, while *6.2*Sb-NT550 had the Orth-MoVO structure. By the comparison between Tri-MoVO and *6.2*Sb-NT400, it was clear that the introduction of Sb significantly improved the acrylonitrile (AN) selectivity. The phase transformation from the Tri-MoVO structure to the Orth-MoVO structure not only enhanced the AN selectivity but also substrates conversion.

Conclusions

Tri-MoVO was collapsed to MoO_3, $(V_{0.07}Mo_{0.93})_5O_{14}$ and $V_{0.95}Mo_{0.97}O_5$ by the heat-treatment at 550 °C under N_2 flow. However, Tri-MoVO containing Sb was transformed to Orth-MoVO structure by the same heat-treatment. The phase transformation was found to proceed as to transform the heptagonal channel of Tri-MoVO structure into hexagonal channel of Orth-MoVO structure. Due to the similarity of the phase transformation between the MoVTe(Sb)NbO precursor to the MoVTe(Sb)NbO and the Tri-MoVO structure to the Orth-MoVO structure, we suggest that Te or Sb being located at the heptagonal channel moiety in the MoVTe(Sb)NbO precursor plays crucial role for the crystal formation of the MoVTe(Sb)NbO as to stabilize the crystal structure during the heating condition.

References

1. R. K. Grasselli, Catal. Today 99 (2005) 23.
2. M. Sadakane, N. Watanabe, T. Katou, Y. Nodasaka, W. Ueda, Angew. Chem. Int. Ed. 46 (2007) 1493.
3. Ishikawa, S.; Murayama, T.; Kumaki, M.; Tashiro, M.; Zhang, Z.; Yoshida, A.; Ueda, W. *Top. Catal.* **2016**, *59*, 1477–1488.

545622 Elucidating the Role of CO2 in the Soft Oxidative Dehydrogenation of Propane over Ceria-Based Catalysts

Tuesday, June 4, 2019: 5:12 PM
Republic ABC (Grand Hyatt San Antonio)

James H. Carter[1], Ewa Nowicka[2], Christian Reece[3], Sultan Althahban[4], Khaled Mohammed[5], Simon Kondrat[6], David J. Morgan[7], Qian He[2], David Willock[8], Stanislaw Golunski[2], Christopher J Kiely[9], Graham J. Hutchings[1] and **Nicholas Dummer**[2], (1)School of Chemistry, Cardiff Catalysis Institute, Cardiff, United Kingdom, (2)Cardiff University, Cardiff, United Kingdom, (3)Harvard University, Cambridge, MA, United States Minor Outlying Islands, (4)Lehigh University, Bethlehem, PA, (5)UK Catalysis Hub, Research complex at Harwell, Harwell, United Kingdom, (6)Loughborough University, Loughborough, United Kingdom, (7)Carfidff University, Cardiff, United Kingdom, (8)Cardiff University, cardiff, United Kingdom, (9)Materials Science and Engineering, Lehigh University, Bethlehem, PA

Elucidating the Role of CO2 in the Soft Oxidative Dehydrogenation of Propane over Ceria-Based Catalysts

Abstract

A mixed oxide support containing Ce, Zr, and Al was synthesized using a physical grinding method and applied in the oxidative dehydrogenation of propane using CO_2 as the oxidant. The activity of the support was compared with that of fully formulated catalysts containing palladium. The Pd/CeZrAlO_x material exhibited long-term stability and selectivity to propene (during continuous operation for 140 h), which is not normally associated with dehydrogenation catalysts. From temperature-programmed desorption of NH_3 and CO_2 it was found that the catalyst possessed both acidic and basic sites. In addition, temperature-programmed reduction showed that palladium promoted both the reduction and re-oxidation of the support. When the role of CO_2 was investigated in the absence of gas-phase oxidant, using a temporal analysis of products (TAP) reactor, it was found that CO_2 dissociates over the reduced catalyst, leading to formation of CO and selective oxygen species. It is proposed that CO_2 has the dual role of regenerating selective oxygen species and shifting the equilibrium for alkane dehydrogenation by consuming H_2 through the reverse water-gas-shift reaction. These two mechanistic functions have previously been considered to be mutually exclusive.

Summary of Results and Discussion

5 % Pd/CeZrAlO_X was identified as an optimal catalyst after comparing its activity with different Pd loadings and various formulations of Ce:Zr:Al supports (data not presented here). Further investigation into the catalyst, its stability and the reaction mechanism were then carried out. The stability of the catalyst is a major issue in the commercial operation of propane dehydrogenation processes. Most current technologies, which are based on direct (non-oxidative) dehydrogenation, are performed at higher temperatures to achieve cost-effective yields of propene, but this can lead to rapid deactivation of the

catalysts through coking. Therefore, in order to maintain a stable yield of propene, the commercial processes (Catofin, STAR and Oleflex) require addition of steam to the gas-feed (to supress coking), frequent regeneration of the catalyst, continuous partial replacement of the catalyst, or a rising profile for the operating temperature. Nonetheless the commercial catalysts generally exhibit relatively short lifetimes, up to a maximum of 3 years.

We have therefore tested our materials for extended periods to observe any changes in catalytic activity. In our investigation, we focused on the selectivity profile since it became quite apparent that the selectivity improves within time-on-stream. The 5% Pd/CeZrAlOx catalyst exhibited high stability over 140 h of reaction at 500 °C, as shown in Figure 1. Selectivity to propene clearly improved during the reaction whereas the C3H8 conversion significantly decreased during the first 1 h of reaction. This is consistent with our interpretation that unselective oxygen is initially consumed from the catalyst, followed by CO2 dissociation on the catalyst surface, resulting in the formation of selective oxygen species which are subsequently consumed in the oxidative dehydrogenation reaction. ADDIN EN.CITE <EndNote><Cite><Author>Nowicka</Author><Year>2018</Year><IDText>Elucidating the Role of CO2 in the Soft Oxidative Dehydrogenation of Propane over Ceria-Based Catalysts</IDText><DisplayText><style
ISI>://WOS:000430154100092</url></related-urls></urls><isbn>2155-5435</isbn><titles><title>Elucidating the Role of CO2 in the Soft Oxidative Dehydrogenation of Propane over Ceria-Based Catalysts</title><secondary-title>Acs Catalysis</secondary-title></titles><pages>3454-3468</pages><number>4</number><contributors><authors><author>Nowicka, E.</author><author>Reece, C.</author><author>Althahban, S. M.</author><author>Mohammed, K. M. H.</author><author>Kondrat, S. A.</author><author>Morgan, D. J.</author><author>He, Q.</author><author>Willock, D. J.</author><author>Golunski, S.</author><author>Kiely, C. J.</author><author>Hutchings, G. J.</author></authors></contributors><added-date format="utc">1530628972</added-date><ref-type name="Journal Article">17</ref-type><rec-number>760</rec-number><last-updated-date format="utc">1530628972</last-updated-date><accession-num>WOS:000430154100092</accession-num><electronic-resource-num>10.1021/acscata1.7b03805</electronic-resource-num><volume>8</volume></record></Cite></EndNote>[1] It is interesting to note that a gradual decrease of activity to the 4% conversion level was observed, and then relatively stable activity was maintained until the end of the 140 h test. This implies that our 5% Pd/CeZrAlOx catalyst underwent changes not only during the first hour of the reaction but also during the remaining period of the test. To further understand this aspect, we performed detailed characterization of the 'fresh' and '140 h used' material.

B)

A)

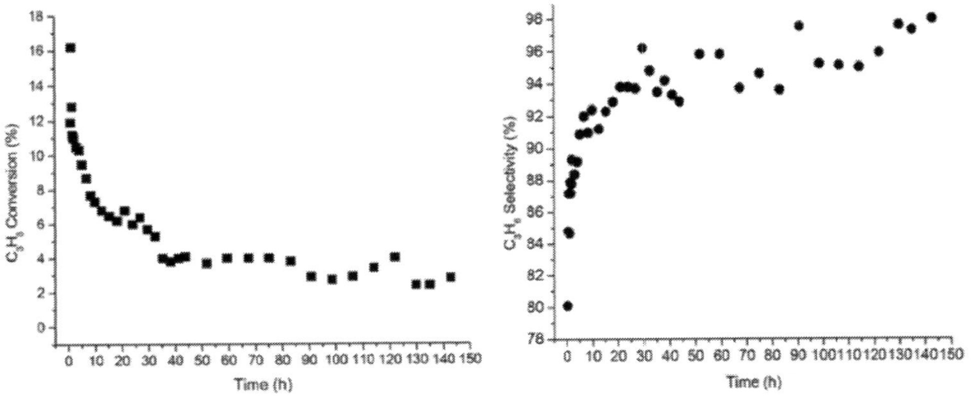

Figure 1. A) Propane conversion and B) propene selectivity during oxidative dehydrogenation over an extended reaction period using the 5% Pd/CeZrAlOx catalyst. Reaction conditions: total flow 15 cm^3 min^{-1} (37% C_3H_8, 37% CO2, 26% He), GHSV = 6000 h^{-1}, T = 500 °C.

We used XAFS to probe the local structures of the fresh and used samples of CeZrAlOx and Pd/ CeZrAlOx, concentrating on the Pd-K and Ce-L3 edge. These clearly show that the PdO present in the fresh Pd-CeZrAlOx catalyst is reduced to metallic Pd during reaction. The evidence for PdO is constituted by the clear Pd-O distance at 2.01 Å, which is entirely absent from the samples analysed after 4 h and 140 h on-line. In addition, the Pd-Pd distances shorten from those characteristic of PdOx in the fresh sample to that of fcc Pd metal in the used samples. Also clear from the EXAFS fitting data is an increase in Pd coordination number with time on-line which is associated with a growth in Pd particle size during reaction. The Ce L3 edge XANES is rich in information regarding the Ce oxidation state. After reaction with propane the Ce underwent gradual reduction with the average valance state dropping to 3.3 after 4 h on-line and further decreasing to 3.2 after 140 h on-line.

To confirm the dissociation of CO2 on the catalyst surface, CO2 was pulsed over reduced CeZrAlOx and 5%Pd/CeZrAlOx catalysts using a TAP reactor. The experiment was first performed at room temperature, and it was found that CO2 irreversibly binds to both materials. It was also found that the 5%Pd/CeZrAlOx was able to dissociate CO2 to form CO, whereas the CeZrAlOx was not, indicating the Pd facilitates the formation of CO at lower temperatures. When the experiment was repeated at a higher temperature (400°C) a stronger CO response was detected in both cases, demonstrating that CO2 was being dissociated more readily at the higher temperature. The increased temperature also caused the CO2 adsorption to become reversible, causing the CO2 response to increase. These data were also supported by complementary DRIFTS experiments.

Interestingly the CO2/CO mass balance was consistently 100% ± 2% during ODH over the Pd/CeZrAlOx, which suggests minimal deposition of carbon on the catalyst during the reaction. This was supported by elemental analysis of the used catalyst after 4 hours of reaction, which showed that the amount of retained carbon was equivalent to only 3.7 % of all the converted propane. These results infer that an additional function of the Pd is to suppress carbon deposition on the catalyst by enhancing the mobility of the surface oxygen or by promoting the reverse-Boudouard reaction.

Conclusions

The role of CO_2 as a soft oxidant in the dehydrogenation of alkanes over reducible metal oxides has previously been attributed to its ability either (i) to reoxidize the reduced surface as part of an MvK reaction mechanism or (ii) to consume the hydrogen released by direct dehydrogenation through the RWGS reaction. These two pathways have previously been thought to be mutually exclusive. It is worth noting that the RWGS reaction may also proceed by a redox mechanism. Our results indicate that, in ceria-based catalysts such as CeZrAlOx, the MvK and RWGS mechanisms can be linked by a common step. We propose that lattice oxygen ions abstract hydrogen from the reactant alkane molecules to form the product alkene and H2O, while the CO2 replenishes these selective oxygen species, releasing CO into the gas phase (see Figure 2). Although the byproducts (CO and H2O) are also the products of the RWGS, this route to alkenes does not require the stepwise formation of H2 as an intermediate. The high selectivity to the desired alkene can be explained by the exclusive formation of fully reduced oxygen ions when CO2 dissociatively adsorbs at surface oxygen vacancy sites. In contrast, the adsorption of O2 is known to lead to the formation of transient electrophilic oxygen species (such as adatoms, O2−, O−) before O2− ions are formed. Although Figure 13. Representative scheme of the reactions involved in the overall ODH process these electrophilic species are generally short-lived, they account for the formation of CO2 (and any CO) when O2 is used as the oxidant. The ability of Pd to increase the oxidation activity of redox metal oxide catalysts is well-known and correlates with the observed enhanced reducibility and re-oxidation of the metal oxide surface. This rapid cycle of creating surface oxygen vacancies and then refilling them from the gas phase prevents the rate of the catalytic reaction becoming limited by the transfer of lattice oxygen ions. However, when O2 is the oxidant, the presence of Pd promotes the rate of both the selective and nonselective oxidation reactions. By way of contrast, our results demonstrate that, when CO2 is used as the oxidant, the presence of Pd has the critical function of increasing conversion without sacrificing selectivity to propene.

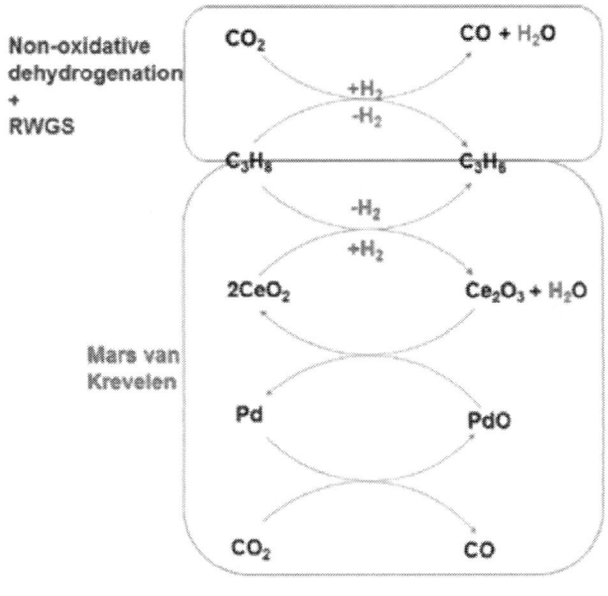

Figure 2. Summary of proposed mechanism of propane dehydrogenation using CO_2

over Pd/$CeZrAlO_x$.

References

ADDIN EN.REFLIST 1. E. Nowicka, C. Reece, S. M. Althahban, K. M. H. Mohammed, S. A. Kondrat, D. J. Morgan, Q. He, D. J. Willock, S. Golunski, C. J. Kiely and G. J. Hutchings, *Acs Catalysis*, 2018, **8**, 3454.

545654 CO_2 Assisted Oxidative Dehydrogenation of Ethane to Ethylene over FeO_x-Based Catalysts

Tuesday, June 4, 2019: 5:36 PM
Republic ABC (Grand Hyatt San Antonio)

Charis Loizidis, Chemical Engineering, Aristotle University Thessaloniki, Thessaloniki, Greece, Eleni Heracleous, School of Science & Technology, International Hellenic University (IHU), Thessaloniki, Greece and **Angeliki Lemonidou**, Chemical Engineering, Aristotle University of Thessaloniki, Thessaloniki, Greece

Introduction

In recent years, significant efforts have been made for utilizing CO_2 in high temperature reactions, such as dry methane reforming, oxidative coupling of methane, in addition to the conversion of CO_2 into fuels such as methanol and DME. From the plethora of processes that are either in industrial scale or under development, promising are those in which CO_2 is used as a mild oxidant instead of O_2 [1]. Characteristic examples are the dehydrogenation of ethylbenzene to styrene and the dehydrogenation of light paraffins to the corresponding olefins [2]. Ethylene is the cornerstone of the petrochemical industry with over 180 million tons production and annual growth of 3-4% [3]. Moreover, the increasing availability of ethane from shale gas has revived the interest in the development of new sustainable, low environmental impact ethylene production processes [4]. Oxidative dehydrogenation (ODH) of ethane is an alternative process for ethylene production with significant advantages over the energy-intensive steam cracking process [5]. The use of CO_2 as mild oxidant in the dehydrogenation of ethane is a promising concept, with the overall reaction being the following: $C_2H_6 + CO_2 \leftrightarrow C_2H_4 + CO + H_2O$

The main advantage of CO_2 over O_2 is the absence of undesirable secondary oxidation reactions to CO_x and the safer reactor operation due to the avoided use of flammable oxygen/hydrocarbon mixtures. In addition, CO_2 is simultaneously converted to CO, a valuable raw material for the production of chemicals (e.g. oxo-alcohols) and liquid fuels through the F-T reaction. The CO_2-assisted ethane dehydrogenation has been mainly studied over redox catalysts that generally have good performance in alkane ODH [6]. Despite the extensive research, the relatively low turnover frequencies, the low ethylene selectivity and the extensive catalyst deactivation still remain major challenges in the pursuit of an active and stable catalyst.

In this study, we present the promising performance of FeO_x-based catalysts supported on (NiO-)ZrO_2-MgO in the CO_2-assisted ethane dehydrogenation to ethylene. Temperature programmed tests provide insight of the optimum temperature window for attaining maximum selectivity. Parametric analysis was performed to explore the effect of the basic operating parameters on the efficiency of the reaction.

Experimental

The FeO_x-based catalysts with 5 and 10 wt% Fe loading were synthesized by combining the methods of sol-gel auto-combustion and wet impregnation on mixed Ni(5%), Mg(70%) and Zr(25%) oxide. The supports were calcined at 800°C for 3h. After impregnation of the Fe precursor salts and drying, the catalysts were further calcined at 600°C for 3h. Characterization included BET, XRD, TPR, TPO and TEM. TP reaction tests were also conducted in a transient unit equipped with a mass analyzer. Testing of the FeO_x catalysts on ZrO_2-MgO with and without 5 wt% NiO was conducted in a lab-scale continuous flow unit with fixed bed reactor. The catalysts were tested at 600°C with CO_2/C_2H_6 feed at 1.3/1 molar ratio in mixture with N_2, and W/F values ranging from 0.013 to 0.043 g.min/cm^3. The products of the reaction were C_2H_4, CO, CH_4, H_2 and H_2O.

Results

The surface areas of all four catalysts are >100 m^2/g. The XRD patterns of the bare supports reveal the presence of cubic crystalline structures of MgO and ZrO_2. No peak corresponding to NiO is detected, probably due to its low concentration and high dispersion. Over the catalysts, a low-intensity peak at 35.5-35.7°, characteristic of ã-Fe_2O_3 and Fe_3O_4 phases, is observed. TEM images of the 5% Fe/NiO-ZrO_2-MgO sample show the presence of ã-Fe_2O_3 nanocrystals (10-15 nm). The TPR of the catalysts show a reduction peak centered at ~400°C, ascribed to the reduction of Fe_2O_3 to FeO. Complete reduction to metallic Fe occurs at temperatures over 700°C, outside the temperature window of the dehydrogenation reaction. No reduction peaks appear in the TPR profiles of the upromoted and NiO-promoted ZrO_2-MgO supports, implying the formation of a solid solution between NiO and MgO that prevents NiO reduction.

TP reaction tests in C_2H_6/CO_2 flow over the bare NiO-ZrO_2-MgO support demonstrate that temperatures >620ïC are necessary for the activation of the reactants. Similar tests with the catalyst show that in the presence of FeO_x the onset temperature drops by 100ïC. In the absence of CO_2, FeO_x is quickly reduced to Fe by C_2H_6 and catalyzes the reforming ethane to CO and H_2 at temperatures as low as 600°C. Therefore, the presence of CO_2 is crucial in maintaining the active phase in oxidized form and drive the dehydrogenation reaction to ethylene. The beneficial effect of CO_2 in reducing coke, possibly via the Boudouard reaction, is also evident.

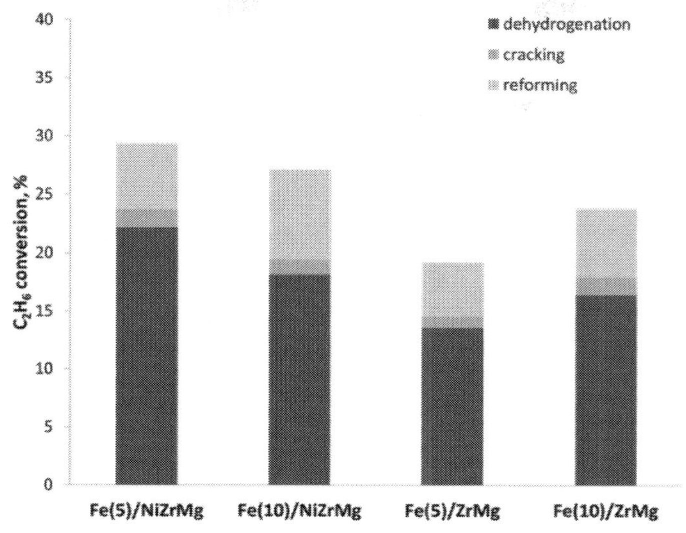

Figure 1. Ethane conversion (T=600°C, W/F=0.026 g.min/cm^3)

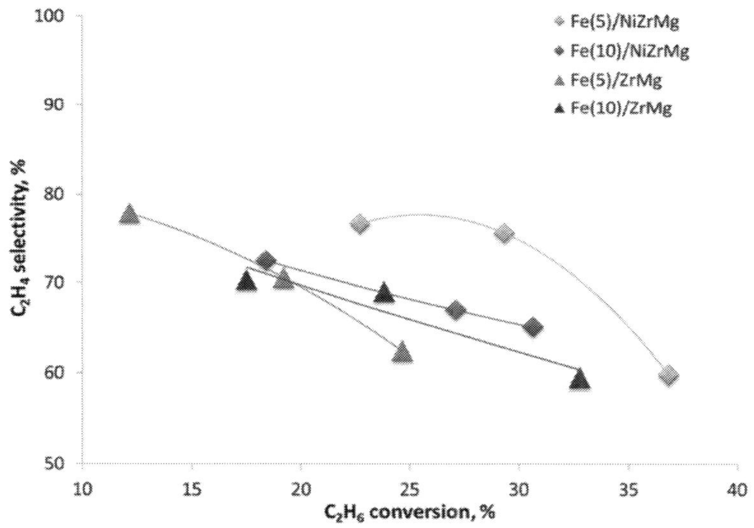

Figure 2. Ethylene selectivity vs ethane conversion (T=600°C, W/F=0.013-0.043 g.min/cm^3)

Testing of the materials at 600°C with CO_2/C_2H_6 molar ratio 1.3 in the lab-scale unit confirm the high activity of the catalysts. Ethane conversions vary in the range of 20-30% (Fig. 1), with similar CO_2 conversion (22-32%). The increase of the FeO_x loading has a positive effect only in the catalysts supported on ZrO_2-MgO. When the support is promoted with NiO, activity is boosted significantly (50% increase), probably masking the loading effect. Ethylene selectivity is very high, in the range of 60-80%. Ethylene selectivity as a function of ethane conversion is presented in Figure 2. Selectivity decreases with increasing conversion due to the increased extent of secondary reactions. The different loading catalysts show similar selectivity-conversion curves, indicating that only the number, and not the nature of the active sites, is altered with increasing loading. On the contrary, incorporation of NiO in the ZrO_2-MgO support enhances selectivity, with the 5 wt% Fe/NiO-ZrO_2-MgO catalyst exhibiting the highest selectivity equal to ~80% at 20-30% conversion. The product distribution allows a first estimation of the reaction network over the FeO_x-based catalysts, which comprises the dehydrogenation of ethane to ethylene, the consecutive oxidation of H_2 to H_2O via the reduction of CO_2 to CO (Reverse Water Gas Shift Reaction), the dry reforming of ethane from CO_2 to CO and H_2 and, to a small extent, the hydrogenolysis of ethane to methane. Figure 1 also shows the contribution of the three main reactions to the total measured ethane conversion. It is apparent that the desired route – dehydrogenation to ethylene – constitutes the main route of ethane consumption, followed by reforming and finally cracking to methane. The dehydrogenation contribution is highest in the 5 wt% Fe/NiO-ZrO_2-MgO catalyst, consistent with its highest selectivity.

The optimum catalyst was further testing at different operating conditions. The effect of temperature on conversion and selectivity is of particular interest. Conversion, as expected, increases significantly with temperature. Ethylene selectivity however, reduces by 10% by only a small temperature increase from 600°C to 625°C at constant conversion (27%). This decrease can be attributed to the side reaction of dry reforming. The apparent activation energy for the consumption of ethane and CO_2, calculated from the Arrhenius plots, is equal to 115 and 123 kJ/mol respectively. These values, although not directly comparable, are consistent with corresponding values over Cr-containing catalysts [7]. A stability test was conducted for 17.5h of reaction to monitor catalyst stability. A 25% activity loss occurs at the early stages of the reaction, with conversion stabilizing at longer reaction times. Characterization of the used catalyst did not show appreciable coke deposition. Loss of activity is probably related to sintering of the active phase and/or partial reduction of Fe_2O_3 to Fe_3O_4. The latter was confirmed by TEM and by visual observations, as the catalyst color changed from orange to green-gray, indicative of iron oxide reduction.

Conclusions

This work demonstrates that FeO_x catalysts on mixed (NiO)-ZrO_2-MgO support constitute a promising catalytic system for the CO_2-mediated ethane conversion to ethylene, exhibiting relatively high ethane and CO_2 conversion with ethylene selectivity up to 80% at mild reaction temperature. Promotion of the ZrO_2-MgO support with NiO enhances both activity and selectivity of the reaction, indicating a participation and/or modification of the FeO_x active sites by NiO. Increase in iron loading increases activity in the FeO_x-based catalysts on unmodified ZrO_2-MgO, without substantial influence on selectivity. This suggests that the active centers remain the same, with the additional loading simply increasing their concentration on the surface. The optimum material is the 5 wt% Fe/NiO-ZrO_2-MgO catalyst that exhibits the highest activity and ethylene selectivity. The stability test reveals moderate catalyst deactivation in the first hours of reaction, attributed to sintering and partial reduction of Fe, with the performance stabilizing at longer operating times. It is worth mentioning that although the research is in early stages, the conversion and selectivity values recorded are among the highest reported in the open literature on Cr-free catalysts.

References

[1] D. Mukherjee, S.-E. Park, B.M. Reddy, Journal of CO_2 Utilization 16 (2016) 301

[2] M.B. Ansari, S.-E. Park, Energy Environ. Sci. 5 (2012) 9419

[3] Steve Lewandowski, IHS Markit, Asia Chemical Conference, Nov 2016

[4] R.A. Kerr, Science 328 (2010) 1624

[5] Z. Skoufa, G. Giannakakis, E. Heracleous, A.A. Lemonidou, Catal Today, 299 (2018) 102

[6] S. Kawi, Y. Kathiraser, Frontiers in Energy Research 3 (2015) 2

[7] N. Mimura, M. Okamoto, H. Yamashita, S.T. Oyama, K. Murata, J. Phys. Chem. B 110 (2006) 21764

545687 Effects of BrøNsted Acid Site Density and Proximity in MFI and CHA Zeolites on Light Alkane Activation and Light Alkene Oligomerization Catalysis

Wednesday, June 5, 2019: 2:42 PM
Texas Ballroom EF (Grand Hyatt San Antonio)

Philip M. Kester, Young Gul Hur, Elizabeth E. Bickel, Fabio H. Ribeiro, Jeffrey T. Miller and Rajamani Gounder, Charles D. Davidson School of Chemical Engineering, Purdue University, West Lafayette, IN

Effects of Brønsted acid site density and proximity in MFI and CHA zeolites on light alkane activation and light alkene oligomerization catalysis

Philip M. Kester, Young Gul Hur, Elizabeth E. Bickel, Fabio H. Ribeiro, Jeffrey T. Miller, and Rajamani Gounder*

Charles D. Davidson School of Chemical Engineering, Purdue University, 480 Stadium Mall Drive, West Lafayette, IN 47907, USA

*rgounder@purdue.edu

The activation of light alkanes to form alkenes, and subsequent oligomerization to heavier molecular weight hydrocarbons, provide a strategy for converting gases into liquids useful as transportation fuels. In the case of Brønsted acid zeolite catalysts (e.g., MFI), the proximity of framework Al atoms, which generate active H^+ sites, has been proposed to influence monomolecular alkane activation turnover rates via polarization of kinetically-relevant carbonium ion transition states [1]. The sizes of primary crystallites and the density of framework Al atoms also influence the rates, selectivity, and deactivation during alkene oligomerization in Brønsted acid zeolites because longer diffusion paths and higher active site densities preferentially increase intracrystalline residence times of larger products [2]. The combined effects of crystallite size and Al content are reflected in a characteristic diffusion parameter, proportional to the volumetric density of H^+ sites and the square of the characteristic diffusion length [2]. Although these structural properties are typically correlated in aluminosilicate MFI zeolite samples synthesized hydrothermally, they can be manipulated independently through alternative synthesis routes, such as discussed here through addition of B heteroatoms into synthesis solutions containing tetra-n-propylammonium (TPA^+) and ethylenediamine (EDA) as structure directing agents (SDAs).

Crystallite sizes (0.3 – 10 μm) and Al density ($0.1 - 0.2 \times 10^{-3}$ mol Al g^{-1}) of B-Al-MFI zeolites are independently varied by the addition of B to synthesis mixtures (Si/B = 2.6 – 13) containing TPA^+ and EDA. The presence of different SDAs during crystallization also influence the siting of Al in zeolite frameworks [3]. Here, we provide evidence that the synthesis of B-Al-MFI zeolites (Si/Al = 50) from mixtures containing only TPA^+ as the SDA contain both proximal and isolated Al, while those made from mixtures containing EDA and TPA^+ contain predominantly isolated Al. Protons compensating framework B are weaker in acid strength (higher deprotonation energy [4]) than those compensating framework Al, requiring methods to discriminate and quantify protons at framework Al prior to interpreting catalytic phenomena on boroaluminosilicates. Temperature programmed desorption (TPD) of ammonia of B-Al-MFI after liquid-phase NH_4^+-exchange quantify all framework B and Al heteroatoms, but TPD performed after NH_4-form B-Al-MFI were purged in flowing He (433 K), or after gas-phase NH_3 adsorption (433 K) onto H-form B-Al-MFI, quantify only protons at framework Al atoms (H^+_{Al}) [5]. Methanol dehydration turnover rates on B-Al-MFI zeolites (415 K, per H^+_{Al}) measured in zero-order kinetic regimes, which are insensitive to acid strength and confinement, are identical to those measured on Al-MFI zeolites, validating these NH_3 site titration protocols. Turnover rates (per H^+_{Al}) measured in first-order kinetic regimes, which are sensitive to differences in both acid strength and confinement [6], are lower on B-Al-MFI zeolites than those previously reported for Al-MFI zeolites [7], suggesting that Al is preferentially sited within larger MFI intersections (~0.7 nm) in boroaluminosilicate compositions. Among CHA zeolites, monomolecular propane cracking rates (748 K, per H^+_{Al}) increase with the fraction of proximal H^+ sites, even among samples of fixed Al composition. We will also discuss the dependences of monomolecular propane activation rates (718-778 K) and of propene oligomerization rates (473-573 K) on H^+_{Al} proximity in MFI zeolites.

References:

[1] Song, C., Chu, Y., Wang, M., Shi, H., Zhao, L., Guo, X., Yang, W., Shen, J., Xue, N., Peng,

L., Ding, W. *J. Catal.* 349 (2017) 163-174.

[2] Sarazen, M. L., Doskocil, E., Iglesia, E. *ACS Catal.* 6 (2016) 7059-7070.

[3] Di Iorio, J. R., Gounder, R. *Chem. Mater.* 28 (2016) 2236-2247.

[4] Jones, A. J., Carr, R. T., Zones, S. I., Iglesia, E. *J. Catal.* 312 (2014) 58-68.

[5] Kester, P. M., Miller, J. T., Gounder, R. *Ind. Eng. Chem. Res.* 57 (2018) 6673-6683.

[6] Carr, R. T., Neurock, M., Iglesia, E. *J. Catal.* 278 (2011) 78-93.

[7] Di Iorio, J. R., Nimlos, C. T., Gounder, R. *ACS Catal.* 7 (2017) 6663-6674.

545373 High and Stable Activity of Dealuminated Zn/H-ZSM-5 for Ethane Dehydroaromatization

Wednesday, June 5, 2019: 2:45 PM
Texas Ballroom EF (Grand Hyatt San Antonio)

Hikaru Saito[1], Satoshi Inagaki[2], Qian Han[2], Ryota Terunuma[1], Tomohiro Yabe[1], Shuhei Ogo[1], Yoshihiro Kubota[2] and Yasushi Sekine[1], (1)Applied Chem., Waseda Univ., Tokyo, Japan, (2)Yokohama National Univ., Yokohama, Japan

We investigated effects of steam treatment on Zn/H-ZSM-5 for ethane dehydroaromatization to suppress coke formation, which deactivates the activity. Steam treatment is a well-known technique to decrease the amount of Brønsted acid sites through converting framework Al species to extra framework ones.
Steam treatment of Zn/H-ZSM-5 prepared by an ion-exchange method was conducted in a fixed bed reactor at 823 K for 1 h under H_2O/Ar flow. The steam-treated samples is denoted as Zn/H-ZSM-5 ST. Catalytic activity tests were conducted in a fixed bed reactor at 873 K and atmospheric pressure under 80vol% ethane and balanced nitrogen.
The initial activity of Zn/H-ZSM-5 ST was comparable to Zn/H-ZSM-5: ethane conversion is about 20%. Although the activity of Zn/H-ZSM-5 was rapidly deactivated in 2 h with time on stream, that of Zn/H-ZSM-5 ST was quite stable for 4 h. Temperature programmed oxidation revealed that the amount of carbon deposition on Zn/H-ZSM-5 ST is one-tenth smaller than that of Zn/H-ZSM-5. During steam treatment, the qualitative and quantitative nature of Zn was not changed from the results of UV-Vis DRS, XRD and ICP-OES. Therefore, we concluded that steam treatment affected Brønsted acidity. Then, we measured 29Si MAS NMR spectra to obtain information on the local structure of H-ZSM-5 supports. In the spectrum of steam-treated H-ZSM-5, the peak intensity of Q4(1Al), which means $(AlO)Si(SiO)_3$, significantly decreased in comparison with parent H-ZSM-5. On the other hand, a little decrease in the peak intensity was observed in the spectrum of Zn/H-ZSM-5 ST. This indicates that ion-exchange of Zn^{2+} with H^+ stabilize framework Al species. In other words, framework Al species related to H^+ (Brønsted acid sites) would be easily converted to extra framework ones. Therefore, the suppression of coke formation was achieved thanks to the optimal amount of Brønsted acid sites.

545403 Paraffin Activation and Conversion Technology (PACT), a New Flexible and Robust Technology for the Conversion of C5+/Natural Gasoline to Higher Valued Products

Wednesday, June 5, 2019: 2:48 PM
Texas Ballroom EF (Grand Hyatt San Antonio)

Jens Michael Poulsen, Haldor Topsoe A/S, Lyngby, Denmark and **Anthony Baldridge**, Phillips 66, Bartlesville, OK

Paraffin Activation and Conversion Technology (PACT), jointly developed by Haldor Topsoe and a U.S.-based development partner is a new, flexible, and robust technology for the conversion of C5+/natural gasoline to higher valued products. The recent increase in the supply of natural gas liquids (NGLs) due to the U.S. shale revolution motivates an opportunity to upgrade surplus C5+/natural gasoline streams to products with larger market and higher value. Existing commercial technologies for NGL conversion are optimized for lighter paraffins, but more cost-effective upgrading of C5+/natural gasoline streams can be realized with novel advancements in catalysis coupled with a unique process design.

Current U.S. EIA estimates show a 60% increase in C5+/natural gasoline supply within the US from 2018 to 2022 with supply continuing to outpace demand. Additionally, global supplies of C5+/natural gasoline/light naphtha are expected to rise at similar rates providing a global overall surplus of these streams. While this supply continues to grow, traditional outlets for C5+/natural gasoline, including gasoline blending, diluent for heavy crudes, and naphtha steam cracking, are likely unable to absorb the growing surplus of C5+/natural gasoline.

The newly developed PACT Process converts C5+/natural gasoline streams into more valuable products including high octane blendstocks, aromatics, olefins, and hydrogen through the utilization of robust operation units to minimize technical risk. Pilot studies of the paraffin activation step of the PACT Process have demonstrated process flexibility for a variety of streams. This presentation will include pilot-scale results highlighting the yield and conversion structure along with key process and economic indicators for this new technology.

545486 Ethene Dimerization Mediated By the Coordination-Insertion Mechanism at Nickel Cations Supported Onto Molecular Sieves

Wednesday, June 5, 2019: 2:51 PM
Texas Ballroom EF (Grand Hyatt San Antonio)

Ravi Joshi[1], Guanghui Zhang[2], Jeffrey T. Miller[1] and Rajamani Gounder[1], (1)Charles D. Davidson School of Chemical Engineering, Purdue University, West Lafayette, IN, (2)Davidson School of Chemical Engineering, Purdue University, West Lafayette, IN

Dimerization is an entry step in oligomerization and chain growth pathways to convert light alkenes derived from shale gas hydrocarbons into chemical intermediates and liquid fuels [1]. Nickel cations supported on various inorganic supports are reported to catalyze alkene dimerization in the absence of externally supplied activators or co-catalysts [2], which are often required in the case of homogeneous nickel-halide complexes [3]. The mechanistic origin of alkene dimerization on Ni sites located on heterogeneous aluminosilicate supports, however, has been ascribed to both coordination-insertion and metallacycle-based cycles. Conflicting conclusions appear to reflect attempts to reconcile observations that externally supplied activators or co-catalysts are not required to initiate the catalytic cycle (which is characteristic of the metallacycle mechanism), and that products are also formed from alkene reaction pathways occurring in parallel (e.g., oligomerization, isomerization) at H^+ sites located on these supports.

Here, ethene dimerization (453 K) was studied in absence of externally supplied activators on Beta zeolites synthesized to contain exchanged Ni^{2+} cations, according to site balances determined by cation exchange, and to CO infrared, UV-Visible, and Ni K-edge X-ray absorption spectroscopies. The catalytic contributions of residual H^+ sites on aluminosilicate supports were suppressed by selectively poisoning them with Li^+ cations or NH_3 base titrants, or by weakening them using a zincosilicate support. Beta zeolites with only H^+ sites form linear butene isomers (1-butene, cis-2-butene, trans-2-butene) in thermodynamically-equilibrated ratios, in addition to isobutene and products of subsequent oligomerization-cracking cycles; thus, isobutene formation serves as a kinetic marker for the presence of H^+ sites. On the other hand, after residual H^+ sites on Ni-zeolites deactivate during initial reaction times or when they are suppressed prior to reaction, linear butene isomers form in non-equilibrated ratios that are invariant with ethene site-time velocity, reflecting primary butene double-bond isomerization events catalyzed at Ni^{2+}-derived active intermediates. Further, *in situ* X-ray absorption spectroscopy showed that Ni cations retain their 2+ oxidation state during ethene dimerization. Also, Ni-zeolites pretreated in oxidative environments (5 kPa O_2, 773 K) show transient activation periods during initial reaction times at dilute ethene pressures (<0.4 kPa), but not at higher ethene pressures (>0.4 kPa) or in the presence of co-fed hydrogen (5 kPa). This behavior is consistent with *in situ* ethene-assisted formation of Ni^{2+}-H intermediates, which isotopically scramble H_2-D_2 mixtures (453 K) and are quantified from surface H/D exchange reactions (453 K).

Taken together, these findings provide unambiguous evidence for the coordination-insertion mechanism as the dominant route for alkene dimerization at Ni^{2+} cations exchanged onto molecular sieves. This implies that kinetic factors determine the isomer distribution in alkene dimers, providing opportunities for altering product isomer selectivity based on changes to the coordination or confining environment around Ni centers.

References -

[1] C.P. Nicholas, Applications of light olefin oligomerization to the production of fuels and chemicals, App. Catal., A, 543 (2017) 82-97.

[2] A. Finiels, F. Fajula, V. Hulea, Nickel-based solid catalysts for ethylene oligomerization - a review, Catal. Sci. Technol., 4 (2014) 2412-2426.

[3] K.P. Bryliakov, A.A. Antonov, Recent progress of transition metal based catalysts for the selective dimerization of ethylene, Journal of Organometallic Chemistry, 867 (2018) 55-61.

545729 Observation of Pt Migration on Pt/ZSM-5 Catalyst for Ethane Aromatization

Wednesday, June 5, 2019: 2:54 PM
Texas Ballroom EF (Grand Hyatt San Antonio)

Hui Wang[1], Junjun Shan[1], Lisa Nguyen[1], Joshua Miles[1], Hua Liu[2], Fu-Kuo Chiang[2] and Jihong Cheng[1], (1)NICE America Research, Mountain View, CA, (2)National Institute of Clean-and-Low-Carbon Energy (NICE), Beijing, China

Observation of Pt Migration on Pt/ZSM-5 Catalyst for Ethane Aromatization

Hui Wang[a, *], Junjun Shan[a], Lisa Nguyen[a], Joshua Miles[a], Hua Liu[b], Fu-Kuo Chiang[b], Jihong Cheng[a]

[a] *NICE America Research Inc, 2091 Stierlin Ct, Mountain View, California USA 94043*

[b] *National Institute of Clean-and-Low-Carbon Energy (NICE), Future Science & Technology City, Changping District, Beijing China 102211*

**Corresponding author: wanghui@nicenergy.com*

Owing to the low cost and increased supply of natural gas in the United States, the direct conversion of natural gas liquids (C2-C4) into aromatics is of great industrial importance.[1] In order to convert light alkanes such as ethane into aromatic products in a single step, a bi-functional catalyst is required where ethane is first activated on metal sites to ethylene, which is then selectively converted into aromatics on zeolite, as illustrated in Figure 1. Therefore, a catalyst with acid-metal intimacy could have synergy that allows for optimal performance in ethane aromatization reaction.

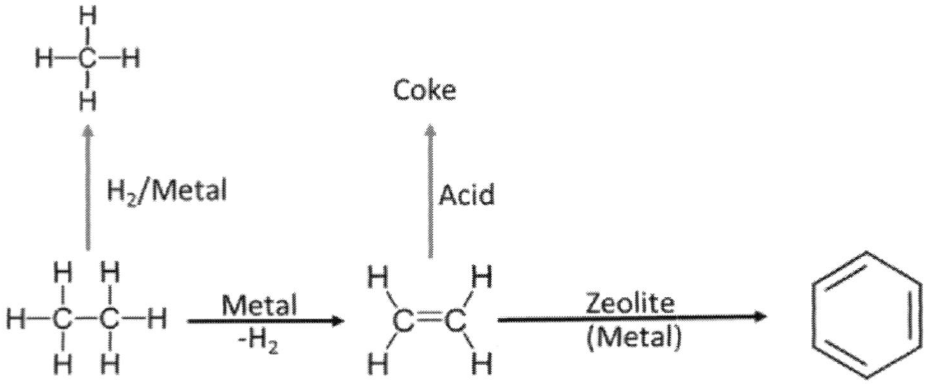

Figure SEQ Figure * ARABIC 1 Simplified mechanism for ethane aromatization

Such a composition can be easily achieved by impregnating or ion-exchanging ZSM-5 with Pt^{2+} during catalyst preparation. In the presence of binder such as alumina,

Pt can be distributed between binder and zeolite, leading to an inferior catalyst. We demonstrated experimentally that under certain conditions, Pt located on alumina binder can migrate onto zeolite crystal. Through advanced characterizations such as dynamic CO adsorption, diffuse reflectance infrared Fourier transform spectroscopy (DRIFTS), STEM, X-ray absorption near edge structure (XANES), Extended X-Ray Absorption Fine Structure (EXAFS), it is evident that Pt was almost atomically dispersed in zeolite micropore in the form of nanoclusters. The stabilization effect of micropore on Pt nanocluster is the primary driving force for Pt migration from alumina binder to zeolite that is almost irreversible.

The synthesis of atomically dispersed metal nanoclusters encapsulated within the zeolite micropores have attracted much attentions in recent decades due to many unique benefits offered by these confined nanoclusters. For example, such high metal dispersion can minimize the metal loading while maximizing metal activity. Furthermore, the encapsulation of such metal clusters within the zeolite pores allows one to take full advantage of shape selectivity of zeolite pore channels. Often time, this is achieved by carefully designed synthesis[2] or inter-structure transformation[3, 4]. Our work presents another way to achieve metal nanoclusters encapsulated within zeolite micopores .

References

[1] Y. Xiang, H. Wang, J. Cheng, J. Matsubu, Catal. Sci. Technol., 8 (2018) 1500-1516.

[2] S. Goel, Z. Wu, S. I. Zones, and E. Iglesia, J. Am. Chem. Soc., 2012 *134* (42), 17688-17695

[3] S. Goel, S. I. Zones, and E. Iglesia, J. Am. Chem. Soc., 2014 *136* (43), 15280-15290

[4] L. Liu, U. Díaz, R. Arenal, G. Agostini, P. Concepción, and A. Corma, Nature Mater., 2017, *16*(1), 132–138.

545402 Origin of Selectivity in the Direct Propylene Epoxidation Reaction Using Molecular Oxygen

Wednesday, June 5, 2019: 3:18 PM
Texas Ballroom A (Grand Hyatt San Antonio)

Janvit Teržan[1], Petar Djinovic[2], Janez Zavasnik[3,4], Iztok Arcon[5], Gregor Zerjav[2], Matjaz Spreitzer[6] and Albin Pintar[1], (1)D05 - Department of Environmental Sciences and Engineering, National Institute of Chemistry, Ljubljana, Slovenia, (2)Department for Environmental Sciences and Engineering, National Institute of Chemistry, Ljubljana, Slovenia, (3)Centre for Electron Microscopy and Microanalysis, Jožej Stefan Institute, Ljubljana, Slovenia, (4)Department of Structure and Nano-/ Micromechanics of Materials, Max-Planck-Institut für Eisenforschung, Dusseldorf, Germany, (5)Laboratory for quantum optics, University of Nova Gorica, Nova Gorica, Slovenia, (6)Advanced Materials Department K9, Jozef Stefan Institute, Ljubljana, Slovenia

Origin of selectivity in the direct propylene epoxidation reaction using molecular oxygen

Janvit Teržan[a, b], Petar Djinovic[a], Janez Zavašnik[c, d], Iztok Arcon[b, e], Gregor Žerjav[a], Matjaž Spreitzer[b, f] and Albin Pintar[a]

[a]*Department for Environmental Sciences and Engineering, National Institute of Chemistry, Slovenia*

[b]*Jožef Stefan International Postgraduate School, Slovenia*

[c]*Centre for Electron Microscopy and Microanalysis, Jožef Stefan Institute, Slovenia*

[d]*Max-Planck-Institut für Eisenforschung GmbH, Germany*

[e]*University of Nova Gorica, Slovenia*

[f]*Advanced Materials Department, Jožef Stefan Institute, Slovenia*

Propylene can be catalytically converted into several useful C_3 oxygenates, such as propylene oxide (PO), propanal and acrolein.[1] Propylene oxide is a highly desired chemical as it is used in the production of propylene glycol ethers, propylene glycol and polyether polyols.[2]

In this work,[3] CuO_x/SiO_2 catalysts

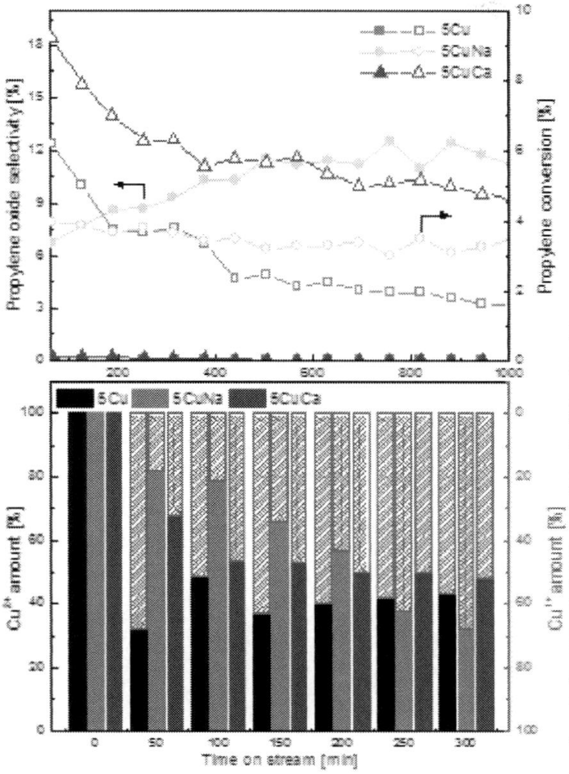

modified with Na, K and Ca were synthesized and characterized by N_2 physisorption, XRD, TEM-EDS, CO_2-TPD, operando UV/Vis DRS, operando XANES and pyridine DRIFT spectroscopy.

After CuOx deposition, a decrease of about 25 % in specific surface area and pore volume is observed, when compared to pure KIT-6 silica support. The shift in pore size distribution towards lower values indicates that the CuO_x phase lies inside the pores of the silica support. Subsequent modification of the CuO_x catalysts with sodium induced only slight changes in surface area, pore diameter and pore volume. X-ray diffraction patterns at low 2θ angles of pure KIT-6 silica support, 5Cu, 10Cu and 10CuNa catalysts indicates a well-ordered mesoporous structure of the KIT-6 support, which remains unaltered during subsequent copper loading and sodium modifications.

The BF-TEM micrographs show a well-ordered mesoporous structure in both the sodium modified and unmodified sample, which confirm the results of the N_2 physisorption and XRD analyses. Additional morphological properties were probed with the UV/VIS DR analysis. They revealed characteristic charge transfer bands between mononuclear Cu^{2+} and oxygen. A broad but less intense peak corresponding to the d-d transition of Cu^{2+} ions in an octahedral environment, which can be attributed to bulk CuO, was also observed. In all samples a shoulder characteristic of $[Cu–O–Cu]_n$-type (oligomeric species) was present. *Ex-situ* and operando UV/VIS DR analyses of the unmodified samples showed a notable absorption red-shift along with the increase of band intensity related to bulk CuO, after the second calcination protocol (to simulate the alkali and earth alkali modification of the samples) and during catalytic conditions. Alkali as well as earth alkali modification of the CuO_x/SiO_2 catalysts noticeably stunted the sintering effect seen in the unmodified samples. Switching to a purely reductive atmosphere results in a progressive increase of light absorption in the whole range of wavelengths for both the modified and the unmodified catalysts. Returning to an oxidative atmosphere causes re-oxidation of the catalyst where the structural changes induced by the reduction are, to a greater extent, reversible in the 5CuNa catalyst, when compared to the unmodified catalyst.

Figure 1: Temporal evolution of propylene conversion and PO selectivity (top) and Cu^{+} fraction as a function of reaction time for the investigated catalysts (bottom).

Acid sites are known to negatively affect selectivity for PO, as these strongly electrophilic surface sites can react with the bridging oxygen in PO, resulting in its further oxidation, or surface blocking and deactivation via strong adsorption.[4] The results of pyridine DRIFTS measurements indicate no Brønsted acid sites on the analyzed samples and a negligible change in the Lewis acidity after copper loading. On the other hand, sodium modification decreases the amount of Lewis acid sites for 44 and 29 % in the 5CuNa and 10CuNa catalysts, respectively. The decrease in Lewis acid site abundance is in line with the sodium content, i.e. 0.8 and 0.5 wt. % in 5CuNa and 10CuNa catalysts, respectively.

Abundance and strength of basic sites in the synthesized catalysts was analyzed using CO_2-TPD analysis, which revealed two separate clusters of peaks for all the catalysts: a low-temperature cluster related to weak basic sites (between 0 and 300 °C), and a high-temperature cluster (comprised of two peaks centered at 450 and 600 °C) connected to strongly basic sites. Deconvolution of CO_2-TPD profiles showed that only in case of Na and K modification, the low-temperature cluster consists of three distinct peaks, centered at 60, 94 and 155 °C.

Abundance and binding strength of hydroxyl groups to the catalyst surface was analyzed using H_2O-TPD-MS technique. It can be seen that Cu addition to KIT-6 silica causes an increase in the density of water and hydroxyls. Their number is decreased after Na and K modification. The largest abundance and strength of hydroxyl binding to the catalyst surface (based on the position of the high-temperature desorption peak) is observed after modification with calcium.

Operando XANES analysis revealed that a substantial fraction of Cu^{2+} is transformed to Cu^{+} during the propylene oxidation reaction. The kinetics of oxygen abstraction and replenishment are substantially different, indicative of modified chemistry of the nucleophilic oxygen species, present in 5CuNa catalyst in contrast to others (5Cu and 5CuCa). Using LCF (linear combination fit) analysis of the 5Cu, 5CuNa and 5CuCa catalysts (at 350 °C in O_2/He atmosphere) shows all copper is present as Cu^{2+} and the majority (~90 %) is subnanometer Cu^{2+} clusters. Only ~10 % is present as nanocrystalline CuO.

Over the non-promoted catalysts, the catalytic reaction products are CO_2, CO, water, acrolein and trace amounts of other oxygenates (acetone, propanal, propanol, ...), however, the latter present less than 5 % of the C containing products. Regardless of the copper content, strong deactivation is observed over all unmodified catalysts: they lost more than 60 % of initial activity during the 17 h of propylene oxidation reaction, which is likely related to sintering of oligomeric $[Cu\text{-}O\text{-}Cu]_n$ species. Modification of 5Cu catalysts with sodium and potassium greatly minimized catalyst deactivation: minimal deactivation is observed during 17 h TOS for the 5CuNa catalyst, whereas catalytic activity of 5CuK and 2.5CuNa stabilizes after 10 h following the initial decrease of about 25 %. After Na and K addition, a notable (~10 %) propylene oxide selectivity was also observed, which increased with prolonging TOS. With calcium modification of the 5Cu catalyst, a dramatic increase in the initial catalytic activity is observed (9 and 6.5 % propylene conversion for 5CuCa and 5Cu catalysts, respectively) with trace (below 0.5 %) PO selectivity.

Acrolein and CO_x selectivities remain very similar to the unmodified 5Cu catalyst.

Based on the research done, we propose that Cu^+ is not crucial for PO formation; instead the electropositive Na^+ and K^+ decrease the nucleophilic strength of oxygen in CuO_x, making its attack on the allylic hydrogen of propylene less invasive. Alkali (K and Na) modification produces finely dispersed alkali metal cations, associated with the CuO_x phase, resulting in a greatly stabilized morphology and catalytic activity. The effect of calcium on decreasing the nucleophilic character of O species in CuO_x is negated by charge compensation by strongly adsorbed hydroxyl groups. As a result, Ca modification for PO selectivity is inefficient.

Key words: propylene, selective oxidation, epoxidation, alkali doping, nucleophilic oxygen.

References

[1] D. Torres, N. Lopez, F. Illas, and R. M. Lambert, "Low-basicity oxygen atoms: A key in the search for propylene epoxidation catalysts," *Angew. Chemie - Int. Ed.*, vol. 46, no. 12, pp. 2055–2058, 2007.

[2] The Dow Chemical Company, "Propylene oxide," 2013.

[3] J. Teržan *et al.*, "Alkali and earth alkali modified CuO x /SiO 2 catalysts for propylene partial oxidation: What determines the selectivity?," *Appl. Catal. B Environ.*, vol. 237, pp. 214–227, Dec. 2018.

[4] G. Mul, "Stability and Selectivity of Au/TiO2 and Au/TiO2/SiO2 Catalysts in Propene Epoxidation: An in Situ FT-IR Study," *J. Catal.*, vol. 201, pp. 128–137, 2001.

580784 The Role of Phosphorus on Vanadia Catalyst for the Selective Oxidation of n-Butane to Maleic Anhydride

Wednesday, June 5, 2019: 3:21 PM
Texas Ballroom A (Grand Hyatt San Antonio)

Jingxiu Xie[1], Christian Schulz[1], K. Knemeyer[2], R. Naumann d'Alnoncourt[2], Benjamin Frank[1], Ralph Kraehnert[1] and Frank Rosowski[1], (1)BasCat – Unicat BASF JointLab, Berlin, Germany, (2)BasCat - Unicat BASF JointLab, Berlin, Germany

The Role of Phosphorus on Vanadia Catalyst for the Selective Oxidation of n-Butane to Maleic Anhydride

J. Xie[1], C. Schulz[1], K. Knemeyer[1], R. Naumann d'Alnoncourt[1], B. Frank[1], R. Kraehnert[1], F. Rosowski[1,2]

[1] *BasCat – Unicat BASF JointLab, Hardenbergstr. 36, Berlin, 10623, Germany*

[2] *BASF SE, Carl-Bosch-Straße 38, Ludwigshafen, 67056, Germany*

Selective oxidation of lower alkanes (ethane, propane, and n-butane) converts these less reactive molecules to valuable chemicals in a resource-efficient and environmentally friendly approach. The selective oxidation of n-butane to maleic anhydride, catalyzed by vanadyl pyrophosphate (VPP), is an important industrial process with a worldwide capacity of more than 500 kton annually. However, its reaction mechanism and active sites remain debatable due to the complexity and dynamic surfaces of VPP under reaction conditions. In this work, the role of phosphorus on the reaction network and mechanism for the selective oxidation of n-butane to maleic anhydride (MAN) is investigated.

The study focuses on two catalysts, namely V_2O_5 and phosphorus-modified V_2O_5, ($PO_xV_2O_5$), which was synthesized by atomic layer deposition (ALD) of phosphorus on the V_2O_5 surface. Parameter field studies of $PO_xV_2O_5$ and V_2O_5 were carried out to compare the reaction networks. Additional co-feed studies of 1-butene and MAN were performed to probe deeper into specific reaction pathways, such as consecutive reactions of 1-butene and MAN.

The elemental loading of P was measured by ICP to be 0.1 %wt., which is 0.7 theoretical monolayer coverage of V surface atoms. From the parameter field study, both catalysts produced MAN and depositing 0.1 w% P via ALD increased MAN selectivity. The combined selectivities to MAN, CO and CO_2 was more than 90 % for both catalysts, and the minor products include 1-butene, acrylic acid, acetic acid, and acetaldehyde. 1-butene and MAN were identified play important roles in the reaction mechanism, thus additional co-feed studies were performed. Complete conversion of the 1-butene co-feed were observed for both catalysts, but n-butane conversion was not significantly affected by the co-feeding of 1-butene. This demonstrates the higher reactivity of 1-butene in comparison to n-butane, and the possibly lack of competitive adsorption between n-butane and 1-butene. From Figure 1a, product selectivities of n-butane and 1-butene were similar, thereby suggesting that P deposition has negligible influence on product selectivities from 1-butene. With MAN co-fed, the consecutive oxidation of MAN was largely suppressed upon P addition (Figure 1b). Without P deposition, MAN was oxidized mainly to CO and CO_2 (Figure 1c). The difference in CO/CO_2 ratio of n-butane. 1-butene and MAN is also used as a descriptor for selective oxidation. The results underline the strong impact of P on the kinetics of the reaction network. This contribution will discuss and compare the reaction networks for both catalysts in order to reveal the role of P in the selective oxidation of n-butane to MAN.

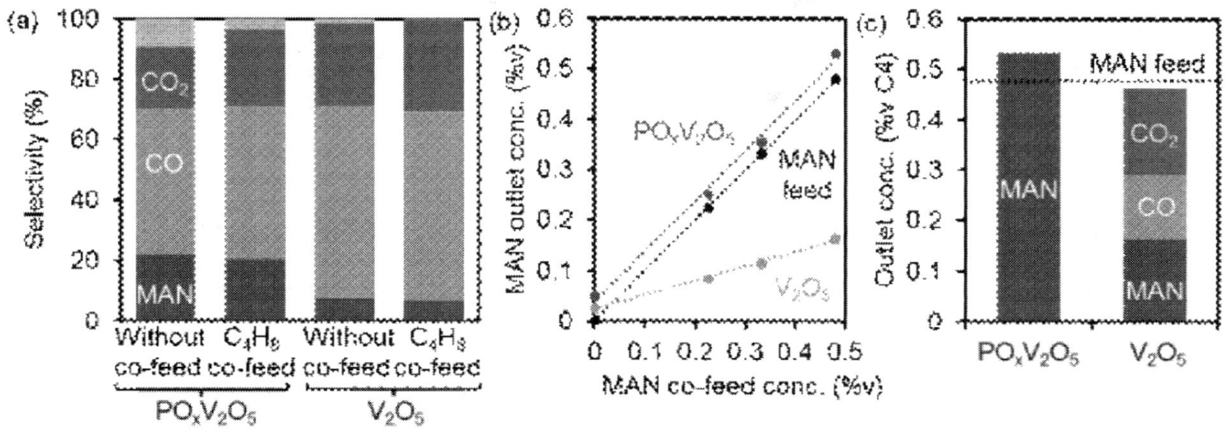

Figure 1. A) Product selectivities of n-butane and 1%v 1-butene co-feed, B) MAN outlet concentrations as a function of MAN co-feed concentrations and C) products from 0.5 %v MAN co-feed at 420 °C, 1 bar, $C_4H_{10}/O_2/H_2O$= 2/20/3 %v, 2000 h^{-1}.

545529 Identifying the Active Site for the V/P Mixed Oxides

Wednesday, June 5, 2019: 3:24 PM
Texas Ballroom A (Grand Hyatt San Antonio)
Gang Fu Sr., Chemistry Department, Xiamen University, Xiamen, China

Identifying the active site is one of the central questions for the multicomponent catalysts. Usually, different components in the multicomponent systems can be divided into active centers and promoters (or linkers). However, recent theoretical works indicated that the boundary between the active and inactive region would be blurred. Taking mixed oxides for example, Guliants et al.[1] suggested that H atom is preferentially adsorbed on Te=O rather than V=O or Mo=O on the M1 phase of $MoVTeNbO_x$ catalysts, while Goddard et al. [2] and our group [3] pointed out that in the binary vanadium phosphorus oxides (VPO) model catalysts, the activity of P=O is favored over V=O in the oxidation of n-butane. In above cases, the activity can be dispersed from the seemingly active sites to the nearest neighbour atoms/sites, which are chemically inert in their pristine form. Accordingly, fundamental questions are emerged as: (i) whether the activities are able to disperse to those sites outside the nearest ones? (ii) If so, what are the factors to determine their activities.

To simplify the question, here we concentrate on the binary VPO system[4]. It was found experimentally that the outermost layer of the VPO is phosphorus rich with the ratio of P/V up to 4, and the valence of V should be +5 rather than +4. However, the surface phosphate species are structurally ill-defined such that there is not a straightforward way to build an exact model to describe the surface structure. Alternatively, crystalline P_2O_5 is of layer structure, composed of a two-dimension network of PO_4. When some P^{5+} ions in P_2O_5 are displaced by V^{5+}, conceptual models for the P rich VPO catalysts can be obtained. Density functional theory (DFT) calculations are carried out to explore the initial activation of n-butane by V^{5+}=O and P^{5+}=O as well. We show that the activation of n-butane by the VPO can be viewed as a PCET process, in which V^{5+} serves as electron acceptor, while both the V=O and the P=O can act as proton acceptor. For P=O abstraction, the activation barrier (ΔE_a) correlate well with the reaction energies (ΔE), which are gradually decay with increasing the separated -OP- chain(s) from the V^{5+}. We find that both the basicity of O and the degree of charge seperation contribute to the reation barrier. For the H abstraction by V=O, the weak basicity of O could be compensated by the negligible charge seperation; whereas when the reaction happens on P=O, the benefit gained from the strong basicity of the P=O could counteract the cost of bond reorganization along the -(OP)- chain(s) and charge seperation as well.

According to our calculations, the center of O lone-pair band (ε_{lp}) is a good descriptor, which can quantitatively describe the tendency of activity. The good linear correlation between ε_{lp} and ΔE can be nicely explained through surface bonding interaction. It is clear that not only the occupied P=O lone-pair electron states but also the empty V $3d_{xy}$ state plays crucial roles. The lone-pair band of P=O would directly interact with H 1s, resulting in one bonding and one antibonding orbitals. The low-lying bonding orbital is doubly occupied, which makes the combined system more stable. On the contrary, the remainder electron could be promoted into the unoccupied V $3d_{xy}$, destabilizing the whole system. Hence, the ΔE would be closely related with the net energy gain/loss of orbital interaction. We expect that such a descriptor is amenable to fast screening the potential active sites over mixed oxide catalysts.

References

1. Govindasamy, A.; Muthukumar, K.; Yu, J.; Xu, Y.; Guliants, V. V. *Phys. Chem. C* **2010**, *114*, 4544-4549.
2. Cheng, M.-J.; Goddard, W. A. *Am. Chem. Soc.* **2013**, *135*, 4600-4603.
3. Fu, G.; Yuan, R.; Wang, P.; Wan, H. *Chinese J. Catal.* **2015**, *36*, 1528-1534.
4. Fu, G.; Wang, P.; Wan, H. *ACS. Catal.* **2017**, *7*, 5544–5548.

545931 Combined Solid- and Gas Phase Kinetic Modelling: Ethylene Oxychlorinationover K-Promoted Cu/Al2O3 Catalyst

Wednesday, June 5, 2019: 3:27 PM
Texas Ballroom A (Grand Hyatt San Antonio)

De Chen[1], Terje Fuglerud[2], Endre Fenes[3], Hongfei Ma[3], Magnus Rønning[4], Samuel Regli[5], Yanying Qi[5] and **Kumar Rout**[6], (1)Chemical Engineering, Norwegian University of Science and Technology, Trondheim, Norway, (2)INOVYN, Porsgrunn, Norway, (3)Norwegian University of Science and Technology, Trondheim, Norway, (4)Department of Chemical Engineering, NTNU, Trondheim, Norway, (5)NTNU, Trondheim, Norway, (6)SINTEF, Trondheim, Norway

Combined Solid- and Gas Phase Kinetic Modelling: Ethylene Oxychlorinationover K-Promoted Cu/Al2O3 Catalyst

Endre Fenes[1], Hongfei Ma[1], Yanying Qi[1], Samuel Regli[1], Kumar R. Rout[2*], Terje Fuglerud[3], Magnus Rønning[1], De Chen[1*]

1 – Department of Chemical Engineering, NTNU, Trondheim, 7491, Norway

2 – SINTEF, Trondheim, 7491, Norway

3 – INOVYN Norway, Porsgrunn, 3936, Norway

*Corresponding author: kumarranjan.rout@sintef.no, de.chen@ntnu.no

<>Introduction

Ethylene oxychlorination is an industrially important process where the $CuCl_2/\gamma$-Al_2O_3 based catalyst undergoes a reduction-oxidation process (Eq.1).[1] The catalyst is disposed to by product formation over alumina acid sites, volatilization of Cu^+ species and hotspot formation due to the exothermicity of the reaction, all of which is industrially abated with K addition. While K simply titrates alumina acid sites, [1] the nature of K interaction with copper species and the subsequent effect on the reduction - and oxidation reaction kinetics is still an open question. Building on our kinetic model of the neat $CuCl_2/\gamma$-Al_2O_3 catalyst, [2] wherein we demonstrated that the nature of the active site depends on the degree of reduction, we extend this approach to K doped catalysts in combination with DFT modelling and operando XAS spectroscopy, the two latter are currently underway.

$$2CuCl_2 \xrightarrow{C_2H_4} C_2H_4Cl_2 + 2CuCl \xrightarrow{1/2O_2} Cu_2OCl_2 \xrightarrow{2HCl} 2CuCl_2 + H_2O$$

(1) <>Experimental

The catalyst were prepared through incipient wetness impregnation of $CuCl_2 \cdot 2H_2O$

and KCl and contains 5wt% Cu and 1.2 wt% K. The catalyst was characterized with *ex situ* XRD, BET and XPS. Step transient reduction (P_{C2H4}=0.065) and oxidation (P_{O2}=0.3) experiments were performed at differ temperatures to obtain the kinetic parameters. During co-feeding experiments, C_2H_4, O_2 and HCl were introduced simultaneously (T=503K, P_{tot}=1Bar, P_{C2H4}=0.009, P_{O2}=0.0045, P_{HCl}=0.018). The effluent gases were monitored my mass spectrometry and Cu oxidation state quantified thorough operando UV-Vis-NIR spectroscopy.[3] <>Results

During the step transient reduction, ethylene gradually removes Cl from the catalyst, reducing Cu^{2+} to Cu^{+} and the Cu coordination from four to two. However, when plotting the reaction rate with respect to the amount of Cl remaining in the catalyst, it is immediately clear that two active sites are necessary in order to properly describe the overall reaction (r_1). The presence of at least two active sites is also supported by temperature programmed reduction and preliminary DFT results. (not shown) Initially only four-coordinated Cu ($r_{1,I}$) is present on the catalyst as $CuCl_2$. Then, as Cl is removed and the Cl concentration of the catalysts drops below a certain concentration (Cl**) defined by Gaussian deconvolution, three-coordinated Cu appear ($r_{1,II}$). As the catalyst is almost completely reduced (Cl*), no more four-coordinated Cu can be sustained and only the three coordinated site remains active until all reducible Cu is converted into the inert, two-coordinated CuCl. The kinetic model describing the reduction reaction is summarized below. (Eq. 2-4)

$$r_1 = r_{1,I} + r_{1,II} \tag{2}$$

$$r_{1,II} = k_{1,II} P_{C_2H_4} (Cl^{**} - C_{Cl}) C_{Cl} \tag{3}$$

$$r_2 = r_{2,I} + r_{2,II} \tag{4}$$

Similar for the oxidation step, there is initially only one active site, two coordinated CuCl. At some point (CuCl**), three and four coordinated Cu coexist and as the catalyst is almost completely oxidized (CuCl*), only three-coordinated Cu sites are active. Thus, a kinetic model similar to the reduction step is proposed (Eq. 5-7) with the addition of a desoprtion term to account for O_2 dissociation. It is noted that the third step (Eq. 1) is generally considered to be much faster than the redox steps and thus assumed kinetically irrelevant.[3]

$$r_{2,I} = \frac{k_{2,I} P_{O_2} (C_{CuCl} - C^{*}_{CuCl})^2}{1 + P_{O_2} K_{eq}} \tag{5}$$

$$r_{2,I} = \frac{k_{2,I} P_{O_2} (C_{CuCl} - C^{*}_{CuCl})^2}{1 + P_{O_2} K_{eq}} \tag{6}$$

$$r_{2,II} = \frac{k_{2,II} P_{O_2} (C_{CuCl}^{**} - C_{CuCl}) C_{CuCl}}{1 + P_{O_2} K_{eq}} \tag{7}$$

The above model was then fitted towards the step transient experiments of the neat and K doped catalyst at different temperatures in order to obtain the activation energy and preexponential factor which are presented in Table 1. From the table, it is evident that K addition lowers the reduction activation energy although the largest difference is in the preexponential factor of the four coordinated $CuCl_2$ which decreased almost three orders of magnitude, resulting in a decreased reduction turnover frequency (TOF) compared to the neat catalyst. This indicates that entropy changes is more significant in K doped catalysts than in the neat catalyst, which could be explained by formation of supported catalyst particles in the K doped catalyst with larger degree of crystallinity compared to a more disorded, partial monolayer coverage of $CuCl_2$ on the neat catalyst as indicated by TEM images. (not shown) Results from combined in situ XAS and UV-Vis-NIR experiments at the Swiss-Norwegian

Table 1. Kinetic parameters determined from step transient experiments.

		Reduction		**Oxidation**	
Catalyst	Parameter	Part **I**	Part **II**	Part **I**	Part **II**
5Cu	A [g_{Cu} / mol atm s]	1.69E+08	4.94E+07	1.7E+05	2.91E+06
	Ea [kJ/mol]	60.25	65.40	41.22	52.27
1.2K5Cu	A [g_{Cu} / mol atm s]	2.76E+05	1.58E+06	3.63E+05	2.04E+08
	Ea [kJ/mol]	45.22	51.89	40.34	70.17

beamline might elucidate the exact nature of the catalyst particles further and especially the tendency of $KCuCl_x$ mixed salt formation. To our knowledge, we are the first to have brought synthesized samples of $KCuCl_3$ as a reference material.

With respect to the oxidation step, the oxidation of the two coordinated site is fairly similar at the investigated K loading while K addition increases both the preexponential factor and activation energy for the oxidation of the three-coordinated site. The overall effect is a slightly increased oxidation rate compared to the neat catalyst, and in combination with the decreased reduction activity, this should result in a lower concentration of volatile CuCl, minimizing Cu volatilization.

The kinetic parameters in Table 1 was applied in a 1D reactor model to simulate conditions were all reactants were introduced to the reactor simultaneously and compared to the actual performance of the laboratory reactor. No further fitting of the kinetic parameters were carried out and the estimated reaction rate (2.23E-04 [mol/mol_{Cu} s]) corresponded well with the experimentally observed one (2.28E-04 [mol/mol_{Cu} s]). The experimental Cu^{2+} concentration was quantified through operando UV-Vis-NIR spectroscopy, and it is noted that the slight discrepancy between the theoretical modelling and experimental data in Figure 2 might be due to carbon formation increasing the abosorption of the visible spectrum. It is though clear that the proposed kinetic model is capable of predicting both the gas phase kinetics as well as the evolution of the active site. While both redox steps were modelled with two active sites in order to account for the changing electronic properties of the active site, resulting from altered Cu coordination as the reacton proceeds, the actual transformation might be continnuous in nature. Although this is a topic we are currently persuing from several angles, the two site approach is a computationally inexpensive approch with physically meningful active sites that are more easily envisioned.

<>Conclusions

A kinetic model, capable of accurately predicting both gas phase and solid state kinetics while taking into account the dynamics of an changing active site is developed from step transient experiments. The kinetic model is capable of accurately predicting the behavior of a laboratory scale reactor during co-feeding conditions. The proposed kinetic model is expected to be applicable in other systems where the catalyst partakes in redox reactions, and highlights the common assumption that the reaction kinetics is a function of temperature and partial pressures alone while ignoring the impact of altering the active site coordination sphere. Both DFT and analysis of combined, operando UV-Vis-NIR and XAS is currently underway and is expected to help elucidate further details about the catalytic system. References

[1] N. B. Muddada, U. Olsbye, L. Caccialupi, F. Cavani, G. Leofanti, D. Gianolio, S. Bordiga, C. Lamberti, Phys. Chem. Chem. Phys. 12 (2010) 5605.

[2] E. Fenes, Y. Qi, M. F. Baidoo, H. Ma, K.R. Rout, T. Fuglerud and D. Chen, In preparation, (2019)

[3] K.R. Rout, M. F. Baidoo, E. Fenes, T. Fuglerud and D. Chen, J. Catal. 352, (2017), 218-228

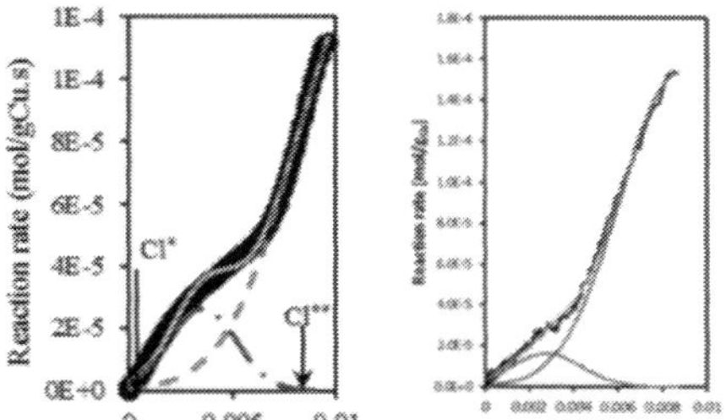

Figure 1. a) Reduction reaction rate with respect to reducible Cl. b) Oxidation reaction rate with respect to available CuCl.

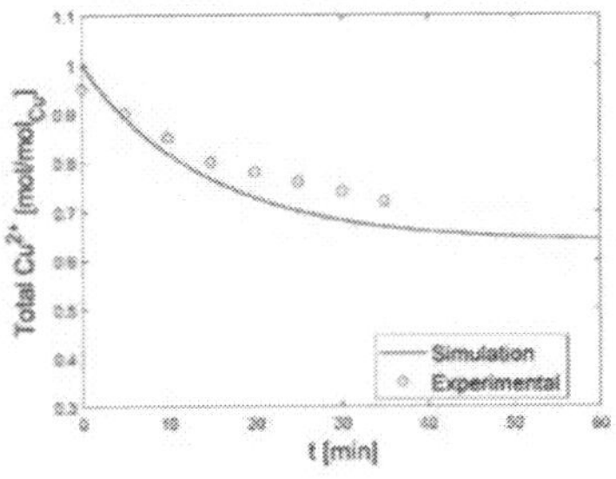

Figure 2. Comparison of the evolution of Cu^{2+} species during co-feeding experiments on the 1.2K5Cu catalyst. (T=503K, P_{tot}=1Bar, P_{C2H4}=0.009, P_{O2}=0.0045, P_{HCl}=0.018).

545404 Understanding the Redox Reactions of Ni and Co during Catalytic Combustion of Methane

Monday, June 3, 2019: 1:30 PM
Republic ABC (Grand Hyatt San Antonio)

Shirley E. Liland[1], Kumar R. Rout[2], Endre Fenes[1] and De Chen[3], (1)Norwegian University of Science and Technology, Trondheim, Norway, (2)SINTEF, Trondheim, Norway, (3)Chemical Engineering, Norwegian University of Science and Technology, Trondheim, Norway

1. Introduction

The catalytic combustion of methane (CCM) is a significant process for energy production and reduction of direct emission from methane[1]. Our study uses an operando setup to explore the synergy between Ni and Co catalysts to understand the oxidation state of the transition metals. The idea is to develop a method where the retrieved in-situ knowledge on the oxidation state is used to improve the CCM activity.

2. Experimental

A series of well-controlled Ni-Co catalysts supported on a hydrotalcite(HT)-like structure have been synthesized using co-precipitation. The catalysts contain a total metal loading of 12wt% with various ratios of Ni and Co content. In addition, the pure Co catalyst was impregnated with 0.1wt% Pd. Several characterization techniques have been utilized to get at further understanding of the catalysts, e.g. hydrogen chemisorption, nitrogen adsorption, XRD, XPS, ICP-MS, TPR and TEM. The catalysts activities were analyzed in a fixed-bed quartz reactor equipped with a fiber optic probe. The probe is pointing directly towards the catalyst bed and connected to a UV-vis-NIR spectrometer giving valuable information on the *in-situ* state of the catalyst. The compositional gas changes were analyzed using a gas chromatograph (GC) and a mass spectrometer (MS). The experiment consists of alternating between O_2 and CH_4 conditions in 3 cycles to compare the oxidation and reduction rates of the catalysts. In addition, steady-state catalytic combustion was performed over 4 h. All analyses were performed at 670 °C to keep it at isothermal conditions.

3. Results and discussions

Figure 1A shows the activity during complete combustion for catalyst with different Ni/Co ratios, it is evident that increasing the amount of Co increases the activity. This could be explained by the oxygen conversion rate, where the results show that cobalt is a much better metal for converting oxygen from the feed compared to nickel. The results for the methane conversion rate show the opposite trend; cobalt has the lowest reduction rate. For all catalysts the oxidation rate is faster than the reduction rate which is also confirmed by the normalized Kubelka-Munk Function (KMF). The KMF is calculated from the reflectance found at 540 nm in the UV-vis spectra, which shows how Co^0 goes to Co^{2+} and Ni^0 goes to Ni^{2+} during combustion, see Figure 1B. It is evident that the activity is related to the oxidation of the catalysts. Due to this the pure Co was impregnated with Pd, which is an oxidative metal. The results clearly show that Pd enhances the methane conversion during combustion, which is found to be due to the prolongation of the oxidation time. A similar technique has also been done for oxychlorination on copper in our group[2-4]. Figure 2 shows in more details the results from the 2nd cycle of the transient reduction of 12Co, here the ion currents of the CO and CO_2 peaks from the MS are combined with the UV-vis results. The results show how the cobalt oxide is reducing from Co^{3+} to Co^{2+} to Co^0, while the CO product is not forming until a certain fraction of the reducible cobalt is in the Co^0 state. This shows that the presence of metallic Co is important for the CO formation, which can be an indication that Co^0 instead of Co^{2+} or Co^{3+} can be applied to enhance the activity towards partial oxidation of methane. This technique is considered valid for several systems where transition metals are undergoing redox reaction. We believe that by utilizing these in-situ data from the UV-vis-NIR spectroscopy, that a method can be developed which relates the activity and selectivity with the oxidation state of the catalysts.

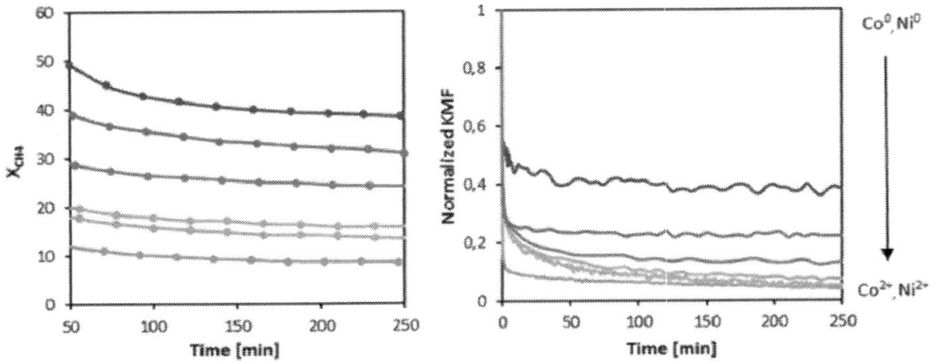

Figure 1: A) Methane conversion over time during complete combustion. B) The oxidation state changes were normalized KMF = 1 is defined as Ni^0, Co^0, normalized KMF = 0 is defined as Ni^{2+}, Co^{2+}. $12Ni_1$ (—), $12Ni_{0.7}Co_{0.3}$ (—), $12Ni_{0.5}Co_{0.5}$ (—), $12Ni_{0.3}Co_{0.7}$ (—), $12Co_1$ (—) and 0.1Pd-$12Co_1$ (—).

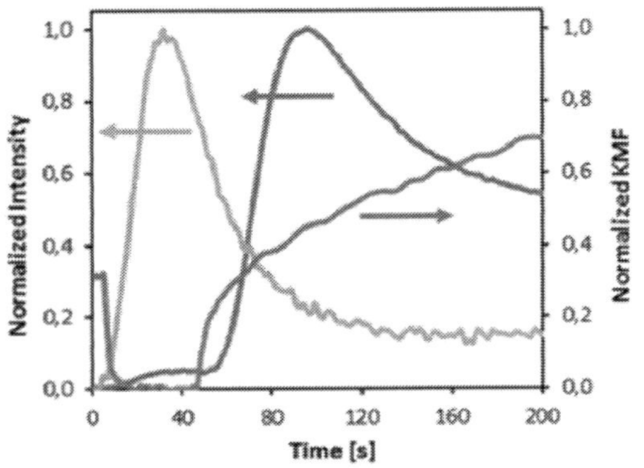

Figure 2: The normalized ion current of CO_2 (—) and CO (—) combined with the Normalized KMF (—) during the transient reduction of 12Co.

4. Conclusions

Based on catalysts with different Ni/Co ratios, the Co has the highest activity, however impregnating Co with Pd has a significant effect on the activity since Pd is an oxidative metal. The high activity is found to be due to the fast oxidation of Co in combination with high amounts of surface oxygen. The UV-vis-NIR spectroscopy combined with the MS signals also show that the selectivity is influence by the oxidation state of the catalyst, where CO is not produced on cobalt oxides.

References

[1] Gélin, P.; Primet, M., *Applied Catalysis B: Environmental* **2002,** *39* (1), 1-37.

[2] Baidoo, M. F.; Fenes, E.; Rout, K. R.; Fuglerud, T.; Chen, D., *Catalysis Today* **2017**.

[3] Rout, K. R.; Baidoo, M. F.; Fenes, E.; Zhu, J.; Fuglerud, T.; Chen, D., *Journal of Catalysis* **2017,** *352* (Supplement C), 218-228.

[4] Rout, K. R.; Fenes, E.; Baidoo, M. F.; Abdollahi, R.; Fuglerud, T.; Chen, D., *ACS Catalysis* **2016,** *6* (10), 7030-7039.

545493 Atomically Resolved *in Situ* investigation of Cu_2o Reduction with Methanol

Monday, June 3, 2019: 1:54 PM
Republic ABC (Grand Hyatt San Antonio)

Hao Chi[1], Meng Li[1], Matthew Curnan[2], Christopher M. Andolina[1], Judith Yang[3], Götz Veser[1] and **Stephen House**[4], (1)Chemical Engineering, University of Pittsburgh, Pittsburgh, PA, (2)Department of Chemical Engineering, Pohang University of Science and Technology (POSTECH), Pohang, PA, Korea, Republic of (South), (3)University of Pittsburgh, Pittsburgh, PA, (4)Chemical and Petroleum Engineering, University of Pittsburgh, Pittsburgh, PA

Catalytic gas-surface reactions are dynamic processes for which changes in structure and composition of the catalysts under working conditions play a pivotal role. Post-reaction examinations of a catalyst often do not accurately represent its active state at reaction conditions [1]. For example, morphology and surface structure of metal particles can change upon cooling and differ from those under reaction conditions [2], and catalytic surfaces are known to reconstruct – often reversibly – upon exposure to adsorbates [3].

Direct *in situ* or *operando* observation of the micro-structural evolution and active sites of a catalyst under reaction conditions aims to address these issues and yield much deeper and more robust insights into catalytic reaction mechanisms [2, 4-6]. In this context, environmental transmission electron microscopy (ETEM) has emerged as a powerful *in situ* technique that allows gathering direct real-time dynamic information on the nanoscale. Using *in situ* TEM techniques, nanoscale morphological evolution and chemical modifications of a catalyst can be visualized directly under gas exposure, providing essential insights into catalytically active phases, transient states, and reaction mechanisms.

Recent *in situ* studies have provided a significant understanding of atomic scale Cu oxidation, especially during early stage oxidation [7,8]. In contrast, the reduction of metal oxides, such as Cu_2O, has been studied much less extensively. However, such reductions play an important role in many technologies, such as catalysis, electrochemistry, and environmental degradation of materials [9]. Here, we present an *in-situ* investigation of the dynamics of the reduction of Cu_2O islands supported on Cu(100) during exposure to methanol (MeOH) using ETEM. Cu_2O nano-islands supported on Cu facets were created via controlled *in situ* oxidation of copper films, followed by reduction in MeOH vapor.

Cu(001) thin films with a thickness of 600 Å were produced by *e*-beam evaporation of 99.999% pure Cu pellets on NaCl(001) substrates. The films were grown in a UHV evaporator system with a base pressure of 1×10^{-9} Torr at a deposition rate of 1.2 Å/s and a substrate temperature of 300 °C. Native oxides (Cu_2O) were removed by injection of H_2 ($p_{H2} = 7.6\times10^{-6}$ Torr) at a temperature of 550 °C, followed by annealing at elevated temperature to forma faceted holes in the Cu films. Cu_2O

nano-islands were then created along Cu(100) facets via controlled *in situ* oxidation of the Cu films (pO_2 = 1.5x10^{-4} Torr, 350 °C). This procedure results in the formation of Cu_2O islands with the epitaxial relation Cu_2O(200)//Cu(200) and typical sizes of ~2-3nm in height and 5-6 nm in diameter. These islands were then exposed to MeOH (p_{MeOH} = 1x10^{-2} Torr, 250°C) and the resulting reduction of the Cu_2O islands was imaged using an ETEM (Hitachi H9500) equipped with a specially designed double-tilt heating holder and gas injection system [10]. The obtained ETEM images and videos were evaluated using Fiji-imageJ.

Our results reveal a complex dynamic of the Cu_2O reduction process, characterized by two distinct temporal phases: During the initial phase of the reduction (~15 s), the Cu_2O islands reduction proceeds strongly anisotropically, i.e. the island radius decreases rapidly while its height remains almost unchanged. Interestingly, once the island radius roughly matches its height, the reduction switches from anisotropic to isotropic island shrinkage as both island height and radius simultaneously decrease until the whole island is completely removed after ~25 s. This shrinking behavior was observed uniformly for a total of eight Cu_2O islands studied on Cu(100), i.e. it appears to be representative of the reduction dynamics of Cu_2O islands on Cu(100) in an MeOH atmosphere. We propose that this behavior is caused by a preferential reaction between MeOH and stepped surfaces on the sides of the Cu_2O islands (*e.g.* Cu(110) with atomic steps) which show substantially higher reactivity than the island top terrace (Cu(100)). To support this hypothesis, we conducted DFT calculations for MeOH adsorption energies on Cu(110) and Cu(100) surfaces, which confirm the preferential adsorption of MeOH on the stepped surface.

The combination of *in situ* observations and computational studies in this work hence revealed deeper atomistic insights into the interactions of methanol with different crystallographic facets of Cu_2O surfaces. More broadly, this work aims to contribute to the understanding and manipulation of supported metal oxide systems widely used in heterogeneous catalysis.

[1] Tao, F. and M. Salmeron, In Situ Studies of Chemistry and Structure of Materials in

Reactive Environments. Science, 2011. 331: p. 171.

[2] Gai, P.L. and E.D. Boyers, Electron Microscopy in Heterogeneous Catalysts. 2003, London: Institute of Physics.

[3] Ertl, G., Reactions on Solid Surfaces. 2 ed. 2009: Wiley.

[4] Sharma, R., et al., In Situ Nanoscale Observations of the Mg(OH)2 Dehydroxylation and Rehydroxylation Mechanisms. Phil. Mag., 2004. 84: p. 2711-2729.

[5] Gai, P.L., Dynamic studies of metal oxide catalysts: MoO3. Phil. Mag., 1981. 43(4): p. 841-855.

[6] Gai, P.L., J.C. Bart, and E.D. Boyes, Electron microscopy of industrial oxidation catalysts. Phil. Mag. A, 1982. 45: p. 531-547.

[7] Zhou, G., Luo, L., Li, L., Ciston, J., Stach, E.A. and Yang, J.C., Physical Review Letters, 109 (2012) 35502.

[8] Zhou, G. and Yang, J.C., Journal of Materials Research, 20 (2005) 1684-1694.

[9] Zhou, G. and Yang, J.C., Phys Rev Lett, 93 (2004) 226101.

[10] Chi, H., Bonifacio, C., Andolina, C., Stach, E.A., Veser, G., and Yang, J.C. Microscopy and Microanalysis, (2017) 23(S1), 2100-2101.

545783 Nickel-Manganese Bimetallic Catalysts for Selective Hydrogenation of CO_x in the Presence of Light Hydrocarbons: Promotional Effect of Manganese

Monday, June 3, 2019: 2:18 PM
Republic ABC (Grand Hyatt San Antonio)

Vahid Shadravan[1], **Matthew Drewery**[2], Meng-Jung Li[1], Eric M. Kennedy[1] and Michael Stockenhuber[1], (1)Chemical Engineering, The University of Newcastle, Newcastle, Australia, (2)University of Newcastle, Callaghan, Australia

Nickel-manganese bimetallic catalysts for selective hydrogenation of CO_x in the presence of light hydrocarbons: Promotional effect of manganese

Vahid Shadravan, Molly Meng-Jung Li, Eric Kennedy, Michael Stockenhuber*

Chemical Engineering, School of Engineering, The University of Newcastle, Callaghan, NSW 2308, Australia

* michael.stockenhuber@newcastle.edu.au

1. Introduction

The worldwide demand for development of new energy storage and resources has reanimated research into carbon oxide hydrogenation reactions. Carbon oxide methanation reaction is a well-known process that has been used for various applications, such as purification of hydrogen used in ammonia synthesis and fuel cells, synthetic natural gas production, and chemical storage of electricity[1]. On the other hand, in the catalytic synthesis of hydrocarbons from oxidative processes, significant quantities of CO and CO_2 usually present as undesirable by-products which hinders commercialization for example in the production of C_2 feedstock for the polymer industry Therefore, effectively removing CO and CO_2 from a synthesized hydrocarbon stream via hydrogenation reaction, while maintaining or increasing the concentration of C_{2+} hydrocarbons is highly desirable for the industrial processes. In this paper, we report on the influence of Ni-Mn/Al_2O_3 catalyst structure and transition metal content on the C_{2+} yield in the hydrogenation of carbon oxides in the presence of hydrocarbons (CH_4, C_2H_4, C_2H_6, C_3H_6, and C_3H_8).

2. Results and discussion

Figure 1a and 1b disclose that the Ni-Mn/Al_2O_3 catalysts with Ni/Mn ratios of 5 and 2 show enhanced CO and CO_2 conversion compared to the Ni/Al_2O_3 benchmark catalyst. However, by decreasing the Ni/Mn ratio to below 2, it is found that the activity was significantly reduced. The monometallic Mn/Al_2O_3 catalyst is inactive for CO conversion during the examined temperature range, instead, CO production is observed at above 300 °C presumably due to dry reforming of methane or reverse water gas shift reactions. On the other hand, Figure 1c shows that the catalysts with the Ni/Mn ratios of 5 and 2 give enhanced C_{2+} yield compared to the unmodified Ni/Al_2O_3 catalysts, whereas a decreased C_{2+} yield observed in the Ni-Mn/Al_2O_3 catalysts with Ni/Mn < 2. Besides, the yield of C_{2+} hydrocarbons increases with increasing temperature and reaches a maximum then subsequently decreases at elevated temperature due to the cracking of C_{2+} hydrocarbons to form methane.

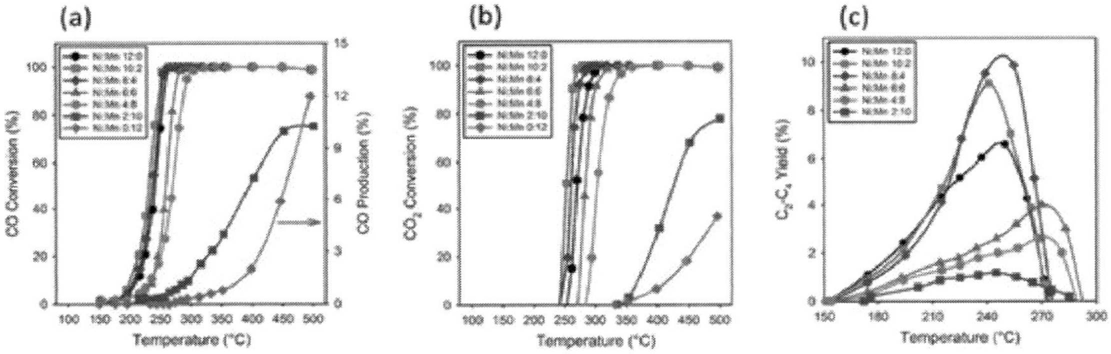

Figure 1. CO conversion (a), CO_2 conversion, and C_2-C_4 yield (c) of the catalysts with varying Ni/Mn ratio.

The observation of the changes in the catalytic performance of Ni-Mn/Al_2O_3 catalysts with different Ni/Mn ratios can be supported by the calculated activation energy value. Table 1 shows the activation energy for the reaction of CO, CO_2, and C_2H_6 with hydrogen over the selected Ni/Al_2O_3 and Ni-Mn/Al_2O_3 (with Ni:Mn = 12:0, 8:4, and 4:8) catalysts. It is found that by adding a moderate loading of manganese (Ni:Mn = 8:4) to the Ni/Al_2O_3 catalyst (Ni:Mn = 12:0), the activation barrier for both CO and CO_2 hydrogenation is reduced. In contrast, adding more manganese (Ni:Mn = 4:8) results in an increase in the activation energy value for CO and CO_2 hydrogenation. As for the ethane hydrocracking reaction, the activation energy value of each catalyst follows the order of Ni:Mn 4:8 > Ni:Mn 8:4 > Ni:Mn 12:0, suggesting that the activation barrier for catalyzing ethane hydrocracking increases as incorporating manganese to the Ni/Al_2O_3 catalyst.

Table 1. Calculated activation energy values for the reaction of CO, CO_2 and C_2H_6 with H_2 over Ni-Mn/Al_2O_3 catalysts.

Catalyst	Reactant	Activation Energy value (kJ/mol)
Ni:Mn 12:0		98.3
Ni:Mn 8:4	CO	93.3
Ni:Mn 4:8		113
Ni:Mn 12:0		80.3
Ni:Mn 8:4	CO_2	75.7
Ni:Mn 4:8		99.4
Ni:Mn 12:0		151
Ni:Mn 8:4	C_2H_6	161
Ni:Mn 4:8		169

In-situ infrared spectroscopy using nitric oxide as a probe molecule (NO-FTIR) was performed to understand the adsorption and structural properties of the transition metal sites with different Ni/Mn ratios. The coordinated nitric oxide molecules, bonded to the catalyst's active sites, have properties which are similar to carbon monoxide on the active sites[2], making NO-FTIR suitable for investigating CO_x hydrogenation over Ni-Mn/Al_2O_3 catalysts. Figure 2 demonstrates the NO-FTIR spectra of the Ni-Mn/Al_2O_3 catalyst with Ni:Mn = 8:4. Sections 'a' and 'b' show the adsorption and desorption spectra, respectively. The peaks presented in the spectra at around 1850 cm^{-1} wavenumbers (see the zoomed-in windows) can be assigned as mononitrosyls arising from the nonreactive adsorption of NO on the Ni-containing samples[3]. Other peaks appearing at wavenumbers around 1300 cm^{-1} and 1600 cm^{-1} arise from the reactive adsorption of NO molecules, which are most likely due to the formation of charged and neutral N_xO_y species such as NO_2 and NO^{2-} surface compounds[2]. It has been reported that mononitrosyl in the form of NO^{d+} coordination (linear geometry) is basically an electron donor to an electron accepting site, whereas the form of NO^{d-} coordination (bent geometry) is an electron acceptor to an electron donating site[4]. Therefore, the peaks assigned as mononitrosyls were deconvoluted to separate the linear-NO (at higher wavenumber) and the bent-NO (at lower wavenumber) peaks, see Figure 3a. The changes of linear/bent ratio on the selected Ni-Mn/Al_2O_3 catalysts are summarized in Figure 3b. According to Figure 3b, decreasing the Ni/Mn ratio results in reducing the value of linear/bent ratio, suggesting that the numbers of the electron donating sites increases as incorporating Mn into Ni/Al_2O_3.

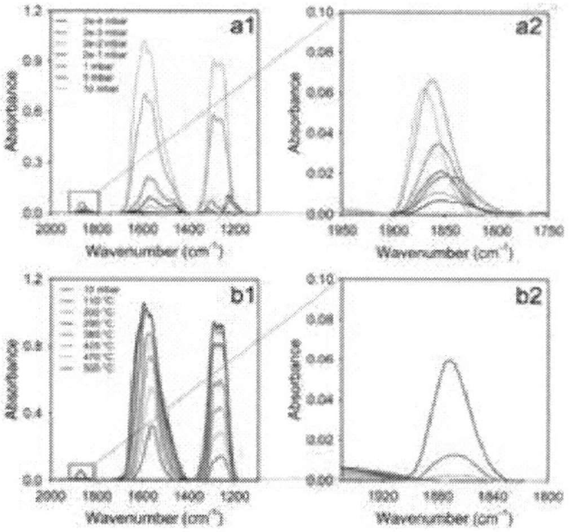

Figure 2. *In-situ* FTIR spectra for a. NO adsorption over bi-metallic Ni-Mn/Al_2O_3 (Ni:Mn 8:4), followed by b. temperature programmed desorption.

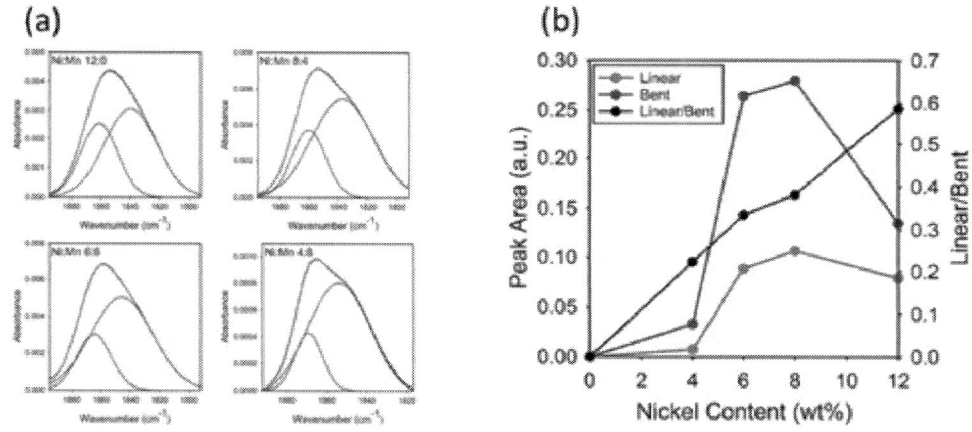

Figure 3. (a) Peak deconvolution for the metal–mononitrosyl species formed on nickel-containing catalysts observed from the *in-situ* NO-FTIR analysis; (b) calculated value of linear/bent ratio of metal–mononitrosyl compounds for the catalysts with different Ni/Mn ratio.

Solid state surface electrostatic potential $V_S(r)$ were computed based on KS-DFT (Kohn-Sham Density Functional Theory) performed at the PW91/pob-TZVP level of the theory. Figure 4 shows the surface electrostatic potential $V_S(r)$ at 0.001 au isodensity level for Ni and Ni-Mn particles containing one conventional unit cell of each crystal (Figure 4a) and 64 crystallographic unit cell (Figure 4b). The redistribution of the electron density in multi-atomic systems leads to the formation of minima ($V_{s,min}$) and maxima ($V_{s,max}$) of the surface electrostatic potential $V_S(r)$. The $V_{s,min}$ and $V_{s,max}$ or areas with different electrostatic values on the surface of the catalyst can be interpreted as different catalytic sites[5]. In this case, the sites with V values can act as electron donors and the ones with V can act as s,min s,max

electron acceptors. The colour-coded maps for Ni and Ni-Mn particles are considerably different, indicating the changes in the electronic structure of the Ni/Al_2O_3 catalyst by the addition of manganese, which is consistent with our findings from *in-situ* NO-FTIR.

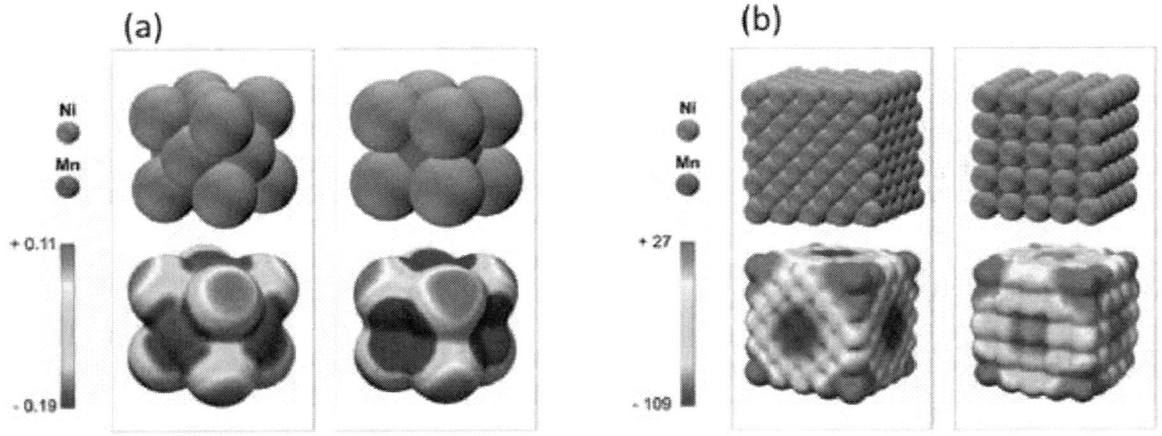

Figure 4. Surface electrostatic potential $V_s(r)$ on 0.001 au isodensity for Ni and Ni-Mn particles with the size of (a) one crystallographic unit cell, and (b) 64 crystallographic unit cell.

Carbide-based mechanism is a widely accepted mechanism for carbon oxide hydrogenation reactions, in which CO_x hydrogenation initiates via the dissociation of C–O bonds. The cleavage of C–O bond generates active surface carbon species which can be followed by their hydrogenation (known as chain growth step) to different C_xH_y species[6]. It has been reported that the changes of the electronic properties of catalysts can affect both chain initiation (C–O cleavage) and chain growth (C_xH_y formation)[7], thus leading to the changes in the catalytic performance. On the other hand, the alteration of electronic properties of catalysts induced by the addition of manganese can also affect the hydrocracking activity (e.g. ethane hydrocracking). It is reported that hydrocracking activity over nickel catalysts depends on the C–C bond rupture. In this case, the C–C bond cleavage happens more readily on catalysts with stronger carbon-metal bonding (more electron accepting sites)[8]. Therefore, the enhancement of CO_x hydrogenation activity and C_{2+} hydrocarbon selectivity by adding manganese to Ni/Al_2O_3 can be explained by the alteration of electronic properties induced by forming bimetallic Ni-Mn catalysts.

3. Conclusion

In summary, it is shown that the addition of manganese to Ni/Al_2O_3 catalyst significantly altered its catalytic performance for hydrogenation of CO and CO_2 in the presence of light hydrocarbons. Adding an optimum amount of Mn to Ni/Al_2O_3 can enhance the catalyst activity and selectivity of C_2-C_4 production. According to our investigation of the catalysts' electronic properties with different Ni and Mn contents, changes in catalytic activity (for CO_x hydrogenation) and selectivity (for light hydrocarbons formation) can be interpreted as being due to the effect of different electronic structure of the catalysts with variety of Ni/Mn ratios.

4. Reference

1. Gao, J. et al., RSC Advances 2015, 5, 22759-22776.

2. Hadjiivanov, K. I. et al., Catalysis Reviews 2000, 42, 71-144.

3. Zecchina, A. et al., Catalysis Today 1996, 27, 403-435.

4. Crabtree, R. H. et al., General Properties of Organometallic Complexes. John Wiley & Sons, Inc.: 2005; pp 29-52.

5. Stenlid, J. H. et al., Journal of the American Chemical Society 2017, 139, 11012-11015.

6. James, O. O. et al., RSC Advances 2012, 2, 7347-7366.

7. Maitlis, P. M. et al., Chemical Communications 2009, 0, 1619-1634.

8. Zeigarnik, A. V. et al., The Journal of Physical Chemistry B 2000, 104, 10578-10587.

545515 The Feasibility of the Thermocatalytic Conversion of CO2 into Synthetic Natural Gas (SNG)

Monday, June 3, 2019: 2:42 PM
Republic ABC (Grand Hyatt San Antonio)
Sogol Mottaghi-Tabar, Robert Currie, Yichen Zhuang and David Simakov,
Chemical Engineering, University of Waterloo, Waterloo, ON, Canada

The synthetic natural gas (SNG) generation process discussed in this talk relies on the thermocatalytic Sabatier reaction:

$$CO_2 + 4H_2 = CH_4 + 2H_2O + \text{heat (165 kJ/mol)}$$

The main idea is to convert the CO_2 contained in waste streams (e.g. cement plant flue gas, landfill gas) into synthetic natural gas (SNG) on site using electrolytic hydrogen generated using renewable or low carbon footprint electricity.

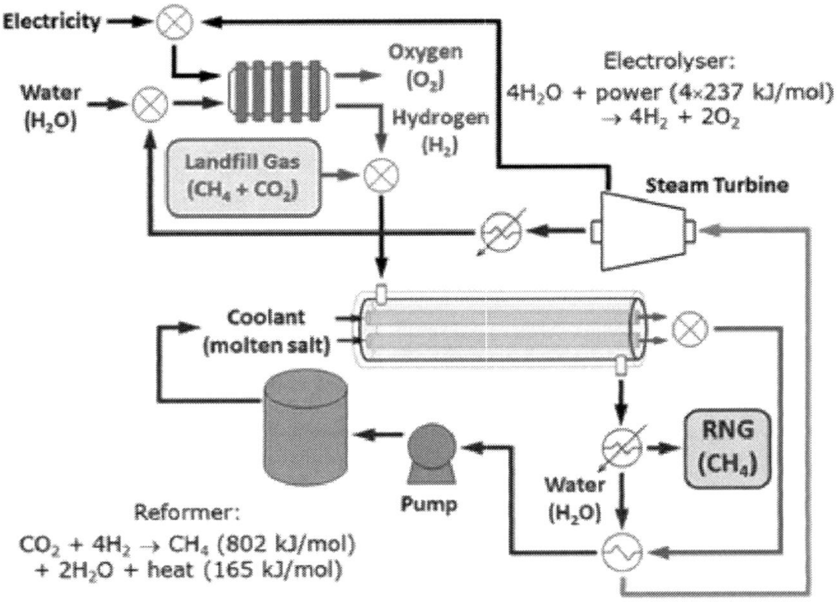

Although CO_2 separation technologies are already commercially available (e.g. pressure swing adsorption (PSA) and membrane separation), the captured CO_2 has to be compressed on site, transported, and eventually stored or utilized somehow. Storing the captured CO_2 in geological formations could be prohibitively expensive. Some of that captured CO_2 can be utilized in the chemical and food industry. However, these markets are limited in their capacity to absorb megatons of separated CO_2 and the capital and operating costs associated with the CO_2 compression and transportation may not be recoverable.

The approach discussed in this talk is an innovative and sustainable alternative that goes beyond conventional CO_2 capture and storage (CCS), relying on CO_2 capture and utilization (CCU) instead. The hydrogen required for the reaction is provided by water electrolysis using renewable or low carbon footprint electricity. The water and heat generated in the process are recuperated to improve the process efficiency. Oxygen (O_2), which is a byproduct of water electrolysis can be utilized as well. With a proper postreatment, such process would produce a pipeline quality SNG which can be directly injected into the existing natural gas infrastructure, replacing the fossil natural gas.

Technological issues related to catalysis, reactor design and system integration are outlined in this talk. The results of the catalytic performance evaluation of Ni- and Ru-based catalysts are briefly presented followed by numerical simulations and a lab-scale experimental proof-of-concept of a novel reactor configuration. Design of the entire SNG production system including upstream and downstream treatment is also briefly discussed. The focus is on the technological and economic feasibility of this conversion pathway.

545549 High-Efficient Chemical/Electrochemical CH_4 Conversion

Monday, June 3, 2019: 3:06 PM
Republic ABC (Grand Hyatt San Antonio)

Yongdan Li, Tianjin Key Laboratory of Applied Catalysis Science and Technology, Tianjin University, Tianjin, China; Chemical Engineering and Technology, Tianjin University, Tianjin, China

Natural gas, which is mainly CH_4, is one of the most important fossil fuels in the world. Nowadays, large amount of natural gas is burned directly to drive turbines to generate electricity, which has a low efficiency and release enormous CO_2. During the last two decades, we have been active and made a number of breakthroughs in high-efficient CH_4 convension and utilization.

Hydrogen is an ideal clean energy source which has raised much attention. Most of the hydrogen supply is produced based on steam reforming and partial oxidation of hydrocarbons and carbonaceous feed stocks such as natural gas and coal. Those techniques also produce CO and CO_2, requiring additional clean and sequestration processes. In 2000, we proposed a process to produce CO_x-free hydrogen and value-added carbon material simultaneously through methane catalytic decompostion (MCD) reaction[1], which has become very attractive due to the simplicity in process flow sheet and high efficiency. The products are easily separated and the clean hydrogen obtained can be used without further purification. Since then, we have focused on the development of novel catalysts and new catalytic processes, and carbons with various structures such as nanotubes[2], nanofibers[3-6] and nonoonions[7] were obtained. We developed a number of catalysts including Fe, Co and Ni suppored by Al_2O_3 and MgO[8-11], and improved their activity and stability with optimize the precusor and with doping promoters such as start with Feitknecht Compound[3, 12], and introduce ZnO[13] and Na_2O[14]. Meanwhile, the influences of the catalyst properities and the reaction conditions on the morphology of the nanocarbons have been revealed[15-19]. Afterwards, we proposed a high-efficient energy conversion system incorporating a MCD reactor with a direct carbon fuel cell consuming the carbon produced and an internal reforming solid oxide fuel cell (SOFC) with hydrogen and unconverted methane as the fuels[20]. In 2011, we summarized the effort and published a review article on the development of methane decomposition on group 8-10 base metal catalysts[21].

Direct CH_4 fuelled SOFC is a very promising clean technique to generate electricity with a high efficiency, and the development of novel anode materials with a high catalytic activity and high coking resistance is a key issue. In 2011, we developed a double-perovskite $Sr_2FeMoO_{6-\delta}$ anode with a high catalytic activity and sufficient mixed ionic-electronic conductivity[22]. A single cell supported by a 300 µm-thick $La_{0.8}Sr_{0.2}Ga_{0.8}Mg_{0.2}O_{3-\delta}$ (LSGM) electrolyte layer exhibited a maximum power density (P_{max}) of 600 mW cm^{-2} at 850 °C with dry CH_4 as the fuel. Another CH_4-fuelled single cell with Ni-$Ce_{0.8}Sm_{0.2}O_{1.9}$ (SDC) cermet anode and 120 μm-thick SDC electrolyte layer showed a P_{max} of 671 mW cm^{-2} at 700 °C[23]. The performance dropped 3.7% after the cell discharged at 600 °C for 72 h. An addition of a small amount of Sn in the anode showed a positive effect on the stability of the anode, and the performance drop of the cell decreased to 1.51%[24]. Besides Sn, we found that some other additives could also improve the resistance to carbon deposition of Ni-SDC anode with hydrocarbon fuels by the modification of crystal and electronic structure of Ni[25-27]. Furthermore, some basic additives such as NbO could adsorb water readily and facilitate carbon removal reactions[28]. Recently, we developed $LaBaMn_{1.6}Fe_{0.2}Co_{0.2}O_{5+\delta}$ as an anode material, and a single cell supported by the LSGM electrolyte layer achieved a P_{max} of 500 mW cm^{-2} with a duration of more than 200 h at 850 °C with CH_4 as the fuel.

Keywords: Methane catalytic decompostion; Carbon nanofiber; Solid oxide fuel cell; Anode

References

1. Y.D. Li, J.L. Chen, Y.N. Qin, L. Chang, Energy Fuels, 2000, 14, 1188-1194.
2. L.Y. Piao, Y.D. Li, J.L. Chen, L. Chang, J.Y.S. Lin, Catalysis Today, 2002, 74, 145-155.
3. Y.D. Li, J.L. Chen, L. Chang, Applied Catalysis A: General, 1997, 163, 45-57.
4. Y.D. Li, J.L. Chen, Y.M. Ma, J.B. Zhao, Y.N. Qin, L. Chang, Chemical Communications, 1999, 1141-1142.
5. J.L. Chen,Y.D. Li, Y.M. Ma, Y.N. Qin, L. Chang, Carbon, 2001, 39, 1467-1475.
6. X.X. Zhang, Z.Q. Li, G.H. Wen, K.K. Fung, J.L. Chen,Y.D. Li, Chemical Physics Letters, 2001, 333, 509-514.
7. C.N. He, N.Q. Zhao, X.W. Du, C.S. Shi, J. Ding, J.J. Li, Y.D. Li, Scripta Materialia, 2006, 54, 689-693.
8. J.L. Chen, X.Z. Zhou, L. Cao, Y.D. Li, Studies in Surface Science and Catalysis, 2004, 147, 73-78.
9. G.W. Wang, Y. Jin, G.J. Liu, Y.D. Li, Energy Fuels, 2013, 27, 4448-4456.
10. Y. Jin, G.W. Wang, Y.D. Li, Applied Catalysis A: General, 2012, 445-446, 121-127.
11. G.W. Wang, J.L. Chen, Y. Tian, Y. Jin, Y.D. Li, Catalysis Today, 2012, 183, 26-33.
12. Y.D. Li, J.L. Chen, L. Chang, Y.N. Qin, Journal of Catalysis, 1998, 178, 76-83.
13. J.L. Chen, Y.H. Qiao, Y.D. Li, Applied Catalysis A: General, 2008, 337, 148-154.
14. W.Z. Qian, T. Liu, F. Wei, Z.W. Wang, D.Z. Wang, Y.D. Li, Carbon, 2003, 41, 2683-2686.
15. D.X. Li, J.L. Chen, Y.D. Li, International Journal of Hydrogen Energy, 2009, 34, 299-307.
16. J.L. Chen, X.M. Li, Y.D. Li, Y.N. Qin, Chemistry Letters, 2003, 32, 424-425.
17. L.Y. Piao, J.L. Chen, Y.D. Li, China Particuology, 2003, 1, 266-270.
18. J.L. Chen, Q. Ma, T.E. Rufford, Y.D. Li, Z.H. Zhu, Applied Catalysis A: General, 2009, 362, 1-7.

19. J.L. Chen, X. Yang, Y.D. Li, Fuel, 2010, 89, 943-948.
20. Q.H. Liu, Y. Tian, H.J. Li, L.J. Jia, C. Xia, L.T. Thompson, Y.D. Li, Journal of Power Sources, 2010, 195, 6539-6548.
21. Y.D. Li, D.X. Li, G.W. Wang, Catalysis Today, 2011, 162, 1-48.
22. Z.M. Wang, Y. Tian, Y.D. Li, Journal of Power Sources, 2011, 196, 6104-6109.
23. Z.M. Wang, Y.D. Li, J.W. Schwank, Journal of Power Sources, 2014, 248, 239-245.
24. P. Li, Z.M. Wang, X.L. Yao, N.J. Hou, L.J. Fan, T. Gan, Y.C. Zhao, Y.D. Li, J.W. Schwank, Catalysis Today, 2018, DOI: 10.1016/j.cattod.2018.04.030.
25. P. Li, B.L. Yu, J. Li, X.L. Yao, Y.C. Zhao, Y.D. Li, Journal of Power Sources, 2016, 320, 251-256.
26. P. Li, L.N. Fang, N.J. Hou, J. Li, X.L. Yao, T. Gan, L.J. Fan, Y.C. Zhao, Y.D. Li, Journal of The Electrochemical Society, 2017, 164, F1142-F1148.
27. G.C. Ding, T. Gan, J. Yu, P. Li, X.L. Yao, N.J. Hou, L.J. Fan, Y.C. Zhao, Y.D. Li, Catalysis Today, 2017, 298, 250-257.

28. X.L. Yao, L.J. Fan, T. Gan, N.J. Hou, P. Li, Y.C. Zhao, Y.D. Li, International Journal of Hydrogen Energy, 2018, 43, 12748-12755.

545605 Effect of Ni-Fe Segregation on the Deactivation Behavior of Co-Precipitated $NiFeAlO_x$ Catalysts in the CO_2 Methanation Reaction

Monday, June 3, 2019: 4:00 PM
Republic ABC (Grand Hyatt San Antonio)

Thomas Burger[1,2], Stefan Ewald[1,2], Klaus Köhler[1,2] and Olaf Hinrichsen[1,2],
(1)Department of Chemistry, Technical University of Munich, Munich, Germany,
(2)Catalysis Research Center, Technical University of Munich, Munich, Germany

Motivation

The CO_2 methanation reaction is one of the key steps in the power-to-gas process chain. For potential commercialization, efficient and cost-effective catalyst materials need to be developed that feature high intrinsic activity to achieve high CH_4 yields at low temperatures. In addition, the catalyst should be able to withstand high reaction temperatures that may evolve due to hotspot formation in this highly exothermal reaction. Catalyst deactivation by temperature-induced effects may further be aggravated by the high partial pressure of steam at high CO_2 conversion levels. Fe was shown to be an efficient promoter for significantly enhancing the activity of Ni-Al-based catalysts [1-3]. However, comprehensive and time-resolved studies on the catalyst deactivation behavior featuring post-characterization studies to draw conclusions on the structure-activity relationship are lacking so far and therefore are addressed in this work.

Experimental

A $NiAlO_x$ and a $NiFeAlO_x$ (Ni/Al = 1, Ni/Fe = 5.5) catalyst were prepared by co-precipitation at pH 9 and 30 °C. After aging for 18 h, the precipitate was dried at 80 °C and calcined in flowing synthetic air at 450 °C for 6 h. After *in situ* reduction at 485 °C for 5 h (5 % H_2 in Ar), the catalysts were aged in thermodynamic equilibrium for 72 h at 8 bar, 36 NL $(g_{cat}\ h)^{-1}$ ($Ar/H_2/CO_2$ = 5/4/1) and temperatures between 350 and 450 °C. In between, the catalysts were periodically cooled down to 230 °C to evaluate their activity under differential conditions (4 bar, 600 NL $(g_{cat}\ h)^{-1}$, $Ar/H_2/CO_2$ = 75/4/1, 230 °C) and to track deactivation away from thermodynamic equilibrium. After definite aging times, the catalysts were thoroughly characterized without exposure to air.

Results and discussion

Material characterization prior to catalyst testing

The catalysts feature a hydrotalcite-derived mixed oxide structure with mesoporous structure. The BET surface area amounts to 293 $m^2\ g_{cat}^{-1}$ for the benchmark $NiAlO_x$ (37 % Ni) and 266 $m^2\ g_{cat}^{-1}$ for $NiFeAlO_x$ (35 % Ni, 5 % Fe). Catalyst activation leads to partial reduction of Ni in $NiAlO_x$ (60 %) and Fe as well as Ni in $NiFeAlO_x$. For the Fe-promoted catalysts, *in situ* X-ray powder diffraction analysis reveals the presence of Ni-Fe alloy particles by the characteristic shifts of the Ni reflexes to higher diffraction angles. Ferromagnetic measurements show a clear increase of all relevant magnetic characteristics for $NiFeAlO_x$, indicating that Fe substantially contributes to the magnetic properties of Ni. Since Fe is known to be a weak adsorbent for H_2 at room temperature [4], a decrease of the metal surface area from 30.1 $m^2\ g_{cat}^{-1}$ for the benchmark $NiAlO_x$ to 20.1 $m^2\ g_{cat}^{-1}$ is observed in static H_2 chemisorption, indicating that significant amounts of Fe are present on the Ni surface after reduction.

Catalyst testing and material characterization after definite aging times

As illustrated in Figure 1A, the initial CH_4 weight time yield over $NiFeAlO_x$ is 1.5 times higher compared to $NiAlO_x$. Despite, the initial apparent activation energy is 76.2±1.4 kJ mol^{-1} for NiAlOx and 89.5±1.4 kJ mol^{-1} for $NiFeAlO_x$, which clearly shows that Fe modifies the active Ni centers and indicates a compensation effect [5] for the methane production rate over $NiFeAlO_x$.

Aging of $NiAlO_x$ at 450 °C leads to a strong decrease of the catalytic activity. During the first 6 h of aging, already 28 % of the initial activity are lost. After 40 h of aging the catalytic activity reaches a stable level at 49 %. The Ni surface area decreases to 15.3 % during 6 h of aging (BET surface area 207 $m^2\ g_{cat}^{-1}$), after 40 h it stays constant at approx. 12 $m^2\ g_{cat}^{-1}$ (BET surface area 131 $m^2\ g_{cat}^{-1}$). Crystallite size estimation from XRD indicates crystallite growth from 3.2 to 4.8 nm during 32 h of aging. Similarly, the CO_2 uptake declines from 199 to 123 µmol g_{cat}^{-1} during 6 h and to 84 µmol g_{cat}^{-1} during 40 h, indicating that besides sintering effects on the Ni particles also the basic site density decreases. Interestingly, all sorption parameters as well as the Ni crystallite size remain constant once the constant activity level has been achieved. However, no direct correlation of any of the sorption parameters to the catalytic activity can be found. It is therefore hypothesized, that a complex interplay of the total surface area, the Ni surface area and the basic site density is responsible for changes in the catalytic activity of the $NiAlO_x$ catalyst. Throughout the aging treatment, the apparent activation energy remains constant, which indicates that the type of the active sites does not change.

In contrast, the activity for $NiFeAlO_x$ rises within the first 6 h, with a maximum value of 122 % of the initial one before apparent catalyst deactivation starts. After 72 h of aging, the activity is still at 90 %. The performance of the $NiFeAlO_x$ aged at different temperatures is illustrated in Figure 1B. When aged at 350 °C, a stable activity level is obtained after 20 h of aging (119 %) and no apparent deactivation takes place. When aged at 400 °C, the maximum activity is reached at 129 % after 16 h of aging, from where on slow deactivation is observed. It is therefore hypothesized that two countervailing aging effects take place on $NiFeAlO_x$. Besides deactivation phenomena similar to $NiAlO_x$ (all relevant sorption properties of $NiFeAlO_x$ are affected similarly to $NiAlO_x$, cf. Table 1), X-ray diffraction analysis indicates that the previous Ni-Fe alloy gradually segregates during aging. Accordingly, the magnetization of the $NiFeAlO_x$ catalyst first rises but then decreases to finally reach the characteristics of the $NiAlO_x$ catalyst after 72 h of aging. This agrees well with the higher loss of metal surface area for $NiFeAlO_x$ (73 %) compared to $NiAlO_x$ (60%), where the segregated Fe species might block H_2 adsorption sites on Ni. The results indicate that the composition of the Ni-Fe particles is dynamically modified under aging conditions, which leads to a temporal improvement of the CH_4 production rate. The impact of Fe on the active sites is further corroborated by a clear change of the activation energy to 104.8±3.8 kJ mol^{-1} already after 6 h of aging. Again, the opposing enhanced activity indicates a compensation effect [5].

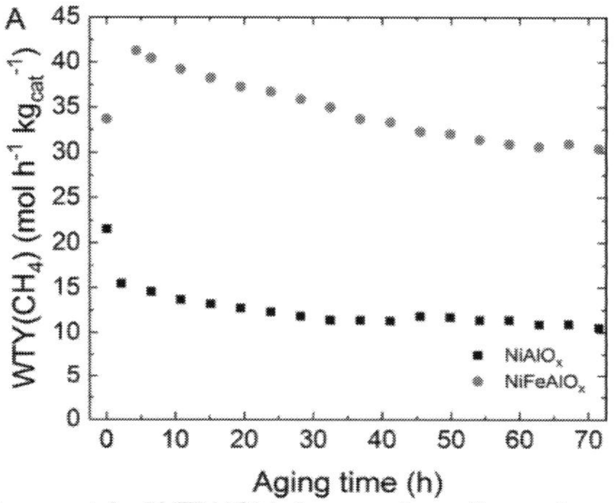

Figure 1A: $WTY(CH_4)$ as a function of aging time for $NiAlO_x$ and $NiFeAlO_x$, aging temperature 450 °C.

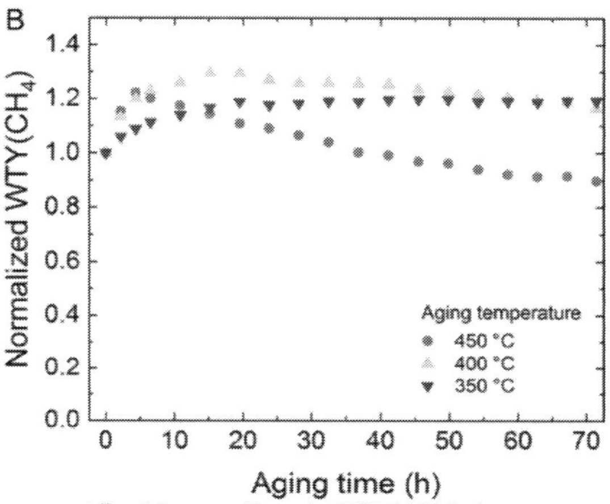

Figure 1B: Normalized $WTY(CH_4)$ as a function of aging time for $NiFeAlO_x$, aging carried out at different temperatures.

Table 1: Catalyst sorption data and activation energies as a function of aging time and temperature.

	t_a	T_a	S_{Ni}	D_{Ni}	d_P [a]	S_{BET}	$U(CO_2)$	E_A
	h	°C	$m^2\ g_{cat}^{-1}$	%	nm	$m^2\ g_{cat}^{-1}$	µmol g_{cat}^{-1}	kJ mol^{-1}
$NiAlO_x$	0	-	30.1	12.2	3.2	293	198.7	76.2±1.4
	6	450	15.3	6.2	4.2	207	123.1	73.9±3.6
	40	450	12.2	5.0	4.8	131	83.8	75.0±2.5
	72	450	12.1	4.9	4.8	132	81.5	73.7±2.1
$NiFeAlO_x$	0	-	20.1	8.6	3.2	266	218.7	89.5±1.4
	6	450	13.4	5.8	3.7	130	121.0	104.8±3.8
	40	450	6.8	2.9	4.4	110	86.9	104.2±2.7
	72	450	5.4	2.3	5.3	108	83.5	101.9±1.2
	72	350	9.0	3.9	4.0	191	139.8	99.7±1.9
	72	400	8.1	3.5	4.7	131	108.7	102.3±0.9

[a] determined by XRD, Ni(Fe) reflex at $2\theta = 51.3–51.7$ °.

Conclusion

It is shown that the high stability of co-precipitated $NiFeAlO_x$ catalysts in the CO_2 methanation reaction is caused by a temporal increase of the catalytic activity, that is linked to (partial) segregation of the Ni-Fe alloy particles, induced by hydrothermal aging. For isothermal low-temperature methanation, the catalytic activity of $NiFeAlO_x$ catalysts therefore can be strongly enhanced by an aging pretreatment. Ambient XPS and high-resolution synchrotron X-ray data might contribute to derive a deeper understanding of the underlying segregation mechanisms and the Ni-Fe particle compositions desirable to maximize the catalytic activity of Ni-Fe-based CO_2 methanation catalysts.

[1] B. Mutz, M. Belimov, W. Wang, P. Sprenger, M.-A. Serrer, D. Wang, P. Pfeifer, W. Kleist, J.-D. Grunwaldt, *ACS Catal.* **2017**, 6802–6814.

[2] T. Burger, F. Koschany, O. Thomys, K. Köhler, O. Hinrichsen, *Appl. Catal., A* **2018**, *558*, 44–54.

[3] C. Mebrahtu, F. Krebs, S. Perathoner, S. Abate, G. Centi, R. Palkovits, *Catalysis Science & Technology* **2018**, *8*, 1016–1027.

[4] H. J. Jung, M. A. Vannice, L. N. Mulay, R. M. Stanfield, W. N. Delgass, *J. Catal.* **1982**, *76*, 208–224.

[5] E. Cremer, in *Advances in Catalysis, Vol. 7* (Eds.: W. G. Frankenburg, V. I. Komarewsky, E. K. Rideal), Academic Press, **1955**, pp. 75–91.

545479 Strong Ni-MnO Synergy in the Sabatier Reaction

Monday, June 3, 2019: 4:24 PM
Republic ABC (Grand Hyatt San Antonio)

Emiel Hensen, Schuit Institute of Catalysis, Department of Chemical Engineering and Chemistry,, Eindhoven University of Technology, Eindhoven, Netherlands and Wilbert Vrijburg, Eindhoven University of Technology, Eindhoven, Netherlands

1. Introduction

The prospects of producing fuels and chemicals via CO_2 methanation may simultaneously contribute to CO_2 emission reduction and clean-energy storage solutions.[1] However, CO_2 activation remains a considerable scientific challenge.[2] Supported Ni and Ru have been the industrial catalysts of choice for CO and CO_2 methanation due to their high activity and methane selectivity.[3] In recent years, the promotion of Ni-based catalysts with a second metal has been studied. Such bimetallic systems usually include expensive noble metals (e.g. Pt, Pd, Ru),[4] therefore cheaper promoters are highly desired. A transition metal promoter that has received limited attention for Ni-based systems is Mn, this despite its importance in Co-based Fischer-Tropsch catalysts.[5] In our study we report a promotional effect of Mn on Ni/TiO_2 catalysts for CO_2 and CO methanation.

2. Experimental

Mn-promoted Ni/TiO_2 catalysts were prepared using wet-impregnation methods. Catalysts were characterized before reaction with TPR, chemisorption, ICP, and XRD. Particle sizes and metal distribution was studied with STEM-EDX. The reduction of unpromoted and Mn-promoted Ni/TiO_2 was investigated using operando quick-EXAFS at both the Ni and Mn K-edges to determine the oxidation state and coordination of each metal. Catalytic activity and stability for gas phase CO_2 and CO methanation were studied in high-throughput at atmospheric pressure between 200-400 °C. Temperature-programmed hydrogenation was used to study catalyst coking. In-situ FTIR was employed to investigate CO_2 adsorption and observe intermediate species under methanation conditions, enabling us to develop a mechanistic understanding for TiO_2 supported Ni and Ni-Mn catalysts. The mechanistic picture was completed by a combination of isotopic transient kinetic analysis, density functional theory (DFT) calculations and first-principles microkinetics simulations.

3. Results & Discussion

Prior to reduction, XRD and XANES/EXAFS measurements confirmed that Ni and Mn were present as highly dispersed oxides. Detailed analysis of the Ni-Mn/TiO_2 samples' reducibility was provided by operando XANES measurements. The addition of Mn to Ni/TiO_2 catalysts led to a delay in the onset of Ni reduction from 300 °C to

400 °C as determined with operando XANES on the Ni K-edge (Figure 1a-b). This confirmed that Ni and Mn were in close proximity. STEM-EDX showed that all active catalysts consisted of bulk Ni nanoparticles, with Mn-promoted catalysts containing highly dispersed Mn, likely in the form of MnO. The decoration of the Ni particles with MnO led to an enhanced CO_2 methanation activity (Figure 1c). Moreover, exposing Ni/TiO_2 and Ni-Mn/TiO_2 catalysts to CO methanation conditions led to the deactivation of all catalysts. However, the catalyst lifetime was clearly enhanced by the addition of Mn. An investigation into the deactivation pathways for these catalysts identified coking as the likely source. An in situ FTIR study verified that promotion with Mn leads to an increased CO_2 adsorption capacity. Furthermore, Mn addition enhanced CO_2 activation, ultimately leading to improved methanation activity.Mechanistically, we find that partial removal of O atoms from MnO in contact with Ni results in active sites for rapid C-O bond cleavage. Microkinetics simulations based on DFT calculations support this and point out C-O bond dissociation in adsorbed CO and CO_2 at the interface between Ni and MnO contributes to the high catalytic performance. These data are consistent with operando IR measurements and ^{12}CO2 " $^{13}CO_2$ isotopic kinetic analysis.

4. Conclusions

The addition of Mn to Ni/TiO_2 catalysts leads to enhanced CO_2 methanation activity. The superior activity originates from the decoration of Ni nanoparticles by highly dispersed MnO, with higher Mn loadings yielding more active CO_2 methanation catalysts, and improving the stability of the catalysts under CO methanation conditions. These findings demonstrate that Ni-based catalysts may be promoted with cheaper and more abundant transition metal oxides such as MnO, enabling better performance while ensuring a sustainable approach.

References

(1) Aziz, M. A. A.; Jalil, A. A.; Triwahyono, S.; Ahmad, A. *Green Chem.* **2015**, *17*, 2647–2663.

(2) Aresta, M.; Dibenedetto, A.; Angelini, A. *Chem. Rev.* **2014**, *114*, 1709–1742.

(3) Wei, W.; Jinlong, G. *Front. Chem. Sci. Eng.* **2011**, *5*, 2–10.

(4) De, S.; Zhang, J.; Luque, R.; Yan, N. *Energy Environ. Sci.* **2016**, *9*, 3314–3347.

(5) Morales, F.; de Groot, F. M. F.; Gijzeman, O. L. J.; Mens, A.; Stephan, O.; Weckhuysen, B. M. *J. Catal.* **2005**, *230*, 301–308.

545675 Preparation and Testing of Ru/Al_2O_3 Catalysts for a Compact-Scale Fixed Bed CO_2 Methanation Reactor

Monday, June 3, 2019: 4:48 PM
Republic ABC (Grand Hyatt San Antonio)

Alessandro Porta[1], Leonardo Falbo[1], Carlo Giorgio Visconti[1], Luca Lietti[1], Claudia Bassano[2] and Paolo Deiana[2], (1)Department of Energy, Politecnico di Milano, Milan, Italy, (2)ENEA, Rome, Italy

INTRODUCTION

In order to reduce the CO_2 emissions and comply with international agreements, a strong increase in the use of renewable energy and carbon capture technologies is foreseen. Renewables, especially solar and wind power, are intrinsically fluctuating sources and this leads to periods of energy overproduction, during which energy needs to be effectively stored so as not to be lost. Nowadays, batteries are still inadequate to provide long term-storage capacity, and a valid solution may be provided by the chemical storage through the so-called power to gas technology (PtG). This process uses the excess renewable power to fuel an electrolyzer that produces gaseous H_2 (and pure oxygen) from the splitting of H_2O. To facilitate its storage and transportation, which are still an issue, according to the PtG concept, H_2 is then used to make CH_4. Methane has the huge advantage to have an already-existing infrastructure for its transport and use [1]. Furthermore, methane production allows the recycling of waste CO_2 (coming from capture, biogas rectification, etc.) through the Sabatier reaction:

$$CO_2 + 4\ H_2 \leftrightarrow CH_4 + 2\ H_2O \qquad DH^0_R = -165\ \text{kJ/mol}$$

This reaction is generally carried out over Ni-based catalysts, but from a process intensification standpoint, also the more expensive Ru-based catalysts are of interest. Indeed, it is known that Ru is the more active metal for the Sabatier reaction [2], and this allows to operate the Sabatier reaction at lower temperatures with respect to Ni-based catalysts. This is of crucial importance, since both CO_2 conversion and CH_4 selectivity of the highly exothermic Sabatier reaction are limited by thermodynamics at high temperatures. The use of Ru-based catalysts paves the way for operations at low temperatures, possibly enabling high per-pass CO_2 conversions even at low pressures.

Aiming at the optimization of the CO_2 methanation process using a compact scale fixed bed reactor and very active supported Ru-catalysts, it is necessary to gather information about the catalyst selectivity, stability and threshold size so to rule-out the onset of internal mass transfer limitations. For this purpose, we have prepared catalysts using γ-Al_2O_3 spherical supports having different diameters (100, 800 and 2300 μm, Sasol Puralox) and using two different Ru precursors ($Ru(NO)(NO_3)_3$ and $RuCl_3$). The obtained catalysts, prepared via impregnation, have been tested (as such or after crushing and sieving) in a fully-automated rig operating 24/7 in a fixed bed reactor configuration, operating in integral conditions to represent at best the industrial process conditions.

RESULTS AND DISCUSSION

The impregnation of supports bigger than 100 µm with a metal loading of 0.5 wt% of Ru led to catalysts with an eggshell configuration, i.e. all the introduced active phase was located on the outer shell of the catalyst pellets. This result was observed using both metal precursors (Ru chloride and nitrosyl nitrate) and confirmed by SEM/EDX analysis. Using the chloride precursor, an active layer thickness as low 4% of the pellet radius was obtained, while using $Ru(NO)(NO_3)_3$ a thicker shell was obtained.

In order to tune the thickness of the impregnated layer, the impregnating solutions containing $Ru(NO)(NO_3)_3$ and $RuCl_3$ have been acidified, and linear correlations have been observed between the active layers thickness and the molar amount of introduced acid, at constant concentration of the metal precursor in the impregnating solution. By strongly increasing the acidity of the solution, homogenously impregnated pellets with a metal loading of 0.5 wt% could be eventually obtained (Figure 1).

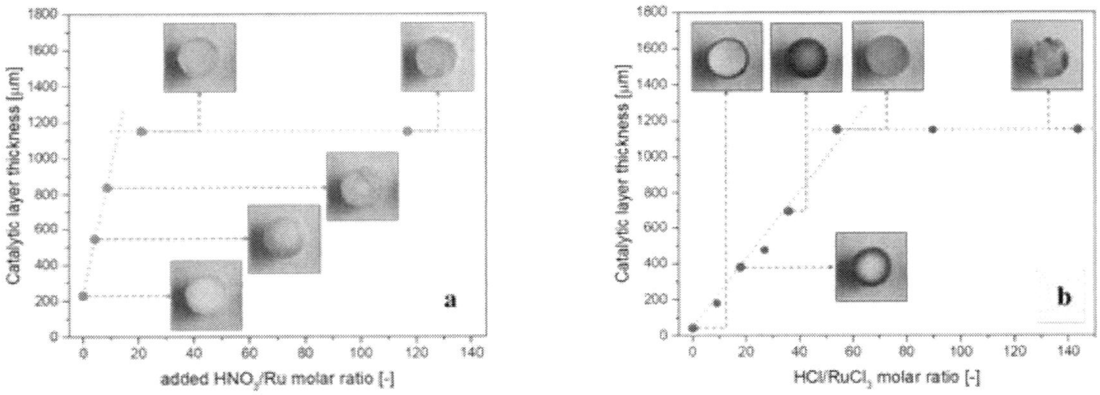

Figure 1. Effect of the acid/metal ratio on the active layer thickness for a 0.5 wt% Ru/Al_2O_3 catalyst of 2300 µm prepared from a) $Ru(NO)(NO_3)_3$ and b) $RuCl_3$

These results have been explained considering the strong ruthenium precursor-support interaction, which results in the fast adsorption of the Ru ions during the impregnation of the outer layers of the alumina support.

At first, the catalytic performance of the two materials prepared using $Ru(NO)(NO_3)_3$ and $RuCl_3$ on the 100 µm powder (and no acid addition) has been evaluated at P = 0 barg, T = 250-430°C, H_2/CO_2 = 3.9, GHSV = 5 L(STP)/h/g_{cat}. The nitrate-based catalyst reached steady performance within the first hours of time on stream, while the chloride-based sample required more than 150 h to reach a stable performance, due to the slow elimination of Cl^- ions from the catalyst surface. For this reason, even though the steady state performance of the two systems after the conditioning transient was comparable, the chloride containing samples have been excluded from the rest of this study. Notably, both samples show an almost complete methane selectivity in all the investigated temperature range, with CO being the most abundant byproduct (always below 2%). Small amounts of C_2H_6 were also detected below 350°C, due to the C-chain growth reactions.

Then, the impregnated pellets prepared using acid-free or acid-containing solutions of $Ru(NO)(NO_3)_3$ and 800 and 2300 µm Al_2O_3 spheres have been tested (after crushing and sieving) so to evaluate the effect of the acid introduction on the catalytic activity. No differences in CO_2 conversion nor in CH_4 selectivity were observed, indicating a negligible effect of the acid on the catalyst performance.

Subsequently, homogeneously impregnated pellets have been tested to verify the absence of internal diffusional limitations. No significant differences in the catalytic activity were observed when varying the dimension of the pellets, indicating the absence of intraporous limitations with a Ru loading of 0.5 wt% up to a characteristic length of diffusion (d) of 383 µm (Figure 2a)

In order to push the system to more stringent conditions, we increased the catalytic activity by increasing the Ru loading of one order of magnitude. The metal loading of 5 wt% was obtained with four successive impregnations and in this case no acid addition was required to obtain a uniform Ru distribution along the pellet radius.

At first the performance of the catalysts at 0.5 and 5 wt% loadings supported onto 100 µm powders has been compared. Interestingly, no differences in CO_2 conversion nor in CH_4 selectivity were observed working under kinetically controlled regime and at constant space velocity per gram of Ru. This indicates that the specific reactivity of the catalyst does not depend on the Ru loading. In line with these results, HR-TEM evidenced that the dispersion of the Ru^0 phase does not vary by increasing the metal loading by one order of magnitude ($d_{p,avg}$ = 1 nm).

Subsequently, these experiments were repeated with for the 5 wt% catalysts supported on 800 and 2300 µm Al_2O_3 spheres (P = 0 barg, T = 250-430 °C, H_2/CO_2 = 3.9, GHSV = 50 L(STP)/h/g_{cat}). In this case, both conversion and selectivity diverge from the performance of the powdered catalyst already in the case of the 800 µm pellets ($d <$ 130 µm), suggesting that the catalyst performance is affected by internal diffusional limitations (Figure 2b). This evidence was confirmed by means of a kinetic study, which showed that the activation energy evaluated for 2300 µm thick catalysts was half the one evaluated for the powdered catalyst with the same Ru loading. Notably, the selectivity to methane drops when using big pellets (though still remaining above 95%) due to the specular increase in the CO formation, which might be due also to a non-isothermal profile along the catalyst pellet generated by the strongly exothermic Sabatier reaction under conditions dominated by transport limitations.

These results are relevant both in terms of mechanistic understanding of the Sabatier reaction and in view of the intensification of the process. A detailed kinetic study (able to describe not only the CO_2 conversion but also the CH_4 selectivity) based on these findings as well as on spectroscopic evidences is currently being developed.

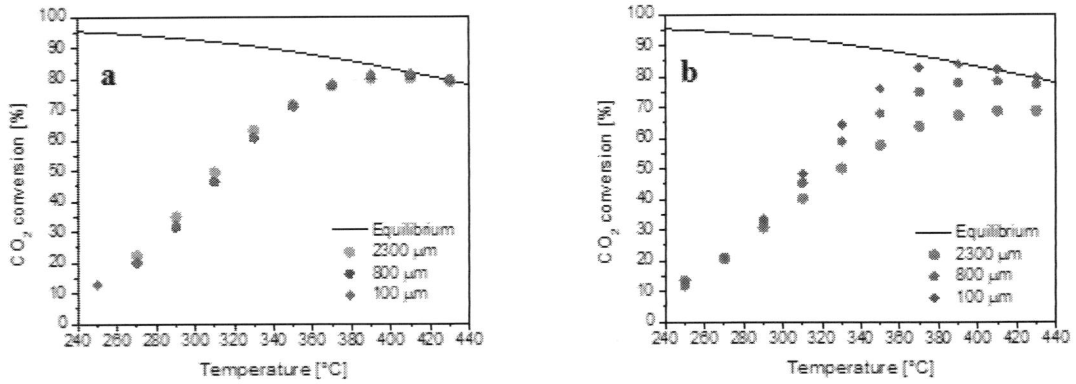

Figure 2. CO_2 conversion as a function of temperature for homogenously impregnated γ-Al_2O_3 spheres of 100, 800 and 2300 µm. a) 0.5 wt% Ru, b) 5 wt% Ru

CONCLUSION

We show that the strong ruthenium precursor-support interaction leads to the formation of eggshell catalysts at metal loadings as low as 0.5 wt%. However, homogeneously impregnated pellets can be obtained by increasing the acidity of the impregnating solution. The reactivity of pellets prepared using acid-free or acidified solutions is similar. Increasing the metal loading by one order of magnitude prevent the formation of eggshell pellets and does not result in an increase of the metallic particle size. As a result, the CO_2 conversion rate per gram of active phase is unaffected by the increase metallic density on the surface. However, in presence of a catalysts with Ru-loadings as high as 5 wt%, intraporous mass/heat transfer limitations strongly affects both the CO_2 conversion rate and CH_4/CO selectivity. An accurate description of these phenomena is pivotal for the process intensification.

REFERENCES

[1] C. Janke, M.S. Duyar, M. Hoskins, R.J. Farrauto, Catalytic and adsorption studies for the hydrogenation of CO_2 to methane, Appl. Catal. B Environ. 152–153 (2014) 184–191. doi:10.1016/j.apcatb.2014.01.016.

[2] P. Frontera, A. Macario, M. Ferraro, P. Antonucci, Supported Catalysts for CO_2 Methanation: A Review, Catalysts. 7 (2017) 59. doi:10.3390/catal7020059.

551390 Advanced Chemical Looping Concept for Energy and Fuels Applications

Tuesday, June 4, 2019: 10:15 AM
Republic ABC (Grand Hyatt San Antonio)

Guy B. Marin and Vladimir V. Galvita, Laboratory for Chemical Technology, Ghent University, Ghent, Belgium

The increasing energy consumption creates the need to investigate new routes for utilization of the available resources, like natural gas and biogas, to produce fuels and chemicals. CH_4 is considered as one of the most affordable carbon feedstocks in the world. However, direct conversion of CH_4 into chemicals results in low yields due to the high C-H bond dissociation energy (436 $kJ \cdot mol^{-1}$). As a result, CH_4 is mainly converted indirectly, by reforming it to syngas, a mixture of H_2 and CO, as an intermediate step. Syngas is a key building block with many downstream applications, as it is used in several synthesis processes for a wide range of chemicals and fuels. Most of the produced syngas is used for the synthesis of ammonia for fertilizers and for hydrogen production, which is exploited in refining processes, while the "gas to liquid" routes for fuels production account for ~8 % consumption.

Currently, different methane reforming processes exist and the most widely studied for syngas production are steam reforming (SRM), dry reforming (DRM), partial oxidation (POM) and autothermal reforming of methane (ARM). During reforming, several side reactions may occur. The water-gas-shift (WGS) reaction leads to a decrease or increase of the H_2/CO ratio. In addition to WGS, CH_4 decomposition (MD) and the Boudouard reaction (BR) may occur. These are responsible for carbon deposition on the active metal sites, leading to catalyst deactivation.

Chemical looping is an emerging technology, capable of low-emission, with application in production of heat, fuels, chemicals, and electricity. The power of the chemical looping concept will be demonstrated in Super-Dry Reforming [1]. Super-dry reforming of methane (SDRM) was developed for enhanced CO production from CH_4 and CO_2. The process uses a CO_2 sorbent (CaO/ZrO_2), an oxygen carrier ($Fe_2O_3/MgAl_2O_4$), and a CH_4 reforming catalyst (Ni-Fe/$MgAl_2O_4$), as well as CH_4 and CO_2 in a 1:3 feed ratio. SDRM combines an exothermic process, $CaCO_3$ formation, with two endothermic processes, CH_4 reforming and Fe_3O_4 reduction, thereby looping the energy utilization. The isothermal coupling of these three different processes results in a higher CO production and free of carbon deposition as compared to conventional DRM. Herein, details of the individual steps and the full process of SDMR, as well as perspectives and economic aspects will be presented [2-5].

References

[1] L.C. Buelens, V.V. Galvita, H. Poelman, C. Detavernier, G.B. Marin, Super-dry

reforming of methane intensifies CO_2 utilization via Le Chatelier's principle, Science 354(6311) (2016) 449-452.

[2] J. Hu, L. Buelens, S.-A. Theofanidis, V.V. Galvita, H. Poelman, G.B. Marin, CO2 conversion to CO by auto-thermal catalyst-assisted chemical looping, Journal of CO_2 Utilization 16 (2016) 8-16.

[3] K. Verbeeck, L.C. Buelens, V.V. Galvita, G.B. Marin, K.M. Van Geem, K. Rabaey, Upgrading the value of anaerobic digestion via chemical production from grid injected biomethane, Energy & Environmental Science 11(7) (2018) 1788-1802.

[4] J. Hu, V.V. Galvita, H. Poelman, C. Detavernier, G.B. Marin, A core-shell structured $Fe_2O_3/ZrO_2@ZrO_2$ nanomaterial with enhanced redox activity and stability for CO_2 conversion, Journal of CO_2 Utilization 17 (2017) 20-31.

[5] J. Hu, V.V. Galvita, H. Poelman, C. Detavernier, G.B. Marin, Catalyst-assisted chemical looping auto-thermal dry reforming: Spatial structuring effects on process efficiency, Applied Catalysis B: Environmental 231 (2018) 123-136.

545641 Conceptual Design of a Dual Fluidized Bed System for an Intensified Steam Methane Reforming Concept Coupled with Ca-Ni Looping

Tuesday, June 4, 2019: 10:54 AM
Republic ABC (Grand Hyatt San Antonio)
Theodoros Papalas, Andy Antzara and **Angeliki Lemonidou**, Chemical Engineering, Aristotle University of Thessaloniki, Thessaloniki, Greece

Introduction

The severe environmental issues along with the foreseen depletion of fossil fuels have accentuated the need of finding alternative fuels and energy production technologies that reduce the dependence on fossil energy and moderate CO_2 emissions. Hydrogen is regarded as an important primary industrial gas as well as a promising and sustainable energy carrier due to its combustion having high energy efficiency and zero carbon emissions. The dominant large-scale production path of hydrogen is steam reforming of natural gas which is considered an energy intensive process and requires multiple steps and complexity to achieve the generation of high-purity hydrogen. Furthermore, due to thermodynamic limitations the process is usually coupled with water-gas shift reactors downstream the reformer to recover additional hydrogen and thus not avoiding the co-production of CO_2. To address the aforementioned issues, a novel technology which provides a path for intensified conversion of natural gas directly to high purity hydrogen in a single step has been developed; the so-called sorption enhanced chemical looping steam methane reforming (SE-CL-SMR) [1]. This concept combines the reformer and WGS reactors in a single unit by introducing a solid sorbent material such as CaO that can in–situ remove the produced CO_2. The exothermic carbonation reaction of CaO with the produced CO_2 enables the performance of both the reforming and water-gas shift reactions in a single vessel, overcoming the thermodynamic limitations of the overall reaction, while the heat generated by the strongly exothermic carbonation is consumed in–situ for the endothermic reforming reaction. When the solid sorbent material reaches saturation, it needs to be regenerated in a separate reactor for it to be re-utilized. The necessary heat for the regeneration is addressed with the introduction of a second chemical loop, where an oxygen transfer material (OTM) such as NiO is also circulating between the two reactors by undergoing continuing redox reactions. During the regeneration stage, a sweep gas such as pure oxygen can be used for the strongly exothermic oxidation of the OTM thus moderating the thermal demands of the regeneration. The oxidized OTM returns to the reformer wherein is reduced by methane and the reformate gases. In addition to the oxygen transfer properties as an OTM, NiO in the reduced form presents excellent catalytic properties for steam methane reforming and water gas shift reactions. A schematic representation of the combined process is presented in Fig. 1.

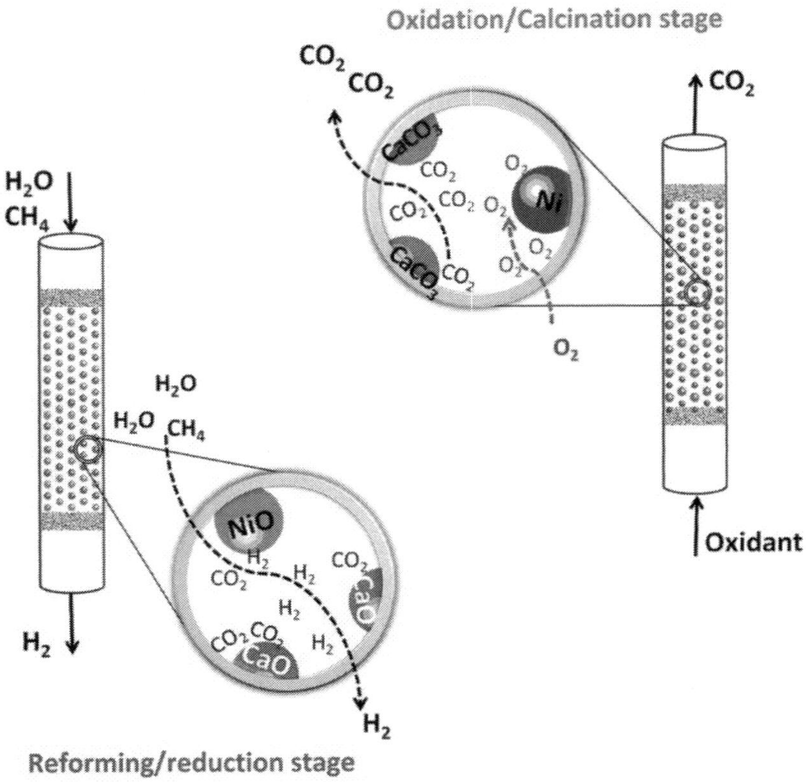

Figure 1: Schematic representation of SE-CL-SMR process

Methodology and Results

This study proposes the implementation of the reduction and regeneration cycles in a system of two interconnected fluidized bed reactors that operate under different hydrodynamic regimes. Based on previous work of our group [2], the solid materials used consist of a CaO-based CO_2 sorbent (66wt% $CaO/CaZrO_3$) and a NiO-based OTM (40wt% NiO/ZrO_2) which form a mixture of Geldart A particles. The design of the reactors is performed with the assistance of the chemical process simulator Aspen Plus® V9 and all the results obtained refer to steady state simulations. The reformer is modeled as a bubbling bed, whose diameter and length are chosen to be equal to 1.5m and 3.1m respectively and thus form a total reactor volume of ~5.3m^3. In order to operate in the bubbling regime, the inlet superficial gas velocity needs to be higher than the minimum bubbling velocity, with the latter being at least three times as high as the minimum fluidization velocity. In this case, the inlet gas velocity is proposed to be nearly 0.07m/s which is ~6.5 times as high as the minimum fluidization velocity. For the simulation of the bubbling bed reactor the two-phase theory is utilized, which indicates the co-existence of an emulsion and a bubble phase [3]. For the estimation of the bubbles' diameter within the bed, a correlation is used that takes into account both the coalescence and splitting frequencies of bubbles inside the reactor [4]. An assumption is made that the bubble diameter remains constant to 11cm for all the length of the reactor, which is the value calculated in the middle of the bed. Bubbles rise with a velocity of ~0.78m/s and perform a plug flow movement within the reactor by occupying ~5% of the total reactor volume. They also contain a small amount of solids that account for the 0.6% of the bubble volume. The rest of the reactor is comprised of the emulsion phase, where it is assumed that perfect mixing between gases and solids occur. The emulsion remains in minimum bubbling condition and thus has a solid fraction and a gas velocity of 0.53 and 0.03m/s respectively. The bed is axially divided into zones, with each being composed of a RCSTR and a RPFR model that are connected in parallel and represent the emulsion and bubble phases respectively [5]. The reactor models solve the mass balance equations for each component by taking into account kinetic equations for the rates of the reactions taking place within the reformer. A Langmuir-Hinshelwood model is utilized for the reforming and water-gas shift reactions, while the reduction of OTM is described through the shrinking core model [6,7]. The carbonation reaction is assumed to reach equilibrium. The equilibrium carbonation constant is calculated from the standard Gibbs free energies of formation of the components that participate in the reaction. The outlet streams of the reactor models of each zone are modified by taking into consideration the mass transport phenomena between the emulsion and bubble phases before entering the next pair of reactor models. The solid phase of the last outlet stream of the bubbling bed enters the regenerator which is assumed to operate as a riser. The dimensions of the reactor are chosen to be 0.17m and 4.50m for the diameter and the length respectively. The main feature of the riser is that solid particles exit from the top of the bed due to the velocity of the gas feedstock being higher than the terminal velocity of the solids. In this case, the superficial gas velocity is estimated to be 0.59m/s which is nearly 79 times as high as the minimum fluidization velocity and surpasses the value of the terminal velocity. The riser is assumed that achieves perfect mixing of gas and solid that allows its axial division into an array of RCSTR models. The kinetic expressions used both for calcination of $CaCO_3$ and Ni reoxidation are based on the shrinking core model [7,8].

The proposed system succeeds in the generation of 210 Nm^3 H_2 /hr with ~95% purity (dry basis). This hydrogen production is achieved with the 92.7% conversion of the methane feedstock and an estimated 79.7% hydrogen yield. Conversions of nearly 100% and 83.5% are attained by the OTM and the sorbent respectively in the reformer. The solid materials return to their initial forms in the regenerator and a recovery of ~55 Nm^3 CO_2/hr with more than 97% purity is attained. Assuming that the reactors operate non-adiabatically, the heat duty of the reformer and the regenerator were calculated around 68kW and 53kW respectively, which sum up to a total ~46 MJ/kmol H_2 thermal demand of the whole process. The results of this study are compared to the ones predicted by the thermodynamic analysis of the combined process for the verification of the proposed model and a small deviation is witnessed between the results for the reformer as shown in Table 1. A deviation was not observed for the degrees of NiO reduction, CaCO since the thermodynamic analysis that all three solid materials achieve almost 100% conversion. The above deviations can be attributed to the amount of gas that enters the reformer with the form of bubbles [5] as well as to the used kinetics data obtained from literature that does not refer to the studied materials. However, further research is required in order to confirm this proposal. Although the concept of two interconnected fluidized beds seems to be an appropriate solution to the continuous hydrogen production through the intensified steam methane reforming process, further parametric studies need to be performed to define the right number of zones of the bubbling bed reactor and the optimized values of all the variables that influence the system's performance.

Table 1: Deviation of this model's results to equilibrium data

	This study	Equilibrium Data [1]	Deviation (%)
CH_4 Conversion (%)	92.7	97.0	-4.4
H_2 Yield (%)	79.7	83.0	-4.0
H_2 Purity (%)	95	94.7	+0.3
CaO Conversion (%)	83.5	81.0	+3.1

References

-

[1] A. Antzara, E. Heracleous, D.B. Bukur, A.A. Lemonidou, Int. J. Greenh. Gas Control 32 (2014) 115-128.

[2] A. Antzara, E. Heracleous, A.A. Lemonidou, Appl. Energy 180 (2016) 457-471.

[3] D. Kunii, O. Levenspiel, Fluidization engineering, 2nd ed., Butterworth-Heinemann, Boston, 1991.

[4] J.-H. Choi, J.-E. Son, S.-D. Kim, Ind. Eng. Chem. Res. 37 (1998) 2559-2564.

[5] R. Porrazzo, G. White, R. Ocone, Fuel 136 (2014) 46-56

[6] T. Numaguchi, K. Kikuchi, Chem. Eng. Sci. 43 (1988) 2295-2301.

[7] A. Abad, J. Adánez, F. García-Labiano, L.F. Diego, P. Gayán, J. Celaya, Energy Fuels 62 (2007) 533-549.

[8] İ. Ar, G. Doğu, Chem. Eng. J. 83 (2001) 131-137.

545682 Methane Oxy-Reforming Coupled with a Water Gas Shift Membrane Reactor for Pure Hydrogen Production

Tuesday, June 4, 2019: 11:18 AM
Republic ABC (Grand Hyatt San Antonio)

Francesco Basile[1], Gabriele Centi[2], Andrea Fasolini[1] and Salvatore Abate[2], (1)University of Bologna, Bologna, Italy, (2)University of Messina, Messina, Italy

The aim of the present work is to prove a full methane to pure hydrogen production by coupling oxy-reforming (OR) of methane and WGS in a Pd thin membrane dense reactor. Oxy-reforming couples the advantages of steam reforming (SR) and catalytic partial oxidation (CPO). Oxy-reforming runs autothermally in relative compact reactors, producing concentrated hydrogen due to the reduced amount of steam. Moreover, a Pd-based membrane on a ceramic support has been employed to purify the hydrogen produced during oxy-reforming and increased by water gas shift (WGS) reaction integrated in the catalytic membrane reactor. The process success is obtained by: i) an oxy-reforming catalyst composed by a $Ce_{0.5}Zr_{0.5}O_2$ *support and a Rh metal phase with improved resistance to sintering and carbon formation at low S/C and O/C ratio. and ii) a thin Pd membrane have been produced to ensure high* H_2 *flux comparable to the* H_2 *production rate, iii) a Pt* $Ce_{0.5}Zr_{0.5}O_2$ *with enhanced WGS activity and low methanation activity even at low S/C and high* H_2*/C ratios.*

1. **Scope**

Although nowadays hydrogen is mainly produced on a large-scale, small-scale delocalized hydrogen production from methane will have an important role in future. In fact, large-scale processes, such as steam reforming, suffer of problems regarding elevated gradients of temperature between the catalytic bed and the walls of the reactor. In addition, large and expensive hydrogen purification units, namely PROXY and PSA, to eliminate co-produced CO and CO_2 are needed. Finally, hydrogen storage and transport is challenging.

Small-scale plants can overcome these issues. However, the integration between small-scale production and separation has to be improved. For this reason, the processes studied in this work are the methane oxy-reforming and the membrane reactor unit for the water gas shift reaction and hydrogen separation. These two have characteristics that make them suitable for an integrated process, such as the possibility to run them on low contact times and reduce some of the limitations found on large-scale. In fact, the oxy-reforming process, which consists in the combination of Steam Reforming (SR) and Catalytic Partial Oxidation (CPO) of methane in a configuration that provides low amount of oxygen (O/C=0.21) and steam (S/C=0.7-1.5) and short residence times (36-150 ms), helps to reduce some of the heat transfer issues found in classical SR and CPO However, the low amount of reagents employed disfavour methane conversion and results in deactivation processes such as carbon formation and sintering. For this reason, an efficient catalyst with good properties must be used to fulfil high hydrogen production for prolonged time. On the opposite, the low amount of reagents employed in the oxy-reforming make it a process suitable for an integration with a hydrogen-separating membrane reactor. In fact, the driving force of this separation is hydrogen partial pressure, which would be badly affected by an over-stoichiometric amount of reagents.

Thus, the process was integrated with a membrane reactor based on an efficient WGS Pt-catalyst and a selective Pd-based membrane that allowed hydrogen separation at a high rate, which was close to the hydrogen production rate over the catalyst.

2. **Results and discussion**

A $Ce_{0.5}Zr_{0.5}O_2$ support for both oxy-reforming and water gas shift was synthesized by reverse microemulsion technique[1]. This support was impregnated with Rh for the oxy-reforming catalyst and with Pt for the water gas shift one and provided stable and efficient catalyst in both cases. The Rh-based catalyst provided methane conversions close the equilibrium in all of the oxy-reforming conditions, showing a high stability, namely a decrease of conversion of only 3% after about 30 h of operation. The best performances were achieved at T=750°C, P=5 atm GHSV=100000 h^{-1} and S/C=1.

To increase hydrogen production and provide it in a pure stream, a hydrogen selective Pd-based membrane filled with a Pt-$Ce_{0.5}Zr_{0.5}O_2$ catalyst for the water gas shift reaction was integrated downstream the oxy-reforming. The Pt/$Ce_{0.5}Zr_{0.5}O_2$, good activity and its performances in the membrane reactor were compared with those of a fixed bed charged with the same catalyst. In the fixed bed operation, the catalyst proved to be selective toward water gas shift reaction, with low methanation activity, even at low S/C and H/C ratios. Though, hydrogen was produced in a good yield and CO was mainly converted through WGS. Decreasing the contact time also helped to reduce the methanation activity, as the WGS reaction was favoured in these conditions. A decrease in conversion was found, but this was to be attributed to a lower consumption of CO in the methanation reaction rather than a decrease of WGS occurrence. In fact, H_2 yield was increased concurrently.

However, several advantages were obtain by substituting the fixed bed with the membrane reactor setup. In particular, the methanation reaction was completely avoided in all the conditions, even at low S/C ratios due to hydrogen removal from the reaction environment. Moreover, for the same reason the equilibrium was shifted, resulting in increased hydrogen yields and carbon monoxide conversions that were higher than those calculated at the equilibrium for the fixed bed process in the same conditions. The membrane provided total hydrogen selectivity and a high permeance of 126 ml/($cm^2 atm^{-0.5}$min) which was comparable to the hydrogen production rate and thus allowed to recover most of the hydrogen produced thanks to the good integration of the oxy-reforming and WGS membrane reactor. The best performances were obtained at T=400°C, P=10 atm, GHSV=100000 h-1 and S/C=1,5 and provided a hydrogen recovery of 90%, with a high permeate flux (731.44 ml/min) of pure hydrogen and a lower one (404.65 ml/min) in the retentate.

As the membrane potential was not completely exploited, with some hydrogen left in the retentate, a sweep gas may be applied in order to further enhance hydrogen separation.

At lower GHSV, The driving force of the separation approached 0, indicating the complete membrane exploitation. However, lower fluxes and recoveries were obtained and a higher amount of unseparated hydrogen. Lowering the pressure to 5 atm, allowed for a better exploitation of the membrane as well but only 325.14 ml/min were found in the permeate, much lower than 690.38 ml/min of the retentate. However, when the results were compared with those obtained in a conventional WGS reactor at same conditions, an improvement of hydrogen production was observed, with values which overtook the equilibrium.

3. Conclusions

Pure hydrogen can be produced from methane in a two-step process consisting of a first oxy-reforming unit followed by a membrane reactor in which the water gas shift reaction and hydrogen separation are carried out simultaneously. This setup proved to be optimal for a small-scale integrated process and performed better than an oxy-reforming unit followed by a fixed bed water gas shift reactor. In particular, by linking oxy-reforming and membrane reactor pure hydrogen was produced in higher yields, with an increased CO conversion that was found to be over the equilibrium value calculated for the fixed bed in the same conditions. Finally, methanation reaction, which consumes the hydrogen produced in the reforming, could be avoided.

4. References

5. A. Martínez-Arias, et al., *Langmuir* **1999,** 15, 4796–4802.

545800 Synergy Effect of Two Sets of Sing-Atom Sites (Ni_1 and Ru_1) on CeO_2 for Reforming CH_4

Tuesday, June 4, 2019: 11:42 AM
Republic ABC (Grand Hyatt San Antonio)

Franklin (Feng) Tao[1], Yu Tang[2], Yu Chen[3], Ben deGlee[3] and Meilin Liu[3], (1)Departments of Chemical Engineering and Chemistry, The University of Kansas, Lawrence, KS, (2)University of Kansas, Lawrence, KS, (3)Georgia Institute of Technology, Atlanta, GA

Heterogeneous catalysis is performed on specific sites on surface of a catalyst even if the specific sites of many catalysts could not be readily identified. Here we report a catalyst consisting of two sets of single-atom sites on surface of CeO_2 nanorods. It is highly active with turn-over rate of producing 73.6 H_2 molecules on each site per second through reforming CH_4 with CO_2 and highly selective (98.5% selectivity) for producing H_2 at a relatively low temperature down to 500°C. The anchored Ni and Ru atoms on CeO_2 surface are singly dispersed and remained at cationic state during catalysis at 500°C. The two types of single-atom sites, Ni_1 and Ru_1 play synergy role in this catalysis, evidenced by lower apparent activation barrier and higher turn-over rate for production of H_2 and CO in contrast to a catalyst with Ni_1 single-atom sites ($Ni_{0.05}Ce_{0.95}O_2$) and one with Ru_1 single-atom sites ($Ni_{0.05}Ce_{0.95}O_2$). Computational studies uncovered the molecular mechanism of the synergy effect which is originated at the differently preferred roles of Ni_1 and Ru_1 in activations of CH_4 and CO_2, respectively and at the sequential roles in first formation of atomic hydrogen from CH_4 on Ni site and then coupling H atoms to form molecular H_2 on Ru_1 site. This demonstration of synergy effect of two sets of single-atom sites on the same surface suggests a new avenue for designing a catalyst with high activity and selectivity at a relatively low temperature.

545612 NANO-Tailoring Nickel Catalyst Surfaces for CO_2 reforming of Methane

Tuesday, June 4, 2019: 12:06 PM
Republic ABC (Grand Hyatt San Antonio)

Anuj Prakash[1], **Shaik Afzal**[1], Anjaneyulu Chatla[1], Patrick Littlewood[2], Nelly Catherin[3], Peter C. Stair[2] and Nimir Elbashir[4], (1)Chemical Engineering Program, Texas A&M University at Qatar, Doha, Qatar, (2)Department of Chemistry, Northwestern University, Evanston, IL, (3)TOTAL Research Center Qatar, Doha, Qatar, (4)Chemical and Petroleum Engineering, Texas A&M University at Qatar, Doha, Qatar

Natural gas reforming, while of critical importance to the gas processing industry has several drawbacks not least of which is the high energy requirement, contributing significantly to a plant's CO_2 footprint. The case of CO_2 or Dry Reforming of Methane (DRM) is therefore of particular interest because of its potential as a viable route to combined CH_4 and CO_2 (Green House Gases) utilization. Commercialization efforts, however, have met with little success, partly owing to the propensity of conventional nickel-based dry reforming catalysts to deactivate rapidly via carbon deposition or 'coking', and sintering. Mitigating these issues has been the objective of several investigations into synthesizing novel yet inexpensive catalysts over the years. One of the more recent approaches is the use of Atomic Layer Deposition (ALD) to deposit nanometer-thick overlayers of either Alumina, Titania or Silica on Nickel oxide with the objective of anchoring down the Nickel whilst retarding the deposition of carbon. Previous studies [1,2] have demonstrated this effect on several catalytic systems, including unsupported nickel oxide nanoparticles.

This work investigates the potency of increasing ALD layers on high-loading, commercial Nickel catalyst activity, and resistance to deactivation. The 20% wt. (nominal) technical catalyst is ALD coated with 1, 5, 10 and 20 cycles of Tri-Methyl Aluminum-H_2O pulses in a static-bed coater unit. Catalyst deactivation is induced by treating the uncoated and coated catalysts in a 1:1:8 CH_4:CO_2:He environment at 650°C, a temperature at which both coking and sintering contribute comparably. Dispersion as a function of available Nickel sites is monitored over the course of the experiment via a combined CO-H_2 chemisorption technique. Uncoated catalyst showed a steep decline in dispersion over 40h of DRM from 4% to less than 1%. After 40h TOS, no nickel active sites were detected by our chemisorption technique. However, of the coated catalysts, the 5 cycle ALD catalyst could maintain a stable, albeit low dispersion (~1%) even after 40 hours on stream. Although the rates of carbon deposition per gram catalyst from Temperature Programmed Oxidation is comparable with uncoated commercial catalyst, the sustained activity of the ALD catalysts hint at the nature and location of carbon deposits themselves not being similar. High-Resolution Transmission Electron Microscopy (HR-TEM) of the spent catalysts highlight these differences, especially the larger particles sizes for commercial catalyst. The effect of regeneration on a 20 cycle ALD coated catalyst during a 500 hour long time on stream run reveals the possibility of fine-tuning the porous ALD overcoat to allow controlled access of reactants to the underlying Nickel.

These findings lends credence to the use of ALD as a relatively inexpensive, scalable method of protecting conventional catalysts to withstand the harsh, deactivating conditions of Natural gas reforming.

Keywords: *Atomic layer Deposition; CO_2; Methane reforming*

References

[1] Baktash, E., Littlewood, P., Schomäcker, R., Thomas, A., & Stair, P. C., *Applied Catalysis B: Environmental* (2015) 122

[2] O'Neill, B. J., Jackson, D. H. K., Lee, J., Canlas, C., Stair, P. C., Marshall, C. L., Elam, J. W., Kuech, T. F., Dumesic, J. A., Huber, G. W., *ACS Catalysis* (2015) 1804

545521 Direct Conversion of Biogas into Synthetic Natural Gas (SNG) over the 0.1-0.5wt% Ru/Al2O3 Catalyst

Wednesday, June 5, 2019: 2:54 PM
Republic ABC (Grand Hyatt San Antonio)
Yichen Zhuang, **Sogol Mottaghi-Tabar** and David Simakov, Chemical Engineering, University of Waterloo, Waterloo, ON, Canada

Introduction

Biogas and landfill gas are the two main sources of greenhouse gases (GHG) emissions which are rich in both carbon dioxide (CO_2) and methane (CH_4). These streams could be potentially utilized as synthetic natural gas (SNG), sometimes also referred to as renewable natural gas (RNG). However, the CO_2 content (up to 50 vol%) has to be eliminated prior to the use. Current commercial solutions rely on CO_2 removal via either membrane separation or pressure swing adsorption (PSA) with the removed CO_2 being vented to the atmosphere. As an alternative way, the thermocatalytic Sabatier reaction can be used to convert the CO_2 contained in biogas and landfill gas into synthetic CH_4, upgrading these waste streams to a pipeline quality SNG. A highly active, selective and stable catalyst is a pre-requisite for implementation of this conversion pathway. In this study, we use the 0.1-0.5wt% Ru/Al_2O_3 catalyst to convert a simulated biogas mixture directly without any upstream CO_2 separation. Our results show a promising new technological avenue for biogas upgrading into SNG.

Methodology

A series of Ru/Al_2O_3 catalysts were prepared using a wet impregnation method. Commercial alumina support was crushed and sieved prior to impregnation. Appropriate amounts of ruthenium chloride were dissolved in acetone and the sieved alumina particles were added to the solutions with different Ru concentrations. The resulted slurries were sonicated and dried. In the flow system setup, mass flow controllers, a back pressure regulator, and a furnace were used to control feed flow rate, pressure, and temperature. Infrared analyzers were used to measure outlet gas concentrations. Catalytic performance was investigated as a function of temperature and space velocity for a range of Ru loadings to identify the optimal active phase content. The catalyst stability was specifically examined. Characterization techniques included ICP-OES, TGA-MS-FTIR and temperature programmed reactions (TPA, TPR, and TPO).

Results

The results of direct conversion of the simulated biogas (1:1 CO_2:CH_4) over the 0.5wt% Ru/g-Al_2O_3 are shown on the left figure. At 350-450 °C, CH_4 selectivity is nearly 100% followed by a gradual decrease at higher temperature. The optimal temperature for this catalyst is around 450 °C with 80% CO_2 conversion and 97% CH_4 selectivity achieved. From the figure on the right it is clearly seen that both CO_2 conversion and CH_4 selectivity were stable over the 65 hour time on stream. The temperature programmed oxidation (TPO) test of the spent catalyst suggested that the extent of coking is negligible. The effect of Ru loading was also investigated. Lowering Ru loading caused the CH_4 selectivity to drop significantly, and optimal operating temperature shifted to higher temperatures as the Ru loading was reduced. The lower limit for maintaining decent methanation performance was identified as 0.1wt%.

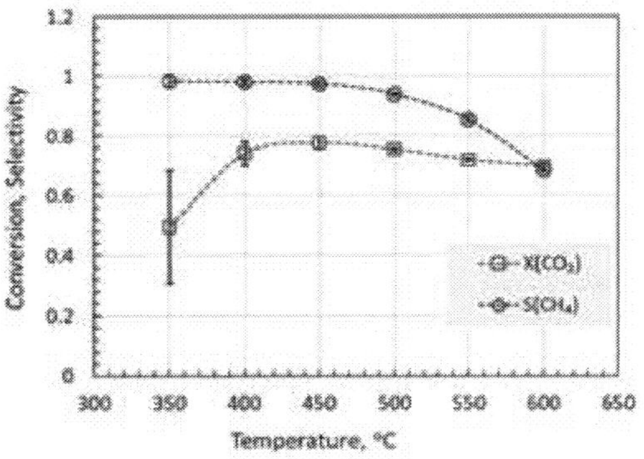

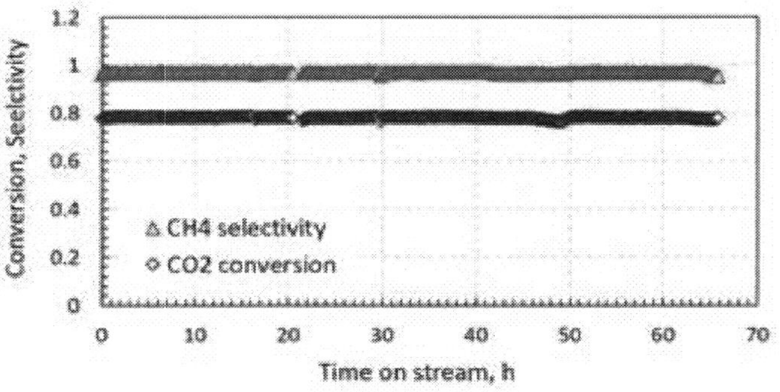

Significance

The results of this study indicate that it is possible to directly convert a mixture of CO_2 and CH_4 (representing biogas) to an upgraded mixture predominantly containing CH_4 with some unreacted H_2, achieving CO_2 conversions above 90%. The proposed technological solution could eliminate the need for the costly CO_2 separation increasing two-fold the rate of RNG production from biogas and landfill gas.

545934 Low-Temperature CO_2 Methanation Catalyst for Power-to-Gas

Wednesday, June 5, 2019: 2:57 PM
Republic ABC (Grand Hyatt San Antonio)

Audrey Waldvogel[1], Sébastien Thomas[2] and **Anne Cecile Roger**[2], (1)ICPEES, Strasbourg, France, (2)ICPEES, Université de Strasbourg, Strasbourg, France

Low-temperature CO_2 methanation catalyst for Power-To-Gas

Audrey Waldvogel, Sébastien Thomas and Anne-Cécile Roger

Institut de Chimie et Procédés pour l'Energie, l'Environnement et la Santé, ICPEES, UMR CNRS 7515, Université de Strasbourg, 67087 Strasbourg, FRANCE

1. Introduction

Renewable energies (RE) are the heart of the main energy transition scenario. In 2014, the objective of 20% share of RE in the total E.U. energy for 2020 was raised to 27% for 2030 [1]. The involvement of France is even more ambitious as it reaches 32% according to the French law on energy transition [2]. This objective corresponds to 40% of RE in the French electricity mix. However, the intermittency of solar and wind electricity can slow down their penetration in the coming decade. The "Power to Gas" (P2G) technology has the potential to overcome this limitation by coupling electric and gas grids, consuming excess of electricity while producing a storable and transportable fuel. The P2G technological chain allows to store the overproduction of renewable energies thanks to their transformation into H_2 by electrolysis based technology (through water conversion by electricity) or in synthetic methane. Then, H_2 or synthetic CH_4 produced can be stored or injected into the natural gas network and supplied in the cities for sustainable mobility means, for electricity and heat needs. Thus, consumers are provided with energy from renewables at any time. Deployment of the P2G technology, currently still in the development stage, is expected from 2025 in France according to the ADEME [3], with the installation of power units of the scale of 1-50 MW.

The CHOCHCO project (Optimization of a flexible technological chain of CO_2/H_2O co-electrolysis and CO hydrogenation into synthetic CH_4) [4], run from 2013 to 2017 in the frame of ANR (French National Research Agency), developed and evaluated an integrated "Power to Methane" chain based on: 1) a high temperature co-electrolyzer for electrochemical reduction of steam and CO_2 into syngas; 2) a modular catalytic methanation reactor and 3) a methane upgrading /drying unit, to reach specifications for further injection into the natural gas grid.

2. Catalyst development

A series of Ni/CZP (10 wt% of Ni° on mixed oxide ceria-zirconia-praseodymia) catalysts has been synthesized in this purpose. The Ni metallic phase has been impregnated on different mixed oxides of same composition $Ce_{1.31}Zr_{2.52}Pr_{0.17}O_{7.97}$, prepared by batch coprecipitation from nitrate salts.

In order to improve nickel dispersion, the way to increase specific surface of the CZP support has been investigated. Thus, the optimization of the coprecipitation synthesis, initially with $NaCO_3$, by the combination of the precipitating salt $(NH_4)_2CO_3$ and addition of a surfactant (CTAB) has been studied.

Table 1. Catalyst characterization.

	Ni/**CZP(Na)**	Ni/**CZP(NH_4)**	Ni/**CZP(NH_4,T)**
S_{BET} (m^2/g)	51	69	84
V_{pore} (cm^3/g)	0.07	0.11	0.18
d_{pore} (nm)	5	5	8
$S_{Ni°}$ **(m^2/g)**	4.3	6.8	7.7

The optimized catalysts exhibited enhanced properties, as presented in Table 1. The use of ammonium carbonates instead of sodium carbonates as precipitating agent is favorable to the specific surface area of the oxide support, allowing a better dispersion of nickel after impregnation. The addition of CTAB as surfactant in the precipitation medium further improves porosity, BET surface and Ni metal surface. The characterization of the surface by thermoprogrammed desorption (TPD) of CO_2 revealed that the use of ammonium carbonates led to a decrease of the mean basic strength : for CZP(NH_4) and CZP(NH_4,S) the main CO_2 desorption occurs around 170 °C whereas it is around 280 °C for CZP(Na). This medium basicity promotes reversible interaction between CO_2 and the catalytic support.

3. Catalytic results

The catalytic activity has been evaluated in a fixed bed reactor on reduced (1h at 400°C under H_2/N_2) catalysts in powder (sieved between 125 and 200 µm), at atmospheric pressure, with a GHSV of 50.000 h^{-1} (STP), under a gas mixture which mimics the composition of the outlet flow of the co-electrolyzer in the P2G process studied in this project ($H_2/CO/CO_2/H_2O/CH_4$ 70/12/8/5/5 in mole).

The increase of the specific surface and Ni metal areas along with a medium basicity, were shown to be responsible for the enhanced catalytic performance (see Figure 1) especially at low temperature (250 °C). The catalyst synthesized with ammonium carbonate in the presence of a surfactant is especially active at low temperature : almost 80 % of CO_x are converted into methane at 250°C, whereas the two other catalysts exhibited almost no activity.

Even at much higher GHSV values (180.000 h^{-1}, STP), the optimized catalyst still exhibits satisfactory methane yields (75% at 250°C).

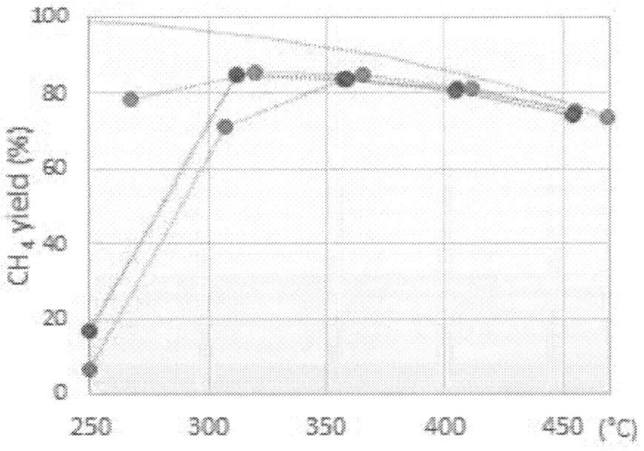

Figure 1. Methane yield vs temperature for the three catalysts. P_{atm}, inlet composition $H_2/CO/CO_2/CH_4/H_2O$ of 70/12/8/5/5, GHSV (STP) = 50.000 h^{-1}.

When submitted to successive 'start and stop' conditions under various atmospheres, the catalyst CZP(NH_4,S) showed a high resistance to carbon deposition and sintering deactivation, which is a key point for the intermittent working conditions of the process.

4. Kinetic study

A kinetic model, of the Langmuir-Hinshelwood type, was developed for the first time on a Ni/CZP type catalyst based on reaction mechanism previously reported [5]. CO methanation, CO_2 methanation, Water Gas Shift and the 3 corresponding reverse reactions were considered. Firstly, activity at low conversions was used to evaluate roughly the kinetic parameters which were then adjusted with data at high conversion by considering plug flow reactor by minimization between experimental and modeled values.

The obtained activation energies are consistent with values reported in the literature [6]. The developed model is accurate to predict the experimental methane yield, as shown in Figure 2.

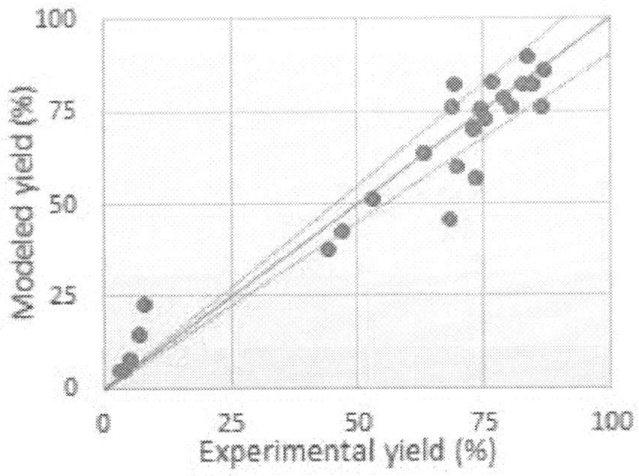

Figure 2. Parity plot for methane yield for the optimized catalyst (Ni/CZP(NH_4,S)).

5. Conclusions

The use of $(NH_4)_2CO_3$ instead of Na_2CO_3 as precipitating agent allows to obtain an oxide with enhanced textural properties, leading to a better dispersion of Ni°. Syntheses with $(NH_4)_2CO_3$ induce more basicity of weak strength, favorable to the reversible adsorption of CO_2. The addition of CTAB as a surfactant further improves the dispersion of nickel.

Their high catalytic efficiency (methane yield > 75% at GHSV 180.000 h^{-1}) at low temperature and atmospheric pressure was demonstrated under representative co-electrolysis exit gas composition, and their stability under intermittent conditions have been proven.

A complete kinetic study has been done. The experimental conditions have been extended from low (250 °C) to high (450 °C) reaction temperatures. The excellent activity was modeled with a reliable kinetic model. The activation energy of the reaction of CO_2 methanation has been shown to be strongly decreased for the optimized catalyst Ni/CZP(NH_4,S).

6. References

[1] Communication of 22/01/2014 of the European Commission.

[2] Law N°2015-992 of 17th August 2015 relative to “la transition énergétique pour la croissance verte”.

[3] ADEME report May 2014, Contribution de l'ADEME à l'élaboration de visions énergétiques 2030-2050.

[4] www.chochco.fr

[5] Ussa Aldana P.A., Ocampo F., Kobl K., Louis B., Thibault-Starzyk F., Daturi M., Thomas S., Roger A.C., *Catalysis Today* 2013, 215, 201 207.

[6] Van Herwijnen T., Van Doesburg H., De Jong W.A., *Journal of Catalysis* 1973, *28(3)*, 391 402.

545601 Advanced Reactor Technology for CO_2 Utilization in Methane Reforming

Wednesday, June 5, 2019: 3:00 PM
Republic ABC (Grand Hyatt San Antonio)

Mohamed Sufiyan Challiwala, The Artie McFerrin Department of Chemical Engineering, Texas A&M University, College Station, TX, Hanif Choudhury, Chemical Engineering, Texas A&M University at Qatar, Doha, Qatar, Nimir O. Elbashir, TEES Gas & Fuels Research Center, Texas A&M at Qatar, Doha, Qatar and **Shaik Afzal**, Chemical Engineering Program, Texas A&M University at Qatar, Doha, Qatar

<>Introduction:

Dry Reforming of Methane (DRM) is one of the enabling processes that converts two greenhouse gases (Carbon dioxide and Methane) to syngas (a mixture of carbon monoxide and hydrogen). Syngas is an important precursor that is used to produce liquid hydrocarbons (or ultra clean fuels) and many other value added chemicals (methanol, Dimethylether, Acetic acid etc). In comparison to the conventional reforming technologies (Steam reforming and Partial Oxidation), the main benefit of this process is that it utilizes carbon dioxide, whereas the other processes require steam and oxygen to be used as oxidant. Equations 1-3 presents the stoichiometric equations for the three reforming technologies:

Dry Reforming (DRM):

$$CH_4 + CO_2 \rightarrow 2CO + 2H_2 \quad \Delta H = 247 \frac{kJ}{mol}$$

(1)

Partial Oxidation (POX):

$$CH_4 + \frac{1}{2}O_2 \rightarrow CO + 2H_2 \quad \Delta H = -36 \frac{kJ}{mol} \qquad (2)$$

Steam Reforming (SRM):

$$CH_4 + H_2O \rightarrow CO + 3H_2 \quad \Delta H = 206 \frac{kJ}{mol}$$

(3)

Although DRM provides a very attractive route towards utilization of CO_2, its successful industrial implementation faces several process limitations. These limitations include high endothermicity of the reaction, low quality of the syngas yield ratio (not suitable for direct use for Fischer Tropsch reaction) and the deactivation of catalyst due to large quantity of carbon formation. Due to the severity of these limitations, DRM is still a grey area that has attracted incredible amount of research both in terms of process advancement (in the way the reaction is carried out) and in terms of the catalytic systems (towards the development of catalyst that are more stable and sustain carbon formation). In our previous publications (Challiwala et al. 2017 [1] and Shaik et al. 2018 [2]), we have demonstrated a different pathway that synergises the benefit of the three reforming technologies, which not only reduces the energy requirements from the overall process, but also targets reduction in the carbon formation during combined reaction. In the present work, we have developed a novel processes that systematically handles the formation of carbon and enhances the CO_2 utilization in the overall process.

<>**Methodology:**

For the preliminary assessment using thermodynamic calculations, equilibrium compositions were evaluated from the solution of overall Gibbs Free Energy (GFE) of the system as shown in equation (4) below:

$$G^{total} = \sum_{i=1}^{N} n_i \left(\Delta G_{fi}{}^{o} + RT\ln y_i + RT\ln\widehat{\varphi}_i + RT\ln\frac{P}{P_0}\right) + n_c \Delta G_{fc(s)}{}^{o} \quad (4)$$

More detailed calculation philosophy with fugacity coefficient calculation using equation of state solution are given in our previous work in [1].

For the experimental proof of concept, we conducted all the tests using Micromeritics 2920 Autochem II instrument. We used a commercial Nickel based catalyst $(20\%\ Ni/\gamma - Al_2O_3)$ to carry out the reaction for our novel reactor configuration termed as carbon generator (CARGEN). For this, we used a mixed gas comprising of $(CH_4, CO_2, O_2, He, Ar)$ and the evolving gas from the CARGEN reactor after the reaction was analysed using a Residual Gas Analyser (RGA) to instantaneously monitor the composition.

<>**Results Highlight:**

The novel process reported in this study, our aim is to produce both syngas and solid carbon, in contrast to conventional approach which targets only the formation of syngas. We propose to utilize a combination of two reactors (rather than one reactor in DRM), with one focussed on production of solid carbon (also known as CARGEN) and the other focussed on production of syngas via combined reforming reaction. The benefit of this process that we have proved from both our thermodynamics calculations and experimental proof of concept indicated significant reduction in energy requirements (about 50% of DRM) in addition to significantly high levels of CO_2 conversions (at least 65% CO_2 conversion). As it is already known that the syngas yield ratio (H_2: CO) obtained from the DRM process is not more that 1:1, the major benefit of this process is that the syngas yield ratio can be tuned as per the process requirements up to 3:1 due to the flexibility in operation provided by the two reactor setup proposed in this work. Due to significant merits obtained from this work, this process was recognized as one of the patentable outcomes from the Qatar National Research Fund (QNRF) Exceptional project. As a part of our continuous efforts in identifying the possibilities for overcoming the challenges associated with DRM process, this work provided a very interesting pathway for CO_2 utilization.

References

1. Challiwala, M.S., et al., *A combined thermo-kinetic analysis of various*

methane reforming technologies: Comparison with dry reforming. Journal of CO2 Utilization, 2017. **17**(Supplement C): p. 99-111.

2. Afzal, S., et al., *Optimization Approach to the Reduction of CO2 Emissions for Syngas Production Involving Dry Reforming*. ACS Sustainable Chemistry & Engineering, 2018. **6**(6): p. 7532-7544.

544333 Photocatalytic Upgrade of CO_2 to Synthetic Natural Gas

Wednesday, June 5, 2019: 3:03 PM
Republic ABC (Grand Hyatt San Antonio)

Steven Bardey[1,2], Audrey Bonduelle-Skrzypczak[1], Antoine Fécant[1], **Valérie Caps**[2] and Valérie Keller[2], (1)IFPEN, Solaize, France, (2)ICPEES (CNRS UMR 7515/ University of Strasbourg), Strasbourg, France

The direct conversion of solar energy into valuable chemical fuels is a great challenge for clean energy production. In this context, the photo-reduction of CO_2 into methane has received increasing attention over the last few years. This way of producing synthetic natural gas would indeed represent a step forward towards sustainability, due to the nature of the carbon source, a major by-product of energy consumption, and its increasing availability. However, the conversion rate of photonic energy into chemical energy by semi-conductor-based photocatalysis is still too low, as current energy carriers make use of only a fraction of the solar spectrum. Hence, the development of photocatalysts with improved and broadened light absorption is needed to efficiently use sunlight as primary energy source. Amongst the possible materials, multi-component systems containing metal nanoparticles (M NPs) in interaction with a semi-conductor (SC) appear promising. In addition to their beneficial role as co-catalysts and electron traps,[1] M NPs may indeed improve the SC photocatalytic performances by plasmon-derived physical phenomena, which requires control over the M-SC interface at the nanoscale[2]. We have recently designed an optimized heterojunction between titania and carbon nitride with improved activity for photocatalytic hydrogen production from water in the presence of Au NPs.[3] Here we show the impact of the nanostructuration of both gold-titania and platinum-titania interfaces on their performances in the solar photocatalytic reduction of CO_2. Results will be discussed in terms of electronic metal-support interactions (eMSI)[4] and Au NPs density, based on TEM, XRD/BET and UV-Vis characterization in particular. Finally, we will present the rational design of highly efficient composites, i.e. materials combining unprecedented quantum yields and selectivities to methane (close to 100%), based on these studies.

[1] Ran, J.; *et al. Adv. Mater.*, **2018**, *30*, 1704649.

[2] Shiraishi, Y.; *et al. Nanoscale*, **2017**, *9*, 834.

[3] Marchal, C.; *et al. Adv. En. Mater.*, **2018**, *8*, 1702142.

[4] Gao, W.; *et al. Nano Lett.*, **2015**, *15*, 2548.

[5] Marchal, C.; *et al. J. Catal.*, **2017**, *352*, 22.

545653 Green Methanation of CO Using New Co-Pt@Flg Nanocomposites

Wednesday, June 5, 2019: 3:06 PM
Republic ABC (Grand Hyatt San Antonio)

Loïc Michel, Cerise Robert, Sécou Sall and **Valérie Caps**, ICPEES (CNRS UMR 7515/ University of Strasbourg), Strasbourg, France

Methanations of CO and CO_2 are reactions of critical importance in the perspective of energy transition [1]. In addition to converting catalytic inhibitors into a useful resource during ammonia synthesis for example, the reaction of CO_2 and hydrogen to methane would provide a strategic way of storing excess electrical power from hydrogen fuel cells into a readily distributable chemical fuel (power to gas). The methanation technology could also potentially be applied to the effluent of the wide-spread cement industry and turn a highly debated greenhouse gas waste into synthetic natural gas on a large scale. Extensive research effort has been devoted to the design of suitable catalysts for this reaction. Typical catalysts involve metal nanoparticles associated with an oxide support. Current processes operate under high pressure, and temperatures above 300°C are required to achieve the targeted conversion levels. Recently, it was shown that reaction rates could be enhanced by using a Co-Pt bimetallic phase [2-4]. In these systems, the reaction was found to be limited by hydrogen spillover across the oxide support [4].

Here we will show that highly efficient Co-Pt-based nanocomposites may be straight-forwardly prepared in one pot by the sonication-assisted $NaBH_4$-reduction of cobalt nitrate and potassium tetrachloroplatinate in the presence of natural graphite (liquid phase process). The resulting Co-Pt@FLG (FLG=few layer graphene) show unprecedented methanation rates at atmospheric pressure and 320°C. The low temperature synthesis, the readily available metal precursors, the low Pt/Co ratios used (0.01-0.05), the short reaction times and the exfoliation of graphite in the course of metal nanoparticles formation make this process an economically interesting alternative to the high temperature, multi-step syntheses and oxide supports used in previous studies. We will present extensive pre- and post-tests characterization (including temperature programmed reduction and oxidation, thermogravimetric analysis, HRTEM and XRD) to discuss the unique role of the few layer graphene matrix in the methanation process, with regard to hydrogen spillover in particular.

[1] Rönsch, S.; Schneider, J.; Matthischke, S.; Schlüter, M.; Götz, M.; Lefebvre, J.; Prabhakaran, P.; Bajohr, S. Review on methanation – From fundamentals to current projects. *Fuel*, **2016**, *166*, 276.

[2] Beaumont, S. K.; Alayoglu, S.; Specht, C.; Michalak, W. D.; Pushkarev, V. V.; Guo, J.; Kruse, N.; Somorjai, G. A. Combining in Situ NEXAFS Spectroscopy and CO_2 Methanation Kinetics To Study Pt and Co Nanoparticle Catalysts Reveals Key Insights into the Role of Platinum in Promoted Cobalt Catalysis. *J. Am. Chem. Soc.*, **2014**, *136*, 9898.

[3] Shin, H. H.; Lu, L.; Yang, Z.; Kiely, C. J.; McIntosh, S. Cobalt Catalysts Decorated with Platinum Atoms Supported on Barium Zirconate Provide Enhanced Activity and Selectivity for CO_2 Methanation. *ACS Catal.*, **2016**, *6*, 2811.

[4] Beaumont, S. K.; Alayoglu, S.; Specht, C.; Kruse, N.; Somorjai, G. A. A Nanoscale Demonstration of Hydrogen Atom Spillover and Surface Diffusion Across Silica Using the Kinetics of CO_2 Methanation Catalyzed on Spatially Separate Pt and Co Nanoparticles. *Nano Lett.*, **2014**, *14*, 4792.

546756 Multifunctional Catalytic Materials with CO_2 Capture and Oxygen Transfer Functionalities for High-Purity H_2 Production Via Sorption Enhanced Chemical Looping Steam Methane Reforming

Wednesday, June 5, 2019
Texas Ballroom Prefunction Area (Grand Hyatt San Antonio)

Andy Antzara[1], Eleni Heracleous[2] and **Angeliki Lemonidou**[1], (1)Chemical Engineering, Aristotle University of Thessaloniki, Thessaloniki, Greece, (2)School of Science & Technology, International Hellenic University (IHU), Thessaloniki, Greece

Introduction

Decreasing availability of fossil resources, together with the environmental issues related to the fossil-derived greenhouse gas emissions, have dictated the necessity for further improvements in the efficiency of energy production processes as well as the transition towards sustainable resources. In this context, hydrogen is heavily supported as an energy carrier, as it is expected to play a key role in the necessary transition from fossil fuels to a sustainable energy future. Likewise, hydrogen demands as a primary industrial gas are continuously increasing in the petroleum refining, mainly driven by clean fuel regulations. Hydrogen however, is produced almost entirely from fossil fuels. Steam reforming of methane (SMR), the dominant large–scale technology for hydrogen production, accounts for almost 50% of the worldwide production. It is evident that a novel technology for large–scale hydrogen production is required.

At this direction, Sorption enhanced chemical looping steam methane reforming (SE-CL-SMR) is considered as an alternative method for the efficient production of pure hydrogen, combining chemical looping steam methane reforming with in-situ CO_2 capture [1,2]. In this process, the reformer contains a CO_2 sorbent and an oxygen transfer material (OTM). In the first step of this cyclic process, the oxide is reduced by CH_4 and serves as the reforming catalyst. The reaction proceeds under near autothermal conditions due to the heat released by the strongly exothermic carbonation. In a second step, the saturated sorbent is regenerated, with the energy provided by the exothermic OTM re-oxidation [3-5]. Aiming at reducing the complexity and cost of the proposed process, we have developed a hybrid multi-functional catalytic material that combines good oxygen transfer characteristics, significant reforming activity and stable CO_2 sorption capacity in a single pellet. Herein, we demonstrate experimentally the feasibility of the SE-CL-SMR process over the NiO/CaO-based hybrid material. The obtained results are compared to the performance of a mechanical mixture of a NiO/ZrO_2 OTM with a CaO/ZrO_2 CO_2 sorbent with the same NiO/CaO composition under identical experimental conditions.

Experimental

Material synthesis: The NiO-based OTM was prepared via wet impregnation using a commercial ZrO_2 support with 40 wt% NiO loading [6]. The CaO-based CO_2 sorbent was prepared via a sol-gel auto-combustion route, using citric acid as a combustion agent and ZrO_2 as structural promoter, with CaO concentration of 66 wt% [7]. The hybrid material was prepared with the same method used for the sorbent synthesis, with ZrO_2 as structural promoter and CaO and NiO concentration of 40 wt% and 25 wt% respectively.

SE-CL-SMR experiments: The SE-CL-SMR experiments were conducted over the hybrid material, as well as over a mechanical mixture of NiO/ZrO_2 with CaO/ZrO_2 at fixed NiO/CaO ratio. For these tests, the materials were initially exposed to CH_4/steam at 650°C, S/C ratio 3 and GHSV 400 h^{-1} until saturation in CO_2. After saturation, the feed was switched to air for reoxidation of Ni and regeneration of $CaCO_3$ at 800°C.

Results and discussion

The composition of the reformate gas on dry basis during the reforming stage for the hybrid material is shown in Figure 1a. During the pre-breakthrough period, where NiO reduction, reforming and water gas shift reactions are carried out simultaneously with in-situ CO_2 capture, very high H_2 concentration was observed (~95%). Similarly high H_2 purity and pre-breakthrough duration were recorded for the OTM/sorbent mechanical mixture (not shown for brevity). Pre-breakthrough period was followed by a transitional period, where the sorbent was gradually saturated (breakthrough period) leading to an increase of CO, CO_2 and CH_4 and decrease of H_2 concentration. Although complete CH_4 conversion is reached from the beginning of the reforming stage, H_2 yield is relatively low (84%) during the first ~10min, indicating that during this short period part of CH_4 is consumed for NiO reduction. After complete NiO reduction, the reactor runs under conditions of simple SE-SMR process with higher H_2 yield (90%), approaching the thermodynamic equilibrium of SE-SMR (Figure 1b). Even after CaO saturation, the reformate gas obtained in the exit of the reactor has composition that approaches that of conventional reforming (post-breakthrough), indicating that the reforming activity of Ni in the hybrid material remains unchanged even after the decrease of surface area due to $CaCO_3$ formation.

After switching to air flow and for almost 20 minutes no additional external heating was applied to the reactor, since the heat generated by Ni reoxidation was more than enough to sharply increase the temperature of the solid, while meeting more than 33% of the heat requirements for the strongly endothermic sorbent regeneration reaction. The use of the hybrid material allowed achieving higher

$CaCO_3$ decomposition without external heating compared to the mechanical mixture for the same NiO/CaO molar ratio (33% vs ~27% – Figure 1c) even though the maximum temperature reached in the reactor during Ni oxidation was lower (~780°C vs 800°C) (Figure 1d). This can be attributed to the direct transfer of heat due to close contact between the two solid phases in the hybrid material, eliminating the conduction losses that exist in the case of the mechanical mixture. It is important to highlight that the use of the hybrid material reduced the volume of the bed by ~25% compared to the OTM/sorbent mechanical mixture. In addition, the hybrid material exhibited excellent stability without deterioration in its performance for 10 consecutive SE–CL–SMR cycles

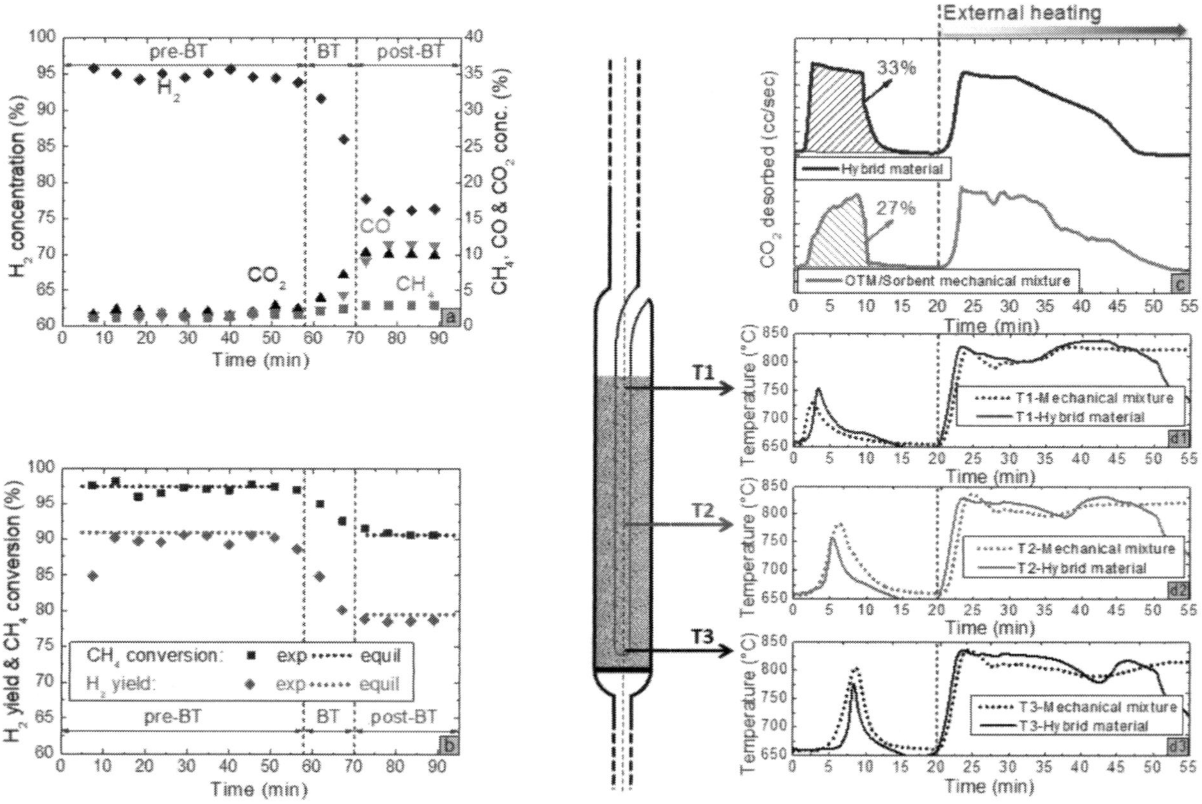

Figure 1. Product concentration **(a)**, CH_4 conversion and H_2 yield **(b)** as a function of time during the reforming stage and CO_2 desorption **(c)** and temperatures profiles **(d1–d3)** as a function of time during the regeneration stage (*Reforming stage:* T=650°C, S/C=3, NiO/CaO=0.5, GHSV=400 h^{-1}, *Regeneration stage:* T=800°C, air flow)

Post-reaction BET measurements of the material after 10 reforming/regeneration cycles showed that the surface area of the spent hybrid material after the reforming carbonation stage of the 10th cycle (with NiO and CaO in their reduced and carbonated form respectively) decreased significantly compare to those of the fresh materials due to formation of $CaCO_3$ (from 12.8 to 5.7 m^2/g). Compared however to the pure CO_2 sorbent at similar state (Table 1), the hybrid material maintains higher surface area for similar carbonation conversion of CaO, indicating that Ni in the hybrid materials serves also as CaO binder. This is related to the high capturing performance of the hybrid material during SE–CL–SMR, similar to that of the OTM/sorbent mechanical mixture. After complete regeneration, the surface area of the materials is restored to even higher value (16.9 m^2/g) than that of the fresh material. This could be ascribed to gradual restructuring of the materials due to the consecutive capture and evolution of CO_2 with cycles.

Table 1: BET surface area and pore volume of CaO/NiO–based hybrid material and CaO–based CO_2 sorbent before and after SE–CL–SMR experiments

CO_2 Sorbent	Surface area (m^2/g)	Pore volume (m^2/g)
66 wt.% CaO/$CaZrO_3$		
fresh–oxidized/calcined	15.3	0.05
used–reduced/carbonated	3.9	0.02
used–oxidized/calcined	14.0	0.04
Hybrid OTM/CO_2 Sorbent		
40 wt.% CaO–25 wt.% NiO/$CaZrO_3$		
fresh–oxidized/calcined	12.8	0.06
used–reduced/carbonated	5.7	0.31
used–oxidized/calcined	16.9	0.14

Conclusions

The synthesis and use of a hybrid material with combined functionalities in the SE-CL-SMR reaction demonstrated the feasibility and high potential of the proposed process. The developed hybrid catalyst demonstrated very high catalytic activity and sorption capacity, resulting in the production of high purity H_2 (~95%), comparable to that obtained over a mechanical mixture of OTM and CO_2 sorbent. During the initial stage of regeneration, the heat generated by Ni reoxidation was adequate to sharply increase the temperature of the solid, causing the decomposition of a substantial part of the saturated $CaCO_3$ without providing external heating. The use of the multi-functional hybrid material significantly reduced the volume of the catalytic bed by at least 25% compared to the mechanical mixture while obtaining similar results. The results clearly demonstrate that the use of a hybrid material in the SE-CL-SMR process decreases the operating costs and renders the process more attractive for commercialization.

References

[1] V. Dupont, Á.B. Ross, I. Hanley, M. V. Twigg, Int. J. Hydrogen Energy **2007**,

32, 67–79.

[2] A. Antzara, E. Heracleous, E., D.B. Bukur, A.A. Lemonidou, Int. J. Greenh. Gas Control. **2015**, *32*, 115–128.

[3] A. Antzara, E. Heracleous, A.A. Lemonidou, Appl. Energy **2016**, *180*, 457–471.

[4] J. Fermoso, M.V. Gil, F. Rubiera, D. Chen, ChemSusChem. **2014**, *7*, 3063–3077.

[5] M. Rydén, P. Ramos, Fuel Process. Technol. **2012**, *96*, 27–36.

[6] L. Silvester, A. Antzara, G. Boskovic, E. Heracleous, A.A. Lemonidou, D.B. Bukur, Int. J. Hydrogen Energy. **2015**, *40*, 7490–7501.

[7] A. Antzara, E. Heracleous, A.A. Lemonidou, Applied Energy. **2015**, *156*, 331–343.

545718 Low Temperature Methane Steam Reforming in a Hydrogen-Selective Membrane Reactor

Wednesday, June 5, 2019
Texas Ballroom Prefunction Area (Grand Hyatt San Antonio)

Francesco Basile[1], Gabriele Centi[2], **Andrea Fasolini**[1] and Salvatore Abate[2],
(1)University of Bologna, Bologna, Italy, (2)University of Messina, Messina, Italy

The present work is focused on pure hydrogen production via Low Temperature Steam Reforming (LTSR) of methane carried out in a membrane reactor. This was composed by a Pd-based membrane deposited on a ceramic support. Different active phases and two process were screened in order to identify the suitable conditions for the membrane reactor, which was able to provide pure hydrogen in a higher yield compared to a classical fixed bed reactor.

1. **Scope**

Pure hydrogen production from methane is a multi-step process, which is usually run on a large scale for economic reasons. However, hydrogen can be produced in a one-pot continuous process suitable for small scale applications, namely the Low Temperature Steam Reforming of methane. In such a setup the Steam Reforming is carried out in a reactor whose walls are composed by a membrane which is selectively permeable to hydrogen. In particular, Pd is the most used material for such application due to its high permeability and selectivity toward hydrogen. However this material deteriorates at temperatures higher than 500°C and for this reason the operative temperature of the Steam Reforming reaction, which usually runs around 900°C, has to be lowered. This has a bad influence on the equilibrium yields that decrease due to thermodynamics. However, the employment of a membrane reactor allows to produce high yields even at these temperatures thanks to hydrogen removal, which shifts the reaction toward the products. Moreover, pure hydrogen can be produced directly in this way thanks to the high selectivity of Pd toward this molecule. On the opposite reducing the reaction temperature can lead to energy saving and opens the possibility of performing also the Water Gas Shift (WGS) reaction together with the Steam Reforming one. Moreover, if Catalytic Partial Oxidation is also conducted together with LTSR, a more energy-efficient process can be obtained as the heat produced by CPO feeds the endothermic steam reforming. However, the former may produce hot spots on the catalytic bed that can provide membrane breaking. For this reason, this process was screened in a fixed bed reactor and compared with the LTSR, with different active phases. Once suitable catalyst and conditions were found, the membrane reactor was run to produce pure hydrogen. As high membrane permeation is a fundamental factor to provide high hydrogen recoveries and enhance methane conversion and hydrogen production, thin dense Pd membrane deposited over an alumina support through electroless plating deposition was employed. This was characterized by a very thin layer of uniformly distributed Pd.

2. **Results and discussion**

The LTSR and oxy-reforming were first conducted in a fixed bed at temperatures between 350 and 500°C, with different pressures, contact times and steam to carbon ratios (S/C). Particular attention was given to the catalytic bed temperature, which could affect membrane performances when too high, methane conversion and hydrogen partial pressure, which is the driving force of the permeation.

A $Ce_{0.5}Zr_{0.5}O_2$ support synthesized by reverse microemulsion technique[1] was selected has it was proved to have good properties in carbon removal thanks to its high oxygen storage capacity[2] and produced efficient catalysts which could withstand the conditions of low temperature oxy-reforming and steam reforming in which carbon formation over the active phase is favoured. This oxide was impregnated with Ru, Rh, Pt and a mix of Pt and Rh and tested in oxy-reforming conditions. Methane conversion was increased at high contact times and was slightly raised by an increase of S/C. On the opposite, lower conversion were obtained when pressure was enhanced or temperature decreased due to the thermodynamics of the process. Between the active phases Rh and one of Pt:Rh mix were found to be the most active ones, although Ru provided interesting results, given its cheaper price. These catalyst reached the equilibrium and provided high hydrogen partial pressures. However, temperatures as high as 600°C were developed in some cases over the catalytic bed due to the occurrence of CPO and exothermic side reactions. Although such high temperatures may provoke membrane breaking, a small part of the catalytic bed can be placed just before the membrane, in order to consume oxygen toward CPO and reduce the development of hot spots inside it. In fact, oxygen is completely consumed before entering the membrane in this way. However, the catalysts was tested even in the low temperature steam reforming and provided high yields and conversion close to the equilibrium value. The methane conversion was positively influenced by low contact times and high S/C ratios, while lower ones were obtained at high pressures. However, the tests conducted at high pressure were able in some conditions to provide an hydrogen partial pressure higher than 1 atm, which can provide permeation if developed inside a membrane reactor. In fact, a membrane reactor was charged with Rh-based catalyst and provided improved results in methane conversion and hydrogen production compared to the fixed bed reactor, while being able to produce a pure hydrogen stream under different conditions with high recoveries thanks to the high permeability of the thin Pd layer. The best results were obtained at 10 atm, given the higer driving force produced in these cases.

3. Conclusions

Pure hydrogen can be produced from methane in a one-step process consisting a low temperature steam reforming membrane reactor in which the reaction and hydrogen separation are carried out together. Different catalyst were tested in both low temperature oxy-reforming and low temperature steam reforming providing high methane conversions. The most promising ones were tested inside a membrane reactor, which helped to increase both hydrogen production and methane conversion with high hydrogen recoveries.

4. References

5. 1. A. Martínez-Arias, et al., *Langmuir* **1999,** 15, 4796–4802.

6. 2. F.Basile et al *Journal of the European Ceramic Society* **2018** Article in press

545746 Oxygen Removal from NG By Methane Oxidation at Low Temperature

Wednesday, June 5, 2019
Texas Ballroom Prefunction Area (Grand Hyatt San Antonio)

Carla Queiroz[1], Valéria Vicentini[1], Holli Garrett[2], Chandra Ratnasamy[2] and Greg Cullen[2], (1)BU Catalysts, Clariant SA, Suzano, Brazil, (2)BU Catalysts, Clariant Corporation, Louisville, KY

Oxygen removal from NG by methane oxidation at low temperature

NGCS 12 Abstract

Carla Queiroz[1], Holli Garrett[2], Valéria Vicentini[1], Greg Cullen[2], Chandra Ratnasamy[2]

[1]*Clariant S.A., Avenue Jorge Bei Maluf, 2163, 08686-000 Suzano, São Paulo, Brazil*

[3]*Clariant Corporation, 1600 West Hill Street, Louisville, KY 40210, United States*

1. Introduction

Oxygen is a virulent contaminant in the natural gas processing, leading to several troubles even at low levels, including degradation of process chemicals, e.g. amine fluids, used to remove acid gases from the natural gas stream and the salts generated as a result eventually precipitate or cause foaming in the treaters of the gas plant. Furthermore, O_2 promotes pipeline corrosion in the presence of a few amount of water and sulphur. In the regeneration of molecular sieves, oxygen can react with hydrocarbons and their residues to produce carbon, which coats the molecular sieves, reducing its life. For underground storage sites, bacterial activity that produces hydrogen sulphide is increased by oxygen. Pipeline specifications are normally set at 10 ppmv, but some practical examples show that this level is constantly reducing, achieving in some industrial plants levels below 2 ppmv. Natural gas customers recognize the problems associated with the presence of oxygen in natural gas and impose a penalty on the price they will pay for gas containing excessive amounts of oxygen. Therefore, it is imperative to employ a system for oxygen removal at low costs, wide oxygen removal range and at low temperature. Various technologies are applicable for oxygen removal, amongst them two general options offer the highest potential from a technical and economic point of view. The first is a continuous adsorption based process, commonly using copper as adsorption materials, which is placed in at least two or more fixed-bed adsorption columns, necessary for alternating operation of loading and regeneration, however, a reducing agent is required, as hydrogen, which is high cost and normally is not available at plants for this purpose. The second commercially and promising alternative for O_2 removal in NG is a heterogeneously catalyzed oxidation reaction of methane to form H_2O and CO_2. Actually, methane acts as a reducing agent for oxygen. However, in this research, under certain operating conditions Pt offer benefits to the removal oxygen by oxidation. The influences of CO_2 and H_2S added separately to the reaction mixture as well as the stability and carbon formation on the catalytic activity were also examined.

2. Results

Figure 1 presents the gas heating temperature for 100% of O_2 conversion for the stream with pure methane for two space velocity studied (Fig.1a) and for the stream containing CO_2 (Fig. 1b). The results indicated the ActiSorb GP 601 catalyst needs a very low temperature to complete removal of O_2 from the NG, when compared to the typical methane combustion catalysts. Furthermore, the behaviour for methane oxidation in Fig 1 is expected, in which the increase the oxygen content in the feed gas requires a higher temperature, necessary for the complete O_2 removal. Likewise, raising the GHSV a slight increase in the gas heating temperature, around 10 °C, was observed. These finding are agreed with the literature [1,2], since the Eley-Rideal mechanism, described for the oxidation of methane, assumes methane to have a limiting influence on reaction rate due to weak absorption interactions with the catalytic surface. On the other hand, O_2 species to be sufficiently absorbed on the catalytic surface at any time and it is reduced easily by methane.

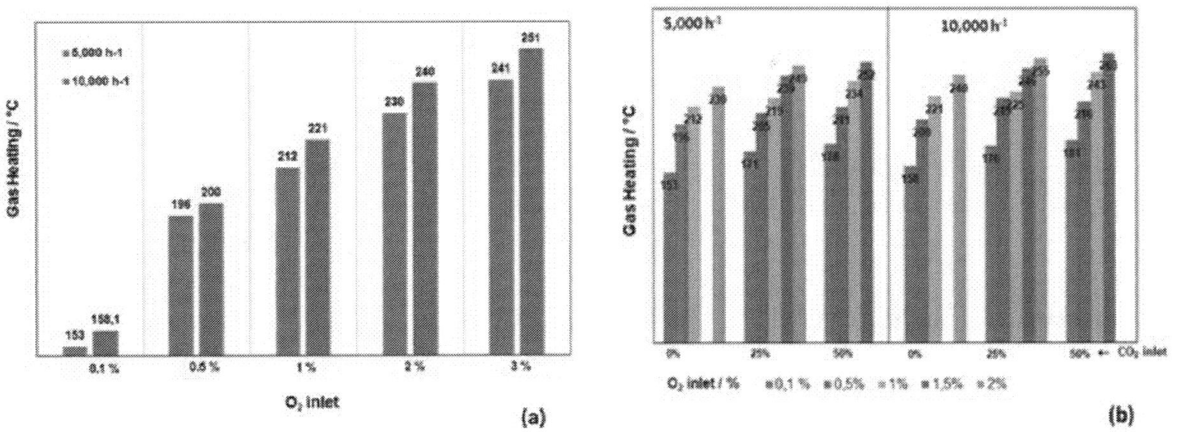

Figure 1: Gas heating temperature for 100 % of O_2 conversion ($O_{2\ outlet}$ > 10 ppmv): (a) O_2 inlet and space velocity effect and (b) CO_2 study.

When CO_2 is contained in the feedgas in 25%vol, the difference in the gas heating temperature to achieve complete oxygen conversion is raised by 20 °C, probably due to a weak competition between the methane oxidation and CO_2 reforming reactions. However, non-expressive effect was observed increasing the CO_2 inlet to 50%, indicating there is no effect related to side reactions or low methane inlet content. Likewise, no-significant effect was found when the space velocity is raised in the conditions employed during the CO_2 effect study.

The results for sulfur evaluation (not showed) revels a strong negative effect on catalytic activity with almost total loss of activity when H_2S is on gas stream. Nevertheless, by performing in situ regeneration, under the same amount of O_2 in the stream, the catalytic activity can be recovered in the same conditions of the fresh catalyst, achieving 100% of O_2 conversion. Regarding to catalytic stability, the results demonstrate high stability for O_2 removal under constant gas heating temperature at low GHSV, with full conversion. Furthermore, it was not identified carbon formation at conditions employed, where we have the same amount of carbon for both, fresh and spent catalyst.

3. Conclusion

The ActiSorb GP 601 Pt-based catalyst operate at low temperature for oxygen removal from natural gas (as function O_2 inlet content), is able to take away oxygen at specified levels and has high thermal stability and low carbon build-up. The investigation showed a strong negative effect of sulfur (as H_2S) on O_2 removal process, but a simple in situ regeneration procedure allows to recover the catalytic activity in the same conditions of the fresh catalyst.

4. Reference

[1] F. Ortloff, F. Graf, T. Kolb. Removal of Oxygen from Biogas via Catalytic Oxidation of Methane *International Gas Union Research Conference*, WP5-26, Copenhagen (2014)

[2] P. Gélin, L. Urfels, M. Primet, E. Tena. *Catalysis Today* 83, 45-57 (2003)

545765 Understanding the Role of Titanium in the Stability of Supported Palladium Catalysts for the Oxidation of Ventilation Air Methane

Wednesday, June 5, 2019
Texas Ballroom Prefunction Area (Grand Hyatt San Antonio)

Matthew Drewery[1], Hadi Hosseiniamoli[2], Meng-Jung Li[1], Eric M. Kennedy[1], Adesina Adesoji A.[3] and Michael Stockenhuber[1], (1)Chemical Engineering, The University of Newcastle, Newcastle, Australia, (2)Western Sydney University, Sydney, Australia, (3)ATODATECH LLC, Pasadena, CA

Understanding the role of titanium in the stability of supported palladium catalysts for the oxidation of ventilation air methane

*M Drewery, H Hosseiniamoli, M. J. Li, E. M. Kennedy, A. A. Adesina and M. Stockenhuber**

*Corresponding author

The growth and rate of increase in greenhouse gas emissions (GHG) represents a serious environmental challenge. The need to decrease emissions to minimise environmental effects, combined with increasing societal pressure has led to a significant increase in research focused on these problems. Ventilation air methane (VAM), the subject of this study, contributes approximately 70% of all coal-mining GHG emissions due to the significantly high flowrates required, however the low methane concentration (<1%) is below the lower flammability limit for methane and is thus unable to be oxidised to decrease the impact of emissions. Additionally, the stability of catalysts for long-term applications is a significant obstacle for the commercial adoption of technology, particularly for the expected humidified feeds.

The current study investigates the catalytic combustion of a humidified lean methane feed stream over Pd/TS-1 and Pd/Silicalite-1 at temperatures below 500°C. A time-on-stream analysis of Pd/TS-1 identified a resistance to deactivation for over 100 hours, with this hydrothermal stability investigated via comparisons with the silicalite catalyst. The silicalite catalyst displayed a gradual deactivation over 30 hours which was not evident in Pd supported on TS-1, which coincided with the agglomeration of Pd on the catalyst surface combined with carbon deposition. Further characterisation and analysis using XAS, XPS and *in situ* IR spectroscopy confirmed that Ti in the TS-1 framework plays an anchoring role, inhibiting the sintering of the catalyst by Pd migration and agglomeration. Despite the lean conditions of the feed, carbonaceous deposits were found to contribute to deactivation.

Introduction

It has been identified that catalytic activity during VAM oxidation is dependent on a number of factors including dispersion, electronic and coordination structure, support properties and interaction of the active metal with the support [1-4]. Palladium is considered the most active metal for the catalytic combustion of methane [3,5,6], however the assessment of Pd catalyst performance is complicated due to the influence of a variety of important factors [2,5,7]. Catalyst deactivation is a major challenge associated with the implementation of catalytic processes, including VAM mitigation, however no clear agreement concerning the cause of catalyst deactivation during VAM oxidation has been reached. Literature has identified two primary causes which affect catalyst activity and stability, namely water vapour inhibition and catalyst sintering.

Previous studies have reported that water vapour competed with surface active sites on Pd-based catalysts via the transformation of palladium oxide into palladium hydroxide in the presence of steam [8-13]. Supporting Pd on zeolites provides a potential mitigation solution for water deactivation. The presence of Al in the zeolite framework leads to a strong affinity for water vapour, with research showing that increasing the Si/Al ratio resulted in the catalyst becoming more hydrophobic leading to higher stability [1,2,14]. Further to this, aluminium free zeolites, such as TS-1 and silicalite-1, exhibit enhanced hydrophobic properties compared to high silica zeolites [15].

Experimental

TS-1 catalyst was prepared adopting the method reported by Taramassoet [16]. Silicalite-1 was prepared using the same procedure while omitting the addition of a titanium source. 1.2 wt% Pd was loaded onto both supports via incipient wetness with Pd(II) nitrate solution and calcined at 550°C prior to reaction. Catalytic combustion of lean methane was performed in a tubular, stainless steel fixed bed micro reactor. The inlet methane concentration was fixed at 0.7% in an air stream containing 3.4% water vapour, reacting at a GHSV of 100 000 h^{-1}. The average water vapour concentration was determined using a humidity probe and was determined to be constant, with an approximate relative humidity (RH) of 80%.

Results

The light off curves of Pd/TS-1 and Pd/Silicalite-1, assessed under the same reaction conditions, shows that Pd/TS-1 displays a higher activity with a 25°C difference at T_{90}. Palladium was identified as the active species, with no methane oxidation noted for unloaded supports.

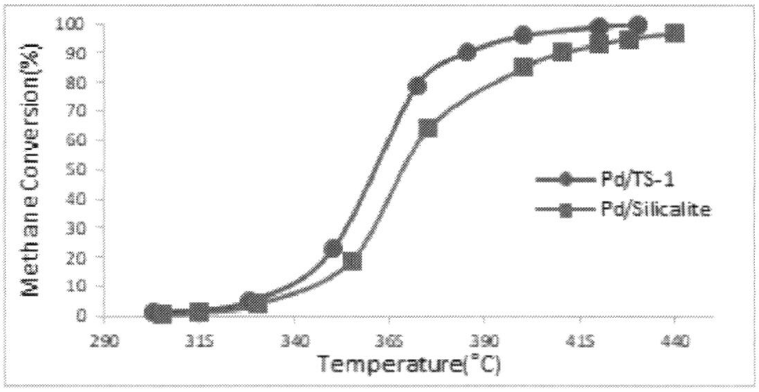

The stability of both catalysts was also evaluated under humid simulated VAM feed at a constant temperature of 400°C, with the water vapour present in the feed stream and produced during oxidation predicted as the primary factors for catalyst deactivation [8].The high stability observed is attributed to the hydrophobicity as a result of the Brønsted acid sites [15,17]. This inhibits the accumulation of hydroxyl groups on the support and increases the amount of available oxygen [5]. Additionally, it has been reported that such high oxygen mobility supports can accelerate hydroxyl desorption impeding accumulation [9]. Furthermore, carbon deposits were shown to contribute to the deactivation of the catalyst.

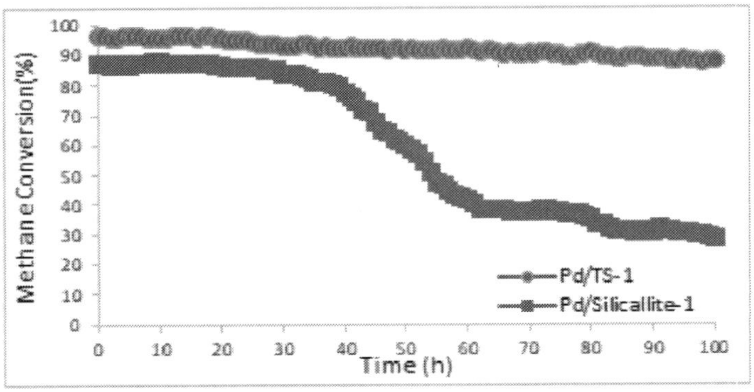

The source of Pd/silicalite deactivation was investigated via a number of characterisation techniques. The process of agglomeration, with concomitant migration of palladium, was identified by analysis with TEM. It can be seen that both the unused Pd/TS-1 and Pd/Silicalite-1 are covered with 2 to 4 nm diameter, highly dispersed palladium particles. While no significant change was identified on Pd/TS-1 after 100 h operation, Pd/Silicalite-1 displayed the agglomeration of palladium particles up to 15 nm, accompanied by much lower dispersion. These results suggest that the deactivation of the silicalite catalyst is a result of agglomeration causing the loss of dispersion, indicating the oxidation reaction is structurally sensitive. This type of deactivation was observed for acid zeolites, which also accumulated carbon deposits under very lean conditions [5,8,18]. The role of acidity and the formation of these deposits is discussed in the paper.

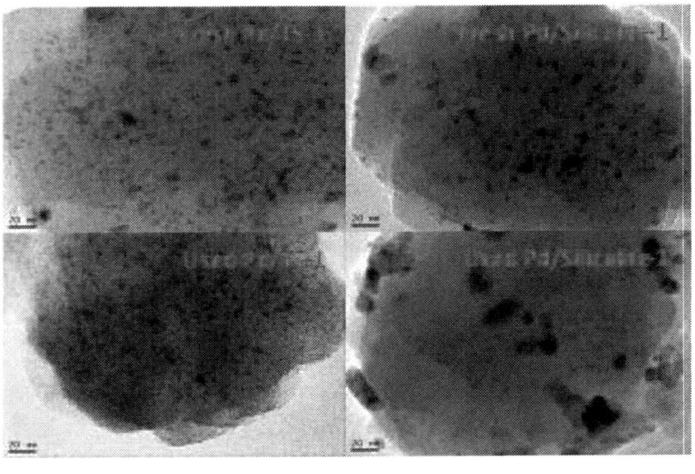

Conclusion

The stability of Pd/TS-1 and Pd/Silicalite-1 was evaluated for catalytic combustion VAM. The long term stability of both catalysts is attributed to the hydrophobicity of these aluminium-free zeolites. However, Pd/Silicalite-1 displayed a gradual deactivation over 30 h of time-on-stream. This deactivation occurred coincident with the agglomeration of palladium atoms on the surface of this catalyst as well as carbon deposition (confirmed by TEM), whereas this phenomenon was not observed for the Pd/TS-1 material. It is concluded that the dispersion of palladium in Pd/Silicalite-1 decreased remarkably compared to the slight change of dispersion observed for Pd/TS-1. It is suggested that the Ti in TS-1 framework form the sites for palladium particles and inhibits the sintering of catalyst through migration and agglomeration of palladium particles.

References

[1] Y. Lou, et al., ACS Catal., 2016.

[2] K. Okumura, et al., Appl. Catal. B, 2003. **40**(2): 151-159.

[3] P. Gelin and M. Primet, Appl. Catal. B, 2002. **39**(1):1-37.

[4] J. Nilsson, et al., ACS Catal., 2015. **5**(4):2481-2489.

[5] A. Setiawan, et al. Phys. Chem. Chem. Phys., 2016. **18**(15): 10528-10537.

[6] J. D. Grunwaldt, M. Maciejewski and A. Baiker, Phys. Chem. Chem. Phys., 2003. **5**(7): 1481-1488.

[7] C. Chen, et al., J. Mater. Chem., 2014. **2**(21): 8126-8134.

[8] A. Setiawan, et al., Catal. Sci. Technol., 2014. **4**(6): 1793-1802.

[9] K. Yasuda, et al., J. Mater. Sci., 2011. **46**(11): 4046-4052.

[10] D. Ciuparu, E. Perkins and L. Pfefferle, Appl. Catal. A, 2004. **263**(2): 145-153.

[11] D. L. Mowery and R.L. McCormick, Appl. Catal. B, 2001. **34**(4): 287-297.

[12] C. H. Bartholomew, Appl. Catal. A, 2001. **212**(1–2): 17-60.

[13] D. Ciuparu, N. Katsikis, and L. Pfefferle, Appl. Catal. A, 2001. **216**(1): 209-215.

[14] K. Okumura, E. Shinohara and M. Niwa, Catal. Today, 2006. **117**(4): 577-583.

[15] D. P. Serrano, Sep. Purif. Technol., 2007. **54**(1): 1-9.

[16] M. Taramasso, G. Perego, and B. Notari, *Preparation of porous crystalline synthetic material comprised of silicon and titanium oxides*. 1983.

[17] D. Ciuparu and L. Pfefferle, Appl. Catal. A, 2001. **209**(1): 415-428.

[18] H. Hosseiniamoli, et al., ACS Catal., 2018. **8**(7): 5852-5863.

545666 Application of Chemical Looping in Fischer-Tropsch Gas-to-Liquids Processes: Performance and Economic Analysis

Monday, June 3, 2019: 5:42 PM
Republic ABC (Grand Hyatt San Antonio)

Vishnu Priya Ramakrishnan and Omar Basha, Chemical, Biological and Bio Engineering, North Carolina A&T State University, Greensboro, NC

Abstract:

The industry of Gas-to-Liquids (GTL) using Fischer-Tropsch (F-T) has always been subject to capricious market and economic factors, typically resulting in project cancellations and delays. This is primarily due to the capital intensive nature of the process, with announced costs reported to range from $18 - $22 billion, and anecdotal reports suggesting even higher cost overruns [1]. Recently, the integration of Chemical Looping with XTL processes has been proposed as a means to reduce the capital requirements, enhance the efficiency and reduce emissions from the process [2, 3]. Theoretically, the integration of Chemical Looping (CL) could potentially eliminate the air separation and the syngas units, which account for over 25% of the F-T GTL EPC cost breakdown [4], however no techno-economic assessment has yet to be performed for GTL applications. Therefore, this work will present detailed process simulations, lifecycle greenhouse gas emissions analyses and cost analysis in a comprehensive analytical framework for multiple alternative CL-GTL system configurations with the aim of quantifying the merit of the CL-GTL process. Moreover, systematic comparisons will be made to alternative GTL options in light of addressing techno-economic challenges.

References:

[1] Biofuels and Emerging Technologies Team, "Gas-To-Liquid (GTL) Technology Assessment in support of AEO2013," Energy Information Administration (EIA)2013.

[2] O. M. Basha, L. Weng, Z. Men, W. Xu, and I. Morsi Badie, "NICE's Indirect Coal-to-Liquid Process for Producing Clean Transportation Fuels Using Fischer-Tropsch Synthesis," Front. Eng, vol. 3, pp. 362-376, 2016.

[3] O. Basha, M. and B. Morsi, I., "Techno-Economic Assessment of Different Tailgas Recycle Strategies for Coal Conversion via Fischer-Tropsch Slurry Technology," in International Energy and Raw Materials Energy Summit (INERMA), Istanbul, 2017.

[4] R. Zennaro, "Fischer–Tropsch Process Economics," in Greener Fischer-Tropsch Processes for Fuels and Feedstocks, ed: Wiley-VCH Verlag GmbH & Co. KGaA, 2013, pp. 149-169.

551467 Innovative Processes Design for Value-Added Products from Natural Gas Liquids in Shale Gas

Wednesday, June 5, 2019: 1:30 PM
Republic ABC (Grand Hyatt San Antonio)

Yiru Li[1], Zewei Chen[1], Wasiu P. Oladipupo[2], Edwin Andres Rodriguez Gil[1], Jeffrey J. Siirola[2], Fabio Ribeiro[3] and Rakesh Agrawal[1], (1)Davidson School of Chemical Engineering, Purdue University, West Lafayette, IN, (2)Charles D. Davidson School of Chemical Engineering, Purdue University, West Lafayette, IN, (3)Purdue University, W LAFAYETTE, IN

Recent advances in hydraulic fracturing and horizontal drilling have led to a surge in shale resource production and transformed the US energy landscape. Compared to natural gas, shale resources contain a substantial amount of condensate and natural gas liquids (NGLs). Many shale basin regions located in remote areas are lacking the infrastructure to distribute the extracted NGLs to other regions—particularly the Gulf Coast, a major gas processing region. A new shale gas valorization process that involves catalytic dehydrogenation followed by catalytic oligomerization has been synthesized. This process of converting natural gas liquids (NGLs) in shale gas into higher molecular weight hydrocarbon has been estimated to be energy-efficient and economically viable[1].

However, due to the highly endothermic and equilibrium limited nature of light alkane dehydrogenation, especially ethane, which is the richest component in NGLs, the process is subjected to challenges such as large recycle ratio, difficulty in reactor heat supply, etc. Here we develop various innovative NGLs upgrading processes to produce value-added products. Processes with new separation, recycle and heat supply strategies are synthesized, designed and simulated. The purpose of the process generation has transformed from designing energy-efficient flowsheet configuration to proposing easy-to-modularize plants considering the locations of remote shale basins.

Reference:

[1] Taufik Ridha, Yiru Li, Emre Gençer, Jeffrey Siirola, Jeffrey Miller, Fabio Ribeiro, and Rakesh Agrawal. Valorization of Shale Gas Condensate to Liquid Hydrocarbons through Catalytic Dehydrogenation and Oligomerization. *Processes* 6, no. 9 (2018): 139.

545431 Effects of Morphology and Dopants on the CO_2 Capacity of Nanofibrous Calcium-Oxide Based Materials for Sorption-Enhanced Steam Methane Reforming

Wednesday, June 5, 2019: 10:30 AM
Republic ABC (Grand Hyatt San Antonio)

Dante Simonetti, Luke Minardi, Faisal H. Alshafei and Zubin Mishra, Chemical and Biomolecular Engineering, University of California, Los Angeles, Los Angeles, CA

Hydrogen production is an essential component of the refining and petrochemical industries. Steam-methane reforming (SMR) is the primary method of hydrogen production in the U.S. Sorption enhanced steam-methane reforming (SE-SMR) is an attractive adaption of SMR that reduces operating temperature and down-stream processing. The primary drawbacks of SE-SMR materials are incomplete carbonation, slow carbonation kinetics, loss in activity upon cycling, and high regeneration temperature.

Nanofibrous calcium oxide, synthesized via electrospinning, was compared against calcium oxide synthesized via calcium acetate decomposition, hydrothermal synthesis with the aid of a surfactant, and natural sources. A thermogravimetric analyzer found that the sorption capacity of calcium oxide from nanofibers, from acetate, via hydrothermal, and from natural sources are 0.79, 0.70, 0.57, and 0.14 $g_{CO2}/g_{sorbant}$ after 1 hour at 600ºC, 1 atm, and 100% CO_2. The capacity of calcium oxide nanofibers and CaO derived from natural sources were measured over repeated carbonation and calcination cycles. Both calcium oxide nanofibers and natural calcium oxide lost ~33% of their initial capacity after ten cycles. Despite that, in the SE-SMR process, the enhanced activity of calcium oxide nanofibers resulted in a breakthrough time, which was twice that of calcium oxide from natural sources. SEM, TEM, XRD, and BET were used to characterize the sorbents to identify the properties that directly influence the sorption properties such as overall capacity, kinetics, and stability.

Various metal nitrates were added to the calcium nitrate electrospinning solution, at various compositional ratios, as dopants to increase the stability of the sorbent. The dopants studied comprised of metals from Groups 2, 3, 4, 12, and 13 on the periodic table of elements. After reacting these doped-sorbents at 600°C and 1 atm with 100% CO_2, the added metals were shown to have various effects on capacity, kinetics, and stability. XRD was used to identify how the dopants affect the crystal structure of calcium oxide, the dopant either forms a pure oxide crystal structure or it forms a mixed metal oxide. SEM was also used to identify morphological changes in the nanofibrous structures. The results demonstrate a promising opportunity for the application of doped electrospun materials in SE-SMR.

545498 PEI Impregnated Mesoporous Carbon Spheres As Ideal Sorbent for Post-Combustion CO_2 Capture from Natural Gas Power Plant

Wednesday, June 5, 2019: 10:54 AM
Republic ABC (Grand Hyatt San Antonio)

Qingjun Chen Sr.[1], De Chen[2], **Kumar Rout**[3] and Siyu Wang[2], (1)Chemimcal Engineering, Norwegian University of Science and Technology, Trondheim, Norway, (2)Chemical Engineering, Norwegian University of Science and Technology, Trondheim, Norway, (3)SINTEF, Trondheim, Norway

PEI Impregnated Mesoporous Carbon Spheres as Ideal Sorbent for Post-Combustion CO_2 Capture from Natural Gas Power Plant

Qingjun Chen,[1] Siyu Wang,[1] Kumar R. Kent,[2] De Chen[1,*]

[1] Department of Chemical Engineering, Norwegian University of Science and Technology (NTNU), 7491 Trondheim, Norway.

[2] SINTEF Industry, 7465 Trondheim, Norway

[*] Corresponding author: Email: de.chen@ntnu.no

1. Introduction:

The development of novel materials and new technologies for CO_2 capture and storage has received great attentions due to the global warming issues. Among the different carbon capture and storage (CCS) technologies, post-combustion CO_2 capture has attracted increasing attentions because this technology can be easily retrofitted into the existing power plants. There are two conventional options that the CO_2 can be captured in existing power plants by amines absorption or solid sorbents. Low temperature CO_2 absorption process with amines absorption has been used in industry, which is a technology adapted from natural gas sweetening. This conventional technology has faced several challenges, such as low energy efficiency and high cost. In addition to this, the amine solvent loss due to regeneration as well as its degradation has resulted in challenges and environmental concerns. The CO_2 capture by solid sorbent may diminish the environmental impact of the CO_2 capture and at the same time result in a lower energy penalty, thus solid adsorption process may be an alternative to achieve CO_2 capture.

The objective of this work is to develop promising innovative solutions for CO_2 capture towards natural gas combined cycle (NGCC) power plant, with a potential to provide significant cost reductions. This work emphasised on the synthesis of PEI impregnated mesoporous carbon spheres with high CO_2 capture capacity, fast kinetics for CO_2 capture. The results showed that mesoporous carbon spheres (MCS) with perfect spherical morphology and control pore structure were prepared by a hard-template assisted reverse emulsion method. After impregnation with PEI, the CO_2 capture capacity was up to 3.22 mmol/g at CO_2 partial pressure of 0.05 bar (5 v%, a typical concentration of the tail gas from nature gas power plant) and temperature of 75 ^{o}C, outperforming the best solid amine sorbents reported at similar condition.

2. Experimental

MCS was prepared by a hard-template assisted reverse emulsion method using formaldehyde and resorcinol as the carbon precursors and silica nanoparticles (LUDOX® SM-30) as the template. Paraffin oil and sorbitan monooleate (Span 80) were used as the disperse phase (oil phase) and surfactant. Formaldehyde, resorcinol and silica nanoparticles were first mixed in deionized water under stirring. After pre-polymerization at 45 °C for 20 min, the solution was transferred to a glass reactor containing a mixture of paraffin oil and Span 80. The hybrid polymer spheres were formed after the stirring with a speed of 200 rpm at 85 °C for 60 min. After further washing, drying, pyrolysis, and removal of silica template, the MCS were obtained. In order to control the pore structure of the MCS, different C/Si ratio (0.75, 1, 1.5) was used in this work. PEI impregnated MCS were prepared by a wet impregnation method. In a typical preparation, the desired amounts of PEI (Mw of 600) were dissolved in 2 g of methanol at 35 ^{o}C under stirring for 30 min, and then 0.3 g of MCS was added to the above solution and further stirred for one night to remove methanol.

CO_2 adsorption at 0.05 bar and dry condition was performed on a TA Instruments Q600 TGA thermogravimetric analyzer. In a typical process, about 10 mg of the adsorbent was placed in a platinum sample pan. Before the adsorption test, the adsorbents were pretreated at 100 ^{o}C under a N_2 flow at 200 mL for 100 min to remove the moisture and adsorbed CO_2 as much as possible; then the temperature was decreased to the design temperature. The gas was then switched from N_2 to CO_2 (5% CO_2, 200ml/min), and the adsorbent was held at that temperature for 100 min for the adsorption study. The CO_2 capture capacity of the adsorbent in mmol/g sorbent was calculated from the weight gain of the adsorbent in the adsorption process. The MCS and PEI impregnated MCS were characterized by SEM, digital camera, N_2-physisorption, TEM, TGA, FTIR, etc.

3. Results and discussion

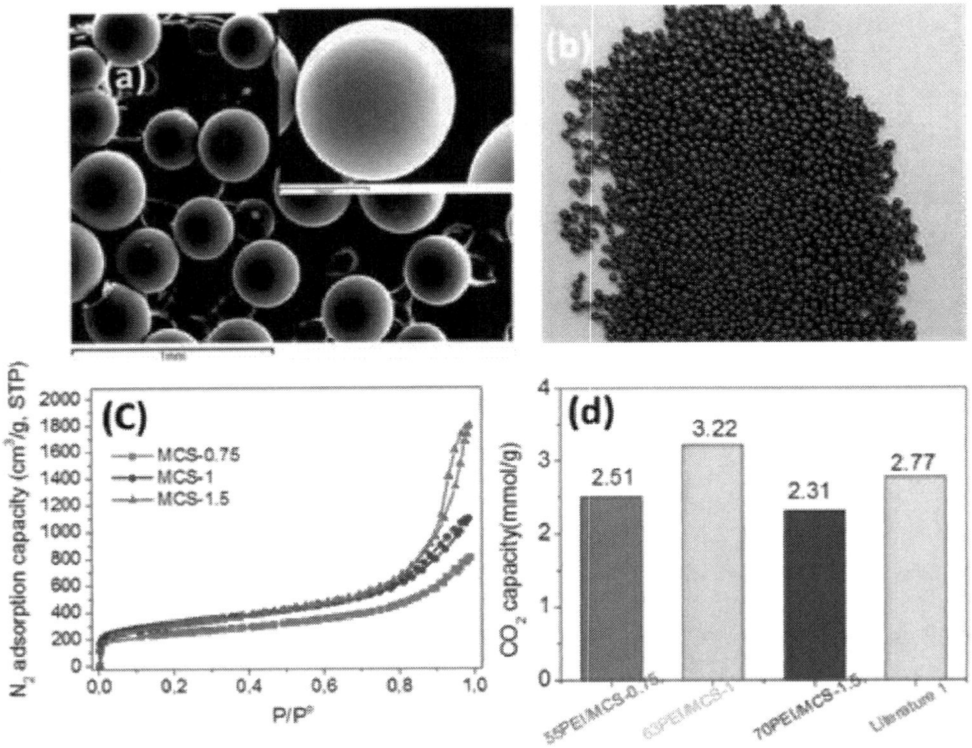

Fig. 1 SEM (a) and digital (b, 5× magnification) images of the MCS-1, pore structure (c) and CO_2 capture capacity (0.05 bar, 75 °C) (d) of PEI/MCS for CO_2 removal

The SEM images of MCS prepared at the C/Si ratio of 1 are shown in Fig.1 (a). Perfect spherical morphology of the MCS is observed with very smooth and clean surface. The diameters of the MCS-1 are very homogeneous and in the region of 0.3-0.4 mm. The digital image in Fig. 1(b) also shows the excellent and homogeneous spherical structures, which make the spheres wonderful mobility and dust-free property. The pore structures of the MCS were measured by N_2 physisorption. The N_2 adsorption-desorption isotherms of MCS with different C/Si ratio are presented in Fig. 1(c). The combination of Type I and IV isotherms indicates the co-existence of micropores and the mesopores/macropores. The micropore volumes (0.16-0.18 cm^3/g) of MCS are similar and mesopore/macroporous volume increased with the C/Si ratio. The total volumes of MCS with C/Si of 0.75, 1, 1.5 are 1.25, 1.61, and 2.68 cm^3/g, respectively. The specific surface areas of the MCS are in the region of 817-1097 m^2/g. The developed pore structure of MCS allows a high amount of PEI loading with good dispersion.

The effect of PEI loading on the CO_2 capture was studied. The results showed that there is an optimal loading for the MCS, where larger or smaller PEI loading can result in a smaller CO_2 capacity. The optimal PEI loadings depending on the pore structure of MCS were 55, 63, and 70wt. % for MCS-0.75, MCS-1, and MCS-1.5, respectively. The adsorption temperature also significantly affects the CO_2 capacity.

There is a balance between the CO_2 adsorption/desorption and diffusion in the PEI film in the sorbent. High temperature improves the CO_2 diffusion, but decreases the CO_2 adsorption. Low temperature is benefit for the adsorption, but it causes a low PEI utilization due to the diffusion limitation. 75 °C is the best temperature for the CO_2 adsorption over PEI/MCS. The CO_2 adsorption capacity measured at CO_2 partial pressure of 0.05 bar and 75 °C are shown in Fig. 1(d). The PEI impregnated MCS with medium pore volume (65PEI/MCS-1) exhibited the highest CO_2 capacity of 3.22 mmol/g, which outperforms the best solid sorbents reported at similar condition.[1,2] The PEI impregnated MCS also exhibited excellent sorption kinetics and regeneration performance (97-99%).

4. Conclusions

Mesoporous carbon spheres (MCS) with perfect spherical morphology and control pore structure were prepared by a hard-template assisted reverse emulsion method. After impregnation with PEI, the obtained sorbent showed excellent CO_2 capacity at low CO_2 partial pressure, which can be used as excellent sorbent for post-combustion CO_2 capture from natural gas power plant.

Reference:

1. J. Wang, M. Wang, W. Li, W. Qiao, D. Long, L. Ling, *AIChE J.* **2015**, 61, 972-980.

2. K. Min, W. Choi, C. Kim, M. Choi, *Nat. Commun.* **2018**, 9, 726.

545489 Future Aspects of Syngas Conversion

Wednesday, June 5, 2019: 11:18 AM
Republic ABC (Grand Hyatt San Antonio)
Frits Dautzenberg, Serenix Corporation, Fort Collins, CO

The production of syngas using SMR can be modified by applying hysteresis heating of catalytically active Ni-Co nano-particles.

Using methanol production as an example, the attractiveness of this approach will be discussed in this paper.

Methanol may also be prepared using a new "Wacker" oxidation system currently under development.

The design and tentative economics of this novel methanol process will be presented.

Mono-ethylene glycol is an imported chemical and new technology is under development in Europe and in Asia based on the conversion of syngas from bio-mass.

Details of the proposed MEG technology will be reviewed in this paper.

551478 Selective Oxidation and Oxidative Dehydrogenation of Ethane and Propane

Wednesday, June 5, 2019: 11:42 AM
Republic ABC (Grand Hyatt San Antonio)

Srikant Gopal[1], Mohammed Al-Hazmi[1] and **Pankaj S. Gautam**[2], (1)SABIC Technology Center, Riyadh, Saudi Arabia, (2)Process Technology, SABIC, Sugar Land, TX

Selective Oxidation and Oxidative Dehydrogenation of Ethane and Propane

Srikant Gopal* and **Mohammed Al-Hazmi**

SABIC Technology Center, Riyadh, 11551, Kingdom of Saudi Arabia

**Corresponding author: +966 (11) 499 9825, gopalsr@sabic.com*

Introduction

Feedstock for the important petrochemicals lies in the range of hydrocarbons with 1 to 4 carbon atoms. The growth in shale gas production in the United States has made petrochemical raw materials such as ethane and propane available at a lower cost. The availability of this feedstock provides opportunities for utilizing newer routes to produce valuable chemicals. The selective oxidation of lower alkanes provides a direct route to produce chemicals such as acetic acid, acrylic acid and acrylonitrile as well as olefins through oxidative dehydrogenation. These routes can provide a process with lower capital and operating costs as well as lower environmental impact. SABIC has been involved in developing technologies leading to C2 compounds and its derivatives including selective oxidation of ethane.

Achieving a selectivity and productivity that is commercially attractive is one of the main challenges of selective oxidation processes. The results reported in recent years for selective oxidation of lower alkanes to carboxylic acids as well as oxidative dehydrogenation to olefins indicate that substantial progress has been made in improving the catalyst selectivity. The progress reported in the patent and journal literature as well as some of our assessments and work in this area will be discussed. The current dominant routes for production of the above chemicals will be compared with the one-step selective oxidation routes. The economic and environmental benefits that selective oxidation routes could potentially provide will be illustrated.

Methods

Comprehensive searches of patents as well as journal articles were carried out on routes for selective oxidation and oxidative dehydrogenation of lower alkanes. The experimental results reported were analyzed to identify the best results achieved over the years for each process. A more detailed study of the patents was also performed to determine the most promising catalysts and the methods that have been employed to improve the catalyst performance. From this assessment, the most suitable catalyst compositions for each process and the current status of the technology can be inferred.

The raw material cost savings is one of the major benefit provided by the selective oxidation process. Based on the best results reported in the literature, and the latest feedstock prices, raw material costs for the selective oxidation route has been calculated and compared with the raw material cost for the current dominant route.

Results and discussion

The most successful catalysts for selective oxidation/ammoxidation and oxidative dehydrogenation of ethane and propane have been complex multimetallic oxides containing several metals, especially the ones containing Mo, V, Nb and Te. Further catalyst performance improvements have been achieved by addition of promoters to the catalyst composition as indicated in Table 1 below, and also through improvements to the catalyst preparation method. The advances made in catalysts for each of the processes will be discussed.

Table 1. Catalyst compositions showing one of the best results reported in literature for selective (amm)oxidation and oxidative dehydrogenation processes.

Process	Catalyst Composition	Selectivity to Target product	References
Ethane to Acetic acid	MoVNbSbCaPd	91	1, 2
Propane to Acrylic acid	MoVTeNbSm	98	3
Propane ammoxidation to Acrylonitrile	MoVSbNbTiLu	69	4
Ethane ODH to ethylene	MoVNbTe	98	5
Propane ODH to propylene	V/CeO2	97	6

The economics of the above processes have been analyzed by estimating the raw material cost per ton of product, considering a range of selectivities and raw material prices. An illustration is shown below for the ethane to acetic acid process in Figure 1. Using the best results reported in the literature, the new processes can be compared with the current dominant industrial process in terms of the raw material costs.

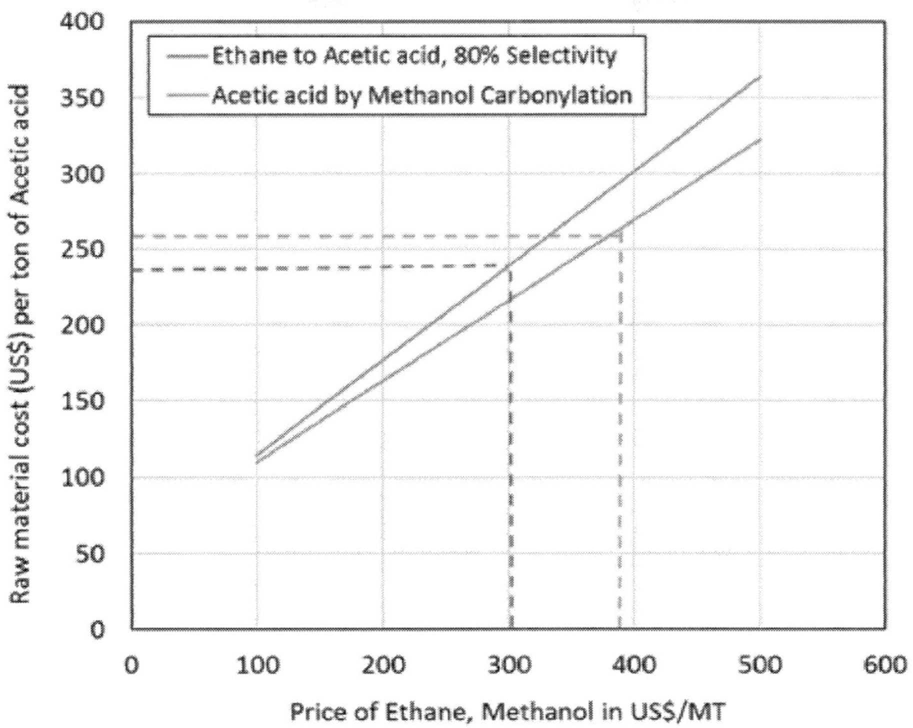

Figure 1. Raw material cost estimate for acetic acid production by ethane oxidation compared to methanol carbonylation. Dotted lines show the raw material cost at current US Gulf coast prices.

The illustration in Figure 1 shows that at current ethane and methanol prices, and 80% selectivity, the SABIC ethane to acetic acid technology can be economically attractive, compared to the methanol carbonylation process. A similar assessment would be presented for the other selective oxidation routes as well. In addition, the other challenges that must be overcome for widespread commercial adoption of the selective oxidation and ODH routes will be discussed in the talk.

References

1. H. Borchert, U. Dingerdissen, US Patent 6399816 (B1).
2. S. Zeyss, U. Dingerdissen, EP 1286942 (B1).
3. A. Gaffney, M. Heffner, R. Song, US Patent 7396955 (B2).
4. C. Lugmair et al., US Patent Application US20080103325(A1).
5. A. Gaffney et al., US Patent 8105972(B2).
6. M. Taylor et al., Top Catal (2009) 52:1660–1668.

545602 Techno-Economic Analysis of Drm+Cosorb Integrated Process for High H_2/CO Syngas Production

Wednesday, June 5, 2019: 12:06 PM
Republic ABC (Grand Hyatt San Antonio)

Shaik Afzal[1], Debalina Sengupta[2], Mahmoud El-Halwagi[1] and Nimir Elbashir[3], (1)The Artie McFerrin Department of Chemical Engineering, Texas A&M University, College Station, TX, (2)Texas A&M University, College Station, TX, (3)Chemical and Petroleum Engineering, Texas A&M University at Qatar, Doha, Qatar

TECHNO-ECONOMIC ANALYSIS OF DRM+COSORB INTEGRATED PROCESS FOR HIGH H_2/CO SYNGAS PRODUCTION

Shaik Afzal[1,2], Debalina Sengupta[2], Mahmoud El-Halwagi[2,3], and, and Nimir Elbashir[1,2,3,4,*]

[1]*Chemical Engineering Program, Texas A&M University at Qatar, P. O. Box 23874, Doha, Qatar*

[2]*Gas and Fuels Research Center, Texas A&M Engineering Experiment Station, College Station, Texas 77843, United States*

[3]*The Artie McFerrin Department of Chemical Engineering, Texas A&M University, College Station, Texas 77843, United States*

[4]*Petroleum Engineering Program, Texas A&M University at Qatar, P. O. Box 23874, Doha, Qatar*

*nimir.elbashir@qatar.tamu.edu

Dry Reforming of Methane process has been studied extensively over various catalytic systems. The main challenges for industrial application are the high energy requirement, severe coking environment and low-quality syngas ($H_2/CO\sim1$) produced by DRM. Optimization work done recently by our group [1], indicates that parallel combinations of a DRM unit with a commercial process like Steam Methane Reforming might not result in an appreciable decrease in overall CO_2 emissions or operating costs. Different process combinations were studied with the aim of leveraging the CO_2 utilization potential of the DRM reaction. Based on the assessment, it was found that a series combination of DRM and COSORB (CO Absorption unit to adjust H_2/CO ratio) can be competitive with existing processes in terms of CO_2 emissions and operating costs. However, it should be noted that the advantage may be achieved only if the captured CO is sold to an external customer. Of all the studied combinations, this particular process combination has significant reductions in CO_2 emissions and operating costs. For a syngas ratio of 2, the DRM+COSORB process would have ~68% reduction in CO_2 emissions and ~21% reduction in operating costs when compared to a Partial Oxidation plant (which utilizes oxygen). Similarly, for a syngas ratio of 3, when compared to an Auto-Thermal Reformer, the reduction in CO_2 emissions is ~72% and reduction in operating costs is ~25%. The comparisons are plotted in Figure 1. These promising results indicate a good potential for the DRM+COSORB process.

The current work builds on the aforementioned findings and a detailed process simulation is carried out in ASPEN Plus V8.8. The objective is to investigate the techno-economic potential of the DRM+COSORB process to produce syngas of high H_2/CO ratio. Data for the COSORB unit has been obtained from literature sources published when the process was being commercialized [2]. The simulation results will be analyzed in terms of overall CO_2 emissions as well as fixed and operating costs of the process. These metrics will be compared with those of currently commercial processes (Partial Oxidation of Methane and Steam Methane Reforming). Figure 1 shows the comparison of these metrics based on our optimization study [1]. The ASPEN simulation will help in validating these estimates for the proposed DRM+COSORB process.

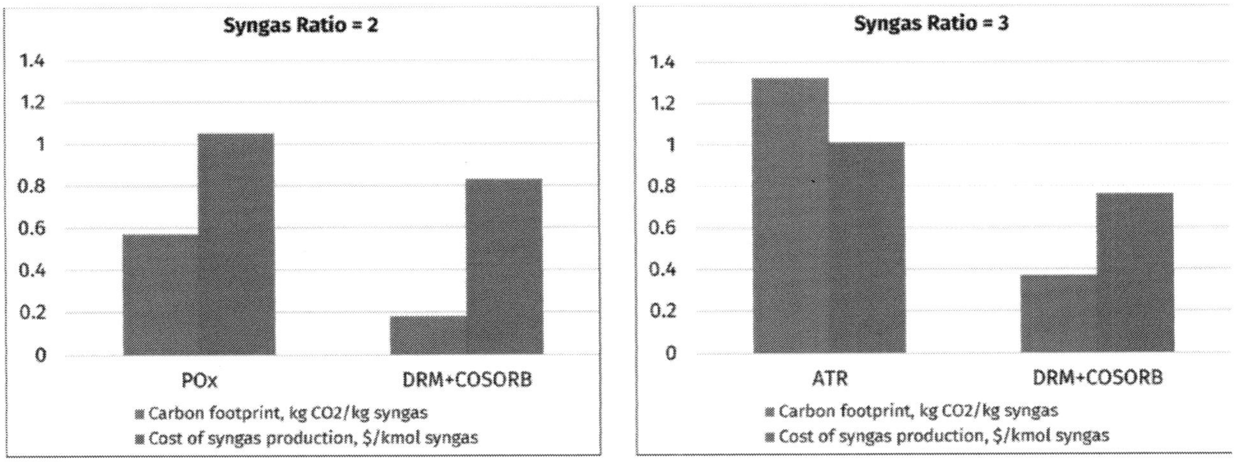

Figure 1 - Comparison of proposed DRM+COSORB Process (captured CO sold externally) with commercial processes for syngas ratio of 2 & 3

Due to the inherent stoichiometry of the Dry Reforming of Methane process, even if coke-resistant catalysts are found to be commercially viable for the DRM conditions, the low-quality syngas problem still persists. The proposed DRM+COSORB process will be able to produce high quality syngas, through a stand-alone DRM unit without the need for external hydrogen blending for syngas ratio adjustment. This simulation work of the DRM+COSORB process is one step towards better understanding of this process which has the potential of producing same quality of syngas as commercial processes ($H_2/CO \geq 2$) with lower CO_2 emissions and competitive operating costs.

Keywords: Dry Reforming, Syngas Ratio, COSORB

References

[1] S. Afzal, D. Sengupta, A. Sarkar, M. M. El-Halwagi, and N. O. Elbashir, "Optimization Approach to the Reduction of CO2 Emissions for Syngas Production Involving Dry Reforming," *ACS Sustain. Chem. Eng.*, vol. 6, p. 7532–7544, 2018.

[2] D. J. Haase and D. G. Walker, "The COSORB Process," *Chemical Engineering Progress*, vol. 70, no. 5, pp. 74–77, 1974.

545704 Systems Analysis of Light Alkane Resources for Fuels and Petrochemicals Manufacture

Wednesday, June 5, 2019: 1:54 PM
Republic ABC (Grand Hyatt San Antonio)

Sean E. DeRosa[1], Yosuke Kimura[2], Gary McGaughey[2], Elena C. McDonald-Buller[2], **David Allen**[3] and Mark A. Stadtherr[3], (1)Sandia National Laboratories, Albuquerque, NM, (2)Center for Energy and Environmental Resources, University of Texas at Austin, Austin, TX, (3)McKetta Department of Chemical Engineering, The University of Texas at Austin, Austin, TX

It has recently been proposed (Ridha et al., 2018) that natural gas liquids (NGLs), which may occur in significant amounts when produced from shale resources, may be converted through dehydrogenation and oligomerization to liquid hydrocarbons suitable as transportation fuels. This will open competing markets for NGLs, which are currently used as feedstock for petrochemicals manufacture (through cracking to ethylene), marketed directly as fuel (propane, butane), or exported. Moreover, the Ridha et al. (2018) technology may potentially be operated to produce significant amounts of olefins for non-fuel applications, thus competing with crackers.

To assess the impact of this technology in the fuels and petrochemicals industry, a network model of the industry has been developed. Models of chemical manufacturing networks originated with Stadtherr and Rudd (1976, 1978) and many iterations of the original industry model have been constructed. Recently, the model framework and material flows have been updated by DeRosa and Allen (2015, 2016) to examine the impacts of shifts in feedstocks from petroleum fractions to NGLs and to study the impact of new aromatics production technology. In this presentation, modifications to this model are described, including the introduction of geospatial resolution. The model is applied to study scenarios involving the introduction of new technologies for processing light alkanes, such as that suggested by Ridha et al. (2018).

References

DeRosa, S. E., Allen, D. T., *ACS Sustainable Chem. Eng.*, **2015**, *3*, 451–459.

DeRosa, S. E., Allen, D. T., *Ind. Eng. Chem. Res.,* **2016**, *55,* 5366–5372.

Ridha, T., Li, Y., Gençer, E., Siirola, J. J., Miller, J. T., Ribeiro, F. H., Agrawal, R., *Processes*, **2018**, 6, 139.

Stadtherr, M. A., Rudd, D. F., *Chem. Eng. Sci.*, **1976**, *31*, 1019–1028.

Stadtherr, M. A., Rudd, D. F., *Chem. Eng. Sci.*, **1978**, *33*, 923–933.

545416 Enhancing Asset Design, Performance and Operation Management through Reservoir-Production-Processing Facility Integrated Modeling

Wednesday, June 5, 2019: 2:18 PM
Republic ABC (Grand Hyatt San Antonio)
Carlos Yengle and Hemant Kumar, Chevron Corporation, Houston, TX

Asset steady state modelling is commonly done by different teams of engineers working independently on asset subsystem design and asset operation management. Design criteria shared between teams are based on a limited set of pre-defined discrete assumptions for each individual subsystem. Currently, the commonly used approach fails to account for subsystem inter-dependencies, and does not enable assessment of changing conditions across the asset lifecycle. This gap often results in sub-optimal facilities design for the asset and cost overruns and/ or lost production. Overall asset management decision-making should be supported with consistent information utilized across the whole enterprise. Integrated modeling approach helps to make informed decisions by considering the combined effect of reservoir uncertainties, well placement, surface network, and process facility on overall asset design, production and management. This paper describes how a full asset integrated model is built and utilized as an effective decision support tool to help optimize overall asset design and operational performance management.

A full asset integrated model is built using typical industry known commercial simulator packages for each individual subsystem and field conditions, the integrated model includes steady state models for Reservoirs, Wells, Production and Injection Network and Processing Facilities. A platform which orchestrates data connectivity and integration while allowing subsystem applications to maintain their dependency in thermodynamic properties and equation of state solving is utilized for transferring data and controlling variables within the subsystems. Data from a major capital project at early stages of development was utilized to pilot test the technique and stress testing of consistency of forecast production across the subsystems, whole asset impact due to a change in an individual subsystem such as reservoir uncertainties, well placement and scheduling, surface network operating conditions, and individual subsystem constraints.

The integrated multi-field network model provided realistic optimal operating conditions and long-term production forecasts of oil production, sales gas and water injection requirements by incorporating the physics at appropriate levels to ensure higher accuracies than discrete models.

Integrated asset modelling utilizes information consistently across the whole asset and includes subsystem interdependencies impact on overall asset steady state operating conditions. Integrated modelling allows independent teams of facilities engineers and reservoir engineers to collaborate holistically while maintaining specific experts needs and resources.

The integrated model technique offers a novel and versatile capability of evaluating diverse operational scenarios to find optimal settings for short-term and the long-term asset production needs.

545565 Model-Based Optimization of an Alkene Oligomerization Process to Produce High Octane Fuels

Thursday, June 6, 2019: 10:30 AM
Republic ABC (Grand Hyatt San Antonio)

Hari S. Ganesh[1], Sergio Vernuccio[2], Michael Baldea[1], Thomas F. Edgar[1], Linda J. Broadbelt[2], Mark A. Stadtherr[1] and **David Allen**[1], (1)McKetta Department of Chemical Engineering, The University of Texas at Austin, Austin, TX, (2)Chemical and Biological Engineering, Northwestern University, Evanston, IL

Ridha et al. [1] proposed a shale gas transformation process to convert shale gas condensate into liquid fuels via catalytic dehydrogenation followed by catalytic oligomerization. Liquid fuels are easier to transport from the remote shale gas base regions to gas processing plants than NGLs. They can also be directly sold in the regional fuel market.

This work focuses on the oligomerization reactor of the Ridha et al. [1] process, using a zeolite (H-ZSM-5) catalyst. The objective of this work is to identify the operating conditions of the reactor that would maximize the high value products. The value of the liquid product stream from oligomerization is measured in terms of barrels of octane. We use a microkinetic model that we developed to predict the concentrations of individual chemical components exiting the oligomerization reactor as a function of the operating conditions such as temperature. The oligomerization products are separated into a light stream, which is recycled, and a heavy product stream. We first calculate the octane numbers of the pure components in the product stream by the structural group contribution method reported in Albhari [2]. Then, we apply appropriate mixing rules to calculate the octane number of the product stream blend. The optimization problem aims to calculate the reactor operating conditions that would maximize the value of the product stream subject to the constraints imposed by the reaction chemistry, and by the size and dynamics of the oligomerization reactor. Operation under optimized operating conditions results in value addition for the overall NGLs to liquid hydrocarbon conversion process.

REFERENCES

[1] Ridha, T., Li, Y., Gençer, E., Siirola, J., Miller, J., Ribeiro, F., & Agrawal, R. (2018). Valorization of Shale Gas Condensate to Liquid Hydrocarbons through Catalytic Dehydrogenation and Oligomerization. *Processes*, *6*(9), 139.

[2] Albahri, T. A. (2003). Structural group contribution method for predicting the octane number of pure hydrocarbon liquids. *Industrial & Engineering Chemistry Research*, *42*(3), 657-662.

551316 Identification of Optimal Gas Conversion Strategies Using a Process Systems Engineering Approach

Thursday, June 6, 2019: 10:54 AM
Republic ABC (Grand Hyatt San Antonio)

Patrick Linke, Dhabia Al-Mohannadi, Shaza Shehab and Razan Ahmad, Chemical Engineering, Texas A&M University at Qatar, Doha, Qatar

Hydrocarbon resources can be monetized through various routes as fuels or other chemicals. The demand for such products and energy is expected to increase with world population increases. At the same time, more stringent environmental issues must be met and a trend to embed elements of circular economies into monetization strategies. Ambitious carbon dioxide emission reduction targets to avoid dangerous climate change effects and ensure sustainability for the future generations is one such example. Monetizing gas under such constraints will require a wide variety of existing and emerging technology options to be considered in determining optimal monetization strategies. Hence, the development of an integrated holistic approach to systematically screen through the many possible strategic options - across multiple feedstock, many conversion routes, products, intermediates, and emissions - is crucial for many countries with natural resource centric economies and as well as for energy companies.

We will present research into a systematic, optimization-based approach for the systematic screening of optimal gas monetization options considering multiple feedstock, conversion processes, product choices, intermediates and emissions, including but not limited to natural gas, energy, value added products, as well as waste such as carbon dioxide. The approach follows a novel problem representation and results in a structural resource management network optimization problem. The work focusses on industrial processing clusters and their integration with surrounding regional energy systems. The proposed method will optimize raw material usage, energy requirements, and power requirements across industrial parks to identify those strategies, which maximizing profitability while meeting emissions constraints. The management of these resources can have multi purposes such as economic diversification or minimizing emissions. The work develops a model that produces the best route to integrate the resources within an industrial city under desired footprint constraints, resource limitation, or an economic value. A case study example will be solved to illustrate the method and outline strategic choices considering sets of feedstock and product choices.

551462 A General Evaluation Framework for Direct Nonoxidative Conversion Strategies of Methane to Aromatics

Thursday, June 6, 2019: 11:18 AM
Republic ABC (Grand Hyatt San Antonio)

Kefeng Huang and **Christos T. Maravelias**, Department of Chemical and Biological Engineering, University of Wisconsin-Madison, Madison, WI

Natural gas is a versatile chemical feedstock, and natural gas conversion to commodity chemicals has recently attracted significant attention because of the shale gas boom in the United States and the low natural gas prices. In addition, direct nonoxidative methane coupling to form aromatics in the absence of O_2 has received significant attention. Yet, high coke yields and catalyst deactivation due to coke formation have presented a significant barrier to develop a commercial methane dehydroaromatization (DHA) process.

In this work, we study nonoxidative natural gas conversion technologies that directly convert methane to aromatics. We first develop an integrated strategy, employing a nonoxidative catalytic system, as well as downstream separation sections for product recovery and purification as well as recovery and recycle of unreacted methane. We then develop a flexible framework that allows us to systematically evaluate numerous technology alternatives and identify the key technology gaps that must be overcome in order to develop commercially attractive processes. The key variables include conversion temperature, methane one-pass conversion, aromatics selectivity, coke selectivity, and natural gas feed rate. We also develop surrogate unit models and a database that allows us to easily calculate key variables from experimental results.

References

Huang, K.; Miller, J.B.; Huber, G.W.; Dumesic, J.A.; Maravelias, C.T.; 2018. A General Framework for Evaluation of Direct Nonoxidative Methane Conversion Strategies. *Joule* 2, 349-365.

Karakaya, C., and Kee, R.J.; 2016. Progress in the direct catalytic conversion of methane to fuels and chemicals. *Prog. Energy Combust. Sci.* 55, 60–97.

551464 Optimal Allocation of Modular and Transportable Methane Processing Units for Use in Oil and Gas Fields

Thursday, June 6, 2019: 11:42 AM
Republic ABC (Grand Hyatt San Antonio)

R. Cory Allen[1,2], Styliani Avraamidou[1,2], Douglas L. Allaire[3], Mahmoud M. El-Halwagi[4] and Efstratios N. Pistikopoulos[1,2], (1)Texas A&M Energy Institute, Texas A&M University, College Station, TX, (2)Artie McFerrin Department of Chemical Engineering, Texas A&M University, College Station, TX, (3)J. Mike Walker '66 Department of Mechanical Engineering, Texas A&M University, College Station, (4)The Artie McFerrin Department of Chemical Engineering, Texas A&M University, College Station, TX

Optimal Allocation of Modular and Transportable Methane Processing Units for Use in Oil and Gas Fields

R. Cory Allen[1,2], Styliani Avraamidou[1,2], Douglas L. Allaire[3], Mahmoud M. El-Halwagi[2], Efstratios N. Pistikopoulos[1,2*]

1. *Artie McFerrin Department of Chemical Engineering, Texas A&M University, College Station, TX 77843, USA*
2. *Texas A&M Energy Institute, Texas A&M University, College Station, TX 77843, USA*
3. *J. Mike Walker '66 Department of Mechanical Engineering, Texas A&M University, College Station, TX 77843, USA*

**stratos@tamu.edu*

Keywords: Two Stage Stochastic Programming, Large Scale Optimization, Process Systems Engineering

Production from tight oil and shale gas wells is expected to account for over two-thirds and three-fourths respectively of the total amount of oil and gas produced in the United States by 2050 [1]. These unconventional wells have steep production decline rates [2] and many tight oil fields lack the proper infrastructure to transport all of the associated natural gas to the market. In order to reduce the capital and operational cost of processing facilities, modular and transportable methane processing units have been suggested as they can be reallocated in the field to meet changes in production, or can be used to abate or cease flaring of the stranded associated gas [3, 4, 5, 6].

This work is based on our earlier work on the capacity planning and scheduling optimization framework for modular processing units in a shale gas field [6]. The framework, which is modeled as a stochastic mixed integer linear program, allows for uncertainty with respect to the influent flow rate into each of the processing facilities in field, which are comprised of modular processing units. In an effort to increase the scale of the problems solved in [6], we introduce cutting planes, which exploit the block diagonal structure of the feasible solution region, as well as employ lagrangian relaxations, which reduces the computational burden of the non-anticipatory constraints. This consequently allows for solutions of large scale problems, that often accompany the development of oil and gas fields due to the number of wells being drilled and well as their accompanying production uncertainties to be found.

References:

[1] United States Energy Information Administration. (2018). *Annual Energy Outlook 2018: with projections to 2050*. Washington, DC.

[2] United States Energy Information Administration. (2018). *Drilling Productivity Report November 2018: for key tight oil and shale gas regions*. Washington, DC.

[3] Baldea, M., Edgar, T. F., Stanley, B. L., & Kiss, A. A. (2017). Modular manufacturing processes: status, challenges, and opportunities. *AIChE Journal*, 63(10), 4262-4272.

[4] Yang, M., & You, F. (2018). Modular methanol manufacturing from shale gas: techno economic and environmental analyses of conventional large scale production versus small scale distributed, modular processing. *AIChE Journal*, 64(2), 495-510.

[5] Tan, S. H., & Barton, P. I. (2016). Optimal dynamic allocation of mobile plants to monetize associated or stranded natural gas, part II: dealing with uncertainty. *Energy*, *96*, 461-467

[6] Allen, R. C., Allaire, D., & El-Halwagi, M. M. (2018). Capacity Planning for Modular and Transportable Infrastructure for Shale Gas Production and Processing. *Industrial & Engineering Chemistry Research*.

550885 Center for Innovation on New Energies - CINE- a New Research Effort for Methane Conversion Using Electrochemical Technologies

Monday, June 3, 2019
Texas Ballroom Prefunction Area (Grand Hyatt San Antonio)

Andre Ferlauto[1], Estevam Spinacé[2], Reginaldo Muccillo[2], Daniel de Florio[1], Almir Oliveira Neto[2], Elisabete Santiago[2], Thiago Lopes[2], Jorge Vaz[2], Eliana Muccillo[2] and **Fabio Fonseca**[2], (1)UFABC, Santo André, Brazil, (2)IPEN-CNEN/SP, São Paulo, Brazil

The Center for Innovation on New Energies (CINE) is an Engineering Research Center sponsored by FAPESP (São Paulo Research Foundation, Brazil) and Shell. CINE was launched on August 2018 to grow as a world-class research hub to promote the transition towards a more sustainable energy system. CINE's four divisions have the mission to perform interdisciplinary research in the areas of new dense energy carriers, energy storage, and devise new pathways to convert methane to products. The Methane to Products (M2P) division is built to grow as a scientific platform for research on advanced electrochemical technologies for methane conversion to high-value products. The M2P division covers a broad range of topics that comprises a sustainable path for transforming methane to a feedstock using photo and electrochemical processes. CINE's M2P division is based on the association of two leading institutions of São Paulo metropolitan area: Instituto de Pesquisas Energéticas and Nucleares (IPEN) and Universidade Federal do ABC (UFABC) teamed up with leading academics from other countries (such as Portugal, France, Denmark, and Germany) to extend the scientific capabilities. The main objective of the RD3 is to explore new possibilities for methane conversion by using advanced electrochemical technologies. Systematic studies of different reactions paths using either sun- or electric-driven electrochemical reactors, from ultra-high temperature (~1,400°C) to room temperature, are planned. Such studies comprise the formation and training of high-qualified students in all levels that will share a multidisciplinary research environment with strong links to both knowledge and technology dissemination. The final goal is to produce and disseminate knowledge and technology to contribute to efficient and practical systems for the sustainable production of high-value chemicals from methane. Specific objectives are summarized by the topics covered by the six initial projects of the M2P division - Composite ceramic membranes for carbon dioxide separation - Photocatalysts for methane conversion to valuable products coupling with hydrogen evolution from water - Methane conversion at ultra-high temperatures by solar driven redox reactions - Catalysts for selective formation of methanol from methane in fuel cell systems - Electrochemical conversion of CH_4 to methanol mediated by CO_2 - High temperature electrochemical CH_4 conversion in solid oxide reactors. Such projects will be briefly introduced to stimulate discussions and new collaborations.

545483 Promotion of Ammonia Synthesis By Surface Protonics at Low Temperature

Monday, June 3, 2019
Texas Ballroom Prefunction Area (Grand Hyatt San Antonio)

Yuta Tanaka, Kota Murakami, Ryuya Sakai, Tomohiro Yabe, Shuhei Ogo and Yasushi Sekine, Applied Chem., Waseda Univ., Tokyo, Japan

Ammonia is an important feedstock used as a raw material for fertilizers, resins, and fibers. Lately, much attention has been paid to the ammonia utilization as a hydrogen carrier due to its high hydrogen content and handling faculty. However, there are still a lot of challenges in terms of harsh reaction conditions because of rigid N_2 bonding. To date, Haber Bosch process using Fe-based catalyst has been utilized under extreme condition of 20-40 MPa and 673-873 K owing to its thermodynamic and kinetic limitation. Thus a more efficient process is required for ammonia synthesis.

Recently, our group reported the activation of ammonia synthesis using Ru-based catalyst by applying the electric field. It is revealed that the surface protonics in the electric field promote ammonia synthesis, especially the cleavage of N_2 bonding, rate determining step of this reaction. By means of *in-situ* IR measurements and first principle calculation, it is clarified that N_2cleavage is considerably enhanced by proton in the electric field, proceeding via novel intermediate (N_2H^+).

Herein, we conducted kinetic analysis for ammonia synthesis in the electric field to investigate the contribution of proton to the N_2 dissociation in detail. Results showed that the apparent activation energy decreased gradually as the temperature decreased. We inferred this unique phenomenon was induced by the increment of proton species on the surface. For the elucidation of the assumption, the temperature dependency for the conductivity of CeO_2 pellets under dry condition (N_2 + H_2) was measured using electrical impedance spectroscopy. Results clarified that the proton conductivity increased at lower temperature, which is the same tendency as the activity. These results suggest that N_2 cleavage in the electric field proceeds through N_2H^+ with proton species produced from lattice oxygen and supplied H_2.

545410 TPSR Studies: The Effects of Pretreatment Techniques on Silica Support for Cobalt Fischer-Tropsch Synthesis Catalysts

Monday, June 3, 2019
Texas Ballroom Prefunction Area (Grand Hyatt San Antonio)
Chike G. Okoye-Chine, IDEAS, University of South Africa, Roodepoort, South Africa

This paper focuses on the use of the Temperature Programmed Surface Reaction (TPSR) studies to investigate the effect of pretreatment techniques on silica support for cobalt Fischer-Tropsch (FT) catalysts. Untreated silica gel and two other silica gel each pretreated with thermal treatment and ethylene glycol (organic solvent) were used to prepare 10%Co/SiO_2 catalysts via incipient wetness impregnation. In this work, various catalyst characterization techniques were used i.e. BET, XRD, Hydrogen chemisorption, Temperature Programmed Reduction (TPR), Temperature Programmed Desorption (TPD), Scanning Electron Microscopy (SEM) and Transmission Electron Microscopy (TEM). Also, the reaction behavior of the treated and untreated catalysts were studied using Temperature Programmed Surface Reaction (TPSR) technique by hydrogenation of pre-adsorbed CO. The characterization of the catalysts showed that the catalyst made with ethylene glycol pretreated silica gel gave a better cobalt metal dispersion and the TPSR studies indicated that this same catalyst exhibited higher CO activity compared to the other catalysts.

546629 Structure-Performance Relationships on Co-Based Fischer-Tropsch Synthesis: The Impact of H_2-CO-H_2 Activation Treatment

Monday, June 3, 2019: 5:48 PM
Texas Ballroom EF (Grand Hyatt San Antonio)

Nikolaos Tsakoumis[1], Eleni Patanou[2], Rune Myrstad[3], Erling Rytter[4] and **Edd Blekkan**[4], (1)Department of Chemical Engineering, NTNU, Trondheim, Norway, (2)NTNU, Trondheim, Norway, (3)SINTEF Industry, Kinetics and Catalysis Research Group, Trondheim, Norway, (4)Chemical Engineering, Norwegian University of Science and Technology, Trondheim, Norway

Structure-Performance relationships on Co-based Fischer-Tropsch Synthesis:
The impact of H_2-CO-H_2 activation treatment

Nikolaos E. Tsakoumis [a], Eleni Patanou [a], Rune Myrstad [b], Erling Rytter [a,b], Edd A. Blekkan [a]

[a] Department of Chemical Engineering, Norwegian University of Science and Technology (NTNU), N-7491 Trondheim

[b] SINTEF Industry, N-7465 Trondheim, Norway <>1. Introduction

Fischer – Tropsch synthesis (FTS) is the heart of the gas-to-liquids (GTL) process which is widely used for the utilization of synthesis gas deriving from natural gas, coal and in the future also biomass[1]. Cobalt based catalysts are the most attractive choice for natural gas based FTS, because of their high activity, good stability and low water–gas shift activity[2]. FT catalyst activity and selectivity depend on several factors such as Co nanoparticle (NP) size, nature of the support, the presence of promoters, preparation methods and pre-treatment conditions[3]. FTS is a structure sensitive reaction[4] where catalyst performance strongly depends on NP size. In addition, the crystallographic phase of cobalt appears to have an influence on catalyst performance[5]. The phase of Co nanoparticles can be altered by the application of multistep preparation procedures that include a carburization step[6].

The present study has been performed on Re/Co/γ-Al_2O_3 and Co/α-Al_2O_3 catalysts and focuses on the effect of pre-treatment conditions on FTS. The pre-treatment includes a sequence of Reduction (H_2) - Carburization (CO) - Reduction (H_2) (RCR) on calcined Co-catalysts and is compared with a conventional one step H_2 reduction. The goal of the work is to unravel relationships between nanoparticle structure and the effect of different types of carbon on the FTS performance. <>2. Results

Catalysts were prepared by insipient wetness impregnation of Co and Re (if added) salts. All kinetic data reported are from lab scale fixed-bed reactors[7]. The materials prepared were extensively investigated using *ex situ* and *in situ* characterization methods. This included synchrotron based X-ray diffraction (XRD) and X-ray absorption near edge structure spectroscopy (XANES), thermo-gravimetric analysis (TGA) coupled with a mass spec (MS), H_2 Chemisorption, Raman spectroscopy (RS), transmission electron microscopy (TEM) and ^{13}C-NMR. The obtained average cobalt nanoparticle size was 12nm, far above the size sensitivity region. All the samples were reduced at 350°C prior to the carburization step. Subsequent carburization was performed at temperatures from 200°C to 300°C and eventually the final reduction step at temperatures from 200°C to 450°C. Air exposure of the carburized sample was evaluated for exploiting potential use of a carburized catalyst as a precursor.

As has been shown previously by several research groups the H_2 reduction of Co_3O_4 NPs, found in the calcined catalyst, leads to a complex metallic Co^0 structure consisting of both fcc-Co and hcp-Co where the fcc phase is dominant. After carburization of the reduced catalyst, under pure CO at 14 bar, the Co NPs adapt mainly a Co_2C orthorhombic structure, while significant amount of carbon is co-produced. After the final H_2 reduction step the Co_2C decomposes to Co NPs with a hcp-Co rich configuration independent of the temperature conditions applied, whereas a small contribution from fcc-Co can be observed (Figure 1). Clearly, the application of RCR has both a positive and negative influence at CO consumption rate and C_{5+} selectivity, depending on the applied temperature conditions of each step and air exposure of the sample. On the contrary, catalyst deactivation behaviour is similar independent on the applied activation procedure (Figure 1). Co NPs retain their structure even after 120 h on stream.

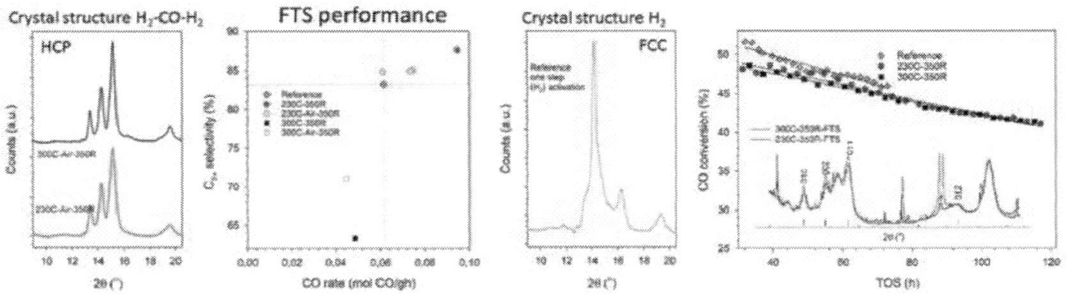

Figure 1. *Effect of the RCR activation protocol CO reaction rate (210 C, 20 bar, GHSV=15000 Nml/h•g, H_2/CO=2.1, TOS=24 h), C_{5+} selectivity (210 C, 20 bar, 46-48% CO conversion, H_2/CO=2.1) and crystal structure. The reference catalyst, which underwent conventional H_2 reduction activation, is denoted with green colour.*

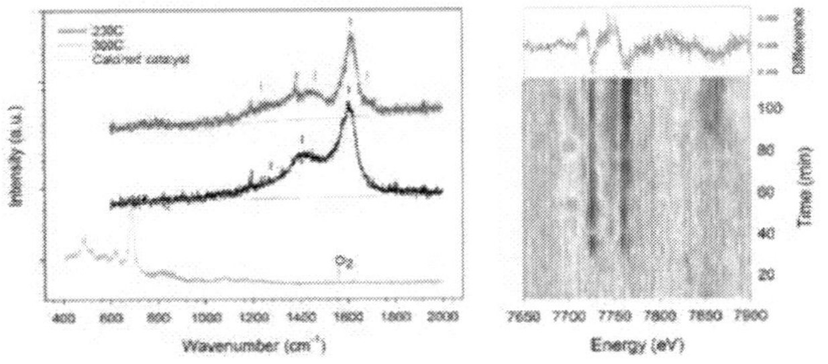

Figure 2. *(a) Raman spectra of the calcined catalyst and carburized catalyst samples at 230°C and 300°C (325 nm excitation and 0.3 mW laser power). (b) Contour plot of differential XANES (ΔXANES) of the normalized spectra shows the slow development of a phase formed on Co_2C; first minus last scan (up).*

Carbon formation is evident after carburization treatments at all temperatures investigated (Figure 2). At 230°C a less H_2 resistant carbon is produced than after carburization at 300 C. More than 5 types of carbon were identified to be present with respect to their structure and resistance to hydrogenation, in addition to the Co_2C that was detected by XRD. Atomic carbon remaining on cobalt lattice due to low temperatures of the final reduction step has a negative effect on FTS performance. The detrimental effect of more stable structures of carbon appears related to the secondary hydrogenation ability of the catalyst however; it merits further investigation since it apparently demonstrates both positive and negative aspects.

The effect of air exposure of the carburized sample was investigated. Apparently, although the bulk crystalline structure of the catalysts doesn't change, minor structural changes occur as observed by in situ XANES (Figure 2). A clear performance loss is observed upon exposure to air for the catalyst carburized at low temperature (230 C). Our results clearly demonstrate that the application of RCR activation protocols combined with O_2 passivation or air exposure with the idea of further use of the material as a precursor in CSTRs or other temperature limited applications probably are not applicable. <>3. Conclusions

· H_2 reduction of a calcined Re/Co/γ-Al_2O_3 catalyst lead to the formation of Co NPs with intergrown fcc-Co and hcp-Co, rich in stacking disorder. Fcc is the dominant phase. On the contrary, a three step RCR activation procedure gives hcp-rich Co NPs. Small fcc-Co is present depending on the applied conditions.

· Overall, the three step RCR activation procedure had a strong impact on the catalyst performance compared to the ordinary H_2 reduction process. FTS results show a rather diverse behavior depending on the pretreatment temperature. Both exceptional and very poor performance was recorded.

· Poor performance can be related partially to remaining carbon species. Carbon formation is evident after carburization treatments at all investigated temperatures 200°C to 300°C. Optimum carburization temperature lies between 200 and 265°C. The H_2 resistant carbon structures formed has primarily a negative influence in catalyst performance, with only one minor positive aspect.

· Air exposure of Co_2C leads to minor phase change of Co, observed by *in situ* XANES, that has an impact on catalyst performance.

· A coverage dependence on the crystallographic phase for the surface species could also be extracted. <>References

(1) Dry, M. E. In *Handbook of Heterogeneous Catalysis*; 2008; pp 2965–2993.

(2) Iglesia, E. *Appl. Catal. A Gen.* **1997**, *161*, 59–78.

(3) Rytter, E.; Tsakoumis, N. E.; Holmen, A. *Catal. Today* **2016**, *261* (2016), 3–16.

(4) Bezemer, G. L. L.; Bitter, J. H.; Kuipers, H. P. C. E.; Oosterbeek, H.; Holewijn, J. E.; Xu, X.; Kapteijn, F.; van Dillen, A. J.; de Jong, K. P. *J. Am. Chem. Soc.* **2006**, *128* (12), 3956–3964.

(5) Ducreux, O.; Rebours, B.; Lynch, J.; Roy-Auberger, M.; Bazin, D. *Oil Gas Sci. Technol. - Rev. l'IFP* **2008**, *64* (1), 49–62.

(6) Hofer, L. J. E.; Peebles, W. C. *J. Am. Chem. Soc.* **1947**, *69* (10), 2497–2500.

(7) Patanou, E.; Tsakoumis, N. E.; Myrstad, R.; Blekkan, E. A. *Appl. Catal. A Gen.* **2018**, *549*, 280–288.

554581 New Insights into the Catalytic Performance of Na-Promoted Fe/Al_2O_3 Catalyst in Fischer-Tropsch Synthesis

Monday, June 3, 2019
Texas Ballroom Prefunction Area (Grand Hyatt San Antonio)

Aliu A. Adeleke[1,2], Xinying Liu[1], Michela Martinelli[2], Diane Hildebrandt[1] and Burtron H. Davis[2], (1)Institute for the Development of Energy for African Sustainability (IDEAS), College of Science, Engineering & Technology, University of South Africa, Johannesburg, South Africa, (2)Center for Applied Energy Research (CAER), University of Kentucky, Lexington, KY

Na in Fe-based catalysts can be used to increase CO conversion and C_2-C_4 olefins and decrease conversion of H_2 and C_1 selectivity, but its behaviour at different reaction temperatures is of importance in Fischer-Tropsch synthesis (FTS). The dependency of reaction temperature on the C_1 formation rate, conversions of H_2 and CO, water-gas shift reaction, olefins and paraffins of C_2-C_4 and C_5-C_{12} hydrocarbons, and C_{13+} hydrocarbons for prepared Fe/Al_2O_3 and $FeNa/Al_2O_3$ catalysts was evaluated in a tubular fixed-bed reactor. This was carried out to investigate the effects of Na in Fe-based catalyst at different reaction temperatures (250 - 310 ^oC). The results show that Na effects in Fe-based catalysts to increase CO conversion and decrease H_2 conversion are dependent on the reaction temperature in FTS. Na-promoted Fe-based catalyst ($FeNa/Al_2O_3$) gave lower C_1 formation rates at some certain lower reaction temperatures (250 ^oC and 270 ^oC) when compared with the unpromoted Fe-based catalyst (Fe/Al_2O_3). The presence of Na in the Fe-based catalyst improved the C_1 formation rates at some certain higher reaction temperatures (290 ^oC and 310 ^oC). Na was found to hinder the selectivities towards C_2-C_4 paraffins and C_{13+} hydrocarbons including the oxygenates and improve the formation of C_2-C_4 olefins and C_5-C_{12} hydrocarbons at various reaction temperatures.

545371 Low-Temperature Catalytic Oxidative Coupling of Methane over Ce-W-O System Catalysts in the Electric Field

Monday, June 3, 2019
Texas Ballroom Prefunction Area (Grand Hyatt San Antonio)

Yuna Takeno, Shuhei Ogo, Ayaka Sato, Tomohiro Yabe and Yasushi Sekine, Applied Chem., Waseda Univ., Tokyo, Japan

Direct catalytic conversion of methane to ethylene by oxidative coupling of methane (OCM) is attracted our attention. However, OCM reaction demands high temperatures over 973 K because of the stable structure of methane. At such high temperature, non-selective oxidation cannot be avoided, which greatly reduces C_2 hydrocarbon selectivity. To solve this problem, we adopted a novel catalytic system, a catalytic reaction in an electric field, to activate methane at low temperature, such as 423 K. We previously reported that CeO_2 supported tetrabutylammonium salt of Keggin-type phosphotungstate (TBA-$PW_{12}O_{40}$/CeO_2) showed high OCM activity in the electric field even at 423 K. The purpose of this report is to investigate active site structure for OCM reaction in the electric field at low temperature, and to evaluate contribution of an electric field to OCM activity.TBA-$PW_{12}O_{40}$/CeO_2 after reaction in the electric filed was characterized with Raman spectroscopy. It showed the peaks of TBA-$PW_{12}O_{40}$ disappeared and the peaks of $Ce_2(WO_4)_3$ and WO_3 appeared. Therefore, catalytic activity tests of Ce-W-O system catalysts were conducted in the electric field. From the activity tests, $Ce_2(WO_4)_3$/CeO_2, $Ce_2(WO_4)_3$ and WO_3/CeO_2 showed OCM activity while WO_3 did not showed OCM activity. Then, the catalysts before and after reaction were characterized by Raman spectroscopy. The $Ce_2(WO_4)_3$/CeO_2, $Ce_2(WO_4)_3$ and WO_3/CeO_2 catalysts after reaction had the $Ce_2(WO_4)_3$ structure, however the WO_3 catalyst after reaction retained the WO_3 structure. In accordance with catalyst activity tests and Raman spectroscopy, $Ce_2(WO_4)_3$ was active site structure for OCM in the electric field. *In-situ* Raman spectroscopy revealed that $Ce_2(WO_4)_3$/CeO_2 structure was distorted and W-O bond weakened by application of the electric field. Consequently, it was suggested that the distorted W-O bond contributed to OCM activity in the electric field.

545558 Tungsten Polyoxide Catalysts for Oxidative Condensation of Methane

Monday, June 3, 2019
Texas Ballroom Prefunction Area (Grand Hyatt San Antonio)

Tolkyn S. Baizhumanova[1,2], **Svetlana A. Tungatarova**[1,2], Zauresh T. Zheksenbaeva[1,2], Kaysar Kassymkan[1], Gulnar N. Kaumenova[1,2] and Rabiga O. Sarsenova[1], (1)Laboratory of Organic Catalysis, D.V. Sokolsky Institute of Fuel, Catalysis and Electrochemistry, Almaty, Kazakhstan, (2)Chemistry and Chemical Technology faculty, al-Farabi Kazakh National University, Almaty, Kazakhstan

Tungsten Polyoxide Catalysts for Oxidative Condensation of Methane

Tolkyn S. Baizhumanova[1,2], Svetlana A. Tungatarova[1,2*], Zauresh T. Zheksenbaeva[1,2], Kaysar Kassymkan[1], Gulnar N. Kaumenova[1,2], Rabiga O. Sarsenova[1]

1 – D.V. Sokolsky Institute of Fuel, Catalysis and Electrochemistry, 142 Kunaev str., Almaty, 050010, Kazakhstan

2 - al-Farabi Kazakh National University, 71 al-Farabi ave., Almaty, 050040, Kazakhstan

* corresponding tungatarova58@mail.ru

Introduction

The increasing need for industry in ethylene and the largest reserves of natural gas make the process of oxidative conversion of hydrocarbons especially attractive. Additional opportunities are opened at embedding the process in the technological chain of production of valuable products from natural gas (motor fuels and their components, products and semi-products of organic and petrochemical synthesis) without intermediate deep cleaning of ethylene. The most active and selective catalysts provide 20% yield of C_2 hydrocarbons. Kazakhstan has also conducted research on the ODM. The obtained results are perspective and allow carrying out an economically and environmentally beneficial recycling of natural gas.

Materials and Methods

Experiments were carried out in flow type system with remove of products from cooling zone. Tubular quartz reactor with fixed bed of catalysts was used for investigation. The catalysts were prepared by impregnation of granulated carriers. Catalytic activity of supported catalysts prepared with using of Mo and W polyoxide compounds was investigated in OCM process. Tests were realized in reaction mixture, containing CH_4, O_2, inert gas with or without water vapor at the temperature of reaction 873-1173 K, contact time $\leq$ 0.5s, and atmospheric pressure.

The catalysts were prepared by impregnation of different carriers (aluminosilicate, alumina, cordierite, active carbon, zeolites, clinoptilolite, etc.) by HPC solutions. The concentration of active component is varied from 1 to 20%. The size of catalytic granules, contact time and relationship of components are varied widely. IRS, ESDR, XRD, TPR, DTA, and isotopic methods were used for physical-chemical investigation.

Results and Discussion

The reaction of oxidative condensation of methane (OCM) to C_2-hydrocarbons (C_2-HC) is the most rapidly developing among the various pathways of methane oxidation. Publications on this problem are much higher than the publication of other possible pathways of partial oxidation. The mechanisms of this process are proposed, but the processes that are embedded in the production do not exist yet. The maximum yield and selectivity of C_2-HC formation from CH_4 ($Y_{C2\text{-}HC}$ = 9.6% and 10.0%, $S_{C2\text{-}HC}$ = 80.0% and 57.6 %, S_{C2H4} = 51.5% and 39.2%, respectively) were achieved on the 5% and 0.5% $H_4SiW_{12}O_{40}$/AlSi catalysts. C_2H_4/C_2H_6 ratio in the products increases sharply toward the formation of C_2H_4 for each of the investigated catalysts with increasing of reaction temperature from 600 to 900°C and the transition from high-percentage to low-percentage catalysts. In some cases it was possible to obtain only C_2H_4 from CH_4 (at T = 800-900°C) and at T = 600°C - only C_2H_6.

The most active catalysts show the highest selectivity for C_2-hydrocarbons: 5% $H_4SiW_{12}O_{40}$/AlSi ($S_{\tilde{N}2H4}$ = 51.5%, $S_{C2\text{-}HC}$ = 80.0%) and 15% $H_3PW_{12}O_{40}$/AlSi (S_{C2H4} = 39.2%, $S_{C2\text{-}HC}$ = 60.2%). Catalysts based on W-HPC with P as central atom in all cases are similar in activity, but less selective in OCM reaction to ΣC_2-hydrocarbons compared to HPC with Si as central atom. The yield of C_2-hydrocarbons on Mo-containing catalysts are lower than on the W-containing catalysts (2.3% sample – 7.8%, 5.0% sample – 9.3%, 10.0% sample – 9.7%, T = 750-900°C). Carriers form series with respect to increasing influence on Y_{C2H4} and S_{C2H4} at 800°C: $S_{\tilde{N}2\acute{I}4}$, %: SiO_2 KSK-2,5 (23,3) < zeocare-2 (37,1) < zeolite NaA (37,2) < zeolite CaX (38,9) < SiO_2 KSM-5 (43,4) < AlSi (49,0) < periclase MgO (57,9) < pentasile CaAZ (58,5). Elements of the I and II groups of the Periodic System (Na, Mg, Cd, Cs) are optimal substituents for the proton in HPC of synthesis of ethylene from methane.

The influence of reaction mixture on the formation of C_2-HC was studied. The yield and selectivity of formation of C_2H_4 increases with the ratio of $CH_4 : O_2$ from 1 : 1 to (1.5-30) : 1. The composition of mixture (mol: CH_4 – 0.0018-0.0031, O_2 – 0.0013, inert gas – 0.008-0.004) is optimal for synthesis of C_2H_4 from CH_4. Introduction of water vapor ($CH_4 : O_2$ = 1 : 0,21, mol) in the reaction mixture has a positive effect on reaction. The results show that the OCM process can be optimized through a rigorous selection of process parameters of reaction and improvement the composition of supported catalysts. The optimal reaction conditions by varying the composition of HPC, the nature of carrier, the ratio of components of reaction mixture and reaction temperature were established. $[PW_{12}]$-HPA at replacement of proton to the Na, Cs as well as $[SiW_{12}]$-HPA at replacement of proton to the Na, Mg, Cd possess the highest activity at 750-850°C and when using the Si-containing oxides as carrier, such as pentasile, SiO_2, or AlSi at supporting of 0,5-15% HPC on carrier and carrying out the experiment in the reaction mixtures: CH_4 – 0.0018-0.0031, O_2 – 0.001, mol, $CH_4 : H_2O = 1 : 0,21$, Ar is residual, t = 0,38-0,46 s. Implementation of OCM process on polyoxide systems in installations with one or two reactors (Table 1), as well as on polyoxide catalysts of the complex composition allowed to increase the conversion of CH_4 to 46.0%, the yield of C_2H_4 - up to 18.5% at $\Sigma Y_{C2\text{-}HC}$ to 22,0%.

Preliminary mathematical analysis of the kinetic results of reaction was conducted with a view to recommending the optimal design of an industrial reactor for the developed OCM process and optimal conditions for the realization of the industrial process on catalysts. It was offered the presumed optimal model of the industrial reactor. Analysis of results showed that the use of industrial-scale multi-shelf (5-7 layers) adiabatic reactor is the best. Productivity of the process according to preliminary estimates may be up to 3.2-3.4 tons of ethylene per day from 1 ton of catalyst.

Thus, the catalysts have been developed from research and the optimal technological parameters of the oxidative conversion of natural and associated gas into ethylene on molybdenum polyoxometalates have been determined.

Thermal stability of supported catalysts based on the 12th series W-HPC in a high temperature gas phase OCM process was examined by XRD and TPR methods under influence of the medium ($O_2 + H_2O + Ar + CH_4$) at T = 20–800°C. It is caused by preservation of fragmented formations of HPC at temperatures up to 800–850°C on a carrier, light renewability under the action of reaction medium containing water vapor, or vapor-air treatment of reactive oxygen-containing fragments of HPA.

The process of OCM into ethylene can be intensified by creating a technology that is close to non-waste technology by additional oxidative dehydrogenation of the product of OCM - ethane to ethylene, and by adding additives to CH_4 of other alkanes in order to increase the conversion of CH_4 to C_2H_4 and C_2H_6.

On the proposed supported catalysts based on 12th series W-HPC (their compositions are close to those used in the OCM) for the concomitant process of oxidative dehydrogenation of C_2H_6 (ODE), the C_2H_4 yield in the presence of water vapor is 41.6-51.3% or 527-641 g C_2H_4/m^3 C_2H_6, selectivity for C_2H_4 reaches 86–94%.

This publication has been made within the project Grant No AP05133881, which is funded by the Ministry of Education and Science of the Republic of Kazakhstan.

545600 Performance of Ce-Added Na_2WO_4 Catalyst for Direct Conversion of Methane to Ethylene

Monday, June 3, 2019
Texas Ballroom Prefunction Area (Grand Hyatt San Antonio)

Heeyeon Kim[1], Jang-won O[1], Guk-hyun Kwon[1], Dae-keun Lee[2] and Ji-haeng, Yu[3], (1)Energy Materials Laboratory, Korea Institute of Energy Research, Daejeon, Korea, Republic of (South), (2)Thermal Energy System Research Laboratory, Korea Institute of Energy Research, Daejeon, Korea, Republic of (South), (3)Separation and Conversion Materials Laboratory, Korea Institute of Energy Research, Daejeon, Korea, Republic of (South)

The production of key petrochemical intermediates using methane as a raw material using direct coupling of methane (DCM) reaction has been studied for the past 30 years as a new technology replacing the existing indirect conversion processes which require tremendous capital investment and result in unfavorable project economics. Recently, as the mining unit price is lowered with the development of shale gas exploitation technology, the DCM becomes more interested by catalysis researchers. However, the optimal catalyst composition and structure have not been clearly presented to date. Also, the development of a catalyst having high performance and durability with a exceeding 30% ethylene yield, which is a commercially viable value, has not been reported yet. The reason is as follows: 1) thermodynamically limited conversion, 2) low selectivity to ethylene (by the oxidation of methane), 3) serious catalyst deactivation via hot spot and coke deposition. Numerous attempts have been made to overcome these problems in DMC reaction, and MnO_2-promoted Na_2WO_4/SiO_2 catalysts have been known to have the highest performance and durability so far. However, even in the case of MnO_2-promoted Na_2WO_4 catalyst, the commercially viable ethylene yield has not been satisfied and the optimal composition and structure of the catalyst and the deactivation mechanism of the catalyst are still under study.

In this study, we have developed catalysts with excellent catalytic performance with superior ethylene selectivity in oxidative coupling of methane (OCM) reaction. Our CeO_2-promoted Na_2WO_4 catalyst has shown high methane conversion with superior ethylene selectivity compared with MnO_2-promoted catalysts. However, as a result of 72 h-reaction, the deactivation rate of CeO_2-promoted catalyst was slightly faster than that of MnO_2-promoted catalyst. Therefore, several techniques have been applied for the long term stabilization of CeO_2-promoted Na_2WO_4 catalysts. Our research group has also been working on the deactivation mechanism of OCM catalysts using computational chemistry and atomic scale characterization techniques in order to secure the long-term stability as well as the high methane conversion in OCM using CeO_2-promoted Na_2WO_4 catalysts.

546391 Na_2WO_4/Mn/SiC Catalyst for Oxidative Coupling of Methane

Monday, June 3, 2019
Texas Ballroom Prefunction Area (Grand Hyatt San Antonio)

Ji Eun Kim[1], Suji Yoon[2], La Hee Park[2,3], Jeong-Myeong Ha[2,3] and **Eun Duck Park**[1,4], (1)Department of Energy Systems Research, Ajou University, Suwon, Korea, Republic of (South), (2)Clean Energy Research Center, Korea Institute of Science and Technology, Seoul, Korea, Republic of (South), (3)Green School (Graduate School of Energy and Environment), Korea University, Seoul, Korea, Republic of (South), (4)Department of Chemical Engineering, Ajou University, Suwon, Korea, Republic of (South)

Since the oxidative coupling of methane (OCM) is a high-temperature and highly exothermic reaction, the hot spot on the catalyst surface is easily formed during the reaction. The control of heat formed during the reaction is an important technique because the hot spot on the catalyst surface can degrade catalytic activity. Herein, Mn_2O_3-Na_2WO_4/SiC was prepared by using SiC with high thermal conductivity and good stability at high temperatures, and the catalyst was applied as a catalyst for OCM reaction. Two different Mn_2O_3-Na_2WO_4/SiC catalysts were prepared by the wet impregnation method using SiC supports with different particle sizes. For comparison, Mn_2O_3-Na_2WO_4/SiO_2 was also prepared by the same method. The catalysts were analyzed by nitrogen adsorption-desorption, XRD, FE-SEM, and HR-TEM. The properties of catalysts were evaluated by OCM reaction. The OCM reaction was carried out at a space velocity of 10,000 h^{-1} in a fixed-bed reactor ($CH_4/O_2/N_2$ = 3/1/1, v/v/v). As a result, Mn_2O_3-Na_2WO_4/SiC catalyst showed 2 % increase in methane conversion and 2.3 % increase in C_2 yield than Mn_2O_3-Na_2WO_4/SiO_2 at 800 .

551302 Tuning Catalyst Electronic Structure & O_2 Activity for OCM Reactions

Monday, June 3, 2019
Texas Ballroom Prefunction Area (Grand Hyatt San Antonio)
Shanti Kiran Nayak, Angelica Benavidez, Lok-kun Tsui and Fernando Garzon, Chemical and Biological Engineering, University of New Mexico, Albuquerque, NM

Natural gas serves as an excellent bridge fuel as we transition to a low carbon economy. However, the high C-H bond strength and the directionality of the sp^3 orbital in the methane molecule make it difficult for methane to form bonds with other species or catalytic surfaces. Hence, high temperatures and effective catalysts are required to activate and transform methane into useful chemicals, products, or electricity. [1]

Tuning heterogeneous catalytic activity by electrochemical methods provides enhanced control of catalyst redox state and nanocatalyst activity. Michaels and Vayenas [2] demonstrated the viability of electrochemical process for dehydrogenation of ethylbenzene to styrene. The presence of an electrical potential can dramatically alter the energy profile of the chemical reaction by lowering the activation barrier for the forward reaction. [1] The oxygen content of non-stoichiometric oxide catalysts can be controlled electrochemically, thus altering catalytic activity. We are developing electrochemical micro reactors to investigate oxidative coupling of methane reactions with better selectivity. By designing cell capacitance and resistivity, the reactor can be switched at high rates and this may change the methane conversion selectivity. The micro reactors are fabricated using ceramic additive manufacturing. We have made progress in the additive manufacturing process for lanthanum manganite, YSZ and Pt for use in the micro reactors. The ceramic additive manufacturing process includes characterization of the starting powder material, milling of powder to attain a uniform particle size, mixing with vehicle and thinner resins, characterization of the rheological properties and sintering of the printed material. The material is also characterized post-sintering and this helps in fine-tuning the sintering temperature parameters.

There is a large body of literature regarding the use of lanthanum perovskites as catalysts for the conversion of methane to CO. Due to their high activity and thermal stability, perovskites have been gaining importance as catalysts for hydrocarbon oxidation over active noble metals such as Pt and Pd which are expensive and do not resist operating temperatures above 850 K. Ciambelli et. al [3] conducted catalytic methane combustion experiments over $AFeO_3$ (A = La, Nd, Sm) and $LaFe_{1-x}Mg_xO_3$ perovskites in a quartz down flow reactor electrically heated in a three zones tube furnace. Their tests showed that all $AFeO_3$ samples gave methane conversion with 100% selectivity to CO_2 below 973 K. The order of activity towards methane combustion was La > Nd > Sm. Further experiments involve the study of the thermodynamic stability and redox behavior of our printed materials using high temperature cyclic voltammetry techniques, heterogeneous conversion of methane to CO using our printed $La(M)MnO_x$ as catalysts and printing and characterization of ceria and rare earth doped ceria films as the electrolyte/catalyst support. The heterogeneous catalysis experiments occur under both steady-state polarization conditions and high frequency switching, investigating the influence of the rapidly changing catalyst activity on the product species.

References:
(1) Gür, T. M. Energy Combust. Sci. 2016, 54, 1–64.
(2) Michaels, J. N.; Vayenas, C. G. J. Catal. 1984, 85 (2), 477–487.
(3) Ciambelli, P.; Cimino, S.; De Rossi, S.; Lisi, L.; Minelli, G.; Porta, P.; Russo, G. Appl. Catal. B Environ. 2001, 29 (4), 239–250.

545162 The Hierarchical Porous W-Mn/SiO_2 Catalyst for Oxidative Coupling of Methane

Monday, June 3, 2019: 5:24 PM
Texas Ballroom EF (Grand Hyatt San Antonio)

Ning Zhang[1], Jian Yang[2], Huahua Zhao[2], Liang Yan[2], Huanling Song[2] and Lingjun Chou[2], (1)State Key Laboratory for Oxo Synthesis and Selective Oxidation, Lanzhou Institute of Chemical Physics, Lanzhou, China, (2)Lanzhou Institute of Chemical Physics, Lanzhou, China

The hierarchical porous W-Mn/SiO_2 catalyst for oxidative coupling of methane

Ning Zhang[ab], Jian Yang[a], Huahua Zhao[a], Liang Yan[a], Huanling Song[a], Lingjun Chou[a*]

[a] State Key Laboratory for Oxo Synthesis and Selective Oxidation, Lanzhou Institute of Chemical Physics, Chinese Academy of Sciences, Lanzhou 730000, P. R. China.

[b] University of Chinese Academy of Sciences, Beijing 100049, P. R. China.

*Corresponding author, Tel: +86 931 4968136; Fax: +86 931 4968129

Email: ljchou@licp.cas.cn

Introduction

Recently, three-dimensionally ordered macroporous and mesoporous (3OM)[1] materials, have captured the interest of scientists and engineers owing to their potential applications in the realms of catalysis, separations and photonic crystals. The macropores could provide interconnected framework and this was beneficial to gas reaction. The reactant molecule could enter into and pass through the channels quickly. The reaction heat could be released by macroporous channels. Until now, few researches investigate about 3OM SiO_2 materials as catalyst support in oxidative coupling of methane (OCM)[2], which has attracted renewed interest. However, it still has great challenges: first, the selectivity of C_2 is low; second, reaction has hot spots at high temperature. Therefore, development of an efficient catalyst for OCM reaction remains to be the most crucial step in the commercialization of this process.

Now, we prepared 3OM SiO_2 as catalyst support to improve OCM catalyst activity. By the comparison of the different support catalysts SBA-15, KIT-6, Tianjin- SiO_2, the 3OM SiO_2 catalyst owns the excellent catalytic performance at 850°C.

Experimental

The 3OM SiO_2 was prepared using a method reported previously. PMMA crystal as macroporous template was synthesized via a soap-free-emulsion polymerization[3]. In a typical synthesis, 15mL monomer MMA was washed three times with a NaOH solution (100mL, 10 wt %) and distilled water respectively to remove any trace of the fire retardant. Then, 11.3g of the washed MMA and 150g of distilled water were added to a 500mL round bottom flask and stirred for an hour. Then, 12mL (0.0005g mL^{-1}) $K_2S_2O_8$ was added to the solution and the mixture was constantly reacted at 70 °C for 7h under a nitrogen atmosphere with mild stirring until a milky solution appeared. The milky solution was centrifuged and dried at 80°C to obtain PMMA. 4g HCl (0.2M) and 6.4g F127 were added into EtOH (30mL) at 40°C for an hour. Then, 8.32g TEOS was added into the above solution to stir 1h to obtain a homogeneous precursor of silicon. The PMMA template was added into Si precursor to soak five day at 40°C. Then, excess solution was removed by filtration. The resulting sample was dried at 120°C overnight and calcined at a heating rate of 1 °C min^{-1} to 600 °C for 5h.

Results and Discussion

Fig.1 shows the typical TEM image of the W-Mn-SiO_2. Fig.2 shows the HAADF image of the W-Mn-SiO_2. It was clearly observed that the catalysts exhibited well developed interconnected networks of the ordered macroporous and mesoporous structures.

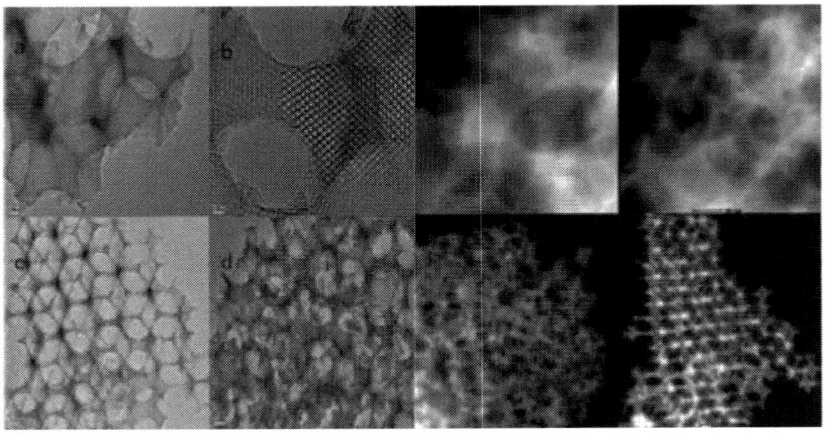

Fig.1 TEM of W-Mn-SiO_2 calcined in 600°C Fig.2 HAADF of W-M-SiO_2 calcined in 600°C

The catalytic performance was listed in table.1. At 850°C, 3OM SiO_2 catalyst showed the optimum catalytic activity. The CO_2-TPD indicated that 3OM SiO_2 catalyst enhanced the moderate-strong basic sites at 566°C. The O_2-TPD revealed the 3OM SiO_2 catalyst owned the highest quantity of lattice oxygen.

Table.1 Catalytic performance of the different support catalysts in 850°C

Catalyst	Conv. (%)	Selectivity (%)		Yield (%)
	CH_4	C_2	C_2H_4	C_2
SBA-15	24.70	61.71	45.19	15.24
3OM SiO_2	27.65	60.37	43.43	16.69
Tianjin SiO_2	24.93	62.57	43.13	15.60
KIT-6	24.49	61.33	41.66	15.02

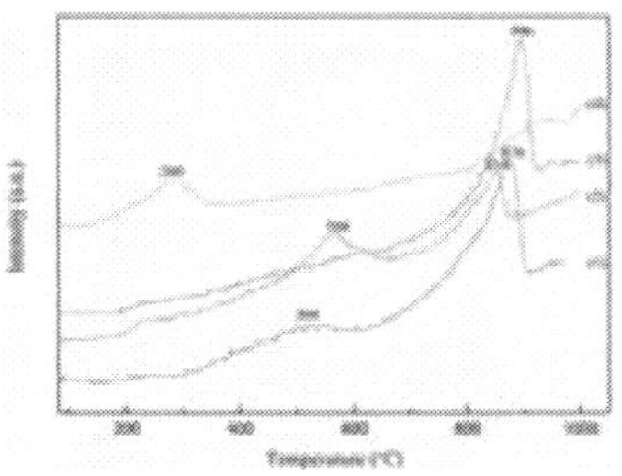
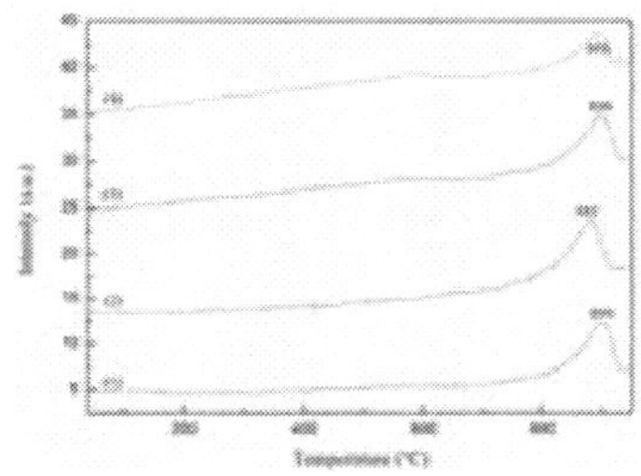
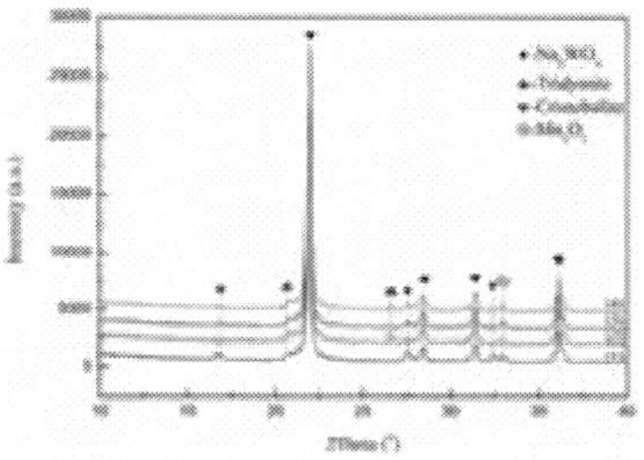

Fig. 3 CO_2-TPD of the catalysts Fig. 4 O_2-TPD of the catalysts Fig. 5 XRD patterns of the catalysts

Conclusion

3OM W-Mn-SiO_2 materials own interconnect macroporous structure and mesopores in macroporous wall, which could make W and Mn more dispersed in channels. The hierarchical porous structure also could enhance the moderate-strong basic sites and O^{2-} quantity of catalyst. This was favourable for OCM reaction. These results indicated that the 3OM SiO_2 catalyst was a promising class of catalysts for the OCM reaction.

REFERENCES

[1] H. M¨ªguez, F. Meseguer, C. L¨®pez, A. Mifsud, J.S. Moya, L. V¨¢zquez, Langmuir : the ACS journal of surfaces and colloids 13 (1997) 6009-6011.

[2] G.E. Keller, M.M. Bhasin, Journal of Catalysis 73 (1982) 9-19.

[3] Z. Wang, A. Stein, Chemistry of Materials 20 (2007) 1029-1040.

544256 The Catalytic Performance of the Mn-Na-W/SiO_2 Catalyst in Oxidative Coupling of Methane: Surface Structure and Further Additives

Monday, June 3, 2019: 5:36 PM
Texas Ballroom EF (Grand Hyatt San Antonio)
Naseem S. Hayek and Oz M. Gazit, Chemical Engineering, Israel Institute of Technology - Technion, Haifa, Israel

The Catalytic Performance of the Mn-Na-W/SiO_2 Catalyst in Oxidative Coupling of Methane: Surface Structure and Further Additives

Naseem S. Hayek* and Oz M. Gazit

Wolfson Faculty of Chemical Engineering, Technion – Israel Institute of Technology, 3200003, Israel

E-mail: naseemha@technion.ac.il; Fax: +97248292850

Oxidative coupling of methane (OCM) has attracted much attention since 1982.[1] This is because OCM is a direct and exothermic route for producing ethylene (Eq. 1), which is a vital building block in the chemical industry. Unfortunately, this process is still industrially inapplicable due to the lack of active and stable catalyst, which overcomes the formation of COx (Eq. 2), the thermodynamically favorable products.

1. $CH_4 + 1/2O_2 \rightarrow 1/2C_2H_4 + H_2O \qquad \Delta H^0_{800^\circ C} = -139\ kJ/mol$

2. $CH_4 + 2O_2 \rightarrow CO_2 + 2H_2O \qquad \Delta H^0_{800^\circ C} = -801\ kJ/mol$

Mn-Na-W/SiO_2 is one of the most studied catalysts for OCM.[2] Its high stability (>500 h) and C_2 selectivity (>70%) make it a good candidate for practical applications. However, the highest C_2 yield reached so far in a single-pass tubular reactor is less than 26%,[3] restricting an economically feasible process.

In the preparation of the Mn-Na-W/SiO_2 catalyst, a phase transition of the impregnated silica from amorphous to -cristobalite occurs during calcination. This phase transition was shown to be critical for obtaining an active catalyst, which is thought to be due to better dispersion and stabilization of the active species. However, despite intensive research on this catalyst, the exact form of the active site and the surface properties that makes this catalyst active and selective are still not completely clear. Also, only a few studies were focused on the addition of more components into the Mn-Na-W/SiO_2 catalyst, but no systematic studies were performed before.

The work here covers two main topics: 1) The effect of the Mn-Na-W/SiO_2 surface parameters on its catalytic performance. 2) The effect of further additives on the catalytic performance of the Mn-Na-W/SiO_2 catalyst. In the first part, we synthesized a set of catalysts with different properties using different silica precursors (i.e. dealuminated β-zeolite, mesoporous SBA-15, and nonporous fumed silica as silica precursors), together with two different Mn precursors that behave differently (i.e. Mn acetate and Mn nitrate).[4,5] Characterizing the catalysts by inductively coupled plasma-optical emission spectroscopy, N_2-physisorption, X-ray diffraction, high-resolution scanning electron microscopy-energy-dispersive spectroscopy, X-ray photoelectron spectroscopy, and catalytic testing enabled us to identify critical surface parameters that govern the activity and C_2 selectivity of the catalyst. Although the current paradigm views the phase transition of silica to α-cristobalite as the critical step in obtaining dispersed and stable metal oxide sites, we show that the choice of precursors is equally or even more important with respect to tailoring the right surface properties. Moreover, surface porosity, metal dispersion, and Na_2WO_4 particle size were shown to be key factors in determining the activity and selectivity of the Mn-Na-W/SiO_2 catalyst. Finally, combining our results with prior studies lead us to single out that the active center is composed of two adjacent Mn_2O_3 and Na_2WO_4 species.

In the second part, set of Mn-Na-W/SiO_2 catalysts were synthesized with different additives (i.e. Ti, Sn, Zr, Ge, Ce, Nb, Fe, Co, and Cu), characterized using the previously mentioned techniques and tested under OCM conditions. Part of the additives promoted the catalytic performance of the conventional catalyst; however, they suffered from severe deactivation, making it questionable whether it is worth adding more components to the conventional catalyst as has been done before or not. In the presentation, interesting results from synthesis, characterization and catalysis will be discussed.

References

(1) Keller G, E.; Bhasin, M. M. Synthesis of Ethylene via Oxidative Coupling of Methane. *J. Catal.* **1982**, *73* (1), 9–19.

(2) Arndt, S.; Otremba, T.; Simon, U.; Yildiz, M.; Schubert, H.; Schomäcker, R. Mn–Na_2WO_4/SiO_2 as Catalyst for the Oxidative Coupling of Methane. What Is Really Known? *Appl. Catal. A Gen.* **2012**, *425–426*, 53–61.

(3) Zavyalova, U.; Holena, M.; Schlögl, R.; Baerns, M. Statistical Analysis of Past Catalytic Data on Oxidative Methane Coupling for New Insights into the Composition of High-Performance Catalysts. *ChemCatChem* **2011**, *3* (12), 1935–1947.

(4) Hayek, N. S.; Lucas, N. S.; Warwar Damouny, C.; Gazit, O. M. A Comparative Study of Precursor Effect on Manganese Post-Synthetic Incorporation into the T-Sites of Dealuminated Beta-Zeolite. *Microporous Mesoporous Mater.* **2017**, *244*, 31–36.

(5) Hayek, N. S.; Lucas, N. S.; Warwar Damouny, C.; Gazit, O. M. Critical Surface Parameters for the Oxidative Coupling of Methane over the Mn–Na–W/SiO_2 Catalyst. *ACS Appl. Mater. Interfaces* **2017**, *9* (46), 40404–40411.

545370 Effect of the Modification of Pd/CeO_2 Catalyst on Steam Reforming of Methane at Low Temperature in an Electric Field

Monday, June 3, 2019
Texas Ballroom Prefunction Area (Grand Hyatt San Antonio)
Ayako Takahashi, Reona Inagaki, Takuma Higo, Shuhei Ogo and Yasushi Sekine, Applied Chem., Waseda Univ., Tokyo, Japan

These days, hydrogen atracts a lot of attention as a clean secondary energy taking the place of petroleum because only water is emitted from the combustion of hydrogen. Hydrogen demand is expected to increase from the viewpoint of environmental protection and an increasing use of fuel cell for vehicles or home systems. Methane steam reforming (MSR) is a primary process to produce hydrogen. Generally, this reaction is conducted at high temperature over 923 K because of the thermodynamic equilibrium and the stability of methane molecule. Such high temperature catalytic processes cause many problems such as sintering of catalyst, damage to plants, necessity for multi-heat exchangers and discarding low-temperature waste heat (under 500 K). Especially, heat loss leads to a decline in energy efficiency. Therefore, it is required to utilize effectively low-temperature waste heat. According to that, the new MSR catalytic process which proceeds at low temperature is demanded.

We have investigated a new reaction process, catalytic reaction in an electric field (*Electreforming*). MSR proceeds even at 473 K by applying electric field. Furthermore, the results indicated that the proton hopping via adsorbed water by *Grotthuss* mechanism emerged on the catalyst in the electric field, and it was revealed that the proton hopping promoted the dissociative adsorption of methane, the rate-limiting step of MSR, by the collision of proton with methane.

In this study, we investigated the effects of support on MSR activity of Pd supported catalysts in the electric field. MSR in the electric field was demonstrated using various CeO_2-based support loaded Pd catalysts. The characteristics of CeO_2 support were controlled by doping or loading of various metal oxides.

545750 Promising Combined Pyrochlore-Perovskite Catalysts to Syngas Production

Monday, June 3, 2019
Texas Ballroom Prefunction Area (Grand Hyatt San Antonio)

Adriana P. Ramon, Engenharia Química, Universidade Federal de São Carlos, São Carlos, Brazil, Elisabete M. Assaf, Instituto de Química de São Carlos/Universidade de São Paulo, São Carlos, Brazil and **José M. Assaf**, Universidade Federal de São Carlos, Engenharia Química, Brazil

The dry reforming of methane (DRM) is a auspicious process to convert biogas or natural gas into $CO + H_2$ syngas. The generation of syngas is of great industrial interest in particular in the production of liquid fuels via Fischer-Tropsch. The major problem in DRM is the catalyst deactivation by the formation of carbon deposits over the metal sites. To overcome this problem the catalyst $La_2Ce_{2-x}Ni_xO_{7-\lambda}$-$LaNiO_3$ was synthesized using coprecipitation (CP) and hydrothermal (HT) methods. These catalysts were characterized and tested catalytically in the DRM at temperatures between 600 to 900°C, and further, tested at 850°C, over 8 hours. The X-ray diffraction analysis showed, in both synthesis methods, the formation of NiO in addition to $LaNiO_3$ and $La_2Ce_{2-x}Ni_xO_{7-d}$. In the temperature programmed reduction analysis, two main reduction peaks characteristic of the perovskite phase were observed with higher reduction temperatures in the HT sample. Furthermore, this catalyst had higher CH_4 and CO_2 conversions at all temperatures, and unlike CP the HT catalyst was active at relatively low temperature of 600°C. In the short term stability at 850°C, the HT catalyst showed very close CH_4 and CO_2 (~90%) conversions suggesting that the reaction is occurring stoichiometrically. In contrast, the CP catalyst showed higher CO_2 (~75%) than CH_4 (~62%) conversion, due the simultaneous occurrence of reverse water gas shift reaction. Regardless of the synthesis method, all catalysts were stable during 8h of time on stream. Furthermore, thermogravimetric analysis after short-term stability tests revealed no significant carbon deposition (<2%). The results indicate that the pyrochlore-perovskite system is active and stable to catalyze the DRM, since it can effectively suppress the deposition of carbon on the surface of the catalyst.

545713 Syngas Generation By CFD Simulation of Kinetic Model and Heat Transfer in Steam Methane Reformer

Monday, June 3, 2019: 5:21 PM
Texas Ballroom A (Grand Hyatt San Antonio)

Saiedeh Arabi, Sandeep Patil and Brian Dennis, Mechanical & Aerospace Engineering, The University of Texas at Arlington, Arlington, TX

This research illustrates the computational fluid dynamics (CFD) simulation to show the effect of inlet Steam/Natural gas flow rate and inlet temperature on efficiency of steam methane reformer (SMR). In syngas production, the approach to get higher ration of H_2/CO is (SMR) which requires a high endothermic reaction and heat source. This work carried out on the reactor design of (SMR) to achieve optimal dimensions for the highest rate of reaction. H_2O/NG, feed temperature, syngas selectivity, and catalyst productivity are the parametric studies in this study. In addition, the length and diameter of chamber are determined as a variable to investigate the temperature distribution through the r and z directions in reactor. To prove the accuracy of developed model, the simulation results are compared with the available plant data at steady state condition. The pilot plant tests were conducted in a reactor chamber at temperatures of 1073–1273 K, steam of 0-3 lb/h, and 3-5 kg Ni catalyst bed. In this study the dropping temperature was observed between 250-300 K.

545744 C-H Activation Chemistry at Metal Phosphide Surfaces

Monday, June 3, 2019
Texas Ballroom Prefunction Area (Grand Hyatt San Antonio)

Jeonghyun Ko, Jessica A. Muhlenkamp, William F. Schneider and Jason C. Hicks, Chemical and Biomolecular Engineering, University of Notre Dame, Notre Dame, IN

Metal phosphides have the potential to combine the intrinsic catalytic reactivity of noble metals with the thermal robustness of ceramic materials. Further, like intermetallics, metal phosphides tend to be ordered and thus their surfaces present well defined and potentially tunable reactive sites. In this work, we combine computational models and experimental synthesis and characterization to evaluate the potential of metal phosphides for catalytic dehydrogenations of light hydrocarbons. Density functional theory (DFT) methods are used to compare ethane dehydrogenation pathways over Ni2P, a representative phosphide known to be active for dehydrogenations, with Ni itself, a material well known to be highly susceptible to coking. Calculations show that the Ni ensembles at the Ni2P surface are metallic and that the adsorption energies of dehydrogenation intermediates are similar to those on Ni. However, computed reaction barriers show that site isolation imposed by the presence of surface phosphorus inhibits reaction pathways that lead toward coke precursors. Ni2P catalysts are prepared on an SiO2 support and are found to exhibit higher selectivity to desired olefin products than do similarly prepared Ni catalysts in both ethane and propane dehydrogenation. The results highlight the promise of these materials and point the way towards further catalyst optimization by control of support, particle size, and composition.

545787 Effect of Ba Addition on the Catalytic Activity of Ga-Al2O3 for Ethane Dehydrogenation

Monday, June 3, 2019
Texas Ballroom Prefunction Area (Grand Hyatt San Antonio)

Yukiko Hosono[1], Hikaru Saito[1], Hirofumi Seki[1], Shun Maeda[2], Shuhei Ogo[1], Kunihide Hashimoto[2] and Yasushi Sekine[1], (1)Applied Chem., Waseda Univ., Tokyo, Japan, (2)Kubota Corp., Hirakata, Japan

Steam cracking of ethane has been an attractive process because ethane got a cheap feedstock. However, this process is non-catalytic and energy-intensive because of high reaction temperatures (> 1073 K) and periodic decoking to remove carbonaceous deposit. Thus, introduction of dehydrogenation catalysts to the reactor is an effective way to lower the reaction temperature and suppress carbon deposition. In our previous study, Ga supported on α-Al_2O_3, which is a model of cracking tubes (AFTALLOY, Kubota Corp.), exhibited high ethylene yield at 973 K. In this study, we examined addition of second metals to Ga/α-Al_2O_3 to enhance the catalytic performance.

Ga-xM/α-Al_2O_3 catalysts were prepared by a co-impregnation method. Note that "M" and "x" denote a second metal and M/Ga molar ratio, respectively. Loading amount of Ga was controlled to be 5wt%. Catalytic activity for ethane dehydrogenation was evaluated in a fixed bed reactor at 973 K and atmospheric pressure. Ethane, water and balanced nitrogen ($C_2H_6 : H_2O : N_2 = 1.0 : 1.4 : 5.5$) were fed at a flow rate of 143 mL min^{-1}.

From the results of activity tests, Ga-0.1Ba/α-Al_2O_3 exhibited the highest selectivity to ethylene (98%) although the activity of Ga-0.1Ba/α-Al_2O_3 was comparable to Ga/α-Al_2O_3. Selectivity to ethylene was improved thanks to the suppression of CO and CO_2 formation. Originally, they are formed through oxidation of carbonaceous deposit by H_2O. Temperature programmed oxidation revealed that the amount of carbon deposition on Ga-0.1Ba/α-Al_2O_3 was smaller than that on Ga/α-Al_2O_3. Then, the electronic state and coordination environment of Ga were measured by Ga K-edge X-ray absorption fine structure spectroscopy. As a result, a part of Ga might form Ga-Ba oxides although Ga mainly existed as β-Ga_2O_3. Therefore, coke formation would be suppressed because active sites of β-Ga_2O_3 for coke formation were poisoned by Ba addition.

545082 The Role of Activated Carbon Supported CeO_x and VO_x Phases during Oxidative Dehydrogenation of Propane to Propylene Using CO2 As a Soft Oxidant

Monday, June 3, 2019
Texas Ballroom Prefunction Area (Grand Hyatt San Antonio)

Petar Djinovic[1], Janez Zavasnik[2,3], Ivo Jerman[4] and Gregor Zerjav[1], (1)Department for Environmental Sciences and Engineering, National Institute of Chemistry, Ljubljana, Slovenia, (2)Department of Structure and Nano-/ Micromechanics of Materials, Max-Planck-Institut für Eisenforschung, Dusseldorf, Germany, (3)Centre for Electron Microscopy and Microanalysis, Jožej Stefan Institute, Ljubljana, Slovenia, (4)Department of Materials Chemistry, National Institute of Chemistry, Ljubljana, Slovenia

The role of activated carbon supported CeO_x and VO_x phases during oxidative dehydrogenation of propane using CO_2 as a soft oxidant

Petar Djinovic[a*], Janez Zavašnik[b,c], Ivo Jerman[d], Ervin Šest[d] and Gregor Žerjav[a]

[a]*Department for Environmental Sciences and Engineering, National Institute of Chemistry,*
Hajdrihova 19, SI-1001 Ljubljana, Slovenia

[b]*Centre for Electron Microscopy and Microanalysis, Jožef Stefan Institute, Jamova cesta 39,*
SI-1000 Ljubljana, Slovenia

[c]*Max-Planck-Institut für Eisenforschung GmbH, Max-Planck-Straße 1, DE-40237 Düsseldorf, Germany*

[d]*Department of Materials Chemistry, National Institute of Chemistry, Hajdrihova 19, SI-1001 Ljubljana, Slovenia*

*petar.djinovic@ki.si

Keywords: CeO_2, VO_x, oxidative dehydrogenation, propylene, CO_2 activation, lattice oxygen

1. Introduction

Propylene is a commodity chemical used for synthesis of acetone, acrylic acid, acrylonitrile, polypropylene plastic, etc. A growing gap between the demand and production of propylene in the last years requires additional propylene production pathways. Catalytic oxidative dehydrogenation of propane using O_2 (O_2-ODH) or

CO_2 (CO_2-ODH) emerge as possible alternatives. The use of a soft oxidant (such as CO_2, reaction 1) could offer the benefits of higher propylene selectivity, less stringent active site architecture (compared to O_2-ODH) and an additional chemical reaction for CO_2 conversion.[1,2]

$$C_3H_8 + CO_2 \rightarrow C_3H_6 + CO + H_2O \qquad H_{298K}=167\ \text{kJ/mol} \qquad (1)$$

The propane CO_2-ODH reaction proceeds with involvement of O species from both CeO_x and VO_x redox phases via the Mars van Krevelen mechanism.[3] The distribution of reaction products is strongly influenced by the parallel RWGS and non-oxidative dehydrogenation reactions.[4] This work focuses on synthesis, characterization and catalytic testing of bulk and supported CeO_x, VO_x and $CeVO_x$ phases over activated carbon. We aim at identifying the morphological, chemical and redox properties that regulate propane and CO_2 activation, as well as their consequences on activity and selectivity in the propane CO_2-ODH reaction.

2. Experimental/methods

Pure CeO_2 was synthesized by dissolving 4.9 g of $Ce(NO_3)_3$ in 84 ml of MQ water; 140 ml of 0.1 M aqueous NaOH was added, mixed for 30 min and transferred to Teflon® lined autoclaves, where it was kept for 24 h at 180°C. Bulk V_2O_5 was synthesized by calcination of vanadium acetylacetonate in air at 600°C for 4 h. Bulk $CeVO_4$ was synthesized by dissolving 92 mg of NH_4VO_3 in water, adding 1 drop of concentrated HNO_3 and 340 mg of $Ce(NO_3)_3$ dissolved in 10 ml of water. The formed suspension was stirred at 80°C for additional 2 h, filtered, dried at 70°C and calcined for 4 h at 600°C in air. Activated carbon (Norit RX3 Extra from Cabot) supported CeO_x and VO_x catalysts were synthesized exactly as $CeVO_4$, with the exception of powdered AC being added together with HNO_3 and final calcination in argon instead of air. Catalysts were characterized by N_2 physisorption, XRD, Raman, H_2-TPR-MS, C_3H_8-TPR-MS, CO_2-TPO-MS, *in-situ* DRIFT and HRTEM techniques. Catalytic tests were performed in a continuous-flow quartz tubular reactor at 550°C using 300 mg of a catalyst and a flow of C_3H_8, CO_2 and He (10 ml/min each, WHSV= 4 L/g_{cat}*h).

3. Results and discussion

The XRD and Raman analyses on CeO_x/AC catalysts showed CeO_2 is formed with the average crystal size of 4 nm, regardless of loading (7-17wt. %). With increasing VO_x loading from 2 to 12 wt. %, a series of VO_x/AC catalysts containing only oligomeric VO_x species, or VO_x in coexistence with crystalline V_2O_5 were synthesized. In the $CeVO_x$/AC catalysts, crystalline $CeVO_4$ phase was formed in coexistence with small fractions of CeO_2 and V_2O_5. Reactivity of lattice oxygen was probed with H_2-TPR and C_3H_8-TPR-MS analyses. Different reactivity and quantity of reactive lattice oxygen was identified in the tested bulk materials: CeO_2 started to reduce at 200 °C and $CeVO_4$ much later at 380°C. No reduction was observed for V_2O_5 in H_2 atmosphere up to 550 °C. The reduction of $CeVO_x$/AC samples was initiated at 330°C.

Distinctly different catalytic performance of the tested materials was observed in the propane CO_2-ODH reaction (Fig. 1A). Activity of CeO_2 is low, stable, with low (15 %) propylene selectivity. Bulk $CeVO_4$ shows no CO_2 conversion; propylene is formed exclusively due to non-oxidative dehydrogenation. Propene selectivity drops to zero within 2h over bulk V_2O_5. VO_x/AC catalysts exibited a propene selectivity decrease from 86 to 80 %, and a propane conversion increase from 3 to 5% when VO_x loading was increased from 2 to 12%, suggesting oligomeric VO_x sites mainly responsible for the propane CO_2-ODH reaction. When CeOx content was increased from 7 to 17 % in CeO_x/AC samples, higher catalytic activity was accompanied by a drop of propene selectivity from 84 to 61%.

By increasing $CeVO_x$ content from 10 to 40 wt. % in $CeVO_x$/AC catalysts, both propane and CO_2 conversions rose twofold (Fig. 1A), along with propene selectivity which reached 78%. Deactivation of all catalysts was observed with time on stream with concentrations of C_3H_6 and CO in the reactor discharge decreasing continuously (Fig. 1B). Contrary, concentrations of H_2, CH_4 and trace amounts of C_2H_4 were stable, indicating two parallel and independent reaction pathways: (i) propane CO_2-ODH and (ii) non catalytic (thermal) propane cracking.

In-situ DRIFT analysis revealed that activation of propane is associated with V_2O_5 and CeO_2 lattice oxygen abstraction and catalyst reduction. Bulk $CeVO_4$ does not react with C_3H_8. The CeO_2 to a lesser extent, but $CeVO_4$/AC and especially V_2O_5 suffer from very limited ability to activate CO_2, therefore lattice O^{2-} vacancy cannot be re-oxidized to close the catalytic cycle. Subsequently, this results in a progressive decline of their CO_2-ODH activity.

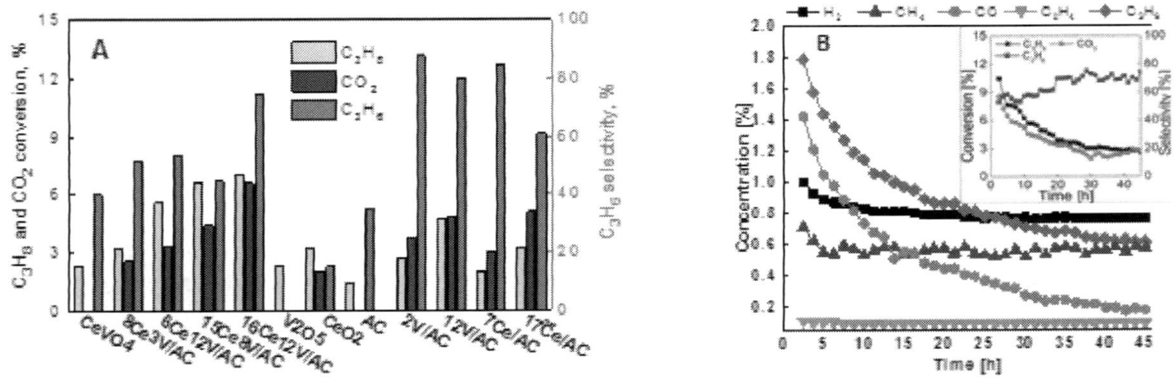

Fig. 1. A) propane and CO_2 conversion, as well as C_3H_6 selectivity for tested

catalysts after 4 h of propane CO_2-ODH reaction, and B) temporal distribution of reaction products over 15Ce8V/AC catalyst (inset shows C_3H_8 and CO_2 conversion along with C_3H_6 selectivity).

4. Conclusions

Two parallel and independent reaction pathways occur: propane CO_2-ODH which is related to participation of lattice oxygen, and propane cracking to methane, H_2 and carbon. Ethylene is formed only in trace amounts suggesting fast C=C bond cleavage. Activation of propane is associated with lattice oxygen abstraction and catalyst reduction. Based on the C_3H_8-TPR and CO_2-TPO results we can postulate that that bulk $CeVO_4$ formation contains negligible activity for propane and CO_2 activation. The ability of bulk crystalline V_2O_5 catalyst to activate CO_2 is very limited leading to a progressive decline of its CO_2-ODH activity. Bulk CeO_2 acts favorably towards catalyst re-oxidation with CO_2 already at 450°C. Presence of vanadium species strongly changes the redox and catalytic performance of CeO_2, hindering its CO_2 activation ability. Finely dispersed, non-crystalline VO_x species appear as the ones mainly responsible for propane activation, leading to propylene formation.

Acknowledgement

The financial support of the Slovenian research agency (ARRS) through research program P2-0150 and project J7-7294 is kindly acknowledged. We thank the Cabot Company for providing the AC sample.

References

(1) Carrero, C. A.; Schloegl, R.; Wachs, I. E.; Schomaecker, R. Critical Literature Review of the Kinetics for the Oxidative Dehydrogenation of Propane over Well-Defined Supported Vanadium Oxide Catalysts. *ACS Catal.* **2014**, *4* (10), 3357–3380.

(2) Cavani, F.; Ballarini, N.; Cericola, A. Oxidative Dehydrogenation of Ethane and Propane: How far from Commercial Implementation? *Catal. Today* **2007**, *127* (1–4), 113–131.

(3) Martínez-Huerta, M. V; Deo, G.; Fierro, J. L. G.; Bañares, M. A. Operando Raman-GC Study on the Structure - Activity Relationships in V5+/CeO2 Catalyst for Ethane Oxidative Dehydrogenation: The Formation of CeVO4. *J. Phys. Chem. C* **2008**, *112* (30), 11441–11447.

(4) Nowicka, E.; Reece, C.; Althahban, S. M.; Mohammed, K. M. H.; Kondrat, S. A.; Morgan, D. J.; He, Q.; Willock, D. J.; Golunski, S.; Kiely, C. J.; et al. Elucidating the Role of CO 2 in the Soft Oxidative Dehydrogenation of Propane over Ceria-Based Catalysts. *ACS Catal.* **2018**, *8* (4), 3454–3468.

545526 Dry Reforming and Oxidative Conversion of Methane on Composite Materials Prepared By Self-Propagating Combustion Synthesis and Impregnation Method

Wednesday, June 5, 2019
Texas Ballroom Prefunction Area (Grand Hyatt San Antonio)

Svetlana A. Tungatarova[1,2], Galina Xanthopoulou[3], Tolkyn S. Baizhumanova[1,2], Zauresh T. Zheksenbaeva[1,2], Manapkhan Zhumabek[1], Gulnar N. Kaumenova[1,2], Rabiga O. Sarsenova[1] and Gulzeinep U. Begimova[1,2], (1)Laboratory of Organic Catalysis, D.V. Sokolsky Institute of Fuel, Catalysis and Electrochemistry, Almaty, Kazakhstan, (2)Chemistry and Chemical Technology faculty, al-Farabi Kazakh National University, Almaty, Kazakhstan, (3)Advanced Ceramics Laboratory, Institute of Nanoscience and Nanotechnology, NCSR Demokritos, Athens, Greece

Dry Reforming and Oxidative Conversion of Methane on Composite Materials Prepared by Self-Propagating Combustion Synthesis and Impregnation Method

Svetlana A. Tungatarova[1,2*], Galina Xanthopoulou[3], Tolkyn S. Baizhumanova[1,2], Zauresh T. Zheksenbaeva[1,2], Manapkhan Zhumabek[1], Gulnar N. Kaumenova[1,2], Rabiga O. Sarsenova[1], Gulzeinep U. Begimova[1,2]

1 – D.V. Sokolsky Institute of Fuel, Catalysis and Electrochemistry, 142 Kunaev str., Almaty, 050010, Kazakhstan

2 - al-Farabi Kazakh National University, 71 al-Farabi ave., Almaty, 050040, Kazakhstan

3 – Institute of Nanoscience and Nanotechnology, NCSR Demokritos, Athens, 15310, Greece

* corresponding tungatarova58@mail.ru

Introduction

The utilization of the greenhouse gases, carbon dioxide and methane, has received increasing attention in the recent years. Particularly, methane catalytic reforming with carbon dioxide can transform two greenhouse gases simultaneously and thereby contribute to reduce global warming. The reaction product, syngas, contains a lower H_2/CO ratio than those available from steam reforming and partial oxidation of methane, which is more preferable as a feedstock for the synthesis of valuable oxygenated chemicals and sulfur free synthetic liquid fuels.

Self propagating high temperature synthesis (SHS) of catalysts offers several benefits with respect to obtaining a homogeneous product, minimizing operating time, simplifying procedure and equipment, much lower energy costs, ease of manufacture and capability for producing materials with unique properties and characteristics. Method is attractive for industrial production: much lower energy consumption than traditional production methods, much lower energy costs, possibility for "just-in-time" manufacturing, high productivity, cheap catalysts, relatively simple process - easily adaptable to industrial scale, controlled physico-chemical properties of the products, large range of new materials which can be used in catalysis, it has wide diapason of structural forms of products - from granules of different size to blocks of honeycomb structure and different geometric forms. In addition, the environmental impact of SHS is very much lower than that of the traditional method, a fact which decreases even further the indirect cost of production. Therefore, the purposes of this work are to produce catalysts by using SHS to characterize the catalyst and experimentally examine their catalytic activity.

Materials and Methods

The SHS catalysts on the base of NiO – Al - α-Al_2O_3 were prepared from powder mixtures consisting of nitrates, metals, and oxides. Cylindrical specimens 10 mm in diameter and about 20 mm long were produced by uniaxial compaction under a pressure of about 10 MPa. Prior to initiation of SHS reaction, the specimens were preheated in an electric furnace at 700 – 900°C for several minutes. The resulting materials were characterized by X-ray diffraction (XRD) using $CuK\alpha_1$ radiation and Scanning Electron Microscopy (SEM) with Energy Dispersive X-ray Spectroscopy (EDX). Further analyses were carried out using TEM and the particle size distributions were estimated by XRD peak broadening analysis. Surface area of the samples was determined by BET method. The catalytic activity of synthesized catalysts for dry reforming of methane was measured using coarsely crushed materials with an average pellet size of 3 mm, as dictated by industrial needs. The tests were carried out in a fixed bed free flow quartz reactor without any pre-reduction. All of the tests were conducted at atmospheric pressure in a flow of CO_2-CH_4-N_2 mixture (1:1:1). The total flow rate of reactants was set at 860 h^{-1} and 3300 h^{-1} and the catalytic reaction was carried out at 700 - 900°C.

Results and Discussion

During the SHS experiments the combustion velocity was measured. The specific surface area of the catalyst system is low and ranges from 0.3 to 2.1 m^2/g. This is due to the high temperatures of combustion during SHS, reaching 1500°C. SHS catalysts on the base of initial batch Al-NiO-Al_2O_3 have a similar qualitative composition, but there are differences in the phase ratio. The increase of NiO concentration and decrease of the Al concentration in the initial charge increases combustion velocity. It is connected with approach the stoichiometric composition, and therefore a greater heat generation affects the increase in the reaction rate.

The study of SHS catalysts structure was carried out using a scanning electron microscope (catalysts with NiO: 24.1 and 34.4 wt% concentration in the initial batch). It was found that the phase analysis by method of chemical analysis is corresponding to the data of XRD analysis: Al, α-Al_2O_3, Al-Ni, Ni, NiO, $NiAl_2O_4$.

Obtained SHS catalysts were tested for catalytic activity in the carbon dioxide dry reforming of methane in the temperature range 750 - 900°C. Hydrogen and CO yields for the studied SHS catalysts on the base of system NiO – Al - α-Al_2O_3 were investigated. Nickel, aluminum and oxygen ratio varies in different areas of catalyst. A high content of nickel, aluminum and oxygen corresponds to the spinel phase. The virtual absence of oxygen at a high content of nickel and aluminum corresponds to NiAl.

The best results for SHS catalysts based on systems NiO – Al - Al_2O_3 are: 93% CH_4 conversion, 100% conversion of CO_2, product yield reaches 92% H_2 and 99 % CO. Ratio hydrogen to carbon monoxide in the reaction product varies in the range of 0.7 – 1.35. Increasing the reaction temperature, in most cases, increases the ratio of H_2/CO due to the amplification of dehydrogenation reaction.

Effect of catalyst composition on the conversion of CH_4, CO_2 and the ratio of H_2/CO also seen in relation to nickel oxide, because from nickel oxide concentration depends content of nickel spinel - active catalyst component. The optimum concentration (maximum conversion of CO_2) of nickel oxide in the starting material is 24 - 30%, and for the conversion of methane - optimum is clearly revealed 29% of nickel oxide. The catalyst with 29% nickel oxide and 51% aluminum in the starting material, after the SHS reaction contains a maximal concentration of NiAl, $NiAl_2O_4$ active phases into carbon dioxide reforming of methane. The optimal lattice parameter for maximum conversion of carbon dioxide and methane are 3.48 - 3.485 Å for aluminum oxide, which plays the role of a catalyst carrier and 1.42 Å - for $NiAl_2O_4$ playing the role of catalyst.

Effect of structural deformation and the composition of catalysts on the conversion of CH_4, CO_2 and H_2/CO (at 900°C) were investigated too. The data indicate a clear structural dependence of the catalyst activity. The optimal lattice parameter for maximum conversion of carbon dioxide and methane are 3.48 - 3.485 Å for aluminum oxide, which plays the role of a catalyst carrier and 1.42 Å - for $NiAl_2O_4$ playing the role of catalyst.

The study of catalysts of similar composition but prepared by the traditional method of wetness impregnation and further their comparison with SHS catalysts was carried out. Investigation of catalysts was carried out under the following conditions: CH_4 : CO_2 : Ar = 1 : 1 : 1, GHSV - 860 h^{-1}. Analysis of the data shows that the conversion of feedstock has similar values both on SHS catalysts, and on traditional supported samples. However, target products yield is significantly higher for SHS catalysts: hydrogen yield is about 48 - 51% on the supported catalysts, and ~ 80% on SHS catalysts; CO yield is about 40 - 42% on supported catalysts, while about 90% on SHS catalysts.

The catalysts of NiO-Al-α-Al_2O_3 series prepared by SHS method and by incipient wetness supporting were tested in the partial oxidation of methane. SHS catalysts show higher activity in CH_4 partial oxidation, H_2 yields (52.0 - 67.0%) are higher compared with supported samples (53.9 - 57.2), and for CO - (21.0 - 27.1) instead of (21.9 - 24.2). Ratio of Í$_2$/ÑÎ = 2.0 – 2.9. The ideal ratio of $H_2/CO = 2$ was obtained for the SHS sample (29.2% NiO + 50.8% Al + 20% Al_2O_3).

Received data indicate a significant advantage of the new composite materials produced by combustion synthesis process. The obtained target products on above-mentioned catalysts are cleaner, which do not require additional treatment.

This publication has been made within the project Grant No AP05132348, which is funded by the Ministry of Education and Science of the Republic of Kazakhstan.

545771 Tri-Reforming of Methane - Operational Studies over a $Ni/CeZrO_2/MgAl_2O_4$ Catalyst

Wednesday, June 5, 2019
Texas Ballroom Prefunction Area (Grand Hyatt San Antonio)
Ananda V. Paladino Lino, Universidade Federal de São Carlos, São Carlos, Brazil, Elisabete M. Assaf, Instituto de Química de São Carlos/Universidade de São Paulo, São Carlos, Brazil and **José M. Assaf**, Universidade Federal de São Carlos, Engenharia Química, Brazil

Methane tri-reforming combines the steam, dry and partial oxidation of methane in a single reactor. The ratio of the syngas (H_2/CO) produced usually ranges from 1.0 to 2.0, and it is the reason for the methane tri-reforming to be considered a process with a great versatility, since syngas quality can be altered by simply changing the feed composition and thus be destined to several applications. An H_2/CO ratio around 2 is usually desired to produce liquid fuels, since it yields hydrocarbons of high chain.

Operational studies related to gas hourly space velocity (GHSV) and feed compositions were carried out in this work aiming to search adequate H_2/CO ratio destined to GTL processes as well as high CO_2 conversion, considering that it competes for the same active sites for water adsorption.

In these studies, Ni (10% wt) supported on $MgAl_2O_4$ spinel promoted with $CeZrO_2$ (Ce/Zr molar ratio= 4) was employed and the catalyst was activated *in situ* under H_2 flow at 750°C. The catalytic tests were carried out at 750°C and atmospheric pressure, for 300 minutes. The ratio $(n_{CO2}+n_{H2O}+ 2.n_{O2})/n_{CH4}$ was kept at 1 for any feed composition, where n_i represents the reactant molar flow rate. The CH_4 volumetric flow rate was 73.4 $NmL.min^{-1}$ for any feed composition. Water was pre-heated before reaching the reactor. The gases effluent composition was analyzed in a gas chromatograph, which was equipped with two Porapak-N® and a 13X molecular sieve columns. Four space velocities were evaluated: 3.0, 4.0, 5.3 and 10.5 $mol.gcat^{-1}.h^{-1}$.

This study showed that below 4 $mol.gcat^{-1}.h^{-1}$ and under stoichiometric feed (3 CH_4:1 CO_2: 1H_2O: 0.5 O_2), the reactants conversion (CH_4= 77% and CO_2=54%) did not change, while CH_4 and CO_2 conversions decreased up to 58% and 33%, respectively, as GHSV increased up to 10.5 $mol.gcat^{-1}.h^{-1}$. Thus, the feed composition studies were made using a GHSV of 4 mol.gcat ^{1}.h ^{1} and varying the O_2/CO_2 and H_2O/CO_2 molar ratios (0, 0.17 and 1.5).

Increasing either O_2/CO_2 or H_2O/CO_2 from 0 to 1.5, the CO_2 conversion decreases from 73% to 30%, despite increasing H_2/CO ratio, once the partial oxidation and steam reforming of methane increase the H_2 production under higher O_2 and H_2O

contents, respectively. O_2 consumption was always complete at any reaction condition, due to O_2 affinity to active sites. The O_2/CO_2 ratio of 0.17 in the feed was considered appropriate for operation, once the achieved conversions were 79% for CH_4 and 60% for CO_2, while the H_2/CO ratio was 1.75.

544628 Indium Oxide As Promoter in Cu/ZrO_2 Catalysts for CO_2 hydrogenation to Methanol

Wednesday, June 5, 2019
Texas Ballroom Prefunction Area (Grand Hyatt San Antonio)

Marco A. Rossi Sr.[1], José M. Assaf[2] and **Elisabete M. Assaf**[1], (1)Instituto de Química de São Carlos/Universidade de São Paulo, São Carlos, Brazil, (2)Universidade Federal de São Carlos, Engenharia Química, Brazil

Indium oxide as promoter in Cu/ZrO_2 catalysts for CO_2 hydrogenation to methanol.

Rossi, M.A.[1]; Assaf, J. M[2]; Assaf, E.M.[1]

[1] University of São Paulo, São Carlos Institute of Chemistry. Av. Trabalhador São Carlense, 400, 13560-970, São Carlos, SP, Brazil. eassaf@iqsc.usp.br

[2] Federal University of São Carlos, Rod. W. Luiz, km 235, 13565-905, São Carlos, SP, Brazil.

The importance of strategies for the capture and utilization of CO_2, that can be found as constituent in biogas and natural gas, is increasing significantly in global context. Among these strategies, the catalytic process of carbon dioxide hydrogenation is one of the most promising with the aim to obtain products with higher added value such as methanol.

The catalysts used in this process are generally based on metallic copper supported on oxides such as zirconia and alumina. Recent theoretical studies have shown the potential of pure indium oxide to obtain methanol with high selectivity, but it was not tested as promoter on catalysts based on copper until the moment.

The objective of this work was to investigate the influence of different molar contents of indium oxide as promoter on catalysts based on metallic copper and zirconia.

Cu/ZrO_2 catalysts without promoter and with 5% and 10% of indium oxide, in molar content, were synthesized by the co-precipitation method at constant pH assisted by the CTAB surfactant, followed by calcination in air. The synthesized materials were characterized by thermogravimetry, ex-situ and in-situ X ray diffraction, energy dispersive x ray spectroscopy, temperature programmed reduction, CO_2 temperature programmed desorption, transmission electron microscopy and N_2 sorption measurements using the BET model.

The catalytic tests were performed in a continuous flow tubular fix-bed PID Microactivity Effi reactor under pressure of 3 MPa, temperature of 250°C and GHSV of 6 $L.g^{-1}.h^{-1}$.

The results allowed to verify that the addition of indium oxide, even in the lowest content, increased the selectivity to methanol from 40% to 55% with decrease in the selectivity for other products such as CO, due to the fact that indium oxide improve the occurrence of the reaction mechanism by the formate route, being this, the most effective pathway to generate methanol with high selectivity.

545681 Sorption-Enhanced Water-Gas Shift Technology Platform for Cost Effective CO_2 Reduction in the Iron & Steel Industry

Wednesday, June 5, 2019
Texas Ballroom Prefunction Area (Grand Hyatt San Antonio)

Leon G.A. van de Water[1], Liliana Lukashuk[1], H.A.J. (Eric) van Dijk[2], Paul D. Cobden[2], Magnus Lundqvist[3], Giampaolo Manzolini[4], Letitia Petrescu[5], Siebe van der Veer[6], Luca Mancuso[7], Jeremy Johns[8] and David Bellqvist[9], (1)Johnson Matthey Technology Centre, Johnson Matthey, Billingham, United Kingdom, (2)TNO, Petten, Netherlands, (3)Swerea Mefos, Lulea, Sweden, (4)Politecnico di Milano, Milano, Italy, (5)Faculty of Chemistry and Chemical Engineering, Babes-Bolyai University, Cluj-Napoca, Romania, (6)Kisuma Chemicals, Veendam, Netherlands, (7)Amec Foster Wheeler Italiana S.r.l., Milano, Italy, (8)Tata Steel UK Consulting Ltd., Newport, United Kingdom, (9)SSAB, Lulea, Sweden

1. Introduction

The interest in CO_2 capture, storage, and utilisation has grown significantly in recent years because CO_2 emissions are the major cause of global warming. A large share of global CO_2 emissions is related to industrial facilities. To avoid major climate change in the long term, it is imperative that each industrial sector looks to improving energy efficiency and decreasing CO_2 output. As an example, the iron and steel industry is responsible for an annual output of ~2.5-3.0 $GtCO_2$/yr, representing 6% of total CO_2 emissions, and 16% of total industrial CO_2 emissions. The STEPWISE project, executed within the European H2020 LCE program, aims at decreasing CO_2 emissions related to the steel-making process from about 2 t_{CO2}/t_{steel} to below 0.5 t_{CO2}/t_{steel} by means of CO_2 removal from the Blast Furnace Gas (BFG) (i.e., an average dry composition of 19% CO, 25% CO_2, 3% H_2, 20 ppm H_2S+COS and 53% N_2) [1-3]. The STEPWISE project has at its heart the Sorption-Enhanced Water-Gas Shift (SEWGS) technology – a solid sorption technology for CO_2 capture from fuel gases in combination with water-gas shift and acid gas removal [4]. SEWGS has advantage over conventional approaches adopted for pre-combustion CO_2 capture requiring at least two water-gas shift reactors with interstage cooling to effect a high conversion of CO to CO_2 (**Figure 1**). Downstream the water-gas shift (WGS) section, further cooling is then necessary to enable the capture of CO_2 by absorption with a solvent. Using the SEWGS process, both the 2nd WGS reactor section and the H_2/CO_2 separation unit are replaced. A combination of bulk conversion of CO in the WGS unit followed by CO_2/H_2 separation in the SEWGS unit enables generation of a high purity CO_2-rich stream intended for storage or further utilisation (such as for production of chemicals) and a H_2-rich stream intended for electricity production. In the STEPWISE project, the SEWGS technology is brought to TRL6 by means of design, construction, operation and modelling a pilot installation in the Iron and Steel industry using actual Blast Furnace Gas. This pilot plant is designed for a 800 Nm^3/h BFG feed (i.e., the capability of processing up to 14 tonnes CO_2 per day) and is located at the facilities of Swerea Mefos in Luleå, northern Sweden next to the SSAB (Sweden) blast furnace from which the pilot receives its BFG feed gas [1-3].

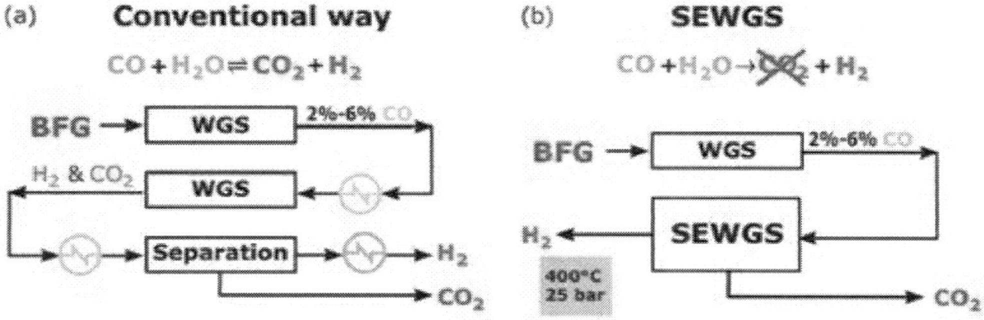

Figure 1. (a) Conventional layout of a shift section and H_2/CO_2 separation for high-pressure CO_2 capture; and (b) layout for improved CO conversion and CO_2 capture section using the SEWGS technology.

2. Experimental

The project covers the design, construction, operation and modelling of a 14 t_{CO2}/day capture unit, consisting of a BFG compression section, an advanced pre-shift section and a single column SEWGS unit. Pilot construction and commissioning was completed Q4, 2017. The 1st pilot plant campaign at Swerea Mefos, Luleå Sweden, using the KATALCO™ 71-6 WGS catalyst for the pre-shift section and the potassium-promoted magnesium-aluminium-based sorbent HT1.5 for the SEWGS unit has been completed. The environmental evaluation of the SEWGS process through Life Cycle Assessment (LCA) and the techno-economic assessment were performed within the STEPWISE project and compared with competitive technologies (reference case consists of post-combustion CO_2 capture via amine scrubbing).

3. Results and discussion

For the WGS section, a commercial iron-based WGS catalyst KATALCO 71-6 was evaluated. The suitability of this catalyst was established during lab-scale testing under BFG conditions, which are considerably different from typical ammonia- or hydrogen-plant conditions. The catalyst showed good resistance towards the low sulphur feed content associated with BFG. Pilot plant testing confirms the catalyst stability of the pre-shift section at lowered steam operation, i.e. $H_2O/CO = 2.0 - 1.5$. A gradual increase in CH_4 production that indicates destructive excessive overreduction of the catalyst was not observed, confirming the catalyst robustness demonstrated earlier during lab-scale testing campaigns. Modelling of the catalyst bed activity in time indicated that the catalyst stability appeared to be good. No significant decrease of the catalyst activity in time was observed throughout the 1st long campaign (i.e., 1000 h operation). Following the campaign, the KATALCO 71-6 catalyst was discharged and samples were taken from different sections of the WGS reactor for characterization. The catalyst appeared to be in very good condition. Post-test characterization indicates that no over-reduction or sulphidation of iron occurred, that the structure of the active catalyst is preserved and that there was no sign for carbon accumulation. The findings agree well with the results of the post-characterization study following lab-scale catalytic testing.

With respect to the SEWGS unit, major achievements are the demonstration of sorbent production at tonnage scale, robust operation of the pre-shift section at lowered steam conditions, and SEWGS performance at >90% H_2S removal efficiency and >90% CO_2 capture rate at <1 mol steam/mol CO_2 and >90% product purity, illustrating that the SEWGS technology provides simultaneous decarbonization and desulphurization.

The preliminary techno-economic evaluation illustrates that the SPECCA coefficient - the energy requirement for capture - is below 2 which is around 30% lower than the amine-based case. The LCA indicates that all impact parameters are lower for the STEPWISE cases compared to post combustion amine scrubbing reference case.

4. Conclusions

In the STEPWISE project, the SEWGS technology for CO_2 capture is brought to TRL6 by means of design, construction, operation and modelling a pilot installation in the Iron and Steel industry using Blast Furnace Gas. This advanced CO_2 removal technology makes use of regenerative solid adsorbents. For the WGS section, a commercial iron-based catalyst was evaluated. The suitability of the commercial catalyst under the atypical conditions for BFG operation was established, illustrating the catalyst robustness at the low sulphur conditions associated with BFG. The LCA and techno-economic assessment indicates that the SEWGS technology has potential to reduce carbon footprint in the Iron and Steel industry.

Acknowledgments

This work is part of the STEPWISE project that has received funding from the European Union's Horizon 2020 research and innovation programme under grant agreement No. 640769.

KATALCO™ is a trademark of Johnson Matthey PLC.

References

[1] http://www.stepwise.eu/project/

[2] https://www.youtube.com/watch?v=7_xbe-EMyRc

[3] H. A. J. (Eric) van Dijk et al., Johnson Matthey Technol. Rev., 62, 4 (2018) 395–402.

[4] M. Gazzani *et al.,* Int. J. of Greenhouse Gas Control 41 (2015) 249-267.

545085 Spectroscopic Characterization of Monomeric and Dimeric Copper Species in Chabazite Zeolites for Partial Methane Oxidation to Methanol

Wednesday, June 5, 2019
Texas Ballroom Prefunction Area (Grand Hyatt San Antonio)

Laura N. Wilcox[1], Ishant Khurana[1], Hui Li[2], Christopher Paolucci[2], William F. Schneider[2] and Rajamani Gounder[1], (1)Charles D. Davidson School of Chemical Engineering, Purdue University, West Lafayette, IN, (2)Chemical and Biomolecular Engineering, University of Notre Dame, Notre Dame, IN

Methane is an abundant and inexpensive alkane that can be sourced from natural and shale gas and upgraded to chemicals of higher value, including methanol. The direct conversion of methane to methanol is challenging, however because of its high C-H bond strength, low polarizability, and low electron and proton affinity [1]. Cu-zeolites have received renewed attention as materials that may facilitate partial methane oxidation (PMO), with many proposed active site motifs, including monomeric and multimeric copper species [2,3,4]. More precise identification of the copper structures that activate CH_4 can be facilitated by studying model heterogeneous supports that contain well-defined copper speciation, as a result of the synthetic and treatment procedures used to prepare them. Chabazite (CHA) zeolites, composed of double 6-membered ring building units that connect to form 8-membered ring windows, are high-symmetry frameworks that contain a single, unique tetrahedral site (T-site). CHA serves as a model support that can be used to simplify the interpretation of experimental characterization data of exchanged Cu ions and complexes, which can be compared more faithfully to structural models studied using computational approaches. Synthetic control of the framework aluminum distribution in chabazite zeolites between isolated and paired Al sites (Al-$(Si\text{-}O)_x$)Al, x>2 for isolated and x=1,2 for paired), influences the speciation of extraframework Cu cations between monovalent $[CuOH]^+$ (ZCuOH) and divalent Cu^{2+} (Z_2Cu) ion sites, respectively, whose concentrations are quantified by ammonia titration after Cu-exchange and through a variety of spectroscopic probes [5]. Monomeric ZCuOH sites in close proximity can condense to form binuclear species of the ZCuOCuZ type, and the combination of H_2 and CO temperature programmed reduction [6] and in-situ UV-visible spectroscopy were used to probe the presence of mononuclear and binuclear Cu site types. Distinct spectral differences between Z_2Cu and ZCuOH sites observed experimentally by UV-visible spectroscopy were corroborated and assigned with ab-initio molecular dynamics (AIMD) combined with time-dependent density functional theory (TD-DFT). Calculations that only consider DFT-optimized single geometries neglect the dynamic changes in structure and coordination present among Cu sites in experiment, and the incorporation of site dynamics was found to be required for proper assignments of spectral features to specific Cu site types. The synthesized model Cu-zeolites were used to probe the oxidation of methane to methanol in a stoichiometric step-wise procedure involving activation at high temperature in air (723-773 K, 0.167 K s^{-1} ramp rate, 1-6 h), methane exposure at low temperatures (473-503 K, 20%-99% CH_4, 0.5-2 h), and then water exposure (473-503 K, 3% in He, 1 h) to form methanol. These results suggest that dimeric copper sites are the dominant precursors to PMO active sites, and we will discuss zeolite synthesis and treatment strategies to generate larger fractions of PMO-active multinuclear copper sites.

[1] Periana R. A. et al., *Science*, **1998**, 280, 560-564.

[2] Ipek B. et al., *ACS Catal.*, **2017**, 7, 4291-4303.

[3] Grundner S. et al., *Nature Comm.*, **2015**, 6, 1-9.

[4] Ravi M. et al., *Angew. Chem. Int. Ed.*, **2017**, 56, 16464-16483.

[5] Paolucci C. et al., *J. Am. Chem. Soc.* **2016**, 138, 6028-6048.

[6] Da Costa P. et al., *Phys. Chem. Chem. Phys.*, **2002**, 4, 4590-4601.

545662 A Study on the Characteristics of Methane Chlorination By Crystalline / Amorphous Carbon

Wednesday, June 5, 2019
Texas Ballroom Prefunction Area (Grand Hyatt San Antonio)

Jip Kim[1,2], Youngmin Kim[3], Kyoung-Su Ha[2] and **Ho-Jeong Chae**[1], (1)Carbon Resources Institute, Korea Research Institute of Chemical Engineering, Daejeon, Korea, Republic of (South), (2)Department of Chemical and Biomolecular Engineering, Sogang University, Seoul, Korea, Republic of (South), (3)Korea Institute of Machinery and Materials, Daejeon, Korea, Republic of (South)

Currently, the petrochemical industry uses naphtha as main raw material. Naphtha has the advantage of being able to produce not only olefin products but also non-olefin products, but it is more expensive than natural gas and coal raw materials. Especially, the domestic petrochemical industry is highly based on basic oil, so it is necessary to reduce the risk through diversification of raw materials. As an alternative plan, C1 gas like methane can be utilized. Methane is a valuable resource which can be used as a basic chemical feedstock or a transportation fuel through chemical conversion processes. And currently, typical and commercialized processes are mostly indirect conversion technologies through synthesis gas, which requires high energy and multi-step process. On the other hand, the direct conversion technology without passing through the syngas has attracted a great deal of attention because it uses low energy and does not require a multi-step process, which is economical. In this study, selective production of methyl chloride was performed by methane chlorination reaction which does not require additional oxygen process. In this study, Crystalline / amorphous carbon was used for the catalysts. The catalysts were characterized by XRD, BET, TEM analysis. The reaction characteristics of the catalysts were invetigated by methane chlorination reaction.

546111 Liquid-Phase Oxidation of Methane into Oxygenates with O_2

Wednesday, June 5, 2019
Texas Ballroom Prefunction Area (Grand Hyatt San Antonio)

Jong Kyu Kang[1] and **Eun Duck Park**[1,2], (1)Department of Energy Systems Research, Ajou University, Suwon, Korea, Republic of (South), (2)Department of Chemical Engineering, Ajou University, Suwon, Korea, Republic of (South)

Methane has been used as a clean energy source because it emits the smallest amount of CO_2 per heat produced among hydrocarbons. Recently increasing attention has been paid to its chemical conversion owing to the upsurge of shale gas production. However, due to the lack of the economic process for the utilization of methane as a chemical feedstock in a small scale, about 35% of methane is just being flared away in the United States. In the convention chemical process, methane is first converted into syngas (a mixture of CO and H_2), which can be further transformed into various chemicals. Since the typical syngas process is energy-intensive, only large-scale methane-based process can be cost-effective compared with petrochemical process to produce the same chemical. To utilize the small gas well efficiently, the efficient direct methane conversion technology into liquids needs to be developed.

Since methane has a strong C-H bond, its functionalization is quite demanding under mild conditions. Therefore, strong oxidants including H_2O_2 have been used to oxidize methane in the liquid phase. Hutchings and his colleagues reported the direct conversion of methane into methanol and formic acid using heterogeneous catalysts such as Cu-promoted Fe/ZSM-5 in the aqueous phase. Because of the high price of H_2O_2, there have been many efforts to use O_2 or air as an oxidant. However, high yields of methane oxygenates could not be accomplished in a single-step process. To compromise these two processes, in situ generation of H_2O_2 from H_2 and O_2 can be proposed in order to oxidize methane selectively under mild conditions.

In this work, Pd/C and ferrous ion was utilized to generate H_2O_2 from a mixture of H_2 and O_2 and to oxidize methane, respectively. The ferrous ion is essential to form two oxygen-radical species (hydroxyl and hydroperoxyl radical) from H_2O_2 to oxidize methane in the aqueous phase. The effect of catalyst compositions on the catalytic activity was examined systematically to find out the role of each component in this catalytic reaction. It can be concluded that methane can be oxidized into formic acid in the co-presence of H_2 and O_2 using the catalyst system composed of Pd/C and ferrous ion.

546124 Ethene Dimerization over Zr-Based Metal-Organic Framework (UiO-67) Functionalized with Ni and Bipyridine

Wednesday, June 5, 2019
Texas Ballroom Prefunction Area (Grand Hyatt San Antonio)

Mustafa Komurcu, Andrea Lazzarini, Gurpreet Kaur, Silvia Bordiga, Karl Petter Lillerud and **Unni Olsbye**, Department of Chemistry, University of Oslo, OSLO, Norway

Ethene dimerization over Zr-based metal-organic framework (UiO-67) functionalized with Ni and bipyridine

Mustafa Kømurcu[1], Andrea Lazzarini[1], Gurpreet Kaur[1], Silvia Bordiga[1,2], Karl Petter Lillerud, Unni Olsbye[1*]

1 – Department of Chemistry, University of Oslo, P.O. Box 1033, Oslo 0315, Norway

2 – Univerisity of Turin, Department of Chemistry, Via P. Giuria 7, Turin 10125, Italy

* corresponding author: Unni.olsbye@kjemi.uio.no

-

The industrial transition from naphta to ethane as feedstock for ethene production has inspired the search for alternative production processes for propene, butene and butadiene. Dimerization of ethene to linear α-butene is one of a few large-scale homogeneously catalyzed reactions [1]. With the intention of making the process more practical and sustainable, recent research efforts have focused on the development of heterogeneous dimerization catalysts. Ni-aluminosilicates is one promising alternative, and recently, metal-organic framework (MOF) based selective oligomerization catalysts are appearing [2, 3]. An important upside of Ni-aluminosilicate catalysts compared to the traditional homogeneous Ni-complexes is that they do not require a co-catalyst (Et_2AlCl). Most Ni-MOF catalysts published to date, however, do require such a co-catalyst [4-6]. Our target is to develop a MOF-based ethene oligomerisation catalyst that does not require a co-catalyst and maintains high linear butene selectivity. In this contribution, we report a series of UiO-67 Zr-MOFs where part of the biphenyl linkers were replaced by bipyridine, as grafting sites for Ni salts. Butene yields up to 1850 mg g_{cat}^{-1} h^{-1} were obtained without the use of a co-catalyst. Detailed characterization of fresh and used catalyst was performed to identify the active site and to confirm the structural stability of the catalyst under reaction conditions.

UiO-67, UiO-67-$bpy_{7\%}$ and UiO-67-$bpy_{12\%}$ (bpy = 2,2'bipyridine-5,5'-dicarboxylate) were functionalized with Ni^{2+} and characterized by XRD, SEM, N_2-adsorption, TEM, UV-VIS and FT-IR. The catalysts were tested as ethene dimerization catalysts in the temperature range from 120 °C - 300 °C, ethene partial pressure from 4 to 26 bar where C.T. = 3.1 $mg_{cat}min\ mL_{ethene}^{-1}$. High temperatures (> 200 °C) and ethene partial pressure (> 4 bar) was required to obtain significant conversion. Only catalysts with high concentration of Ni and bpy showed significant activity, with linear butenes as main product. An initial increase in ethene conversion with time on stream was observed, suggesting that the active state of the catalyst is formed during reaction. The catalysts deactivated over the course of the experiments, most likely due to long chain alkenes retained in the structure or due to formation of nickel nanoparticles. Catalytic activity increased with activation time, where longer duration at elevated temperature led to a higher conversion and faster achievement of maximum conversion. TEM images and FTIR measurements showed the presence of Ni/NiO nanoparticles on spent catalyst. Ex-situ CO-FTIR measurements on 2% Ni-UiO-67-bpy$_{12\%}$ and 4% Ni- UiO-67-bpy$_{12\%}$ used for 0, 100 and 700 min TOS showed that Ni/NiO nanoparticles formed only on 4% Ni- UiO-67-bpy$_{12\%}$ after 700 min TOS (Figure 1), eliminating the possibility of nanoparticles as the single active site for the reaction. Ni^{2+} grafted on bpy linkers is suggested as the active site for this reaction. Switching the feed to 1-butene, no isomerization activity was observed over the bpy-UiO-67. Considering the lack of strong acid sites and taking the product distribution into account we suggest that the catalyst operates through the Cossee-Arlman mechanism.

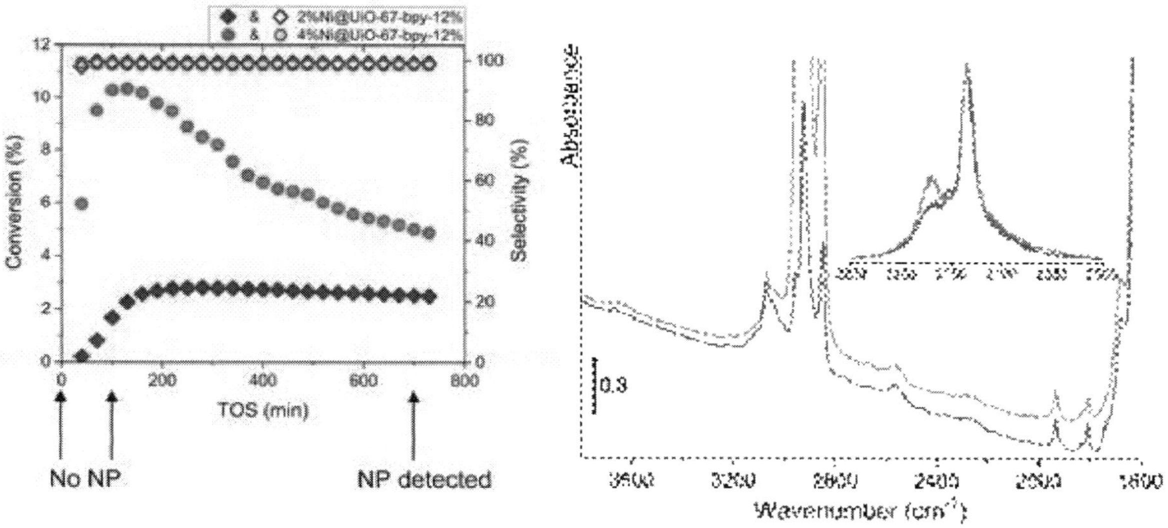

Figure 1: Left: Ethene conversion (solid symbols) and linear butene selectivity (hollow symbols) at 250 °C for the catalysts (shown in legend) versus time on stream. Reaction conditions: m_{cat} = 0.100 g, 30 bar total pressure where P_{ethene} = 26 bar and P_{inert} = 4 bar, contact time: 3.1 mg·min/mL at STP. Right: FT-IR spectra of sample 2% Ni-UiO-67-bpy (blue curve) and 4% Ni-UiO-67-bpy (red curve) collected at the end of reaction. It is possible to notice the ν(C-H) from retained products in the 3000 – 2800 cm^{-1} region. Inset shows the subtracted spectra in the CO spectral region after its adsorption at 77K. The signal at 2170 cm^{-1} evidence the difference in Ni amount present in the two samples, while the broad peak centered at 2060 cm^{-1} for 4% Ni-UiO-67-bpy is ascribed to CO coordinated on Ni/NiO NPs.

1. McGuinness, D.S., *Olefin Oligomerization via Metallacycles: Dimerization, Trimerization, Tetramerization, and Beyond.* Chemical Reviews, 2010. **111**(3): p. 2321-2341.

2. Hulea, V., *Toward Platform Chemicals from Bio-Based Ethylene: Heterogeneous Catalysts and Processes.* ACS Catalysis, 2018. **8**(4): p. 3263-3279.

3. Finiels, A., F. Fajula, and V. Hulea, *Nickel-based solid catalysts for ethylene oligomerization - a review.* Catalysis Science & Technology, 2014. **4**(8): p. 2412-2426.

4. Madrahimov, S.T., et al., *Gas-Phase Dimerization of Ethylene under Mild Conditions Catalyzed by MOF Materials Containing (bpy)NiII Complexes.* ACS Catalysis, 2015. **5**(11): p. 6713-6718.

5. Metzger, E.D., et al., *Selective Dimerization of Ethylene to 1-Butene with a Porous Catalyst.* ACS Central Science, 2016. **2**(3): p. 148-153.

6. Mlinar, A.N., et al., *Selective Propene Oligomerization with Nickel(II)-Based Metal–Organic Frameworks.* ACS Catalysis, 2014. **4**(3): p. 717-721.

555113 Experimental and Computational Fluid Dynamics Study of Two-Stage Natural Gas Combustion Pyrolysis Reactor in High-Valued Chemical Production: The Effect of Feedstock-to-Fuel Ratio and Feedstock Compositions

Wednesday, June 5, 2019
Texas Ballroom Prefunction Area (Grand Hyatt San Antonio)

Lei Chen[1], Sreekanth Pannala[1], Balamurali Nair[2], Istvan Lengyel[3], Byeongjin Baek[3] and David West[1], (1)Corporate Research & Development, SABIC, Sugar Land, TX, (2)CRD, SABIC, Sugarland, TX, (3)SABIC Global Corporate Research, Sugar Land, TX

The abundance of shale gas discoveries has prompted research into conversion of natural gas into high-valued chemical intermediates as compared to historical use as fuels or feedstock for syngas or methanol. Direct routes from CH_4 to C_2 products have been extensively studied but economically viable processes have been elusive. In this study, we used an integrated experimental and modeling approach to optimize natural gas combustion pyrolysis process. Computational Fluid Dynamics (CFD) approach was used to capture the complex interaction between turbulence and mixing, heat and mass transfer, and extremely stiff kinetics for CH_4 oxy-combustion and subsequent hydrocarbon pyrolysis. Following our previous validation study on natural gas pyrolysis[i], this work further introduces recent experimental and computational studies under varying operating conditions, specifically the effects of the feedstock-to-fuel ratio, and composition of the feedstock on the reactor performance metrics (conversion and yield). Both pilot-scale experiments and CFD simulations were conducted under a wide range of operating conditions and the results showed monotonic decrease of feedstock carbon conversion and increased selectivity with the increasing feedstock-to-fuel ratio. The tested feedstocks include natural gas, associated gas (AG), and natural gas liquids (NGL). Experimental and simulation results showed higher conversion and C_2 yields when using AG and NGL as feedstock as compared to conventional natural gas feed, indicating potentially better economic values with flexible feedstock solutions.

[i] L Chen, S Pannala, B Nair, et al., Experimental and numerical study of a two-stage natural gas combustion pyrolysis reactor for acetylene production: The role of delayed mixing, Proceedings of the Combustion Institute, 37(4), 5715-5722, 2019

AUTHOR INDEX

Aasberg-Petersen, Kim 46
Abate, Salvatore 449, 473
Abdelsayed, Victor 293
Adeleke, Aliu A. 512
Afzal, Shaik 65, 453, 462, 494
Agarwal, N. 297
Agarwal, Nishtha 296
Agrawal, Rakesh 484
Ahmad, Razan 501
Akbariyeh, Ashkan 229
Alcala, Ryan 346
Alexopoulos, Konstantinos 256
Al-Hazmi, Mohammed 491
Allaire, Douglas L. 503
Allam, Djaouida 327
Allen, David 2, 497, 500
Allen, R. Cory 503
Al-Mohannadi, Dhabia 501
Almquist, Catherine 37
Alshafei, Faisal H. 70, 485
Alshihri, Saeed 324
Althahban, Sultan 296, 386
Althahban, Sultan M. 297
Altwasser, S. 195
Alvarez, Ramon 2
Ambrosetti, Matteo 336
Andolina, Christopher M. 419
Antzara, Andy 445, 468
Arabi, Saiedeh 25, 129, 130, 229, 529
Arcon, Iztok 404
Arges, Christopher G 257
Argyle, Morris D. 365
Arstad, Bjornar 305
Assaf, Elisabete M. 528, 540, 542
Assaf, Jose M. 528, 540, 542
AuYeung, Nick 325
Avalos-Borja, Miguel 62
Avraamidou, Styliani 503
Ay, Hale 182
Azzam, Sara 70
Bae, Jong Wook 80, 102
Baek, Byeongjin 555
Bahri, Mounib 200
Bai, Shi 315
Baizhumanova, Tolkyn S. 514, 536
Baldea, Michael 500
Baldridge, Anthony 399
Balzarotti, Riccardo 336
Bao, Xinhe 94, 170, 261, 347
Bardey, Steven 465
Barton, John 2
Basha, Omar 483
Basile, Francesco 449, 473
Bassano, Claudia 439
Bastos, Pedro H. C. 85
Batiot-Dupeyrat, Catherine 38
Beato, Pablo 305
Bellqvist, David 544
Benavidez, Angelica 519
Bengaouer, Alain 105, 222
Bensaid, Samir 208, 222
Beretta, Alessanda 336
Berlier, Gloria 305
Bermejo-Deval, Ricardo 379
Bezemer, Leendert 98
Bhan, Aditya 277
Bickel, Elizabeth E. 359, 396
Bielenberg, James 4
Blekkan, Edd 120, 508
Boaz, Nicholas 368
Bodi, Andras 376
Boleubayev, Yerzhan 30
Bonduelle-Skrzypczak, Audrey 465
Bonura, Giuseppe 158, 208
Bool III, Lawrence 21
Bordiga, Silvia 305, 552
Borfecchia, Elisa 305
Borg, Anne 183
Borg, Øyvind 115
Botha, Thys 140
Bouchoul, Nassim 38
Brady, Casper 5
Brandão, Soraia 34, 85
Brandin, Jan 120
Brandl, Marco 353
Brennecke, Joan F. 101
Bridge, Alec 101
Broadbelt, Linda J. 359, 500
Bukowski, Brandon C. 339, 341
Bullard, Denny 2
Buras, Zachary J. 326
Burger, Thomas 433
Cain-Borgman, Craig 195
Canizales, Kelsey 365
Cannilla, Catia 158
Canning, Griffin A. 346
Caps, Valérie 465, 466
Carlsson, Per-Anders 299
Carpenter, John R. 15
Carrero, Carlos A. 84, 351
Carter, James H. 386
Catherin, Nelly 453
Centi, Gabriele 449, 473
Ch, Surbahmanyam 33
Chae, Ho-Jeong 330, 550
Chaise, Albin 105, 222
Challiwala, Mohamed Sufiyan 462
Champon, Isabelle 105
Chang, Xiaoxia 5
Chatla, Anjaneyulu 453

AUTHOR INDEX

Chawdhury, Piu 33
Chen Sr., Qingjun 486
Chen, De 57, 369, 412, 417, 486
Chen, Lei....555
Chen, Wei 104, 177, 218
Chen, Yu 452
Chen, Yuxiang....94, 170
Chen, Zewei 484
Cheng, Jihong....402
Cheng, Jun....109, 372
Cheng, Kang....173
Chernavskii, P.A. 200
Chi, Hao....419
Chiang, Fu-Kuo....104, 402
Chinta, S. 320
Chintamani, Siddarth 25, 229
Chojecki, Adam....242
Chou, Lingjun 521
Choudhury, Hanif....462
Christensen, Earl 230
Chu, Te-Chun....326
Claeys, Michael 112, 140, 164
Cobden, Paul D. 544
Collin, Ryan....325
Costa, Fabio....158
Creci, Simone 299
Cronauer, Donald 145
Cullen, Greg 476
Curnan, Matthew....419
Currie, Robert....428
d'Alnoncourt, R. Naumann....408
Da Silva Jr., Roberto Batista....85
Da Silva Sr., Breno Cerqueira 85
Damstedt, Brad 21
Dankworth, David....93
Datye, Abhaya 320, 346
Dautzenberg, Frits 490
Davilla, J.C. 101
Davis, Burtron H. 145, 205, 512
De Florio, Daniel 505
De Jong, Krijn....141, 192, 202, 237
De Jongh, Petra E. 141, 237
De Wilde, Juray....55
Debuschewitz, Jonas 353
Dedecek, Jiri 307
deGlee, Ben....452
Deiana, Paolo....439
Delgass, W. Nicholas....341
Deng, Dehui 109
Deng, Jiao 109
Dennis, Brian....25, 129, 130, 229, 529
DeRosa, Sean E. 497
Deshlahra, Prashant 381
DeWilde, Joseph....242
Dhingra, Sandeep S. 18
Dimick, Paul S. 7
Dinca, Mircea....304
Ding, Dong 355
Ding, Yunjie....328
Dinh, Kimberly....304
Distaso, Monica 353
Djinovic, Petar....404, 532
Do, Nga Thi Quynh 206
Doherty, Michael F. 49
Donazzi, Alessandro....67
Drewery, Matthew....273, 422, 478
Du, Shoucheng....5
Duarte, Alysson 38
Dugulan, A. Iulian....104
Dumitrescu, Cosmin....91
Dummer, Nicholas....296, 386
Duyckaerts, Nicolas....252
Dyballa, Michael M. 305
Edgar, Thomas F. 500
Egelske, Benjamin T 43
Ehlig, Christine....2
Elbashir, Nimir....65, 453, 462, 494
El-Halwagi, Mahmoud....494
El-Halwagi, Mahmoud M. 503
Enger, Bjørn Christian 115
Erhard, Jannis....353
Ersen, Ovidiu....200
Ewald, Stefan....433
Falbo, Leonardo 439
Fan, Jie....287
Fan, Liang-Shih....79
Fang, Guangzong....261
Fang, Yuan....216
Fang, Yuxin....257
Farberow, Carrie A. 230
Farstad, Mari Helene....183
Fasolini, Andrea 449, 473
Fécant, Antoine....465
Fenes, Endre 369, 412, 417
Feng, Bo....104
Feng, Xiaobo....247
Ferlauto, Andre 505
Fischer, Nico 112, 164
Fitzsimons, Joseph....2
Flake, John C. 257
Flytzani-Stephanopoulos, Maria....321
Fonseca, Fabio....505
Fortman, George 368
Fourre, Elodie....38
Frank, Benjamin 357, 408
Fratalocchi, Laura....248
Freakley, S. 297
Freakley, Simon....296
Freeman, Benny D. 6, 101
Frusteri, Francesco....158, 208

AUTHOR INDEX

Fu Sr., Gang 410
Fu, Ross 368
Fuglerud, Terje 369, 412
Furukawa, Shinya 29
Gaffney, Anne 331
Galvita, Vladimir V. 443
Ganesh, Hari S. 500
Gao Sr., Peng 212
Gao, Xu-Dong 315
Garcia, Omar 2
Garcia-Moncada, Nuria 258
Garrett, Holli 476
Garzon, Fernando 519
Gascon, Jorge 152, 269
Gautam, Pankaj S. 491
Gavrilovic, Ljubisa 120
Gazit, Oz M. 26, 524
Ge, Qingjie 213, 217
Ghosh, Richa 70
Ghouri, Mohammed Minhaj 65
Gil, Edwin Andres Rodriguez 484
Glaum, Robert 357
Goddard III, William A. 368
Goetze, Joris G. 225
Goldberg, Jonathan 368
Goldman, Mark J. 326
Golunski, Stanislaw 386
Gong, Xueqing 286
Gopal, Srikant 491
Gordillo, A. 195
Gordon, Michael 42, 49
Gorling, Andreas 353
Gorny, Martin 206
Gounder, Rajamani 359, 396, 400, 548
Grabau, Mathias 353
Graham, Uschi M. 205
Greeley, Jeffrey 339, 341
Green, William H. 326
Gronemann, Veronika 206
Groppi, Gianpiero 245, 248, 336
Groves, John 368
Gu, Bang 200
Gu, Tian 319
Guerrero, Patricia S Garcia 152
Guffanti, Simone 245
Gunnoe, T. Brent 368
Guo, Lisheng 217
Guo, Pu 109
Guo, Xiaoguang 261
Guo, Xiaoyang 57
Ha, Jeong-Myeong 518
Ha, Kyoung-Su 330, 550
Haag, Stéphane 206
Haas, Alfred 168
Hamdeh, Hussein H. 205
Hammache, Sonia 293
Han, Qian 398
Han, Yunxing 90
Hao, Jianqi 261
Harjowinoto, Danny 29
Harrison, Matthew 2
Hasegawa, Jun-ya 263
Hashimoto, Kunihide 531
Hauber, Christoph 168
Haumann, Marco 353
Hayek, Naseem S. 26, 524
He, Lei 287
He, Mingyuan 241
He, Qian 296, 386
Hemberger, Patrick 376
Hensen, Emiel 104, 177, 218, 267, 437
Hensley, Alyssa 186
Hensley, Jesse E. 230
Heracleous, Eleni 391, 468
Hermans, Ive 349
Hernández, X. Isidro Pereira 186
Hester, Tracy 2
Hibbitts, David 182
Hicks, Jason C. 344, 530
Higo, Takuma 28, 29, 527
Hildebrandt, Diane 512
Hillestad, Magne 120, 137
Hinrichsen, Olaf 433
Hock, Rainer 353
Holmen, Anders 115, 120
Holstein, William L. 62
Hosono, Yukiko 531
Hosseiniamoli, Hadi 478
House, Stephen 419
Hsieh, Tien-Lin 79
Hu, Jianli 9, 91
Huang, Kefeng 502
Hur, Young Gul 396
Hutchings, G.J. 297
Hutchings, Graham J. 164, 296, 386
Iglesia, Enrique 1, 62, 182
Iguchi, Shoji 263
Inagaki, Reona 527
Inagaki, Satoshi 398
Inukai, Shoma 383
Ishikawa, Satoshi 383
Itkulova, Sholpan 30
Ito, Kazuharu 28
Jacobs, Gary 145, 205
Jeong, Yu-ri 323
Jerman, Ivo 532
Jeske, Kai 252
Ji, Zhonghai 347
Jiang, Feng 204
Jiang, Zheng 109

AUTHOR INDEX

Jiao, Feng....94, 170
Jiao, Wenqian....241
Jocz, Jennifer....298
Johns, Jeremy....544
Joos, Jochen....252
Joshi, Ravi....400
Jovanovic, Goran....325
Kalman, Steven....368
Kang, Dohyung....49
Kang, Jincan....173, 372
Kang, Jong Kyu....224, 323, 551
Kapteijn, Freek....152, 269
Karwacki, L.....195
Kassymkan, Kaysar....514
Kataria, Atish....15
Kathe, Mandar....79
Katheria, Sanjay....55
Kaumenova, Gulnar N.....514, 536
Kaur, Gurpreet....552
Keels, Jayson....43
Keller, Valérie....465
Kennedy, Eric M.....273, 422, 478
Kester, Philip M.....396
Khatib, Sheima J.....295
Khodakov, Andrei....50, 200
Khurana, Ishant....548
Kiely, C.J.....297, 386
Kiely, Christopher....296
Kim, Heeyeon....517
Kim, Ji Eun....224, 518
Kim, Jip....330, 550
Kim, Jonglack....252
Kim, Youngmin....330, 550
Kimura, Yosuke....497
Kirchmann, Marius....168
Kirilin, Alexey....242
Kloosterman, Jeffrey....56
Knemeyer, K.....408
Ko, Jeonghyun....344, 530
Köhler, Klaus....433
Kolb, Peter....131
Komurcu, Mustafa....552
Kondo, Junko N.....309
Kondrat, Simon....386
Kong, Fanhe....79
Koninckx, Elsa....359
Koros, William J.....100
Kosinov, Nikolay....218, 267
Kovarik, Libor....148
Kraehnert, Ralph....357, 408
Kreuter, Urs....2
Kropf, A. Jeremy....145
Kruse, Norbert....148
Ksenia, Parkhomenko....160
Kubota, Yoshihiro....398
Kulkarni, Ambarish....260
Kumar, Anurag....277
Kumar, Hemant....498
Kunitake, Yusuke....309
Kwon, Guk-hyun....517
Lalsare, Amoolya....91
Lamberti, Carlo....305
Lazzarini, Andrea....552
Le, Ha V.....311
Le, Thien An....224
Lee, Dae-keun....517
Lee, Sae Ha....323
Leelavathi, Annamalai....381
Lefferts, Leon....258
Lemishka, Mariia....307
Lemonidou, Angeliki....391, 445, 468
Lengyel, Istvan....555
Lercher, Johannes A.....379
Leung, Samuel L.....62
Li, Fanxing....78, 350
Li, Gen....94, 170
Li, Hui....548
Li, Luanyi....319
Li, Meng....419
Li, Meng-Jung....273, 422, 478
Li, Mengwei....321
Li, Na....94, 170
Li, Yiru....484
Li, Yongdan....430
Li, Yuting....329
Li, Zhe....341
Li, Zhinian....287
Liang, Wugeng....319
LiBretto, Nicole....346
Lietti, Luca....248, 439
Liland, Shirley E.....417
Lillerud, Karl Petter....552
Linck, Martin....97
Linke, Patrick....501
Lino, Ananda V.P.....540
Littlewood, Patrick....453
Liu, Hua....402
Liu, Jianwei....182
Liu, Meilin....452
Liu, Mengjie....326
Liu, Shanfu....318
Liu, Su....241
Liu, Xiaoliang....173
Liu, Xinying....512
Liu, Ye....363
Liu, Yi....283
Liu, Yilang....381
Loizidis, Charis....391
Lomachenko, Kirill A.....305
Lopes, Thiago....505

AUTHOR INDEX

López, Núria 376
Louwerse, Manuel J. 202
Loveless, Brett 182
Lu, Yong 363
Luemunkong, Thanaphat 22
Lukashuk, Liliana 544
Lundqvist, Magnus 544
Luo, Lin 356
Lyu, Yimeng 298
Ma, Chao 109
Ma, Hongfei 369, 412
Ma, Jianyu 57
Ma, Qingxiang 90
Ma, Tao 341
Maeda, Shun 531
Maestri, Matteo 67
Mahmoodinia, Mehdi 183
Maier, Florian 353
Malek, Andrzej 242
Mancuso, Luca 544
Maneeintr, Kreangkrai 22
Manoj, Sopariwala Vishnu 124
Mantripragada, Hari C. 74
Manzano, Manuel Nunez 334
Manzolini, Giampaolo 544
Maravelias, Christos T. 502
Marin, Guy B. 332, 334, 443
Marker, Terry 97
Marks, Tobin J. 318
Marquart, Wijnand 164
Marshall, Christopher 145
Martinelli, Michela 145, 512
Martinez, Daniel 237
Martini, Andrea 305
Martino, Annabelle 145
Martos, Carmen 237
McCool, Benjamin A. 255
McDonald-Buller, Elena C. 497
McEwen, Jean-Sabin 186
McFarland, Eric 42, 49
McGaughey, Gary 497
Mehrbod, Mohammad 145
Mejía, Carlos Hernández 141
Memon, Naim 124
Men, Zhuowu 104
Mendes, Pedro S.F. 332
Metiu, Horia 42, 49
Meyer, Randall 255, 304
Mezzapica, Aldo 158
Miao, Ping 104
Miao, Yu 325
Michel, Loïc 466
Miles, Joshua 402
Miller, Brendon 154
Miller, Jeffrey T. 341, 346, 396, 400
Milligan, Cory 341
Minardi, Luke 485
Minette, Florent 55
Mishra, Zubin 485
Mlekodaj, Kinga 307
Mohammed, Khaled 386
Moncada, Jorge 351
Monnier, John R. 43
Moodley, Denzil 140
Moonen, Roel 232
More, Amey 74
Morgan, David J. 164, 386
Morker, Hiralkumar 124
Moroni, Gianluca 67
Morosanu, Eduard Alexandru 222
Mortensen, Peter M. 46
Mottaghi-Tabar, Sogol 428, 455
Muccillo, Eliana 505
Muccillo, Reginaldo 505
Muhlenkamp, Jessica A. 344, 530
Müller, Ulrich 195
Murakami, Kota 28, 128, 506
Myrstad, Rune 120, 508
Nadgouda, Sourabh 79
Nagy, Anton J. 10
Nair, Balamurali 555
Najfach, Aaron 37
Nakaya, Ayumi 263
Nakayama, Akira 263
Nash, Connor 230
Nayak, Shanti Kiran 519
Neiss, Christian 353
Nese, Valentina 252
Neto, Almir Oliveira 505
Neurock, Matthew 182
Nguyen, Lisa 402
Ni, Lingli 379
Nielsen, Robert 368
Nishikawa, Yuta 263
Nowicka, Ewa 386
Oar-Arteta, Lide 152
Obradovic, Ana 332
Oelmann, Tobias 206
Ogihara, Hitoshi 263
Ogo, Shuhei 28, 29, 89, 128, 398, 506, 513, 527, 531
Ohtsuka, Yuki 263
Okoye-Chine, Chike G. 507
Oladipupo, Wasiu P. 484
Olsbye, Unni 19, 305, 552
Ordomsky, Vitaly 50, 200
Ortega, Carlos 232
Ortiz-Toral, Pedro 97
Oßwald, Patrick 326
Ostadi, Mohammad 137
Østberg, Martin 46

AUTHOR INDEX

Osuga, Ryota 309
Ould-Chikh, Samy 269
Paalanen, Pasi....................202
Palmer, Clarke 42, 49
Palou-Rivera, Ignasi 4
Pan, Xiulian 94, 170, 261, 347
Panditha-Vidana, D. G. 57
Pannala, Sreekanth 555
Paolucci, Christopher 548
Papalas, Theodoros....................445
Papp, Christian 353
Pappas, Dimitrios K. 305
Park, Eun Duck....................224, 323, 518, 551
Park, Jaesung 6
Park, Justin 365
Park, Kyung Soo 80
Park, La Hee....................518
Parsons, Lindsey 351
Parvathikar, Sameer....................15
Parvulescu, A. 195
Patanou, Eleni 508
Patil, Sandeep 25, 129, 130, 229, 529
Paunovic, Vladimir 366, 376
Perez, Hector 319
Pérez-Ramírez, Javier 366, 376
Pestman, Robert....................104, 177
Petrescu, Letitia 544
Petty, Giusy 152
Peukert, Wolfgang 353
Pfänder, Norbert....................252
Pham, Hien N. 320
Pintar, Albin....................404
Pirone, Raffaele....................208
Pirro, Laura....................332
Pistikopoulos, Efstratios N. 503
Plaisance, Craig....................257
Plessers, Dieter 301
Pohl, Felix....................357
Porta, Alessandro....................439
Potgieter, Jana....................140
Poulsen, Jens Michael....................399
Prakash, Anuj 453
Pretz, Matt....................356
Prieto, Gonzalo....................171, 252
Puértolas, Begoña....................376
Qi, Yanying....................412
Queiroz, Carla....................476
Quiroga, Cesar 2
Rahimi, Nazanin....................49
Rahman, Mustafizur....................295
Ramakrishnan, Vishnu Priya....................483
Ramon, Adriana P. 528
Ras, Erik-Jan....................232
Ratan, Sanjiv 13, 55
Ratnasamy, Chandra 476
Ray, Debjyoti 33
Razdan, Neil....................277
Reddick, Ian....................325
Reece, Christian....................386
Regalbuto, John R. 43
Regli, Samuel 412
Reible, Danny....................2
Ren, Yang....................341
Ribeiro, Fabio 359, 484
Ribeiro, Fabio H. 339, 341, 346, 396
Rimer, Jeffrey D. 320
Robert, Cerise....................466
Roger, Anne Cecile 105, 160, 457
Roman, Yuriy 304
Rønning, Magnus 412
Rosowski, Frank....................357, 408
Rossi, Marco A. 542
Rout, Kumar....................412, 486
Rout, Kumar R. 120, 369, 417
Ruddy, Daniel 230
Rytter, Erling....................115, 120, 137, 508
Sagar, Sarsani 319
Saito, Hikaru....................398, 531
Sakai, Ryuya 128, 506
Sall, Sécou 466
Salomone, Fabio 208
Santiago, Elisabete 505
Santos, Vera 242
Sarsenova, Rabiga O. 514
Sasaki, Kyuro....................22
Sato, Ayaka 513
Schaidle, Joshua A. 230
Scharfe, Matthias 376
Schlichting, Holger 206
Schlögl, Robert 317
Schneider, William F. 344, 530, 548
Schomäcker, Prof. Dr. Reinhard....................317
Schreiber, Moritz 379
Schuhmann, Timm 206
Schulz, Christian 357, 408
Schwach, Pierre 261
Schwartz, Nichole....................368
Scieranka, Kinga 242
Seemakurthi, Ranga Rohit 339
Seifert, Gotthard....................186
Seki, Hirofumi 531
Sekine, Yasushi 28, 29, 54, 89, 128, 398, 506, 513, 527, 531
Selleri, Tommaso 336
Sengupta, Debalina....................494
Senn, Megan....................346
Serna, Pedro....................304
Serra, Christophe 160
Shadravan, Vahid....................422
Shafer, Wilson D. 205

AUTHOR INDEX

Shan, Junjun 321, 402
Shareghi, Adam 325
Sharma, Pradeepkumar 15
Shehab, Shaza 501
Shekhawat, Dushyant 293
Shen, Zebin 109
Sievers, Carsten 298
Siirola, Jeffrey J. 484
Simakov, David 26, 428, 455
Simmons, Mark. J 154
Simonetti, Dante 70, 485
Sivri, Ali 91
Sklenak, Stepan 307
Skoglundh, Magnus 299
Smart, Simon 42
Smith, Justin 84
Smith, Mark W. 293
Sobalik, Zdenek 307
Song, Huanling 521
Song, Yuanjun 104
Soni, Abhi 124
Sorensen, Esben 61
Spinacé, Estevam 505
Spreitzer, Matjaz 404
Sridhar, Apoorva 295
Stadtherr, Mark A. 497, 500
Stair, Peter C. 453
Stanis, Ron 97
Stavitski, Eli 298
Steinman, Eric 5
Steinruck, Hans-Peter 353
Stitt, E. Hugh 154
Stockenhuber, Michael 273, 422, 478
Stowers, Kara 365
Strømsheim, Marie Døvre 183
Stuckert, Ines 21, 60
Stump, Brian 2
Su, Junjie 241
Suarez, Alma I. Olivos 269
Sullivan, Mark 304
Sun, Jian 213, 217
Sun, Lanlan 364
Sun, Xiaohui 152
Sun, Yuhan 212
Susman, Mariano 320
Svelle, Stian 305
Svenum, Ingeborg-Helene 183
Swarup, Vijay 17
Szentmiklósi, László 376
Tabor, Edyta 307
Taccardi, Nicola 353
Tafjord, Joakim 133
Takahashi, Ayako 527
Takeno, Yuna 513
Tan, Eric 230
Tanabe, Toyokazu 383
Tanaka, Yuta 128, 506
Tanchoux, Nathalie 38
Tang, Yu 329, 452
Tao, Franklin (Feng) 283, 329, 452
Tathod, Anup 26
Tatibouët, Jean-Michel 38
Taylor, Melinda 2
Terunuma, Ryota 398
Teržan, Janvit 404
Thakur, Raj 84
Theodori, Gene 2
Thomas, Sébastien 105, 457
Thum, Lukas 317
Thybaut, Joris W. 332
Tian, Fangyuan 73
Toko, Kenta 28
Tong, Andrew 79
Toraman, Hilal Ezgi 256, 362
Torimoto, Maki 29
Tranca, Diana 186
Traverso, Andrew 325
Triesscheijn, Amy 232
Tronconi, Enrico 248, 336
Trunschke, Annette 317
Tsakoumis, Nikolaos 115, 508
Tseng, Han-Ting 359
Tsubaki, Noritatsu 216, 217, 234, 247
Tsui, Lok-kun 519
Tungatarova, Svetlana A. 514, 536
Turek, Thomas 10
Ueda, Wataru 383
Umrigar, Vaishali 124
Upham, David Chester 42
Valentin, L'hospital 160
Valishevsky, Kirill 30
Van De Water, Leon G.A. 154, 544
Van Deelen, Tom W. 192, 202
Van Der Hoeven, Jessi E.S. 141
Van Der Linden, Bart 269
Van Der Sande, Christian 152
Van Der Veer, Siebe 544
Van Dijk, H.A.J. 544
Van Geem, Kevin M. 334
Van Steen, Eric 140
Van Vreeswijk, Sophie H. 225
Vandegehuchte, Bart D. 332
Vandewalle, Laurien A. 334
Vanhaecke, Estelle Marie M. 57
Vaz, Jorge 505
Vemuri, Balakotaiah 319
Venvik, Hilde 57, 120, 183
Vernuccio, Sergio 359, 500
Veser, Götz 74, 419
Vicentini, Valéria 476

AUTHOR INDEX

Vinçon, Ilka 186
Visconti, Carlo Giorgio....245, 248, 439
Vizcaino, Arturo....237
Vlachos, Dionisios G. 256, 362
Vollmer, Ina....269
Von Jouanne, Annette 325
Vrijburg, Wilbert....437
Vullum, Per Erik....57
Waldvogel, Audrey 457
Walmsley, John C. 57
Wang, Dawei 79
Wang, Dong....241
Wang, Hui....212, 402
Wang, Lucun....355
Wang, Mengheng 173
Wang, Peng....104
Wang, Shenliang 287
Wang, Siyu....486
Wang, Tianyun....283
Wang, Wei 315
Wang, Yangdong....241
Wang, Ye 109, 173, 372
Wang, Yong....186
Wang, Zhiqiang....286
Wangerow, Jim 97
Wasserscheid, Peter 353
Watson, Michael J....154
Weber, J. Lennart....237
Weckhuysen, Bert M. 202, 225
Wei, Jian....213
Wei, Junmei....62
Wei, Li....294
West, David 319, 555
Weststrate, Kees-Jan 183
Wezendonk, Tim A. 152
Wilcox, Laura N. 548
Wilkinson, Sam 154
Williams, Bryce A. 206
Willock, David....386
Wittich, Knut....357
Wolf, Moritz 112, 353
Wu, Jian-Feng....315
Wu, Long-Min....315
Wu, Wei 355
Wu, Yue 341
Wu, Zhenwei 341
Xanthopoulou, Galina....536
Xiang, Yizhi....148
Xiao, Liping....287
Xie, Jingxiu....202, 357, 408
Xie, Quanhua....372
Xie, Shunji 109
Xie, Zaiku 241
Xiong, Haifeng....346
Xu, Bingjun 5
Xu, Dikai 79
Xu, Hualong....197
Xu, Jianguo....55
Xu, Rui 298
Xu, Wayne....104
Xu, Yinan 339
Yabe, Tomohiro....28, 29, 89, 128, 398, 506, 513
Yamada, Kensei....28, 89
Yamanaka, Ichiro....263
Yan, Liang....521
Yang, Guohui....216, 247
Yang, Jian 521
Yang, Judith 419
Yang, Junhao....94, 170
Yang, Weimin....241
Yao, Jinsong....104
Yao, Ruwei....213
Yarulina, Irina....269
Yengle, Carlos....498
Yokochi, Alexandre....325
Yokoi, Toshiyuki 309
Yoneyama, Yoshiharu....216, 247
Yoon, Hee Wook....6
Yoon, Suji....518
Young, Michael....2
Yu, Miao....218
Yu, Si-Min....315
Yu, Xiang 50
Zavasnik, Janez 404, 532
Zeng, Jiren 49
Zerjav, Gregor....404, 532
Zhang, Guanghui....341, 400
Zhang, Haikun....109
Zhang, Huamin....372
Zhang, Jianghao 186
Zhang, Jianli....90
Zhang, Kui 104
Zhang, Mingsen....294
Zhang, Ning 521
Zhang, Qinghong....109, 173, 372
Zhang, Yi....283
Zhang, Yitao 79
Zhang, Zhiqiang....363
Zhao, Guangyu 273
Zhao, Guofeng 363
Zhao, Huahua....521
Zhao, Qingrui 294
Zhao, Tiansheng 90
Zheksenbaeva, Zauresh T.....514, 536
Zhou, Haibo 241
Zhou, Lin....341
Zhu, Jian 363
Zhuang, Yichen 428, 455
Zhumabek, Manapkhan 536
Zichittella, Guido....366, 376
Zou, Shihui 287